AIR POLLUTION

THIRD EDITION

VOLUME I

Air Pollutants, Their Transformation
and Transport

ENVIRONMENTAL SCIENCES

An Interdisciplinary Monograph Series

Editors: Douglas H. K. Lee, E. Wendell Hewson, and Daniel Okun

A complete list of titles in this series appears at the end of this volume.

AIR POLLUTION
THIRD EDITION

VOLUME I

Air Pollutants, Their Transformation and Transport

Edited by

Arthur C. Stern
Department of Environmental Sciences and Engineering
School of Public Health
University of North Carolina at Chapel Hill
Chapel Hill, North Carolina

ACADEMIC PRESS New York San Francisco London 1976

A Subsidiary of Harcourt Brace Jovanovich, Publishers

ACADEMIC PRESS, INC.
111 Fifth Avenue, New York, New York 10003

United Kingdom Edition published by
ACADEMIC PRESS, INC. (LONDON) LTD.
24/28 Oval Road, London NW1

Library of Congress Cataloging in Publication Data

Stern, Arthur Cecil.
    Air pollution.

    (Environmental sciences)
    Includes bibliographical references and index.
    1.   Air—Pollution.   I.   Series.
    TD883.S83   1976 v. 1   363.6   75-13098
    ISBN 0–12–666601–6

PRINTED IN THE UNITED STATES OF AMERICA

To Dorothy, in loving memory

Contents

PART A   THE POLLUTANTS

1. Classification and Extent of Air Pollution Problems

Leslie A. Chambers

2. The Primary Air Pollutants—Gaseous Their Occurrence, Sources, and Effects

Paul Urone

PART B   THE TRANSFORMATION OF AIR POLLUTANTS

PART C   THE TRANSPORT OF AIR POLLUTANTS

11. Meteorological Measurements

E. Wendell Hewson

12. Air Pollution Climatology

Robert A. McCormick and George C. Holzworth

List of Contributors

Numbers in parentheses indicate the pages on which the authors' contributions begin.

Leslie A. Chambers (3), Environmental Sciences, School of Public Health, University of Texas Health Science Center at Houston, Houston, Texas

Morton Corn* (77), Department of Occupational Health, Graduate School of Public Health, University of Pittsburgh, Pittsburgh, Pennsylvania

Merril Eisenbud (197), Institute of Environmental Medicine, New York University Medical Center, New York, New York

A. J. Haagen-Smit (235), Division of Biology, California Institute of Technology, Pasadena, California

E. Wendell Hewson (563), Department of Atmospheric Sciences, Oregon State University, Corvallis, Oregon

George C. Holzworth (643), Meteorology and Assessment Division, Environmental Sciences Research Laboratory, U.S. Environmental Protection Agency, Research Triangle Park, North Carolina

Alvin R. Jacobson (169), Environmental Health Sciences, Office of Allied Health Professions, Illinois State University, Normal, Illinois

Warren B. Johnson (503), Atmospheric Sciences Laboratory, Stanford Research Institute, Menlo Park, California

William P. Lowry (327), Department of Geography, University of Illinois, Urbana, Illinois

* Present address: Assistant Secretary of Labor for Occupational Safety and Health, United States Department of Labor, Washington, D.C.

Robert A. McCormick (643), Smith-Singer Meteorologists, Inc., Raleigh, North Carolina

Samuel C. Morris (169), Brookhaven National Laboratory, Associated Universities, Inc., Upton, New York

R. E. Munn (289), Atmospheric Environment, Environment Canada, Downsville, Ontario

Ralph C. Sklarew (503), Xonics, Inc., Van Nuys, California

Gordon H. Strom (401), 35 Oak Street, Tenafly, New Jersey

D. Bruce Turner (503), Meteorology and Assessment Division, Environmental Sciences Research Laboratory, U.S. Environmental Protection Agency, Research Triangle Park, North Carolina

Paul Urone (23), Environmental Engineering Department, University of Florida, Gainesville, Florida

Raymond C. Wanta (327), P.O. Box 98, Bedford, Massachusetts

Lowell G. Wayne (235), Pacific Environmental Services, Inc., Santa Monica, California

D. M. Whelpdale (289), Atmospheric Environment, Environment Canada, Downsville, Ontario

Preface

This third edition is addressed to the same audience as the previous ones: engineers, chemists, physicists, physicians, meteorologists, lawyers, economists, sociologists, agronomists, and toxicologists. It is concerned, as were the first two editions, with the cause, effect, transport, measurement, and control of air pollution.

So much new material has become available since the completion of the three-volume second edition that it has been necessary to use five volumes for this one. Volumes I through V were prepared simultaneously, and the total work was divided into five volumes to make it easier for the reader to use. Individual volumes can be used independently of the other volumes as a text or reference on the aspects of the subject covered therein.

Volume I covers two major areas: the nature of air pollution and the mechanism of its dispersal by meteorological factors and from stacks. Volume II covers the effect of air pollution on plants, animals, humans, materials, and the atmosphere. Volume III covers the sampling, analysis, measurement, and monitoring of air pollution. Volume IV covers two major areas: the emissions to the atmosphere from the principal air pollution sources and the control techniques and equipment used to minimize these emissions. Volume V covers the applicable laws, regulations, and standards; the administrative and organizational strategies and procedures used to administer them; and the energy and economic ramifications of air pollution control. The concluding chapter of Volume II discusses air pollution literature sources and gives guidance in locating information not to be found in these volumes.

To improve subject area coverage, the number of chapters was increased from 54 of the second edition (and 42 of the first edition) to 71. The scope of some of the chapters, whose subject areas were carried over from the second edition, has been changed. Every contributor to the second edition was offered the opportunity to prepare for this edition either a revision of his chapter in the second edition or a new chapter if the scope of his work had changed. Since 8 authors declined this offer and one was deceased, this edition includes 53 of the contributors to the second edition and 46 new ones.

The new chapters in this edition are concerned chiefly with aspects of air quality management such as data handling, emission inventory, mathematical modeling, and control strategy analysis; global pollution and its monitoring; and more detailed attention to pollution from automobiles and incinerators. The second edition chapter on Air Pollution Standards has been split into separate chapters on Air Quality Standards, Emission Standards for Stationary Sources, and Emission Standards for Mobile Sources. Even with the inclusion in this edition of the air pollution problems of additional industrial processes, many are still not covered in detail. It is hoped that the general principles discussed in Volume IV will help the reader faced with problems in industries not specifically covered.

Because I planned and edited these volumes, the gap areas and instances of repetition are my responsibility and not the authors'. As in the two previous editions, the contributors were asked to write for a scientifically advanced reader, and all were given the opportunity of last minute updating of their material.

As editor of this multiauthor treatise, I thank each author for both his contribution and his patience, and each author's family, including my own, for their forbearance and help. Special thanks are due my secretary, Patsy Garris, and her predecessors, who carried ninety-nine times the burden of the other authors' secretaries combined, and Jean Myers and Juanita Jones for preparing the Subject Index for this volume. I should also like to thank the University of North Carolina for permitting my participation.

Arthur C. Stern

Contents of Other Volumes

Part **A**

THE
POLLUTANTS

1

Classification and Extent of Air Pollution Problems

Leslie A. Chambers

I. Natural History of the Air Pollution Problem

The fundamental characteristics of the air pollution problem have not changed significantly in the most recent decade, but there have been important changes in control technology, in understanding of atmospheric processes and interactions among airborne pollutants, and in administrative and legislative instruments for regulation and abatement. Much effort has been expended toward understanding of the exposure–response relationships between observed concentrations of air pollutants and human receptors, but the conclusive evidence needed as a firm basis for control standards remains elusive to a disquieting degree.

Meanwhile a remarkable shift in public and governmental awareness of the potential for damage to health and general welfare posed by increasing deterioration of the air resource has occurred. Whereas regulatory measures were formerly implemented with great difficulty except for acutely affected cities such as Los Angeles and London, popular demand has now resulted in the adoption of rigorous standards at national levels, and the exploration of possibilities for international codes. There is a tendency to act on the presumption that any change in atmospheric composition is to be prevented—that its damaging effect, if not already documented, must be assumed until proved innocuous. It is therefore timely to reconsider the nature of the threat.

There is implicit in the words "air pollution" an assumption of some sort of atmospheric norm from which variance can be observed. While relative constancy in composition of the earth's gaseous envelope during the brief span of recorded history may be accepted, it is helpful, in a consideration of man's relationship to his supporting medium, to consider the nature of the interaction between living organisms and the atmosphere in times spanning the evolution of our species. Most certainly air composition has undergone great qualitative changes in the two billion years or more since the first anaerobic self-reproducing units of matter came into being.

The first primitive living cells could not have occurred or continued to exist in an atmosphere composed of the chemical constituents we now regard as normal and requisite to life. In particular, the primeval gaseous environment probably contained almost no free oxygen; indeed, all existing evidence indicates that the oxygen in more recent air has accumulated as a result of photosynthetic processes utilized by early nonoxygen-dependent species. For these latter the discharge and accumulation of oxygen was no less a catastrophic form of atmospheric pollution than is the emission of CO_2, CO, NO_2, and other metabolically difficult chemical compounds resulting from activities of currently dominant life forms, especially man.

In addition to the "oxygen revolution" there is evidence that the earth's atmosphere has undergone a sequence of qualitative changes, e.g., substantial variations in CO_2 content, during the long ages of prehistory.

Much of the body of facts on which estimates of such changes are based derive from analysis of fossil records of adaptive changes in plants and animals apparently in conformance with environmental variations. Thus there is exhibited a remarkable adaptive flexibility in persistent life forms with relation to very gradual changes in atmospheric content.

On the other hand, there is equally persuasive evidence that species, and even dominant orders of plants and animals, have often been reduced to insignificance by environmental changes brought about too quickly to relate well with available rates of adaptive response. Herein lies the hard core of man's current conflict with environmental change. The magnitudes he now represents have become significantly large in relation to the air resource; his existence in it is causing effective changes within a few years while his machinery for adaptive adjustment requires many generations.

Fortunately, the human species possesses cognitive properties and capabilities for apprehensive actions designed to forestall cataclysmic feedback of the products of his biological and economic existence. But the implementation of these potentials is in itself a kind of adaptation which requires time; a brief review of the record of human concern for, and rate of, avoidance reactions may or may not prove reassuring.

II. Historical Perspective

A. Air Pollution prior to the Industrial Revolution

The quality of the atmosphere, on which existing terrestrial forms of life are dependent, has been recognized as an important variable in the environment only during the past few decades. It can be supposed that smoke and fumes from forest fires, volcanoes, and crude "domestic" heating and cooking arrangements were troublesome or lethal in discrete localities even before our human ancestors became organized into fixed communities and that the odors of decaying animal and vegetable refuse, attested to by existing residues of prehistoric garbage dumps in and near stone age dwellings, were cause for protesting comment in such language as may have been available to the temporary residents.

But it is unlikely that such circumstances can have been regarded as more than incidental to devastating natural cataclysms or as reasons for transfer to another dwelling site, until social evolution reached the husbandry level involving association of family units into more or less fixed communities. Only then could human activities in the aggregate have produced sufficient effluvia to affect an occupied neighborhood. To what extent they did so is entirely conjectural with respect to all of prehistory and can be guessed only by tenuous inference with respect to most of the ancient and medieval periods. The embodiment in folk knowledge of the middle ages, and in prescientific belief, of the concept of "miasmas," or poisonous airs, as etiological agents of certain diseases may indicate a deduction from accumulated survival experience related to recognized sources of unwholesome air, but is more likely a mistaken association of "malarias" with the odors of swamps rather than the mosquitoes which they supported.

Writers on air pollution occasionally have cited classic references to blackened buildings and monuments as evidence that the smoke nuisance has a reality spanning thousands of years. But the grime of antiquity, while a reasonable expectation, does not suffice to indicate a contemporary recognition of its impact on ancient communities or their members. In fact, accumulated knowledge of domestic heating practices and of the available primitive metallurgical and other limited industrial processes utilized during the first thirteen or fourteen centuries of the Christian era leads to the inference that generalized air pollution could not have been an important problem in the villages and towns of the time; cities, in the sense of modern magnitudes, were nonexistent. The frequently cited references to deaths caused by toxic atmospheres, e.g, the suffocation of Pliny the Elder by volcanic fumes as recorded by Tacitus, seem not to be pertinent except in the sense of demonstrating that the human species was then, as it is now, physiologically responsive to anoxia or to poisonous gases.

Throughout the earlier periods of history wood was the prime source of energy; dependence on it undoubtedly slowed the evolution of industrial processes and eventually limited the per capita availability of heat as depletion of nearby forests proceeded. The discovery of the energy potential of coal and its gradual displacement of wood occurred in Europe about the time of Marco Polo's return from his travels through the more technologically advanced civilizations of Asia. But in spite of its abundance in the West, and its retrospectively apparent advantages, the European adaptation to its use which culminated in the Industrial Revolution proceeded slowly and against all the resistance normal to major economic readjustments. Coal was an "unnatural" fuel; its sulfurous combustion products confirmed its suspected association with anticlerical forces at a time much too closely related to the ascendance of strict orthodoxy; and above all, as a matter of record, it caused neighborhood "action committees" to protest against its evident pollution of the atmosphere.

In England, Germany, and elsewhere, various limitations and prohibitions relative to the use, importation, and transport of coal were proclaimed officially, and in isolated instances there is evidence that capital penalties were imposed. Nevertheless, the overriding demands for domestic heat and industrial power made these efforts useless and assured their disposal in the limbo of unenforceable law. Coal made possible the Industrial Revolution; and then there was smog.

B. *Air Pollution as Related to Coal Smoke and Gases*

From the beginning of the fourteenth century to the early part of the twentieth, air pollution by coal smoke and gases occupied the center of the stage almost exclusively, and in many industrialized areas of the world it is still the dominant concern. That it remains a community problem in spite of repeatedly demonstrated technological capability for its control would be surprising if public and official hesitance to pay the price were not so characteristic a factor in the evolution of all types of health protective programs. Positive action has seldom been anticipatory; instead it has occurred only after dramatic disasters or large-scale sensory insults have aroused public clamor based on fear. We build levees only after floods have devastated whole regions; we abate pollution of water supplies only after typhoid epidemics or similarly impressive episodes; and we take necessary action to control air pollutants only after their killing or irritating potentials have been realized on a large scale as in London in 1952 or in Los Angeles around 1945.

In no case is the very early recognition of a public health problem and the failure to take any effective action until it threatened personal survival better illustrated than in the case of air pollution produced by the unrestricted use of coal in Great Britain. During the reign of Edward I (1272–1307) there was recorded a protest, by the nobility, against the use of "sea" coal; and in the succeeding reign of Edward II (1307–1327) a man was put to the torture ostensibly for filling the air with a "pestilential odor" through the use of coal.

Under Richard III (1377–1399) and later under Henry V (1413–1422), England took steps to regulate and restrict the use of coal, apparently because of the smoke and odors produced by its combustion. The earlier action took the form of taxation, while Henry V established a commission to oversee the movement of coal into the City of London.

Other legislation, parliamentary studies, and literary comments appeared sporadically during the following 250 years. In 1661 a notable pamphlet was published by royal command of Charles II. It consisted of an essay entitled "Fumifugium; or the Inconvenience of the Aer and Smoke of London Dissapated; together with Some Remedies Humbly

Proposed,"* written by John Evelyn, one of the founding members of the Royal Society. It is unfortunate that the author's seventeenth century style has attracted more attention in the twentieth century than has the content of his paper. Evelyn clearly recognized the sources, the effects, and the broad aspects of the control problem to an extent not far surpassed at the present time except for detail and for technological terminology. Thus it is clear not only that the London of 1661 was plagued by coal smoke, but also that the problem and its content were recognized by at least one of the scientific leaders of the period.

By the beginning of the nineteenth century the smoke nuisance in London and other English cities was of sufficient public concern to prompt the appointment (1819) of a Select Committee of the British Parliament to study and report upon smoke abatement. The effect of the study is suspected to have been similar to that of dozens of other committee recommendations during the ensuing years. The gradual development of the smoke problem culminated in the action-arousing deaths, in a few days, of 4000 persons in London in December, 1952.

Records of lethal air pollution concentrations during the nineteenth century are not definitive; in fact, recognition of their occurrence has resulted largely from retrospective examination of vital records and contemporary descriptive notes. In 1873 an episode having the characteristics of the 1952 event occurred in London and more or less severe repetitions have affected metropolitan life at irregular intervals.

The term "smog" originated in Great Britain as a popular derivation of "smoke-fog" and appears to have been in common use before World War I. Perhaps the term was suggested by H. A. Des Voeux's 1911 report to the Manchester Conference of the Smoke Abatement League of Great Britain on the smoke-fog deaths which occurred in Glasgow, Scotland in 1909. During two separate periods in the autumn of that year very substantial increases in the death rate were attributed to smoke and fog, and it was estimated that "1063 deaths were attributable to the noxious conditions."

With few isolated exceptions, the extreme atmospheric concentrations of pollutants produced by coal burning in Britain have not been duplicated elsewhere. Nevertheless, coal-based industrial economies on the continent of Europe and in the United States have caused discomfort, public reaction, and regulatory action. A generation before the dramatic incident which killed 20 and made several hundred ill in the industrial town of Donora, Pennsylvania, in 1948, public protest groups had appeared in several American cities. In some, such as St. Louis, Cincinnati,

* "The Smoake of London—Two Prophecies," selected by J. P. Lodge, Jr. Maxwell Reprint Co., Elmsford, New York, 56 pp. (1969).

and more recently in Pittsburgh, popular movements resulted in substantial elimination of the smoke nuisance, by substitution of less smoky fuels, and by enforced employment of combustion practices designed to eliminate smoke. It has thus been demonstrated that high smoke densities are preventable, although the cost may be large. London, although handicapped in its smoke abatement effort by lack of low-volatility coal supplies, relative dependence on imports for other fossil fuels, and a centuries-old pattern of inefficient household heating, has achieved substantial improvement because of willingness to pay the high price.

No rigorous identification of the constituents of coal smoke responsible for the respiratory illnesses with which it has been associated has been produced, although the effects have been generally attributed to sulfur dioxide and trioxide. Recently the probability of a role of tar, soot, and ash particles in the total irritative effect has been the inspiration for several investigations. But the information available to us on the relationship of coal smoke to human health has been insufficient to explain fully the death and discomfort it has caused.

Smoke and gases from the burning of coal have been the chief atmospheric pollutants in all parts of the industrialized world for more than 400 years. In spite of the recent rapid shift to petroleum and natural gas, coal smoke still is a major contributor to poor air quality in most urbanized areas.

C. Pollution by Specific Toxicants

While pollutants resulting from use of the dominant energy sources—coal and petroleum products—generally arise from a large number of points within a community and therefore often cause a general deterioration of the air supply over large areas, more restricted regions closely adjacent to individual sources may be even more seriously affected. Many localized events have emphasized that critical concentrations of pollutants, other than smoke, having proved toxic properties can adversely affect air quality. A large number of substances used in manufacturing and commerce have been recognized officially as hazards to industrial workers, and maximal limits of acceptable concentration for 8-hour exposures have been established. While these limits are not applicable where intermittant exposure of an unselected population is the concern, they do indicate the classes and species of substances that are potentially hazardous.

Perhaps the most publicized example of serious air pollution by an identified toxicant was the episode at Poza Rica, near Mexico City, in which numbers of people were affected and a few died from exposure to hydrogen sulfide. Metallic fumes and acid mists from metallurgical processing have occasionally rendered downwind regions wholly uninhabitable for plants as well as man. Fluorides escaping from aluminum processing and other industrial sources have been the cause of losses to cattle farmers. Malodorous pollutants from a wide variety of source types have produced responses ranging from public irritation to overt and wholesale illness.

But such unquestionable local reactions to specific pollutants are relatively infrequent. Usually dispersive processes reduce the concentrations of emitted toxic materials to levels below the possibility of immediate or acute biological response. Under such circumstances the pollutants may provide a more or less continuous low dosage to occupants of an extended area, providing the possibility of slow accumulations of substances such as lead, or a continuum of low-grade insults which may eventually overpower physiological defenses. The effects of low dosages of pollutants long continued remain in need of intensive study; and the possibilities of synergism among two or more substances simultaneously breathed at subacute concentrations for extended periods of time have been suggested often but infrequently explored.

D. The Emergence of Petroleum Products

It is possible that future historians may recognize a second industrial revolution born in the years following completion of Drake's first oil well in Pennsylvania. Subsequent release of a flood of fossilized energy in the form of petroleum and natural gas not only has transformed industrial and domestic heating practices, but has made possible wholesale changes in transportation and provided the raw materials for a great variety of petrochemical products.

Combustion of oil and gas has diminished the coal smoke nuisance and hazard to the extent that the use of these fuels has displaced coal. With more than one type of mineral energy source available, the magnitude of the change has varied markedly among geographic regions, in a manner closely related to propinquity of oil and gas fields, extension of pipeline networks and other transport facilities, relative local costs of delivered fuel, and other logistic factors. Within the United States, for example, large areas of the Southwest now consume negligible quantities of coal, while portions of the eastern seaboard, the Southeast, and the midwestern industrial complexes exhibit mixed patterns of coal, oil, and natural gas use.

Throughout the United States and many other parts of the world, use of petroleum products in the forms of gasoline and oil has been tremendously accelerated, especially since World War II, by the almost exclusive employment of internal combustion engines in highway, railway, and marine transport. Thus, even in those regions unfavorably situated for the rapid adoption of oil and gas for heating and manufacturing, the combustion residues of petroleum products have become factors in community air pollution.

The contribution of automobile engine exhausts to the atmosphere was pointed out as a potential hazard as early as 1915, and the objectionable fumes from diesel power plants have been a matter of concern at least as long, but it was not until about 1945 that the first acute commmunity air pollution problem definitely attributable to petroleum products and their use forced itself into public and official recognition. The Los Angeles type of air pollution, hereafter referred to as smog in deference to local usage deriving from an unfortunate transposition of the term from its conceptual counterpart in Great Britain, has become the infamous prototype of similar developments appearing with increasing frequency in metropolitan areas of the United States and other countries.

As a matter of fact, the physical system underlying the obvious manifestations of "smog" in Los Angeles includes neither smoke nor fog. Early in the hastily organized effort to abate the air pollution which became irritatingly evident during the wartime industrialization of southern California, Professor A. J. Haagen-Smit demonstrated that the eye irritation, damage to green leaves, and light-scattering characteristic of smog could be produced by ultraviolet irradiation of hydrocarbon vapors together with nitrogen dioxide. This and much subsequent work have proved that the "new" kind of air pollution results from exposure to sunlight of mixtures of olefins and other reactive products of petroleum manufacture and use and oxides of nitrogen. The variety of intermediate and terminal products formed under different conditions of relative concentration, humidity, temperature, solar radiation intensity, and admixture with other reactive gases and particles is certainly very large. Among them are ozone, organic hydroperoxides, peroxyacyl nitrates (PAN), several aldehydes, and other irritants which have been positively identified; a wide variety of free radicals not experimentally demonstrated but necessary intermediates in the photochemical transitions from primary reactants to more stable products; and a number of possibly troublesome substances whose occurrence is still hypothetical. Subsequent chapters of this volume deal specifically with these reaction mechanisms and their consequences.

Most of the pollutants related to petroleum production, processing, and use have intrinsic toxic or irritative potentials of a rather low order. By contrast, their photochemical reaction products may affect biological systems at extremely low concentrations. Thus, the control of the primary reactants must be based on their identification and their regulation to atmospheric levels incapable of generating effective amounts of secondary products. Ozone, for example, is not known to be produced in significant quantity from direct sources within the Los Angeles area; yet it occurs frequently at levels greater than 0.25 ppm by volume of air as the result of photoenergized reactions involving hydrocarbons at the level of a part per million or less, and NO_2 in the same range of concentration. To prevent toxic accumulations of ozone, it is necessary to control sources of both hydrocarbon vapor and nitrogen oxides which except for the circumstance of reaction in the general atmosphere would be harmless.

Current economic trends and knowledge of proved world petroleum resources indicate that air pollution due to hydrocarbons, petrochemical products, and engine exhausts will become increasingly evident in most metropolitan areas for some years to come. Eventually, of course, retarding pressures generated by depletion of supplies will become operative but the present generation will find it necessary to protect local air supplies against contamination by gasoline vapors and exhaust gases.

E. Relationship to Energy Sources

What a biological system utilizes as its source of energy determines the characteristics of its waste products. Similarly, the fuel used by a community governs the kinds, amounts, and properties of its refuse. The aerial excreta of a city may be modified by local patterns of industry, solid waste disposal practices, or occasional counteracting "perfumes" such as paper mill mercaptans, but the products generated in energy transformations constitute the core of the community air pollution problem.

It is not necessary to discount the importance of localized nuisances in order to accept this primary thesis. Odors, toxic dusts and fumes, and corrosive acid mists are of great importance to the locality directly affected. But the primary threat to the air resources of modern cities can be firmly attributed to the kinds of materials they use for fuel and the ways in which they use it.

A major change in the quality of the air pollution problem could occur only with a major change in energy sources. It is interesting to speculate on what may occur as nuclear power or direct utilization of solar energy becomes practicable and economically competitive. In the one case a totally different kind of air pollution may require careful control; radioactive by-products of nuclear fuels could be troublesome to an extent not foretold by any previous experience with products of fossil fuels. In the present state of nuclear power technology, the magnitude and quality of the potential air pollution problem cannot be precisely defined; but it can be hoped that power packages will be so constructed as to minimize emission of active wastes.

Since the beginning of the twentieth century, worldwide atmospheric concentrations of carbon dioxide have been increasing steadily in a manner related to the increased global use of fossil fuels. Carbon dioxide is not often considered to be an air pollutant, since it produces adverse physiological effects only at relatively high concentration, and because biological and geochemical processes are known to provide a sufficient natural disposal system. Its atmospheric increase apparently reflects an accelerating disparity between the CO_2 production rate and the rate of approach to equilibrium with marine and terrestrial sinks. Unchecked increase in the rate of combustion of carbon fuels apparently would increase general CO_2 levels eventually to meteorologically and physiologically significant levels. Perhaps it may within a few generations compete with radioactive wastes for the dubious distinction of being a worldwide air polluter.

Any substantial shift of energy dependence from fossil fuels to nuclear or solar power plants will tend to reestablish the planetary CO_2 equilibrium. It is especially exciting to consider the air conservation potential of solar energy and other nonpolluting power sources now being explored. Should these prove capable of displacing current combustive transformations, the community air pollution problem would be reduced to more or less routine policing of localized sources.

III. Primary Concepts of Air Pollution

A variety of definitions of air pollution have been devised, each expressing more or less completely the individual philosophical, theoretical, practical, or protective motivation of its author. Any circumstance which adds to or subtracts from the usual constituents of air may alter its physical or chemical properties sufficiently to be detected by occupants of the medium. It is usual to consider as pollutants only those substances added in sufficient concentration to produce a measurable effect on man or other animals, vegetation, or material.

Pollutants may therefore include almost any natural or artificial composition of matter capable of being airborne. They may occur as solid particles, liquid droplets, gases, or in various admixtures of these forms.

Pollution of the air by a single chemical species appears to be a most unusual event; certainly most community problems involve a very large number of kinds and sizes of substances.

In an effort to classify the pollutants thus far recognized, it is convenient to consider two general groups: (a) those emitted directly from identifiable sources, and (b) those produced in the air by interaction among two or more primary pollutants, or by reaction with normal atmospheric constituents, with or without photoactivation. Any taxonomic system based on available sampling and analytical methods is almost certain to fall short of a complete description of the qualities of a polluted air supply. This is true because few, if any, of the polluting entities retain their exact identities after entering the atmosphere. Thermal and photochemical reactions, sometimes catalytically facilitated by gases, or on solid or liquid surfaces, provide a dynamic, constantly changing character to the total system and to its individual constituents. Eventually it may be possible to define a polluted air mass in space and time, by a complex integration of reaction pathways and rates as governed by fluctuating free-energy levels. That capability is only a dream at present.

A. Primary Emissions

It is usually possible to determine the kinds and amounts of primary pollutants emitted from each source in a community. Much information is available as to the chemical species and physical states of discharges from most types of artificial and natural generators. While the end effect of the emissions cannot be predicted with certainty from these data alone, they do define the primary reactants, and after other troublesome reaction chains have been identified, enable retroactive abatement with respect to the primary species contributing to the chains.

Primary emissions are often categorized, quite illogically because of our imperfect knowledge, under a mixture of headings defining chemical properties, physical phases, and magnitudes.

For purposes of generalization a listing of the following type is probably as inclusive as any:

Fine particles (less than 100 μm in diameter)
Coarse particles (greater than 100 μm in diameter)
Sulfur compounds
Organic compounds
Halogen compounds
Radioactive compounds

In one form or another each of these groups of pollutants will be considered in detail in the following chapters.

The finer aerosols include particles of metal, carbon, tar, resin, pollen, fungi, bacteria, oxides, nitrates, sulfates, chlorides, fluorides, silicates, and a host of other species obviously overlapping all of the more specific categories. As particles, they scatter light in conformance with well-established physical laws relating wavelength and particle size. As suppliers of large specific surfaces they afford opportunity for catalysis of normally slow interactions among adsorbed pollutants. As charged entities they govern to a substantial degree the condensation and coalescence of other particles and gases. As chemical species *per se* some of them exhibit high orders of toxicity to plant and animal species or are corrosive to metals and other materials. To the extent that they are radioactive they increase the normal radiation dosage and are suspected to be factors in abnormal genetic processes. And finally, as plain dust deposited in accordance with the physical laws governing precipitation and electrostatic attraction, they soil clothing, buildings, and bodies to constitute a general nuisance.

The coarser particles, upward from 100 μm in diameter, present the same types of problems in greatly diminished degree. This is true because their mass assures rather prompt removal from the air by gravitational attraction, because anatomical defensive mechanisms prevent their penetration into human or animal lungs, and because the same mass of substance in such large units affords substantially less opportunity for interaction with other components of the polluted air supply. On the other hand, their soiling effect may be more evident simply because after leaving a source they are readily deposited without opportunity for wide dispersal.

Interest in the sulfur compounds has been prolonged and intense because of their suspected role in the London disasters of 1952 and other years and because of the extreme toxicity of hydrogen sulfide. Combustion of sulfur-containing fuels contributes large amounts of SO_2 and some SO_3; many industrial processes and waste disposal practices generate H_2S; and the nauseous odors of organic sulfur compounds are well recognized associates of some pulp manufacturing and petrochemical processes. All of these affect plants and animals adversely at different, but generally low concentrations. There is substantial evidence that the full air polluting potential of SO_2 is realized only after it has reacted with other substances in the atmosphere.

Organic compounds released to typical community air supplies include a very large number of saturated and unsaturated aliphatic and aromatic hydrocarbons together with a variety of their oxygenated and halogenated derivatives. They are emitted principally as vapors but the less volatile compounds may occur as liquid droplets or solid particles. Some have odors which are characteristic and often objectionable. A number, notably the polynuclear aromatics, have been associated with carcinogenesis. But the majority have relatively low potential for serious air pollution effect so long as they retain their specific identities. Outstanding exceptions can be found, e.g., formaldehyde, formic acid, acrolein, and some compounds containing phosphorus and fluorine.

The nitrogen compounds most abundantly generated and released are nitric oxide, nitrogen dioxide, and ammonia. The first two of these are produced in high-temperature combustion and other industrial operations by the combination of normal atmospheric oxygen and nitrogen. While NO_2 is irritating to tissues at relatively low concentrations, the major interest in both the oxides is related to their participation in atmospheric photochemical reactions.

Carbon dioxide and carbon monoxide arise in huge amounts, from the complete and the incomplete combustion of carbonaceous fuels respectively. In Los Angeles County, the daily production of CO is estimated to exceed 10,000 tons with more than 80% of it resulting from incomplete utilization of the carbon content of gasoline in automobile engines. The ability of CO to impair the oxygen-carrying capacity of hemoglobin gives it special status as a primary pollutant. Carbon dioxide in very high concentration affects the human vascular control mechanism, but the quantity required is too great to be of much concern. Mention has already been made of the possible long-range influence of the general rise in atmospheric CO_2 on worldwide meteorological phenomena.

Certain inorganic halogen compounds, among them HF and HCl, are produced from metallurgical and other industrial processes. Both are corrosive and irritating *per se*, and the metallic fluorides have toxic properties which have precipitated some costly legal actions among operators of producing factories and neighboring residents whose crops and cattle have been severely damaged.

It is not within the scope of this discussion to elaborate on the very specialized nature, sources, or properties of radioactive pollutants. Except for fallout of nuclear weapon residues, these materials have not yet presented a major practical problem beyond the vicinities of nuclear reactor operations. That they will do so with increasing use of nuclear power and industrial applications of isotope techniques is very possible.

B. Secondary Pollutants

It was suggested earlier that the total polluted air mass over a populated area is chemically and physically unstable. As a whole the system tends, like everything else in nature, to approach a state of minimal free energy. The rates, reaction routes, and intermediate steps involved in the process are influenced by many factors such as relative concentration of reactants, degree of photoactivation, variable meteorological dispersive forces, influences of local topography, and relative amounts of moisture.

In the simplest case two species may react thermally, as in the formation of a halide salt by combination of acid mists with metallic oxides. When water droplets are airborne, solution reactions may occur, as in the formation of acid mists by reaction of dissolved oxygen and SO_2.

The formation of sulfuric acid in droplets has been shown to be enormously accelerated by the presence of certain metallic oxides such as those of Mn and Fe in droplets. This illustrates the well-established role of catalytic processes in affecting step rates in the overall system.

Surfaces of liquid and solid particles contribute variously to the energy degradation processes. They may be able to adsorb gases from very dilute mixtures, thereby accelerating normal reactions by providing discrete sites of high reactant concentration. In the adsorbed form the retention of toxic gases in the respiratory system of man may be enhanced, and the apparent irritative effect of the gas may be increased. Some species of particles provide sites for surface catalysis of simple and complex reactions, and at least a few cases have been studied in which semiconducting metallic oxide surfaces are active in the catalysis of photoenergized events.

Photochemical reactions involved in air pollution have been analyzed enough in recent years to prove their major role in smog manifestations of the type experienced in Los Angeles. The primary photochemical event appears to be the dissociation of NO_2, providing NO and O radicals which are able to initiate sustained free radical reaction chains. The number and kinds of transient radicals and semistable compounds formed are then governed by the relative abundance and susceptibility of other chemical species in the system and by environmental energy factors.

The secondary pollutants produced during events of this type are among the most troublesome that air pollution control agencies are required to abate. They include ozone, formaldehyde, organic hydroperoxides, PAN, and other very reactive compounds, as well as potentially damaging concentrations of short-lived free radicals so long as photoactivation is maintained in the presence of a sufficient supply of primary and secondary reactants. It will be recognized that free radical mechanisms do not preclude the participation of O_2, H_2O, or other normal atmospheric constituents in the formation of end products.

To unravel so complex and temporally variable a system continues to challenge air pollution research, and the precise prediction of the characteristics of the system an hour into the future may never be more reliable than a probability function. On the other hand, it is clear that the simple process of collecting and analyzing stable chemical species and physical entities, as now practiced, cannot provide sufficient knowledge of the continually changing assemblage of transient groupings which are prime factors in the effects produced by air pollution.

C. Recognized Atmospheric Processes

In addition to chemical recombinations, several other major factors regulate the impact of primary and secondary pollutants. Principal among these are processes of nucleation and condensation, sedimentation, and other air-cleansing phenomena which tend to remove substances from the atmosphere, and meteorological processes which may dilute the reactants or tend to concentrate them.

Condensation nuclei released from many sources, both natural and artificial, under appropriate circumstances can induce the accumulation of vapors into aerosols. These in turn may coalesce with other particles to an extent great enough to permit their eventual deposition on exposed surfaces. As in the case of chemical reaction, such physical processes are rate related to concentrations; it is not clear that aggregation and sedimentation play a significant role in air purification except under unusual circumstances or with respect to coarse particles.

The energy-degrading mechanisms discussed in the previous section can be regarded as natural purification processes in the sense that their end products are less reactive, and therefore usually less troublesome than the primary or intermediate pollutants. The mechanism is analogous in its effect to the biological oxidation of organic pollutants in sewage and water supplies; in each case the oxidized products are relatively ineffective physiologically. Thus the development of a large excess of ozone in an air mass usually signals an ensuing rapid decline in eye irritation, presumably because the primary organic reactant in the air mass is near exhaustion and the ozone itself assures final oxidation of the irritating intermediate compounds.

Much closer analogies between water pollution and air pollution problems are apparent when physical dispersive factors are considered. In each case the volume of medium available for dilution of contaminants and the speed of mixing are dominant in determining the capability of the stream or the air mass to accommodate a given output without presenting localized or general affronts to users of the water or air.

Air supplies are affected in this quantitative sense by the degree of containment beneath inversions, the magnitude of horizontal and vertical wind movements, and by the degree of turbulence induced by convection and nonlinear flow. As is the case with all types of meteorological phenomena, these factors are governed by both external synoptic forces and by localized topographic and thermal influences.

It is possible, given sufficient data, to establish wind direction and velocity frequencies, local and regional thermal variations, and other pertinent factors and to develop equations expressing the most probable concentrations of pollutants likely to occur in relation to a source of known characteristics. Hypothetically it is also possible to relate these influences to future air pollution events in and adjacent to a large community of different sources. Some suggestive elementary models have been proposed. But as practical tools for the regulation of regional problems, meteorological analyses have not been used effectively except in relation to localized emissions and their effects in the immediate vicinity.

The existence of *usual* patterns of air movement over specific geographical areas has suggested to many the possibility of affecting favorably the quality of air supplies by some form of zoning or regulated placement of sources. Sites in deep valleys and elsewhere, subject to frequent inversion entrapment, can be recognized as unfavorable to maintenance of good air quality. However, it is unfortunately true that the acute air pollution episodes in affected localities occur when the meteorological pattern is *not usual*. For this reason a question may be raised as to the ultimate usefulness of regional zoning as a means of controlling community air pollution. The whole matter of the relationship between meteorological probability and relatively infrequent atypical pollutional occurrences associated with unusual air movement, or lack of it, is intensely interesting. But it is currently a mathematical exercise involving equations, the solutions of which may well remain indefinite.

Meteorological factors are the chief diluters and dispersers of pollution. Where and when they fail to perform these functions adequately, the sources of pollution must be controlled.

IV. Types of Effects Associated with Air Pollution

In the following chapters detailed attention is devoted to the kinds and magnitudes of effects known and suspected to be produced by air pollution. They may be grouped under five general headings.

A. Visibility Reduction

Historically the earliest noted and currently the most easily observed effect of air pollution is the reduction in visibility produced by the scattering of light from the surfaces of airborne particles. The degree of light obstruction is related to particle size, aerosol density, thickness of the affected air mass, and certain more subtle physical factors. Particulates responsible for the phenomenon may be either primary pollutants, e.g., coal smoke, or secondary, e.g., photochemical smog. At times London and eastern American cities have been so seriously affected by pollutional reduction in visibility as to have experienced severe curtailment of transport and other municipal activities. The attenuation of ultraviolet and other radiations reaching the surface through layers of aerosols may be associated with adverse physiological effects in man and vegetation.

B. Material Damage

Direct damage to structural metals, surface coatings, fabrics, and other materials of commerce is a frequent and widespread effect of air pollution. The total annual loss from these and the incidental increase in cleansing and protective activities in the United States are not accurately known, but have been estimated at several billion dollars. This destruction is related to many types of pollutants, but is chiefly attributable to acid mists, oxidants of various kinds, H_2S, and particulate products of combustion and industrial processing. Secondary pollutants contribute a substantial share. For example, O_3 is known to cause rapid and extensive damage to many kinds of rubber goods and textiles.

C. Agricultural Damage

A large number of food, forage, and ornamental crops have been shown to be damaged by air pollutants. Curtailed value results from various types of leaf damage, stunting of growth, decreased size and yield of fruits, and destruction of flowers. Some plant species are so sensitive to specific pollutants as to be useful in monitoring air quality. Annual bluegrass, the pinto bean, spinach, and certain other forms have been so employed.

Substances thus far identified as responsible for the damage include ethylene, PAN, SO_2, acid mists, fluorides, O_3, and a number of organic oxidants. Research on the etiology, physiology, and biochemistry of air pollution pathologies in plants has contributed to knowledge of related phenomena in man.

D. Physiological Effects on Man and Domestic Animals

Donora, Poza Rica, London, and the Meuse Valley of Belgium have given dramatic proof that air pollution can kill; and, together with other evidence, they have implied less shocking but more extensive effects of air pollutants on the health of affected populations. Long, continued exposure to sublethal concentrations of many substances and combinations thereof are suspected to have physiological effects, but in most cases the quantitative aspects of the relationships remain undefined.

The high incidence of "chronic bronchitis" in British cities, nasopharyngeal and optic irritation in Los Angeles, and the rapid rises in lung carcinoma among metropolitan populations appear to be closely associated with air pollution. Fluorosis in cattle exposed to fluoride-containing dusts has been proved to be related to emissions from certain industrial operations. More subtle physiological effects of air pollution are suggested by laboratory observations of suppression of ciliary action, alterations in pulmonary physiology, specific enzymic inhibitions, and changes in blood chemistry.

E. Psychological Effects

Since fear is a recognizable element in public reactions to air pollution, the psychological aspects of the phenomenon cannot be ignored. Psychosomatic illnesses are possibly related to inadequate knowledge of a publicized threat. Little effort has been directed toward evaluation of such impacts in relation to general mental health of affected groups or determination of their role in individual neuroses. Only in practical politics has any significant action been based on recognition of the psychological attitudes induced by periodic public exposure to an airborne threat.

V. Air Pollution as a Problem of the Future

In spite of its long history of development, community air pollution must be looked upon as a problem of the future. Only a few of the largest population concentrations of the present day are occasionally using their air supplies faster than natural processes can replenish them. Such overuse must be expected to occur with increasing frequency as populations increase, since per capita demands for air cannot be expected to decline.

So long as the air resource was almost infinitely large in relation to daily withdrawal and use, its pollution caused discomfort and illness only in areas immediately adjacent to individual sources. As regional and world populations increase, a time must come when human occupation of the medium will threaten the quality of the total air resource. On a world scale, residues of nuclear weapon testing and huge outpourings of CO_2 from fossil fuel combustion have already demonstrated the extent to which human activity can affect the total gaseous milieu.

Air pollution shares with all other threats to public health and welfare the certainty of becoming more and more severe as long as the population increase remains unchecked. Unless some effective population control is permitted to intervene, the monetary cost of maintaining an acceptable air quality can be expected to rise in some exponential relationship to the numbers of people and associated activities requiring it. This will be true regardless of the speed with which fossilized energy sources are replaced by thermonuclear or solar power plants. Nevertheless, such substitutions can be expected to delay, perhaps for generations, the development of large-scale, completely intolerable, and economically uncontrollable situations.

The emergence of air pollution as a regional or even global phenomenon has already had significant impacts on governmental and administrative procedures. Air masses recognize no political jurisdictions, and in their movements frequently do violence to democratically evolved concepts of local autonomy. As the geographic breadth of air resource problems increase there will appear administrative mechanisms designed to deal with them as regional, national, or international entities. Since they involve aspects of transportation, refuse disposal, industrial zoning, and power utilization, it is difficult to believe that jurisdictional adjustments to meet the regulatory need will conform with traditional governmental concepts.

Water, food, and air must forever constitute the survival bases of human and other populations; we will pay for them whatever they cost in time, money, and effort, since without them we die. It is therefore shortsighted to consider the air resource as a competitively priced commodity; it is priceless. Today it may be appropriate to consider which of several alternative air pollution control measures can be imposed without affecting the public purse unduly; this is true because there are still not enough of us to pose more than a marginal threat to air quality other than locally. But a quadrupled population, if realized, may force consideration of basic resources as fundamental limits to survival, rather than as dollar-valued items affecting the cost of comfort.

2

The Primary Air Pollutants—Gaseous
Their Occurrence, Sources, and Effects

Paul Urone

I. Introduction

A. The Atmosphere

The earth is completely enveloped in a layer of a gaseous mixture composed largely of nitrogen and oxygen and called the atmosphere. It is held by gravity but moves with relative ease in complex patterns over the face of the earth. It has its highest density at the earth's surface but thins rapidly with increasing height. Although traces of atmospheric gases are found rotating with the earth as high as 6000 miles, more than half of the mass of the atmosphere lies below 3.5 miles and 99% below 18 miles (29 km): a very thin layer when compared to the 8000 miles of the earth's diameter [Fig. 1; (*1*)].

1. The Homosphere and Heterosphere

Based upon *molecular composition,* the atmosphere is divided into two general regions: the homosphere and the heterosphere. The homosphere extends outward to some 55 miles (90 km) from the earth. Except for water vapor which varies over a wide range of concentrations, the homosphere is distinguished by the uniformity of its composition. This is a

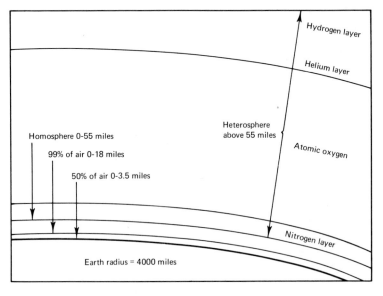

Figure 1. The homosphere and heterosphere relative to earth's radius (*1*).

result of the continuous turbulent movements of the air and winds in the lower regions of the atmosphere. As a result the homosphere, on a dry basis, is often treated as a single gas with a definite set of physical properties.

Above the homosphere is the heterosphere. Because of gravitational forces and a low rate of mixing with the lower regions of the atmosphere as well as the intensity of the higher energy photons coming from the sun, the heterosphere has developed into four distinguishable layers (2). The lowest layer of the heterosphere exhibits a preponderance of molecular nitrogen. It extends from approximately 55 to 125 miles from the earth. Above it is the atomic oxygen layer which extends from 125 to 700 miles. The third layer is enriched in helium and extends from 700 to 2000 miles. The fourth layer is the hydrogen layer, and it extends from 2000 to 6000 or more miles from the earth's surface. Although the heterosphere contains less than 0.01% of the mass of the atmosphere, its presence is highly important for life on earth. It helps filter out the highly energetic portions of the sun's rays which otherwise would ionize or burn up living organic matter. The heterosphere forms a very strong oxidative reservoir for impurities that diffuse upward, and man uses it for the propagation and reflection of radio waves.

2. The Temperature-Related Regions

The atmosphere is also classified into four characteristic temperature regions [Fig. 2; (3)]. In the lowest portion of the atmosphere, called the troposphere, the air temperature decreases with height up to 5–10 miles (8–17 km) depending upon latitude and the season of the year. The upper boundary of the troposphere is closest to the earth at the poles in winter, and furthest at the equator during the summer. The rate of temperature change with altitude is called the lapse rate. For a "normal" atmosphere the lapse rate is —3.5°F/1000 ft (0.64°C/100 m). The "dry adiabatic" lapse rate thermodynamically is —5.5°F/1000 ft (1°C/100 m). The actual temperature lapse rate for a given locality varies with ambient weather conditions and is important with respect to the effect it has on the turbulent vertical movements of the air (4).

Above the troposphere the lapse rate decreases rather sharply. The air temperature becomes stabilized and starts to increase slightly with altitude to a height of approximately 20 miles (32 km). This layer of temperature-stabilized air forms a stratified cover over the troposphere and furnishes the basis for its name: the stratosphere. The transition region between the troposphere and the stratosphere is called the tropopause.

The stratosphere is characterized by its even temperature (—60° to

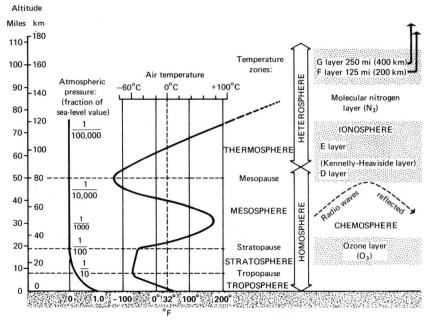

Figure 2. Structure of the atmosphere (3). (A. N. Strahler, "The Earth Sciences," 2nd ed., pp. 18–62. Copyright 1971. Reprinted by permission of Harper and Row, New York, New York.)

—90°F) and by the absence of convective currents of air. Consequently, diffusion through the stratosphere is quite slow. A residence time of 2 years has been estimated for sulfate particles (5). Concern has been expressed for use of the stratosphere for supersonic airplane flights (6). Emissions of such flights could create slowly diffusing high altitude contrails of relatively high concentrations of water vapor and carbon dioxide. Any nitric oxide emitted would react rapidly with ozone with a consequent lowering of the atmosphere's shielding ability for ultraviolet radiation below 3000-Å wavelength.

Above the stratopause, the mesosphere increases in temperature to about 30 miles (48 km) after which the temperature falls to less than —100°F at the mesopause. Above the mesopause the mesosphere blends into the thermosphere where the temperature again rises indefinitely. At this altitude the air is extremely thin. The molecules and atoms acquire high kinetic energy, hence their high "temperature." However the actual heat content is quite small.

3. The Chemosphere and Ionosphere

Another broad classification of the atmosphere is one that is based on its chemical and physical properties. In this case, the atmosphere is by definition divided into the chemosphere and the ionosphere (*2*). The chemosphere exhibits the chemical properties of the molecules, atoms, and free radicals of which it is composed. It includes the ozone layer which has a maximum concentration at about 20 km. Other active species include atomic oxygen and hydrogen and hydroxyl (OH) and hydroperoxide (HO_2) radicals (*6*).

The ionosphere contains relatively large numbers of ions and is recognized by its ability to reflect radio waves. Several layers have been identified. Their height and thickness depend upon the angle of the sun and the longevity of the ions. The lowest layer is known as the Kennelly–Heaviside, or D, layer. The others have been named E, F_1, F_2, and G. The number of ions in the layers vary from 10^3 to 10^6 per cubic centimeter (*7*).

4. Composition of the Homosphere

Figure 3 graphically portrays the volume percent composition of the air in the homosphere on a dry basis. Mole percent relationships and volume percent relationships are numerically equal. Partial pressures of the individual components of a gaseous mixture are proportional to either the volume or mole percent.

Except for slight variations in the trace components, the composition of dry air in the homosphere is uniform enough to be assigned an average "molecular" weight of 28.96 (*8*). This holds to an altitude of about 50

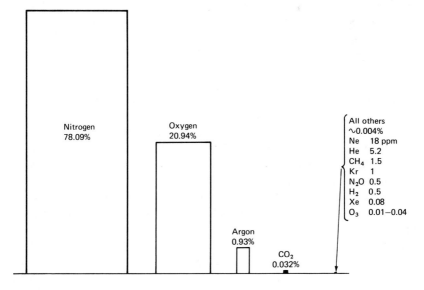

Figure 3. Composition of dry air (% v/v).

miles (80 km). Above this the composition changes gradually toward the lighter gases, and the "molecular" weight at 400 miles (640 km) is about 16.8. At lower altitudes, the presence of water vapor which has a molecular weight of 18.02 renders the air lighter and more buoyant.

Nitrogen and oxygen are by far the most abundant atmospheric gases, and when combined with argon, account for approximately 99.96% of the air. All other components are minor and their concentrations are more easily expressed as parts per million (ppm) rather than percent. Carbon dioxide is the most abundant of the minor gases. Its concentration varies with time and location averaging about 0.032% (320 ppm). The remainder of the atmospheric gases are present in trace amounts totaling between 0.003 and 0.004% (30–40 ppm).

Nitrogen, the most abundant atmospheric gas, is a relatively inert, diatomic gas with a molecular weight of 28.01. It occupies 78.09% of the volume of the atmosphere. The two atoms of the molecule are tightly bound together. Nitrogen enters chemical reactions only under special conditions, such as in the presence of lightning, fixation by certain bacteria, or at high temperatures or pressures, such as in forest fires, combustion sources, and the automobile engine.

Oxygen, the second most abundant gas in the atmosphere, comprises some 20.94% of the atmosphere by volume and 23.21% by weight. When considered in both its free and combined forms it is the most abundant of all the elements and accounts for 49.2% of the combined mass of the earth's crust, including the oceans and the atmosphere. Considering the molecular structure of water (H_2O), quartz (SiO_2), limestone ($CaCO_3$), etc., oxygen occupies more than 90% of the volume of the atmosphere, hydrosphere (liquid water realm), and lithosphere (solid crust realm) combined. Like nitrogen, oxygen in its free state is diatomic. It has a molecular weight of 32.00 and a density of 1.43 gm/liter at 0°C and 1 atm pressure. Unlike nitrogen, oxygen is very reactive. It is essential to life, being the necessary ingredient for metabolic processes. Chemical reaction with oxygen (called oxidation) is an ever present phenomenon involving living and inert matter. The rate of oxidation ranges from extremely slow to explosively fast, depending upon the type of substance, the temperature, and the concentration of oxygen present. Oxygen is several times more soluble in water than is nitrogen and is effective in oxidizing substances in solution and in thin films of water adsorbed on nonporous and porous surfaces.

Argon, the third most abundant gas in the atmosphere, is present at slightly less than 1% (9300 ppm). This amounts to a significant 5×10^{13} tons on a global basis. It is essentially inert and is generally ignored in atmospheric processes. It is a monoatomic gas and is relatively heavy, with an atomic weight of 39.95. It has a boiling point near that of liquid oxygen ($-186°$ vs $-183°C$ for oxygen). It can be assumed to be present in liquid or compressed oxygen unless special precautions have been taken to remove it. In the presence of an electric arc, such as in welding, it emits intense ultraviolet rays which can cause severe eye and skin damage in the absence of proper shielding.

Helium, neon, krypton, and xenon are the other noble gases found in trace amounts in air. Like argon they are essentially inert, forming a few rare compounds under special conditions. All have special industrial, commercial, and research uses. Krypton is of interest because of its long-lived radioactive isotope (^{85}Kr), a by-product of nuclear power generation (9, 10).

B. Water Vapor

The amount of water vapor in the troposphere (lower homosphere) is highly variable. It depends upon geographic location, nearness of water bodies, wind directions, and ambient air temperatures. It may be present from a low of 0.02% (200 ppm) in arid regions to as much as 6% (60,000 ppm) in warm humid climates.

The water vapor content of air is generally measured as percent of the saturation vapor pressure of water at that temperature and is expressed as percent relative humidity. This does not describe the absolute amount of water in the air as is shown in Figure 4 where the relative humidity of a given amount of water vapor is shown to vary widely depending on the ambient air temperature. The density of water vapor is less than two-thirds that of dry air, and it has an important effect on its buoyancy and turbulence. The heat capacity of water vapor is twice that of dry air and is important in atmospheric movements caused by temperature lapse rates. The adsorption of water vapor, on the other hand, is a function of the relative saturation vapor pressure, or relative humidity, as is shown in Figure 5 (11). Highest adsorption is observed at high relative humidities irrespective of the absolute amount of water vapor present.

The water molecule is small and highly polar. It adsorbs strongly on many substances and forms at least a monomolecular layer on most polar solids. It has been shown to significantly increase the amount of light scatter of suspended particles in air (12–14). The effect is most noticeable at high relative humidities where multiple molecular layer adsorption and capillary condensation is encountered. Corrosive gases, such as sulfur dioxide, dissolve in the water attached to, or adsorbed on, building materials and works of art to react slowly with time.

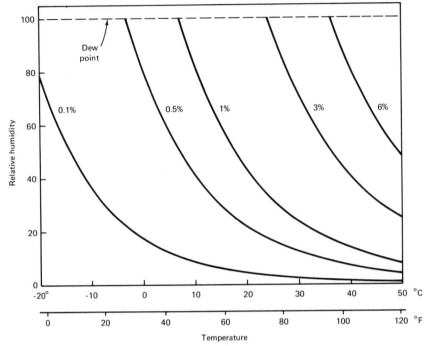

Figure 4. Variation of relative humidity with temperature for given percentages of moisture in air.

Water vapor plays an important role as solvent and catalyst for many thermal and photochemical reactions in the atmosphere. It can be assumed to be present to a significant extent in all air pollution situations. The nucleation of water vapor into raindrops and snowflakes and their precipitation clears the air of unknown megatons of pollutants per year.

The following equation is used to convert measurements of atmospheric composition (and pollutant) concentrations from a dry to a wet basis (and vice versa):

$$\% \text{ Wet basis} = \% \text{ dry basis} \times \frac{100 - \% \text{ H}_2\text{O}}{100} \tag{1}$$

C. The Gaseous State

1. Ideal Gas Law

At ambient temperature, the major gases of the atmosphere are removed enough from their critical temperatures that the ideal gas laws

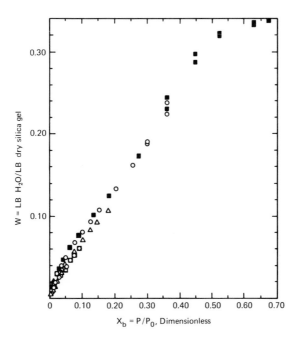

Figure 5. Amount of water adsorbed on silica gel at different temperatures as a function of relative humidity (P/P$_o$) (*11*). ■: 82°F; ○: 100°F; △: 125°F; □: 150° F; ◆: 175°F; ▲: 200°F.

can be applied without serious error. For most practical purposes, the trace gases and vapors (a vapor is the gaseous form of a substance that is liquid or solid at ambient temperature and pressure, i.e., alcohol vapor, methane gas) are also considered to behave like ideal gases because they are so dilute and so far removed from their saturation vapor pressures or dew points.

The ideal gas law

$$PV = nRT \qquad (2)$$

represents the pressure, volume, and absolute temperature relationships of a gas under consideration. R is the gas constant and is calculated from

$$R = (PV/T)_{\text{STP}} \qquad (3)$$

where $(PV/T)_{\text{STP}}$ represents the standard pressure, volume, and absolute temperature of a mole of gas in the units being used (English or metric).

The number of moles n may be calculated from

$$n = \frac{\text{weight of substance}}{\text{molecular weight}}$$

$$= \frac{\text{volume of substance}}{\text{mole volume at same temperature and pressure}} \qquad (4)$$

2. Chemical Dynamics

The trace gases in air retain the physical and chemical properties of their pure state and react accordingly. They are, however, highly dilute in a complex medium of relatively large amounts of nitrogen, oxygen, argon, and water vapor. Carbon dioxide is generally two or three orders of magnitude greater. A wide variety of other chemical species and particles, viable and nonviable, are also present. During daylight hours the atmosphere is bathed with photons of energies varying from those of the near ultraviolet (90 kcal/mole) to the far infrared (<1 kcal/mole) and capable of forming free radicals or energetically excited species as primary photochemical reactants. Secondary photochemical reactions involving the free radicals and excited species are common. Photochemical reactions in air tend to be "oxidative," with a considerable amount of polymerization and condensation—resulting in the formation of haze or smog.

The mass action law and equilibria relationships hold. The high state of dilution ensures that only the more rapid and persistent reactions can occur to any significant extent. Reactions with oxygen will tend to be pseudo first order because of the relatively large concentration of oxygen in the air. Reactions can be photochemical, thermal, homogeneous in a gaseous or liquid medium, or heterogeneous. The complexity of polluted atmospheres combined with wind movements, diurnal variations in sunlight and dark, and the difficulty of measuring trace substances make it difficult to study and understand atmospheric chemical reactions.

3. Expressions of Gaseous Concentrations

For various reasons, the amounts and concentrations of trace gases in air are expressed in a number of ways involving units of weight or volume per unit weight or volume of air in either the English or metric systems. To those not familiar with such expressions some confusion can occur. The relationships are really quite simple. Just as percent equals the part divided by the whole and then multiplied by 100, then

$$\text{Percent } (\%) = \frac{\text{part}}{\text{whole}} \times 10^2 \tag{5}$$

$$\text{Parts per million (ppm)} = \frac{\text{part}}{\text{whole}} \times 10^6 \tag{6}$$

$$= \text{percent} \times 10^4$$

$$\text{Parts per hundred million (pphm)} = \frac{\text{part}}{\text{whole}} \times 10^8 \tag{7}$$

$$\text{Parts per billion (ppb)} = \frac{\text{part}}{\text{whole}} \times 10^9 \tag{8}$$

The units used may be the same: i.e., volume ÷ volume (v/v), weight ÷ weight (w/w), or they may differ as weight ÷ volume (w/v), etc. Most atmospheric ratio expressions, e.g., parts per million, are assumed to indicate a volume–volume relationship, unless otherwise noted.

The concentration of a pollutant whether it be gaseous or particulate is at present commonly expressed on a weight per unit volume basis, usually as micrograms of pollutant per cubic meter of air; i.e., $\mu g/m^3$. The conversion between $\mu g/cm^3$ and ppm is achieved by the following equations:

$$\mu g/m^3 = \frac{\text{weight of substance, gm}}{\text{volume of air sampled, m}^3} \times 10^6 \tag{9}$$

$$\mu g/m^3\,* = \text{ppm} \times \frac{\text{molecular weight}}{24{,}450} \times 10^6$$

$$= \text{ppm} \times \text{molecular weight} \times 40.90 \tag{10}$$

$$\mu g/m^3 = \text{grain/ft}^3 \times 2.29 \times 10^6 \tag{11}$$

$$\text{ppm}\,* = \mu g/m^3 \times \frac{2.445 \times 10^{-2}}{\text{molecular weight}} \tag{12}$$

$$\% \text{ Weight basis} = \% \text{ volume basis} \times \frac{\text{molecular weight}}{28.96} \tag{13}$$

4. Averaging Time

Concentrations of pollutants in air may be expressed as instantaneous values or as time-averaged values depending upon the type of analytical instrumentation used and the statistical basis for presentation of the data. Some analytical measurements require 5 minutes or more of sam-

---

* At 25°C, 1 atm pressure.

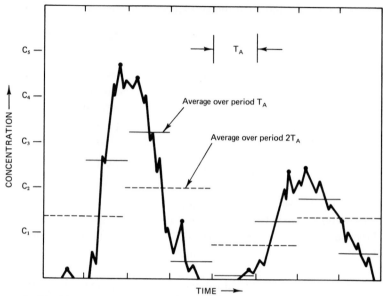

Figure 6. Graphic representation of effect of instantaneous measure of concentration and time averaging of concentration (15). ●: peak value in T_A.

pling time. In such cases the measurement represents an integrated value for the period of the test. Daily, monthly, and/or yearly averages are often more illustrative of a pollution exposure level than a large number of data points.

Figure 6 shows the relationship of the instantaneous values to the time-averaging values (15). Table I shows the relationship among various

Table I   Generalized Relationships among
Commonly Used Averaging Times (16)

Averaging time employed	Relative maximum average concentration in the atmosphere
1 month	0.5
24 hr	1.0
8 hr	1.2
2 hr	1.8
1 hr	2.2
30 min	2.4
15 min	2.7
Single measurement[a]	3.3

[a] Encompasses range up to 10 minutes.

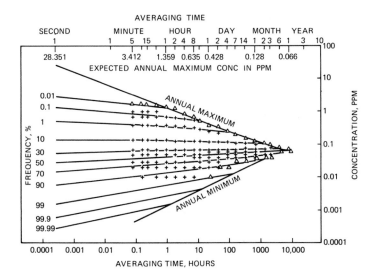

Figure 7. Concentration vs averaging time and frequency for nitrogen oxides in Washington, D.C., from 12/1/61 to 12/1/64 (*17*).

averaging times relative to a 24-hour average concentration taken as 1.0 (*16*). Larsen [Fig. 7; (*17*)] found that the concentrations of pollutants at a given point follow a log-normal distribution. Depending on the degree of exposure; the slopes and the variances of such a distribution would vary for individual cases. Plots of cumulative frequency measurements for various averaging times, on a log basis, develop an "arrowhead" profile showing maximum and minimum concentrations for the various averaging times. Maximum concentrations (C) can be found by

$$C = M_g \sigma_g{}^z \tag{14}$$

where M_g is the geometric mean of a number of measurements (N) obtained during a given period of time (such as 1 or 2 years); σ_g is the geometric deviation of the measurements for the same time period; and z is the "standard normal deviate" of the measurements which is equal to the probability calculated for a single occurrence out of $1.67\,N$ measurements having the given σ_g for the measurements (*17, 18*).

II. The Polluted Atmosphere

Polluted atmospheres generally are associated with man's industrial and domestic activities. However, many of the major gaseous pollutants

Table II   Comparison of Trace Gas Concentrations (ppm) (19)

	Clean air (20)	Polluted air (19)	Ratio polluted-to-clean
CO_2	320	400	1.3
CO	0.1	40–70	400–700
CH_4	1.5	2.5	1.3
N_2O	0.25	(?)	—
$NO_2(NO_x)$	0.001	0.2	200
O_3	0.02	0.5	25
SO_2	0.0002	0.2	1000
NH_3	0.01	0.02	2

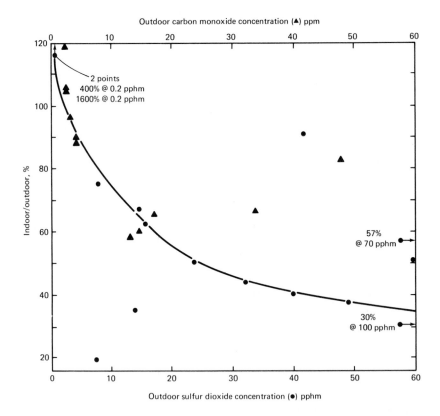

Figure 8. Indoor concentrations of sulfur dioxide and carbon monoxide as a function of outdoor concentrations (21).

are also emitted by nature. Taken on a worldwide basis, the total mass of trace gases emitted by nature exceed those emitted by man by several orders of magnitude. Nonetheless, man's activities do adversely affect the quality of the atmosphere, particularly in dense urban areas and near large emission sources. For many of the pollutants, serious long-term worldwide effects are feared. The effects may be immediate and obvious, such as poor visibility, eye irritation, and objectionable odors; or the effects may be noticeable only through longer periods of observation, such as in corrosion. More subtle effects require sophisticated statistical studies to determine such things as human health effects and changes in the earth's energy balance.

Table II compares typical concentrations of pollutants (19) with those found in uncontaminated areas. It can be seen that the ratio of concentration of polluted air to clean air ranges from fractional to a 1000-fold. Table III by Robinson and Robbins (20) summarizes the worldwide sources, atmospheric concentrations, residence times, and removal reactions for eight principal gaseous air pollutants. Except for sulfur dioxide, emissions from natural sources exceed those from pollution sources. Figures 8 and 9 show the relationship between outdoor and indoor pollution levels for sulfur dioxide and carbon monoxide. Measurements such as these indicate serious penetration into homes near strong pollution sources (21).

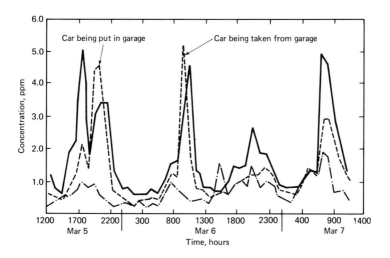

Figure 9. Carbon monoxide concentrations in house with gas range and furnace and with attached garage (21). Solid line, kitchen; dashed line, family room; dot-dashed line, outside.

Table III Summary of Sources, Concentrations, and Major Reactions of Atmospheric Trace Gases (20) [E. R. Robinson and R. C. Robbins, in "Air Pollution Control" (W. Strauss, ed.), Part II, pp. 1–93. Copyright 1972, Wiley (Interscience), New York, New York. Reprinted by permission of John Wiley & Sons, Inc.]

Contaminant	Major pollution sources	Natural sources	Estimated emissions (tons)		Atmospheric background concentrations	Calculated atmospheric residence time	Removal reactions and sinks	Remarks
			Pollution	Natural				
SO₂	Combustion of coal and oil	Volcanoes	146×10^6	No estimate	0.2 ppb	4 days	Oxidation to sulfate by ozone or, after absorption, by solid and liquid aerosols	Photochemical oxidation with NO_2 and HC may be the process needed to give rapid transformation of $SO_2 \rightarrow SO_4$
H_2S	Chemical processes, sewage treatment	Volcanoes, biological action in swamp areas	3×10^6	100×10^6	0.2 ppb	2 days	Oxidation to SO_2	Only one set of background concentrations available
CO	Auto exhaust and other combustion	Forest fires, oceans, terpene reactions	304×10^6	33×10^6	0.1 ppm	<3 years	Probably soil organisms	Ocean contributions to natural source probably low
NO/NO_2	Combustion	Bacterial action in soil (?)	53×10^6	NO: 430×10^6 NO_2: 658×10^6	NO: 0.2–2 ppb NO_2: 0.5–4 ppb	5 days	Oxidation to nitrate after sorption by solid and liquid aerosols, hydrocarbon photochemical reactions	Very little work done on natural processes
NH_3	Waste treatment	Biological decay	4×10^6	1160×10^6	6 ppb to 20 ppb	7 days	Reaction with SO_2 to form $(NH_4)_2SO_4$ oxidation to nitrate	Formation of ammonium salts is major NH_3 sink
N_2O	None	Biological action in soil	None	590×10^6	0.25 ppm	4 years	Photodissociation in stratosphere, biological action in soil	No information on proposed absorption of N_2O by vegetation
Hydrocarbons	Combustion exhaust, chemical processes	Biological processes	88×10^6	CH_4: 1.6×10^9 Terpenes: 200×10^6	CH_4: 1.5 ppm non CH_4: <1 ppb	4 years (CH_4)	Photochemical reaction with NO/NO_2, O_3; large sink necessary for CH_4	"Reactive" hydrocarbon emissions from pollution = 27×10^6 tons
CO_2	Combustion	Biological decay, release from oceans	1.4×10^{10}	10^{12}	320 ppm	2–4 years	Biological adsorption and photosynthesis, absorption in oceans	Atmospheric concentrations increasing by 0.7 ppm/year

III. The Gaseous Compounds of Carbon

The gaseous compounds of carbon found in natural and polluted atmospheres comprise a broad spectrum of the compounds of organic chemistry. Because carbon can form bonds with elements such as hydrogen, oxygen, nitrogen, and sulfur and at the same time combine with itself to form a series of straight and branched chain, cyclic, and combined cyclic–chain systems, an almost infinite number of compounds are possible. Many gaseous carbon compounds such as methane (marsh gas), carbon dioxide, carbon monoxide, the terpenes [Table IV; (22)], and other plant volatiles are emitted in nature through biological processes, volcanic action, forest fires, natural gas seepage, etc. In areas inhabited by man, the emissions of commerce, industry, and transportation are largely concentrated in urban areas and generate high local concentrations of volatile solvents and fossil fuel combustion products.

A. The Hydrocarbons

Table V shows the emissions of hydrocarbons in the United States since 1940 (23). Transportation is by far the principal emitting source, and its emissions seem to have peaked off slightly starting in 1968. Table VI gives the average concentration for some 30 hydrocarbon compounds identified and measured in Los Angeles, California air (24). More than

Table IV   Worldwide Terpene Emission Estimates (22)

Investigator	Method	Estimate in tons
Went[a]	Sum of sagebrush emission and terpenes as percentage of plant tissues	175×10^6
Rasmussen and Went[b]	1. Bagging foliage 1 liter/10 cm^2	23.4×10^{6d}
	2. Enclosure forbs 0.65 m^3/m^2	13.5×10^{6d}
	3. Direct in situ ambient conc.	432×10^6
Ripperton, White, and Jeffries[c]	Reaction rate O_3/pinene	2 to 10 $\times$ previous estimates

[a] F. W. Went, Proc. Nat. Acad. Sci. **46**, 212 (1960).
[b] R. A. Rasmussen and F. W. Went, Proc. Nat. Acad. Sci. **53**, 215 (1965).
[c] L. A. Ripperton, O. White, and H. E. Jeffries, "Gas Phase Ozone-Pinene Reactions," pp. 54–56. Div. of Water, Air, and Waste Chemistry, 147th Nat. Meeting Amer. Chem. Soc., Chicago, Illinois, 1967.
[d] Not corrected for vertical foliage area over ground area.

Table V   Estimates of Hydrocarbon Emissions, 1940–1970 (10⁶ tons/year) (United States) (23)

Source category	1940	1950	1960	1968	1969	1970
Fuel combustion in stationary sources	1.4	1.3	1.0	1.0	0.9	0.6
Transportation	7.5	11.8	18.0	20.2	19.8	19.5
Solid waste disposal	0.7	0.9	1.3	2.0	2.0	2.0
Industrial process losses	3.3	5.2	4.3	4.4	4.7	5.5
Agricultural burning	1.9	2.1	2.5	2.8	2.8	2.8
Miscellaneous	4.5	4.2	4.4	4.9	5.0	4.4
Total	19.1	25.6	31.6	35.2	35.2	34.7
Total controllable[a]	14.7	21.4	27.2	30.3	30.2	30.3

[a] Miscellaneous sources not included.

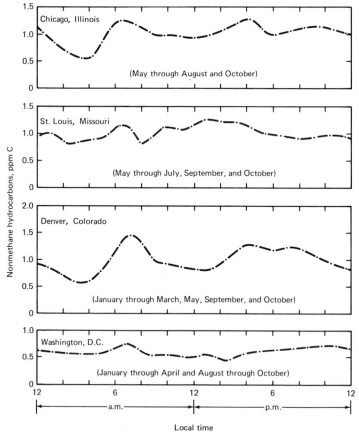

Figure 10. Nonmethane hydrocarbons by flame ionization analyzer, averaged by hour of day over several months for various cities (25).

Table VI   Average Hydrocarbon Composition, 218 Ambient Air Samples, Los Angeles, California (24)

Compound	Concentration	
	ppm	ppm (as carbon)
Methane	3.22	3.22
Ethane	0.098	0.20
Propane	0.049	0.15
Isobutane	0.013	0.05
n-Butane	0.064	0.26
Isopentane	0.043	0.21
n-Pentane	0.035	0.18
2,2-Dimethylbutane	0.0012	0.01
2,3-Dimethylbutane	0.014	0.08
Cyclopentane	0.004	0.02
3-Methylpentane	0.008	0.05
n-Hexane	0.012	0.07
Total alkanes (excluding methane)	0.3412	1.28
Ethylene	0.060	0.12
Propene	0.018	0.05
1-Butane + isobutylene	0.007	0.03
trans-2-Butene	0.0014	0.01
cis-2-Butene	0.0012	Negligible
1-Pentene	0.002	0.01
2-Methyl-1-butene	0.002	0.01
trans-2-Pentene	0.003	0.02
cis-2-Pentene	0.0013	0.01
2-Methyl-2-butene	0.004	0.02
Propadiene	0.0001	Negligible
1,3-Butadiene	0.002	0.01
Total alkenes	0.1020	0.29
Acetylene	0.039	0.08
Methylacetylene	0.0014	Negligible
Total acetylenes	0.0404	0.08
Benzene	0.032	0.19
Toluene	0.053	0.37
Total aromatics	0.085	0.56
Total	3.7886	5.43

60 hydrocarbons have been identified, but the total number possible is very large and is limited only by the sensitivity and selectivity of the analytical method used (25). The compounds are classified into four major functional types: alkanes (paraffins), alkenes (olefins), acetylenes, and aromatics. The concentrations are expressed in both parts per million

Table VII   Ozone Levels Generated in Photooxidation[a] of Various Hydrocarbons with Oxides of Nitrogen (25, 27)

Hydrocarbon	Ozone level, ppm	Time, min
Isobutene	1.00	28
2-Methyl-1,3-butadiene	0.80	45
trans-2-Butene	0.73	35
3-Heptene	0.72	60
2-Ethyl-1-butene	0.72	80
1,3-Pentadiene	0.70	45
Propylene	0.68	75
1,3-Butadiene	0.65	45
2,3-Dimethyl-1,3-butadiene	0.65	45
2,3-Dimethyl-2-butene	0.64	70
1-Pentene	0.62	45
1-Butene	0.58	45
cis-2-Butene	0.55	35
2,4,4-Trimethyl-2-pentene	0.55	50
1,5-Hexadiene	0.52	85
2-Methylpentane	0.50	170
1,5-Cyclooctadiene	0.48	65
Cyclohexene	0.45	35
2-Methylheptane	0.45	180
2-Methyl-2-butene	0.45	38
2,2,4-Trimethylpentane	0.26	80
3-Methylpentane	0.22	100
1,2-Butadiene	0.20	60
Cyclohexane	0.20	80
Pentane	0.18	100
Methane	0.0	—

[a] Hydrocarbon concentration (initial) 3 ppm; oxide of nitrogen (NO or NO_2, initial) 1 ppm.

(ppm) and parts per million as carbon (ppm C). The latter is calculated by multiplying the former by the number of carbon atoms in the respective compound. Parts per million as carbon is considered to be more representative of the hydrocarbon burden of the air.

In themselves, the hydrocarbons in air have relatively low toxicity. They are of concern because of their photochemical activity in the presence of sunlight and nitrogen oxides (26, 27). They react to form photochemical oxidants of which the predominant one is ozone (Table VII). Oxidants, including peroxyacyl nitrate (PAN), are responsible for much of the plant damage and eye irritation associated with smog. Methane has very low photochemical activity. As a consequence, hydrocarbon concentrations are often measured separately as methane on the one hand

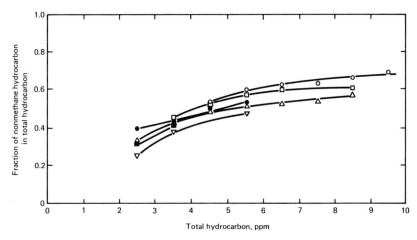

Figure 11. Nonmethane hydrocarbon fraction to total hydrocarbon for selected locations *(28)*. ○: Los Angeles, California, 1967; □: Azusa, California, 1967; △: Los Angeles, California, 1968; ▽: Los Angeles, California, 1968—Sundays; ◆: Brooklyn, New York, 1969; ■: Bayonne, New Jersey, 1968.

and nonmethane hydrocarbons on the other (Fig. 10). Methane will vary from 40 to 80% of the total hydrocarbons in an urban atmosphere [Fig. 11; *(28)*].

Strictly speaking, the hydrocarbons are the compounds of hydrogen and carbon. At least two of the techniques used for measuring "total" hydrocarbons in air include many other classes of organic compounds. The nondispersive infrared method (NDIR), for example, measures compounds containing carbon–hydrogen bonds. This includes most organic compounds. The flame ionization method measures anything that reacts to form ions in a hydrogen flame. Pure hydrocarbons give higher specific responses, but without prior separation; the longer chain alcohols, aldehydes, esters, acids, etc., also give responses.

B. The Oxygenated Hydrocarbons

The oxygenated hydrocarbons, like the hydrocarbons, include an almost infinite number of compounds. They are classified as alcohols, phenols, ethers, aldehydes, ketones, esters, peroxides, and organic acids *(29)*.

Some minor amounts of oxygenated hydrocarbons are emitted as solvent vapors from the chemical, paint, and plastics industries. The greater quantities of primary emissions are more usually associated with the automobile. Table VIII *(30)* lists some typical oxygenates found in auto-

Table VIII Oxygenates in Exhaust from Simple Hydrocarbon Fuels (30)

Oxygenate	Concentration range, ppm[a]
Acetaldehyde	0.8–4.9
Propionaldehyde (+ acetone)[b]	2.3–14.0
Acrolein	0.2–5.3
Crotonaldehyde (+ toluene)[c]	0.1–7.0
Tiglaldehyde	<0.1–0.7
Benzaldehyde	<0.1–13.5
Tolualdehyde	<0.1–2.6
Ethylbenzaldehyde	<0.1–0.2
o-Hydroxybenzaldehyde (+ C$_{10}$ aromatic)[d]	<0.1–3.5
Acetone (+ propionaldehyde)[b]	2.3–14.0
Methyl ethyl ketone	<0.1–1.0
Methyl vinyl ketone (+ benzene)[e]	0.1–42.6
Methyl propyl (or isopropyl) ketone	<0.1–0.8
3-Methyl-3-buten-2-one	<0.1–0.8
4-Methyl-3-penten-2-one	<0.1–1.5
Acetophenone	<0.1–0.4
Methanol	0.1–0.6
Ethanol	<0.1–0.6
C$_5$ alcohol (+ C$_8$ aromatic)[f]	<0.1–1.1
2-Buten-1-ol (+ C$_5$H$_8$O)	<0.1–3.6
Benzyl alcohol	<0.1–0.6
Phenol + cresol(s)	<0.1–6.7
2,2,4,4-Tetramethyltetrahydrofuran	<0.1–6.4
Benzofuran	<0.1–2.8
Methyl phenyl ether	<0.1
Methyl formate	<0.1–0.7
Nitromethane	<0.8–5.0
C$_4$H$_8$O	<0.1
C$_5$H$_6$O	<0.1–0.2
C$_5$H$_{10}$O	<0.1–0.3

[a] Values represent concentration levels in exhaust from all test fuels.
[b] Data represent unresolved mixture of propionaldehyde + acetone. Chromatographic peak shape suggests acetone to be the predominant component.
[c] Toluene is the predominant component.
[d] The C$_{10}$ aromatic hydrocarbon is the predominant component.
[e] Benzene is the predominant component.
[f] The aromatic hydrocarbon is the predominant component.

mobile exhaust. The aldehydes are the preponderant oxygenates in emissions but are emitted in minor amounts when compared to hydrocarbon, carbon dioxide, carbon monoxide, and nitrogen oxide emissions. Many oxygenated compounds are formed as secondary products from photochemical reactions (26).

C. The Oxides of Carbon

1. Carbon Dioxide

Carbon dioxide is not generally considered an air pollutant. It is non-toxic, and immense quantities of it ($\sim 10^{12}$ tons) are cycled through the biosphere annually (20). It is an essential ingredient of plant and animal life cycles. Through photosynthesis it is converted to plant tissues; oxygen is produced as a by-product. Without photosynthesis, the world's supply of oxygen would reduce drastically to that formed by lightning and photolytic processes acting on water (2, 7).

The concentration of carbon dioxide in air is variable and depends upon whatever sources or sinks are present and such factors as the growing season when plants tend to deplete the amounts present. Callendar (31) studied carbon dioxide measurements from 1870 to 1955 (Fig. 12). A nineteenth century base value of 290 ppm was established and is generally accepted. Present day values have been set at 320 ppm with an annual growth rate of about 0.7 ppm (20).

Worldwide combustion of fossil fuel is a primary cause of the relatively rapid increase in carbon dioxide in the atmosphere. Robinson and Robbins have reviewed the sources, sinks, and effects of carbon dioxide (20). Table IX shows carbon dioxide emissions projected to the year 2000. A relative increase of nearly 300% in emissions over those of 1965 is predicted. Robinson and Robbins assume that half the carbon dioxide

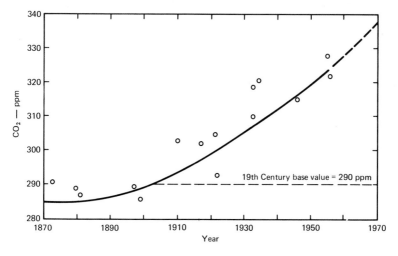

Figure 12. Average CO_2 concentration in North Atlantic region 1870–1956 (31).

Table IX Projected CO_2 Emissions: 1965–2000 (20) [E. R. Robinson and R. C. Robbins, *in* "Air Pollution Control" (W. Strauss, ed.), Part II, pp. 1–93. Copyright 1972, Wiley (Interscience), New York, New York. Reprinted by permission of John Wiley & Sons, Inc.]

	Emissions, 10^9 tons/year				
	1965	*1970*	*1980*	*1990*	*2000*
Coal	7.33	7.40	7.55	7.70	7.85
Petroleum	4.03	5.28	8.57	13.90	22.50
Natural gas	1.19	1.62	2.79	4.80	8.27
Incineration	0.46	0.51	0.61	0.73	0.88
Wood fuel	0.68	0.68	0.68	0.68	0.68
Forest fires	0.39	0.39	0.39	0.39	0.39
Total	14.08	15.88	20.59	28.20	40.57
Relative change	100 %	113 %	146 %	200 %	288 %

emitted remains in the atmosphere. This would result in an increase to about 370 ppm (20).

Carbon dioxide contributes to what is called a "greenhouse" effect in the atmosphere. As in a greenhouse, radiation penetrates the atmosphere and is absorbed by the earth. The earth also radiates energy into space at a reduced level and at longer wavelengths; otherwise the earth would increase in temperature indefinitely. A balance is maintained between the incoming and outgoing energy. Figure 13 (33) shows two radiation envelopes; one at 6000°K to indicate the radiation coming in from the sun; the other at 300°K to indicate the energy radiating out from the earth at longer wavelengths. Carbon dioxide absorbs radiation strongly from this envelope and consequently contributes to a warming, or greenhouse, effect. The temperature increase theoretically resulting from an increase of concentration to 370 ppm would be 0.5°C (32). In reality the earth's energy balance is much more complicated. Water vapor, which absorbs strongly in the infrared, the amount of cloudiness which reflects sunlight, and global atmospheric circulation patterns play important parts (20, 33). An increase in the reflectivity of the earth's atmosphere caused by an increase in the suspended particulate matter (34) or an increase in cloud cover could offset the warming tendency of carbon dioxide.

2. Carbon Monoxide

Carbon monoxide is a colorless, odorless, and tasteless gas slightly lighter than air. It is considered a dangerous asphyxiant because it combines strongly with the hemoglobin of the blood and reduces the blood's

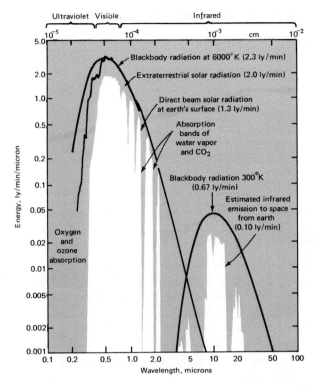

Figure 13. Spectra of solar and earth radiation (33). (Copyright 1965. Reprinted by permission of The Univ. of Chicago Press, Chicago, Illinois.)

ability to carry oxygen to cell tissues. Untold numbers of deaths have been caused by carbon monoxide in coal mines, fires, and closed places. A healthy working man can work 8 hours a day, 40 hours a week, without noticeable adverse effects at 25 ppm (the threshold limit value).

Carbon monoxide is a product of incomplete combustion of carbon and its compounds. It is emitted by fossil fuel combustion sources in greater quantities than all other pollutant sources combined. Table X summarizes the estimates of emissions in the United States (23). The automobile is by far the largest single pollution emission source. Figure 14 shows the maximum carbon monoxide concentrations found at eight Continuous Air Monitoring Program (CAMP) stations in the United States (35).

Recent carbon isotope studies conducted at Argonne National Laboratory (36) showed that nature produces huge quantities of carbon monoxide: from 3 to 640×10^9 tons/year as compared to 0.275×10^9 tons/year from worldwide pollution sources (Table III). The principal natural source is believed to be the result of the photochemical oxidation of methane through an OH radical mechanism (36, 37). Other natural

Table X Estimates of Carbon Monoxide Emissions (United States) 1940–1970 (10^6 tons/year) (23)

Source category	1940	1950	1960	1968	1969	1970
Fuel combustion in stationary sources	6.2	5.6	2.6	2.0	1.8	0.8
Transportation	34.9	55.4	83.5	113.0	112.0	111.0
Solid waste disposal	1.8	2.6	5.1	8.0	7.9	7.2
Industrial process losses	14.4	18.9	17.7	8.5	12.0	11.4
Agricultural burning	9.1	10.4	12.4	13.9	13.8	13.8
Miscellaneous	19.0	10.0	6.4	5.0	6.3	3.0
Total	85.4	103.0	128.0	150.0	154.0	147.0
Total controllable[a]	66.4	92.9	121.0	145.0	148.0	144.0

[a] Miscellaneous sources not included.

sources include the decomposition of chlorophyll to give relatively high concentrations of carbon monoxide particularly in the fall (0.2 to 0.5×10^9 tons/year). Volcanoes, natural gas, forest fires, bacterial action in the oceans (0.15×10^9 tons/year) are other sources. The estimated total amount of carbon monoxide emissions from natural sources in Table III are consequently low by 30- to 50-fold, and the residence time of carbon monoxide in air needs to be reduced somewhere between 0.1 to 0.3 year (37, 38).

The background concentration of carbon monoxide is estimated from data gathered in the Pacific (20, 39) to be approximately 0.1 ppm. Table XI shows the range of maximum hourly average values for the years of 1962–1967 for eight major United States cities (40, 41). The theoreti-

Table XI Carbon Monoxide Concentrations in Representative United States Cities Hourly Maxima, ppm, 1962–1967 (41) [W. L. Faith and A. A. Atkisson, Jr., "Air Pollution," 2nd ed., pp. 163–205. Copyright 1972, Wiley (Interscience), New York, New York. Reprinted by permission of John Wiley & Sons, Inc.]

	Yearly maxima		Theoretical geometric mean (17, 51)
	Highest	Lowest	
Chicago, Illinois	59	28	13.2
Cincinnati, Ohio	34	20	4.8
Denver, Colorado	55	40	6.7
Los Angeles, California	47	35	9.7
Philadelphia, Pennsylvania	54	37	6.9
St. Louis, Missouri	29	25	5.5
San Francisco, California	38	22	4.8
Washington, D.C.	41	25	3.5

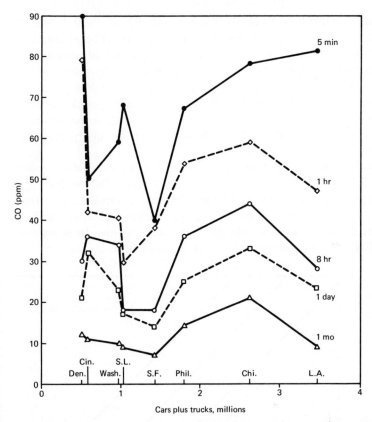

Figure 14. Maximum CO concentrations at Continuous Air Monitoring Program (CAMP) stations. 1962–1968 maxima vs cars plus trucks. Denver (Den.), Colorado; Cincinnati (Cin.), Ohio; Washington (Wash.), D.C.; St. Louis (S.L.), Missouri; San Francisco (S.F.), California; Philadelphia (Phil.), Pennsylvania; Chicago (Chi.), Illinois; Los Angeles (L.A.), California (35).

cal geometric mean hourly concentrations for the entire period are also shown. CO concentrations are more than ten times the level of concentrations of other major pollutants.

IV. The Gaseous Compounds of Sulfur

A. The Sulfur Oxides

Sulfur forms a number of oxides (SO, SO_2, S_2O_3, SO_3, S_2O_7) but only sulfur dioxide (SO_2) and sulfur trioxide (SO_3) are of any importance as gaseous air pollutants. The peroxide, S_2O_7, has been suggested as existing in the lower stratosphere where a layer of sulfate particles has been found (42, 43).

Sulfur trioxide is generally emitted with SO_2 at about 1–5% of the SO_2 concentration (*44, 45*). A few industries such as sulfuric acid manufacturing, electroplating, and phosphate fertilizer manufacturing may emit higher relative amounts (*46*). Sulfur trioxide rapidly combines with water in air to form sulfuric acid (H_2SO_4) which has a low dew point. An aerosol or mist is easily formed, and SO_3 or H_2SO_4 is frequently associated with haze and poor visibility in air (Fig. 15). The analysis for SO_3 or H_2SO_4 in air is quite difficult, and the data have to be interpreted with some care (*46*).

Sulfur dioxide is a colorless gas with a pungent, irritating odor. Most people can detect it by taste at 0.3 to 1 ppm (780 to 2620 $\mu g/m^3$). It is highly soluble in water: 11.3 gm/100 ml as compared to 0.169 gm/100 ml for carbon dioxide, forming weakly acidic sulfurous acid (H_2SO_3). In clean air it oxidizes slowly to sulfur trioxide. It is oxidized more readily by atmospheric oxygen in aqueous aerosols. Heavy metal ions

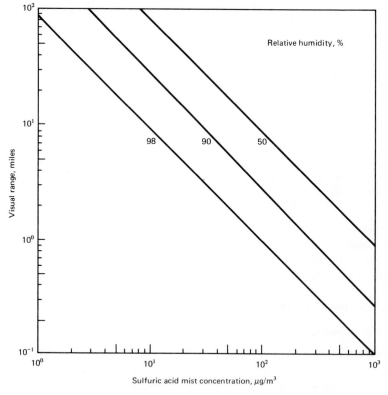

Figure 15. Calculated visibility (visual range) in miles at various sulfuric acid mist concentrations and different relative humidities (*18*).

in solution catalyze the reaction which stops when aerosols become acidic. Atmospheric ammonia neutralizes the acid to form ammonium sulfate, and ammonium sulfate is commonly found in atmospheric particles (47). In moist air and in the presence of nitrogen oxides, hydrocarbons, and particulates, sulfur dioxide reacts much more rapidly (48, 49).

Today, sulfur dioxide remains one of the major atmospheric pollutants. Its worldwide emissions have been estimated at 146 megatons/year by Robinson and Robbins (Table III) and more recently as 100 (150 as sulfate) megatons per year by Kellogg et al. (5) who predict emissions of about 275 megatons per year for the year of 2000. Estimated United States sulfur dioxide emissions for 1970 were 33.9 megatons (Table XII). Steam electric and industrial emissions accounted for 70 and 18% of the

Table XII Estimates of Sulfur Oxide Emissions (United States) 1940–1970 (10^6 tons/year) (23)

Source category	1940	1950	1960	1968	1969	1970
Fuel combustion in stationary sources	16.8	18.3	17.5	24.7	25.0	26.5
Transportation	0.7	1.0	0.7	1.1	1.1	1.0
Solid waste disposal	Neg[a]	0.1	0.1	0.1	0.2	0.1
Industrial process losses	3.8	4.2	4.7	5.1	5.9	6.0
Agricultural burning	Neg	Neg	Neg	Neg	Neg	Neg
Miscellaneous	0.2	0.2	0.3	0.3	0.2	0.3
Total	21.5	23.8	23.3	31.3	32.4	33.9
Total controllable[b]	21.3	23.6	23.0	31.0	32.2	33.6

[a] Negligible (less than 0.05×10^6 tons/year).
[b] Miscellaneous sources not included.

Table XIII Sulfur Dioxide Concentrations in Representative United States Cities Hourly Maxima, ppm, 1962–1967 (18)

	Yearly maxima		Theoretical geometric mean (17, 51)
	Highest	Lowest	
Chicago, Illinois	1.69	0.86	0.111
Cincinnati, Ohio	0.57	0.41	0.018
Denver, Colorado	0.36	0.17	0.014
Los Angeles, California	0.29	0.13	0.014
Philadelphia, Pennsylvania	1.03	0.66	0.060
St. Louis, Missouri	0.96	0.55	0.031
San Francisco, California	0.26	0.11	0.006
Washington, D.C.	0.62	0.35	0.042

latter, respectively (*23*). Intensive efforts are being made to control sulfur dioxide emissions by either removing sulfur from coal and oil or removing sulfur dioxide at the combustion source (*50*).

Ambient air concentrations of sulfur dioxide are routinely measured in many cities and have been the subject of a large number of studies. Table XIII gives typical data obtained from the United States Continuous Air Monitoring Program (CAMP). Figure 16 shows the frequency distribution of sulfur dioxide measurements made in selected United States cities. An approximate log-normal distribution is shown by the

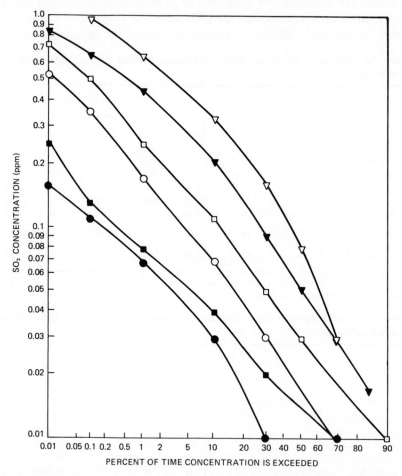

Figure 16. Frequency distribution of sulfur dioxide levels in selected United States cities, 1962–1967 (*18*). ▽, Chicago, Illinois; ▼, Philadelphia, Pennsylvania; □, St. Louis, Missouri; ○, Cincinnati, Ohio; ■, Los Angeles, California; ●, San Francisco, California.

straight portions of the lines. This confirms to some extent the model developed by Larsen and others (*17, 18, 51*).

B. Reduced Sulfur Compounds

1. Hydrogen Sulfide

Hydrogen sulfide (H_2S) is a toxic, evil smelling gas well known for its rotten egglike odor. It can be detected at concentrations as low as 0.5 ppb (7 $\mu g/m^3$) (*52*). Its natural emission sources include anaerobic biological decay processes on land, in marshes, and in the oceans. Volcanoes and natural hot water springs emit hydrogen sulfide to some extent. A total of approximately 100 megatons (268 when expressed as sulfate) is estimated to be emitted in nature [Table III; (*5*)]. However this estimate has been made with strong reservations. The analysis of very low concentrations in air is subject to error, because some of the hydrogen sulfide is oxidized to sulfur dioxide during the sampling process (*5*).

Approximately 3 megatons per year are estimated to be emitted by pollution sources (*20*) (Table III). One of the larger single sources is the kraft pulp industry which uses a sulfide process to extract cellulose from wood (*53*). Because of the strong odor of sulfides such plants can be detected by their odor 40 miles or more downwind, unless emissions are carefully controlled. Other hydrogen sulfide pollution sources include the rayon industry, coke ovens, and the oil refining industry. The processing of "sour" crude oil results in the emission of hydrogen sulfide and other volatile organic sulfides. Hydrogen sulfide emissions from industrial processes are sometimes used as fuel for boilers or are released in burning flares. In either case they are burned to sulfur dioxide and emitted to the air as such. Today, many modern refineries recover their sour gases and process them to form sulfuric acid or elemental sulfur (*54*).

Hydrogen sulfide concentrations in urban air are rarely higher than 0.1 ppm (140 $\mu g/m^3$). Cholak (*55*) analyzed Cincinnati air over a period of 5 years and rarely found hydrogen sulfide to exceed 0.01 ppm (14 $\mu g/m^3$). A survey in Houston, Texas showed average values of 0.02 ppm in the most highly polluted section of the city. The highest level measured was 0.28 ppm (390 $\mu g/m^3$) (*41, 56*). Katz (*57*) found relatively high levels in Windsor, Ontario with a mean concentration of approximately 0.1 ppm and a maximum of 0.6 ppm (835 $\mu g/m^3$). Hydrogen sulfide blackens lead-based paints. A level of 0.1 ppm is said to produce blackening of such paints within 1 hour (*41*). In air, hydrogen sulfide is oxidized to sulfur dioxide within hours, adding to the ambient sulfur dioxide level (*5*).

2. Mercaptans and Sulfides

Other sulfur compounds that are of interest in air pollution, principally because of their strong odors, are methyl mercaptan (CH_3SH), dimethyl sulfide (CH_3SCH_3), dimethyl disulfide (CH_3SSCH_3), and their higher molecular homologs (*53*). They have odors similar to those emitted by skunks and rotting cabbage. Total emissions of these compounds are unknown. A number of studies have been concerned with their evaluation (*58*) and their measurement in air [Fig. 17; (*59*)].

V. The Gaseous Compounds of Nitrogen

Nitrogen forms the very stable diatomic gas, N_2, which makes up over 78% of the atmosphere and, fortunately, helps temper the oxidative power of atmospheric oxygen. It also forms a large number of gaseous and nongaseous compounds, many of which are essential to living matter.

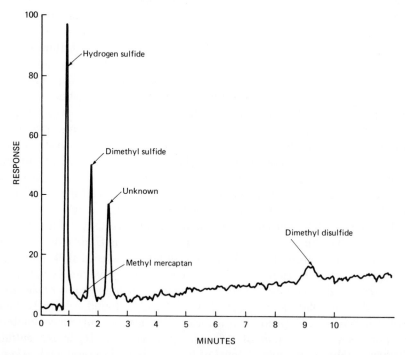

Figure 17. Sulfur gases in ambient air, *in situ* analysis, January 4, 1972, 1424 hours, freeze-out of 1000 ml of air. Column temperature 49°C. Reprinted from *American Laboratory* (*59*) with permission of International Scientific Communications, Inc.

They are produced by such natural processes as bacterial fixation, biological growth and decay, lightning, and forest and grassland fires. To a lesser extent but in higher local urban concentrations, nitrogen compounds are produced by man through a wide number of agricultural, domestic, and industrial activities. In the reduced state, nitrogen forms such compounds as ammonia, amides, amines, amino acids, and nitriles. In the oxidized state, it forms seven oxides and a large number of nitro, nitroso, nitrite, and nitrate derivatives (60).

A. The Oxides of Nitrogen

The oxides of nitrogen include nitrous oxide (N_2O), nitric oxide (NO), nitrogen dioxide (NO_2), nitrogen trioxide (NO_3), nitrogen sesquioxide (N_2O_3), nitrogen tetroxide (N_2O_4), and nitrogen pentoxide (N_2O_5). They and two of their hydrates, nitrous acid (HNO_2) and nitric acid (HNO_3), can exist in air. However, only three, nitrous oxide (N_2O), nitric oxide (NO), and nitrogen dioxide (NO_2) are found in any appreciable quantities. The latter two, NO and NO_2, are often analyzed together in air and are referred to as "nitrogen oxides" and given the symbol "NO_x." Nitrous oxide (N_2O) is not included in the "NO_x" measurement, but it is possible for the higher oxides to be included if they happen to be present (61).

Nitrous oxide (N_2O) is a colorless, slightly sweet, nontoxic gas present in the natural environment in relatively large amounts (0.25 ppm) when compared to the concentrations of the other trace gases except carbon dioxide, methane, and the noble gases. It is used as an anesthetic in minor surgery and dentistry. When mixed with air and inhaled it produces a loss of feeling. Its effects are not severe and soon pass off. It is commonly called "laughing gas" because under some conditions it can cause those who inhale it to laugh violently. The major natural source of nitrous oxide is due to biological activity in the soil and possibly in the oceans. A worldwide production rate of 10^9 tons per year and a residence time of 4 years has been estimated (20, 62). Nitrous oxide has been associated with photochemical reactions in the upper atmosphere (63), but because of its low reactivity in the lower atmosphere it is largely ignored in air pollution studies. There are no known significant pollution sources (20).

Nitric oxide (NO) is a colorless, odorless, and tasteless gas. It is produced in nature by biological action and by combustion processes. It is suspected as being formed and rapidly oxidized in closed silos where dangerous concentrations of nitrogen dioxide have been found (64). In air it is oxidized rapidly by atmospheric ozone and photochemical processes and more slowly by oxygen to form nitrogen dioxide (NO_2). Worldwide natural emissions are estimated by Robinson and Robbins to be 430×10^6 tons per year. Background concentrations are variable and difficult to measure. They are estimated to range from 0.25 to 6 ppb. The residence time in air is about 5 days (20).

As a pollutant, nitric oxide is produced largely by fuel combustion in both stationary and mobile sources such as the automobile. In the high temperatures of the combustion zone, nitrogen reacts with oxygen to form nitric oxide:

$$N_2 + O_2 \rightleftharpoons 2\, NO \tag{15}$$

The reaction is endothermic and proceeds to the right at high temperatures. At low temperatures the equilibrium lies almost completely to the left, but the rate of recombination is extremely slow. Consequently, the

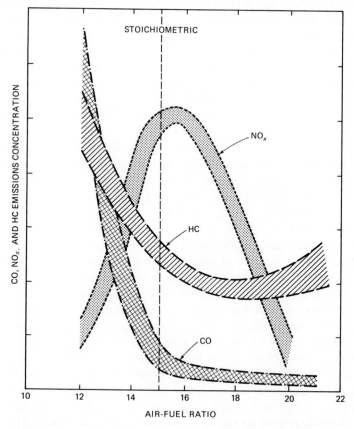

Figure 18. Effects of air–fuel ratio on exhaust composition (approximate ranges, not to scale) (65).

amount of NO emitted is a function of the flame structure and temperature as well as the rate at which the combustion mixture cools. If the cooling rate is rapid, equilibrium is not maintained and the NO concentration, although thermodynamically unstable, remains high (*65, 66*). The proper catalyst can, of course, expedite its decomposition to nitrogen and oxygen. In exhaust gases where higher concentrations and temperatures prevail, some of the nitric oxide is oxidized to nitrogen dioxide. This generally varies from 0.5 to 10% of the nitric oxide present (*67*).

Figure 18 shows the relative amounts of nitrogen oxides, hydrocarbons, and carbon monoxide in the exhaust of an automobile as a function of the ratio of the air-to-fuel mixture used for the engine. At low air-to-fuel ratios ("rich" mixtures), flame temperatures are low, combustion is incomplete, hydrocarbon and carbon monoxide emissions are high, and nitrogen oxides emissions are low. At higher air-to-fuel ratios ("lean" mixtures) the temperature of the combustion flame becomes hotter, the nitrogen oxides increase until the air–fuel ratio is greater than the stoichiometric point and then decrease rapidly as the excess air cools the flame (*65*).

Worldwide pollution sources emit approximately 53×10^6 tons per year of NO and NO_2 combined (NO_x). Table XIV gives United States estimates for NO_x emissions expressed as NO_2. Fuel combustion in stationary sources and transportation account for more than 95% of the 22.7×10^6 tons emitted per year in the United States. Table XV shows maximum and minimum hourly averages for NO_x in several United States cities.

Table XIV   Estimates of Nitrogen Oxide (NO_x) Emissions (United States), 1940–1970 (10^6 tons/year) (*23*)

Source category	1940	1950	1960	1968	1969	1970
Fuel combustion in stationary sources	3.5	4.3	5.2	9.7	10.2	10.0
Transportation	3.2	5.2	8.0	10.6	11.2	11.7
Solid waste disposal	0.1	0.2	0.2	0.4	0.4	0.4
Industrial process losses	Neg[a]	0.1	0.1	0.2	0.2	0.2
Agricultural burning	0.2	0.2	0.3	0.3	0.3	0.3
Miscellaneous	0.8	0.4	0.2	0.2	0.2	0.1
Total	7.9	10.4	14.0	21.3	22.5	22.7
Total controllable[b]	7.1	10.0	13.8	21.1	22.3	22.6

[a] Negligible (less than 0.05×10^6 tons/year).

[b] Miscellaneous sources not included.

Table XV   Nitrogen Oxide (NO$_x$) Concentrations in Representative United States Cities Hourly Maxima, ppm, 1962–1968 (67)

	Yearly maxima		Geometric mean
	Highest	Lowest	
Chicago, Illinois	1.06	0.69	0.75
Cincinnati, Ohio	1.42	0.45	0.83
Denver, Colorado[a]	0.72	0.56	0.62
Los Angeles, California	1.35	0.98	1.24
Philadelphia, Pennsylvania	1.79	0.97	1.53
St. Louis, Missouri[b]	0.92	0.44	0.57
Washington, D.C.	1.30	0.68	0.83

[a] 1965–1968
[b] 1964–1968

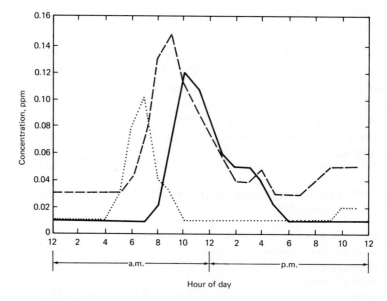

Figure 19. Typical diurnal variation of NO, NO$_2$, and O$_3$ concentrations in Los Angeles, California (67). Solid line, ozone; long dashed line, nitrogen dioxide; dotted line, nitric oxide.

In a polluted atmosphere, nitric oxide is oxidized to nitrogen dioxide primarily through photochemical secondary reactions. Figure 19 shows the diurnal variations of NO, NO$_2$, and O$_3$ in a typical photochemical pollution situation. Nitric oxide reaches a maximum during the early morning traffic rush hours. The rising sun initiates a series of photochemical reactions which convert the nitric oxide to nitrogen dioxide. Within a few hours the nitrogen dioxide reaches a maximum during which it photochemically reacts to form ozone and other oxidants. Both the nitrogen dioxide and the ozone eventually disappear through the formation of nitrated organic compounds, peroxides, aerosols, and other terminal products. The cycle is repeated the following day. If the air mass is not swept away or is brought back by a reversing wind, the residual gases add to the new day's pollutants (26).

Nitrogen dioxide is a reddish-brown gas with a pungent, irritating odor. At concentrations higher than those found in the atmosphere, it forms a colorless dimer, nitrogen tetroxide (N_2O_4). Natural emissions are due primarily to biological decay involving nitrates being reduced to nitrites, followed by conversion to nitrous acid (HNO_2), decomposition to nitric oxide, and oxidation to nitrogen dioxide. Natural emissions are estimated to be 658×10^6 tons per year.

Nitrogen dioxide is one of the more invidious pollutants. It is irritating and corrosive in itself, but more importantly it serves as an energy trap by absorbing sunlight to form nitric oxide and atomic oxygen:

$$NO_2 + h\nu \rightarrow NO + O \tag{16}$$

The atomic oxygen is very reactive forming ozone with oxygen and initiating a number of secondary photochemical chain reactions. Nitrogen dioxide absorbs light strongly in the yellow to blue end of the visible spectrum and the near ultraviolet. Figure 20 (66) shows the absorption spectrum of nitrogen dioxide, and Figure 21 (67) indicates the amount of light absorbed in terms of parts per million—mile concentrations. A mile thick layer of air containing 0.1 ppm of NO_2 reduces the ultraviolet light reaching the ground by more than 25%. Viewed through a horizontal layer of 10 miles, the same concentration reduces the blue and ultraviolet light more than 90% (Fig. 21). The yellow-brown haze often seen hovering over a city is in a large part due to nitrogen dioxide and the aerosols it helps generate (68).

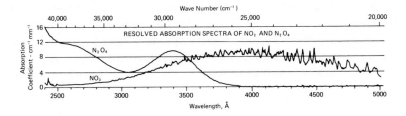

Figure 20. Absorption coefficients ($1/pl \log_{10} I_0/I$) of NO_2 and N_2O_4 vs wavelength and wave number, measured at 25°C (66).

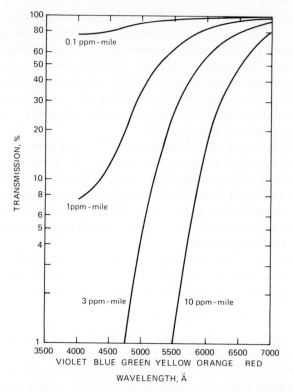

Figure 21. Transmittance of visible light at different NO₂ concentrations and viewing distances (67).

Nitrogen trioxide (NO_3) and nitrogen pentoxide (N_2O_5) have been postulated as intermediates in the photochemical oxidation of hydrocarbons and sulfur dioxide (48, 69–71). They are not commonly observed; their concentrations are expected to be small and difficult to measure in air in the presence of NO, NO_2, and their various photochemical reaction products. The pentoxide hydrolyzes readily with water vapor in the air to form nitric acid vapor (HNO_3) which has been detected in the stratosphere by spectroscopic means (19). Peroxyacetyl nitrate (PAN), an eye-irritating photochemical reaction product from hydrocarbons and nitrogen oxides, has been identified and measured in air (66, 71). Atmospheric concentrations as high as 0.1 ppm (500 $\mu g/m^3$) have been reported (67).

B. Ammonia

Ammonia (NH_3) is considered to be a relatively unimportant pollutant gas. Huge quantities of it are formed in nature principally by bacterial breakdown of amino acids in organic waste materials. Background concentrations have been variously measured from 1 to 20 ppb (*72–74*), and the atmospheric residence time is estimated to be about 7 days. It has a sharp odor and can be detected by taste at the relatively high concentration of 20 ppm. There are no known harmful effects at ambient air concentrations. Pollution sources are estimated at only 4 million tons on a worldwide basis compared to 1169 million tons from natural emissions. Ammonia is one of few basic gases released in any quantity by nature, and it plays an important role in the reactions and fate of many gaseous pollutants. It forms ammonium (NH_4^+) salts with sulfuric, nitric, and hydrochloric acids. Sizable fractions of atmospheric particles have been identified as ammonium compounds (*75*).

VI. The Gaseous Halogens

Gaseous halogen pollutants include the elements and compounds of fluorine, chlorine, bromine, and iodine. Of these, chlorine, hydrogen fluoride, hydrogen chloride, the Freons, and the halogenated pesticides and herbicides are encountered most frequently. A number of the halides and their compounds, in particular the fluorides, herbicides, and pesticides, are of importance because of the strong effects even low concentrations have on plant and animal life (*76–78*). The Freons are of concern because of their potential in destroying the ozone layer (*79–81*).

A. The Fluorides

The presence of fluoride in air is worldwide. Its concentrations are generally low [Table XVI; (*82*)], but higher concentrations are frequently found in areas surrounding pollution sources. Only a few studies have attempted to measure gaseous fluoride as such. More commonly, filterable aerosols, settled particulates, and residues on or in plant and animal tissues are analyzed for their fluoride content. Table XVII (*83*) shows the amounts of combined gaseous and particulate atmospheric fluoride found in urban and nonurban areas in the United States.

The gaseous forms of fluorine emitted to the air include fluorine (F_2), hydrogen fluoride (HF), silicon tetrafluoride (SiF_4), and hydrofluosilicic acid (H_2SiF_6). The principal sources include the aluminum, steel, glass, brick, tile, and phosphate fertilizer industries (*84, 85*). When fluoride minerals such as fluorite (CaF_2) and cryolite ($3NaF \cdot AlF_3$) are present either as trace or major components in the raw materials of a manufacturing process in which high temperatures are used, one or more of the volatile fluoride compounds are emitted to the atmosphere. Elemental

Table XVI   The Concentration of Fluoride Ion in Parts per Billion in the Atmosphere of a Number of Communities (82)

Location	Average or mean	Range
Ft. William, Scotland	48[a]	
Claggan Farm, Scotland	33[a]	
Cincinnati, Ohio (1957)	0.42	0.04–1.20
New York, New York	2.0	
Yonkers, New York (summer)		0.2–0.4
Spokane, Washington	1.6[a]	
Portland, Oregon		0.6–0.8
Rural	0.3	
Salt Lake City, Utah		0.05–1.0
Logan, Utah	0.02	
Canadian–Washington Border	0.3[a]	
San Francisco, California area		0.2–0.4

[a] One determination only.

Table XVII   Summary of Analyses of Atmospheric Fluoride in the United States, 1966–1968 (83)

	No. of stations	Samples	Number of samples with F^- content in $\mu g/m^3$ in the ranges shown[a]				
			<0.05	0.05–0.09	0.10–0.99	≥1.00	Max.
Urban							
1966	100	2521	2161	152	206	2	1.89
1967	122	2967	2612	134	212	9	1.74
1968	147	3687	3287	103	290	7	1.65
1966–1968	—	9175	8060	389	708	18	—
Nonurban							
1966	29	711	687	24	0	0	0.09
1967	30	729	721	5	3	0	0.16
1968	29	724	724	0	0	0	<.05
1966–1968	—	2164	2132	29	3	0	—
CAMP[b]							
1967	6	814	645	108	61	0	0.92
1968	6	909	560	180	169	0	0.55

[a] Minimum detectable level is 0.05 μg F/m³ of air.
[b] Continuous Air Monitoring Program stations, United States.

fluorine is too reactive to remain uncombined in air for very long but it may sometimes be found in the immediate vicinity of high temperature–high energy processes such as electrolytic reduction cells in the manufacture of aluminum metal. In the presence of silicate compounds, SiF_4 is formed and emitted to the atmosphere. It can subsequently be hydrolyzed to form hydrogen fluoride which then can act as a gaseous pollutant. It can react further to form fluorite (CaF_2), a stable, insoluble substance which settles on the surrounding vegetation to be ingested by grazing animals:

$$2CaF_2 + SiO_2 \xrightarrow{\text{heat}} SiF_4 + 2CaO \tag{17}$$

$$SiF_4 + 2H_2O \xrightarrow{\text{hydrolysis}} SiO_2 + 4HF \tag{18}$$

$$2HF + CaO(CaCO_3) \xrightarrow[\text{in air}]{\text{reaction}} CaF_2 + H_2O(CO_2) \tag{19}$$

In the phosphate fertilizer industry, fluorapatite $[3Ca_3(PO_4)_2 \cdot CaF_2]$ is often present. Fluoride content of the ore is generally over 3% (86). When the ore is treated with sulfuric acid to release phosphoric acid, gaseous HF, SiF_4, H_2SiF_6, and other similar substances are formed and emitted to the atmosphere unless removed by some control process. These compounds may act directly as gases, or they may act through their hydrolyzed or reaction products.

Ambient air concentrations for residential and rural areas have been reported by Cholak (82). His data (Table XVI) indicate worldwide background levels of a few tenths of a part per billion. As indicated above, most measurements of atmospheric fluorides have been made on filterable particles which are sometimes assumed to adsorb gaseous fluorides (83). Cholak did not indicate the method of sampling used. Meetham (87) quoted values of 0.02–0.22 mg fluoride/m^3 (24–270 ppb as HF) near an aluminum plant with gaseous fluorides ranging from 15 to 81% of the total (41). Leonard and Graves (88) measured fluoride levels in the vicinity of a phosphate fertilizer plant. They found monthly averages ranging from 3.9 to 5.65 ppb expressed as hydrogen fluoride. The highest 24-hour average was 14.2 ppb and the lowest was 0.6 ppb. Hodge and Smith studied air quality criteria for fluoride effects on man (89) and reported that urban air contains from less than 0.2 $\mu gF/m^3$ to as much as 1.9 $\mu g F/m^3$ (0.24 to 2.1 ppb as HF).

B. Chlorine, Hydrogen Chloride, and Chlorinated Hydrocarbons

The gaseous forms of chlorine and its compounds most frequently encountered in polluted atmospheres include the element (Cl_2), hydrogen chloride (HCl), and the vapors of a number of chlorinated hydrocarbon solvents, pesticides, and herbicides. The chlorinated hydrocarbons solvent vapors include those of chloroform ($CHCl_3$), carbon tetrachloride (CCl_4), trichloroethylene (C_2HCl_3), perchloroethylene (C_2Cl_4), etc. Freon vapors ($CFCl_3$, C_2FCl_3) widely used for refrigeration and as aerosol spray propellants have been found to be building up background concentrations in the air (*79*). Except in the vicinity of major pollution sources, the atmospheric concentrations of the gaseous chlorine compounds are very low.

Fear has been expressed that the Freons will eventually drift into the stratosphere and destroy a significant fraction of the protective ozone layer (*80, 81*). Figure 22 (*81*) shows the calculated effects that the Freons would have on the ozone layer using computerized models and various assumed production rates and times of cessation of production.

Elemental chlorine (Cl_2) is widely used in the chemical and plastics industries, in water and sewage treatment plants, in household bleaches in the form of hypochlorite which releases chlorine readily, and in swimming pools (*90*). Chlorine is a heavy, yellowish colored gas, with a strong pungent odor. It is very reactive and highly irritating to the mucous

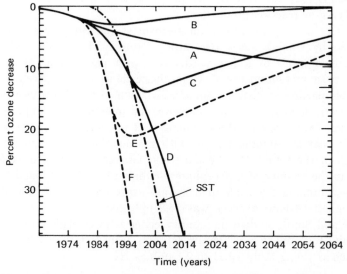

Figure 22. Calculated effects of Freons on global ozone (*81*). A: Production held constant at present rate. B: Production ceases in 1978. C: Production increases 10% per year, ceases in 1995. D: Production increases 10% per year. E: Production increases 22% per year, ceases in 1987. F: Production increases 22% per year. SST: Estimate for upper bound on NO injection future supersonic fleets. (Copyright 1975 by the American Association for the Advancement of Science.)

membranes. Accidental releases of the gas which is shipped in large barge and railroad tank car quantities can be very dangerous to a community. Normal industrial emissions are not excessively high (*91*).

Although a large number of studies have been published on the presence and behavior of particulate chlorides in air, only a few studies have made an effort to measure gaseous elemental chlorine. Nordlie (*92*) analyzed the magmatic gases of Kilauea. Concentrations varied from 0 to 0.76% reported as Cl_2 but probably included a high fraction of HCl. Bartels (*93*) estimated that volcanic contributions of chloride to the atmosphere amounted to 7.6×10^6 metric tons per year as well as 7.3×10^6 metric tons per year of fluoride, but he did not state what chemical forms were involved. Cholak (*94*) and Katz (*57*) have indicated chloride concentrations in a number of American cities ranging from 0.016 to 0.095 ppm.

Hydrogen chloride is a common air pollutant (*95*) emitted by a large number of industrial and domestic activities (such as the incineration of chlorinated plastics). Concentrations of 2–8 $\mu g/cm^3$ (1.3 to 5 ppb) were found at 10 m above the surface of the North Atlantic (*96*). Again measurements of HCl as a gas in air are not common, both because it is not highly toxic and because it is generally present in relatively low concentrations when compared to sulfur dioxide and other acidic gases. Analyses for halogenated hydrocarbons have shown them to be generally less than 0.1 ppm (*41*).

There has been a considerable amount of interest in the presence of pesticides and herbicides in air. In the majority of cases these are sampled and measured as particulate matter. Stanley *et al.* (*97*) sampled air using a combination of filtration with glass fiber cloth, impingement in hexylene glycol, and adsorption with alumina. Their results are shown in Table XVIII. Gaseous fractions were about 50% of the total except for one very high *p,p'*-DDT sample where the gaseous portion was about 15%. Miles *et al.* (*98*) used impingers with ethylene glycol to measure pesticides. Tessari and Spencer (*78*) exposed coated nylon chiffon to measure pesticides. Woodwell *et al.* (*99*) and Cramer (*100*) modeled the circulation of DDT on earth and found that transport through the air was an important dispersion factor. This is not surprising since most of these compounds have finite vapor pressures. The work of Stanley *et al.* (*97*), in addition to showing the presence of gaseous pesticides, also suggests that particulate pesticides evaporate while being sampled through fiber filters. It was estimated that the background concentration of DDT was 84×10^{-3} $\mu g/m^3$. Although the amounts of pesticides present in air are considerably less than the amounts ingested from residues on food (*97*), there is grave concern that the buildup of these substances in the oceans could adversely affect many of the microorganisms that form important

Table XVIII Maximum Pesticide Levels Found in Air Samples (Levels in ng/m³)ᵃ (97) [Reprinted with permission from C. W. Stanley, J. E. Barney, II, M. R. Helfon, and A. R. Yobs, *Environ. Sci. Technol.* 5, 430 (1971). Copyright by the American Chemical Society.]

Pesticidesᵇ	Baltimore, Maryland	Buffalo, New York	Dothan, Alabama	Fresno, California	Iowa City, Iowa	Orlando, Florida	Riverside, California	Salt Lake City, Utah	Stoneville, Mississippi
p,p'-DDT	19.5 (89)	11.0 (40)	177.0 (88)	11.2 (62)	2.7 (56)	1560 (99)	24.4 (85)	8.6 (62)	950 (98)
o,p'-DDT	3.0 (59)	2.9 (24)	88.0 (72)	5.5 (28)	2.1 (21)	500 (95)	6.2 (44)	1.4 (29)	250 (98)
p,p'-DDE	2.4 (4)		13.2 (32)	6.4 (3)	3.7 (10)	131 (29)	11.3 (6)		47 (76)
o,p'-DDE			3.9 (13)			9.6 (7)			1.9 (25)
α-BHC	4.5 (27)			4.5 (4)	4.4 (9)			9.9 (30)	
Lindane	2.6 (4)				0.1 (1)			7.0 (24)	
β-BHC	2.2 (4)							1.8 (3)	
δ-BHC								9.9 (5)	
Heptachlor					19.2 (37)	2.3 (7)			
Aldrin					8.0 (1)				
Toxaphene			68.0 (11)			2520 (9)			1340 (55)
2,4-D								4.0 (1)	
Dieldrin						29.7 (50)			
Endrin									58.5 (25)
Parathion						465 (37)			
Methyl parathion			29.6 (9)			5.4 (3)			129 (40)
Malathion						2.0 (4)			
DEFᶜ									16 (12)
Total samples	123	57	90	120	94	99	94	100	98

ᵃ Numbers in parentheses indicate number of samples containing detectable amounts of the pesticide.
ᵇ Abbreviations used
 DDT = Dichlorodiphenyltrichloroethane
 DDE = Dichlorodiphenyldichloroethylene
 BHC = Benzenehexachloride
 2,4-D = 2,4-Dichlorophenoxyacetic acid
ᶜ A cotton defoliant.

links in the nitrogen, phosphorus, sulfur, and oxygen cycles essential for life on earth.

C. Bromine

Interest in bromine as a pollutant is chiefly centered in its role as a lead scavenger (65). Ethylene bromide ($C_2H_2Br_2$) is added to gasoline containing tetraethyllead to form the relatively volatile lead bromide ($PbBr_2$) which is then exhausted to the air. In most cases interest is in the lead bromide as a particle. However, not all lead from the auto is associated with bromine; nor is it logical to expect all lead bromine compounds to be nongaseous. Moyers et al. (101) analyzed ambient air samples for gaseous and particulate bromine in Cambridge, Massachusetts, using charcoal adsorption for the gas and filters for the particles. Concentrations ranged from 0.12 to 0.45 $\mu g/m^3$ (0.05 to 0.2 ppb). Moyers (102) also found that the marine environment contains approximately 0.05 $\mu g/m^3$ of gaseous bromine. Bowman et al. (103) have also reported on the presence of bromine in air, but their results represent mostly particulate matter.

VII. Ozone and Oxidants

Ozone, O_3, is a bluish gas about 1.6 times as heavy as air and highly reactive. It is formed at high altitudes by photochemical reactions involving molecular and atomic oxygen (60). Its concentration in the atmosphere depends upon the altitude; being greatest in the stratosphere. At 20 km it is 0.20 ppm. Its concentration in rural areas away from pollution sources is approximately 0.02 ppm (104). Very minor amounts of ozone are formed during lightning and thunderstorms. Ozone strongly absorbs ultraviolet light in the wavelength region of 2000–3500 Å and very weakly at about 6000 Å. Its absorption of the energetic portion of the ultraviolet light prevents serious damage to living tissues (104).

Ozone and other oxidants such as PAN (peroxyacetyl nitrate, 105) and hydrogen peroxide (106) are formed in polluted atmospheres as a result of a rather wide variety of photochemical reactions (26, 107). The overall effect is a stinging of the eyes and mucous membranes first noticed in Pasadena, California, a suburb of Los Angeles. Shortly thereafter, polluted atmospheres were labeled as "Los Angeles" type because of their general oxidative character. "London" (England) type smogs (i.e., smoke plus fog) were reductive in nature because of their higher concentrations of sulfur dioxide and soot from the burning of coal.

Figure 19 shows the diurnal variation of nitrogen oxides and ozone in a typical Los Angeles type of photochemical pollution. However, since London has cleared its air with a vigorous smoke abatement program, it is experiencing Los Angeles type of pollution as is shown by Figure 23 (*108*).

To prevent possible serious health effects, an ambient air quality standard maximum 1-hour concentration of 160 $\mu g/m^3$ (0.08 ppm) has been adopted. Alert levels were set at 200 $\mu g/m^3$ (0.1 ppm). Figure 24 shows the number of times that the alert level was exceeded in Los Angeles, California for 1967 thru 1971 (*109*). A study of oxidant levels in the San Francisco, California, Bay Area show a trend to lower annual oxidant levels (*100*). However greater efforts are needed to reduce these values. Two studies have shown that indoor air follows outdoor air concentrations rather closely (*111, 112*).

A number of metals and their compounds have vapor pressures high lems have been measuring total oxidant and ozone concentrations above the alert levels (*113*). There is reason to believe that the "oxidative" conditions in these instances are not the same as those found in the larger cities. Ripperton *et al.* (*114*), for example, have found evidence for tropospheric photochemical production of ozone.

Chesick (*115*) and others (*116*) have been concerned over the effect that high-flying jet planes would have on the upper atmosphere. Water vapor and nitrogen oxides emitted from the jet exhausts conceivably could react with ozone and reduce its insulating quality for strong ultraviolet rays.

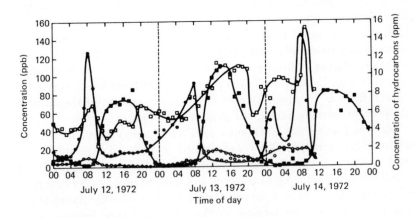

Figure 23. Diurnal variations of air pollutants measured in London, England from July 12 to July 14, 1972. ■, Ozone, ppb; ●, nitric oxide, ppb; □, nitrogen dioxide, ppb; ○, hydrocarbons, ppm (*108*).

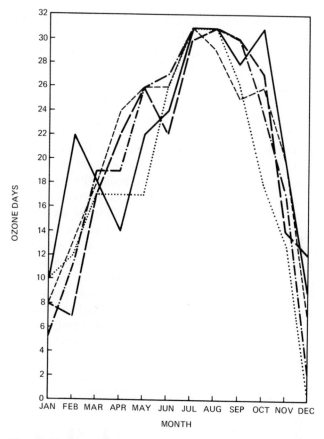

Figure 24. The number of days each month in Los Angeles County, California during which the ozone concentration has risen to 0.1 ppm or above (*109*). Solid line: 1967; short dashed line: 1968; long dashed line: 1969, dashed-dotted line: 1970; dotted line: 1971. [Reprinted with permission from R. H. Sabersky, D. A. Sinema, and F. H. Shair, *Environ. Sci. Technol.* **7,** 347 (1973). Copyright by the American Chemical Society.]

VIII. Mercury

A number of metals and their compounds have vapor pressures high enough to exist as gases in air even if only in trace quantities. Mercury, in particular, is one of them. At room temperature it has a vapor pressure of 0.002 mm. This could lead to a concentration of about 2.6 ppm (21,000 $\mu g/m^3$). However, most studies for trace metals limit the sampling procedures to filtering systems where only particles are collected (*117*). Some of the gaseous metals may adsorb on the particles and be analyzed. However there is also the probability that the metals would desorb into the stream of air moving through the filter.

Knapp (*118*) reviewed the mercury problem. Mercury consumption in the United States exceeded 6 million pounds in 1969, and it is not surprising that unsafe amounts of it have been found in certain water sediments and aquatic life. Goldwater (*119, 120*) described the role of the atmosphere in the mercury cycle (Fig. 25).

The concentration of mercury in air over nonmineralized areas range from 0.6 to 50 μg/m^3. The average at 20 miles offshore over the Pacific is between 0.6 and 0.7 μg/m^3. Table XIX shows maximum mercury concentrations measured in the Western United States (*121*). Many climatic factors affect the amount of mercury to be found in air. These include in particular temperature and wind circulation. The type of mercury compound is important. The rate of evaporation decreases Hg $>$ Hg$_2$Cl$_2$ $>$ HgCl$_2$ $>$ HgS $>$ HgO. The most common form of mercury is cinnabar, HgS. Organic mercury shows a very high tendency to vaporize (*121*).

Man-made pollution sources include the following activities: (a) Mercury production and consumption. Evaporation takes place as the ores are mined, transported, processed, and used. (b) Exhausts from metal smelters. Many metal ores contain mercury from percentage quantities to trace amounts. (c) Combustion of fuels. Coal in particular contains significant quantities of mercury. Power plants using thousands of tons of coal per day can release large quantities of mercury even if present in only trace amounts (*122*).

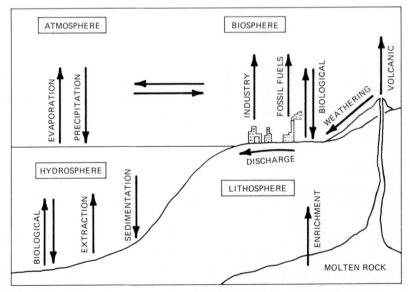

Figure 25. The mercury cycle (from *119*). (From L. G. Goldwater, *Sci. Amer.* **224,** 15. Copyright 1971 by Scientific American, Inc. All rights reserved.)

Table XIX   Maximum Mercury Concentration in Air Measured at Scattered Mineralized and Nonmineralized Areas of the Western United States (121)

Sample location	Maximum Hg concentration $(ng/m^3)^{a,b}$	
	Ground surface	400 ft above the ground[c]
Mercury mines		
Ord mine, Mazatzal Mountains, Arizona	20,000 (50)	108 (4)
Silver Cloud mine, Battle Mountain, Nevada	2000 (50)	24 (8)
Dome Rock Mountains, Arizona	128 (6)	57 (20)
Base and precious metal mines		
Cerro Colorado Mountains, Arizona	1500 (5)	24 (2)
Cortez gold mine, Crescent Valley, Nevada	180 (60)	55 (4)
Coeur d'Alene mining district, Wallace, Idaho	68 (40)	nd
San Xavier, Arizona	nd[d]	25 (3)
Porphyry copper mines		
Silver Bell mine, Arizona	nd	53 (3)
Esperanza mine, Arizona	nd	32 (3)
Vekol Mountains, Arizona	nd	32 (4)
Ajo mine, Arizona	nd	30 (3)
Mission mine, Arizona	nd	24 (3)
Twin Buttes mine, Arizona	20	22 (3)
Pima mine, Arizona	nd	13 (3)
Safford, Arizona	nd	7 (2)
Unmineralized areas		
Blythe, California	nd	9 (20)
Gila Bend, California	nd	4 (2)
Salton Sea, California	nd	3.5 (2)
Arivaca, Arizona	nd	3 (2)

[a] ng/m^3 = nanograms (10^{-9} gm) per cubic meter of air. 1 ng/m^3 = 10^{-4} ppb.
[b] Number of measurements shown in parentheses.
[c] Samples taken from single-engine aircraft.
[d] nd means no data available.

REFERENCES

1. A. N. Strahler and A. H. Strahler, "Environmental Geoscience," pp. 29–58. Hamilton Publishing Co., Santa Barbara, California, 1973.

2. B. Mason, "Principles of Geochemistry," 3rd ed., pp. 208–223. Wiley, New York, New York, 1966.
3. A. N. Strahler, "The Earth Sciences," 2nd ed., pp. 18–62. Harper, New York, New York, 1971.
4. S. Petterssen, "Introduction to Meteorology," 3rd ed. McGraw-Hill, New York, New York, 1969.
5. W. W. Kellogg, R. D. Cadle, E. R. Allen, A. L. Lazrus, and E. A. Martell, *Science* **175**, 587 (1972).
6. A. J. Broderick, ed., "Proceedings of the Second Conference on the Climatic Impact Assessment Program," U.S. Dept. of Transportation No. DOT-TSC-OST-73-4. Nat. Tech. Inform. Serv., Springfield, Virginia, 1973.
7. H. Riehl, "Introduction to the Atmosphere," 2nd ed. McGraw-Hill, New York, New York, 1972.
8. COSPAR International Reference Atmosphere (1965).
9. A. W. Klement, Jr., C. R. Miller, R. P. Minx, and B. Schleien, "Estimates of Ionizing Radiation Doses in the United States: 1960–2000," pp. 47–49. Office of Radiation Programs, United States Environmental Protection Agency, Rockville, Maryland, 1972.
10. "Effects on Populations of Exposures to Low Levels of Ionizing Radiation," pp. 15–24. Biol. Effects Comm., Med. Sci. Div., Nat. Acad. Sci., Washington, D.C., 1972.
11. S. H. Jury and H. R. Edwards, *Can. J. Chem. Eng.* **49**, 663 (1971).
12. C. Orr, Jr., F. K. Hurd, and W. J. Hurd, *J. Colloid Sci.* **13**, 472 (1958).
13. D. A. Lundgren and D. W. Cooper, *J. Air Pollut. Contr. Ass.* **19**, 243 (1969).
14. D. S. Covert, R. J. Charlson, and N. C. Ahlquist, *J. Appl. Meteorol.* **11**, 968 (1972).
15. R. C. Wanta and A. C. Stern, *Amer. Ind. Hyg. Ass., Quart.* **18**, 156 (1957).
16. A. C. Stern, *J. Air Pollut. Contr. Ass.* **14**, 5 (1964).
17. R. I. Larsen, *J. Air Pollut. Contr. Ass.* **19**, 24 (1969).
18. "Air Quality Criteria for Sulfur Oxides," United States Environmental Protection Agency No. AP-50. Research Triangle Park, North Carolina, 1969.
19. R. D. Cadle and E. R. Allen, *Science* **167**, 243 (1970).
20. E. Robinson and R. C. Robbins, *in* "Air Pollution Control" (W. Strauss, ed.), Part II, pp. 1–93. Wiley (Interscience), New York, New York, 1972.
21. F. B. Benson, J. J. Henderson, and D. E. Caldwell, "Indoor-Outdoor Air Pollution Relationships: A Literature Review," United States Environmental Protection Agency No. AP-112. Research Triangle Park, North Carolina, 1972.
22. R. A. Rasmussen, *J. Air Pollut. Contr. Ass.* **22**, 537 (1972).
23. J. H. Cavender, D. S. Kircher, and A. J. Hoffman, "Nationwide Air Pollution Trends," United States Environmental Protection Agency No. AP-115. Research Triangle Park, North Carolina, 1973.
24. Los Angeles County Air Pollution Control District, "Laboratory Data." Los Angeles, California, 1970–1972.
25. "Air Quality Criteria for Hydrocarbons," United States Environmental Protection Agency No. AP-64. Research Triangle Park, North Carolina, 1970.
26. C. S. Tuesday, ed., "Chemical Reactions in Urban Atmospheres." Amer. Elsevier, New York, New York, 1971.
27. R. J. Gordon, M. Mayrsohn, and R. M. Ingels, *Environ. Sci. Technol.* **2**, 1117 (1968).
28. A. P. Altshuller, W. A. Lonneman, and S. L. Kopczynski, *J. Air Pollut. Contr. Ass.* **23**, 597 (1973).

29. J. D. Roberts and M. C. Caserio, "Basic Principles of Organic Chemistry." Benjamin, New York, New York, 1967.

30. D. E. Seizinger and B. Dimitriades, *J. Air Pollut. Contr. Ass.* **22**, 47 (1972).

31. G. S. Callendar, *Tellus* **10**, 243 (1958).

32. S. Manabe and R. T. Wetherald, *J. Atmos. Sci.* **24**, 241 (1967).

33. W. D. Sellers, "Physical Climatology," p. 20. Univ. of Chicago Press, Chicago, Illinois, 1965.

34. R. A. McCormick and J. H. Ludwig, *Science* **156**, 1358 (1967).

35. T. Y. Chang and B. Weinstock, *J. Air Pollut. Contr. Ass.* **23**, 691 (1973).

36. C. M. Stevens, L. Krout, D. Walling, and A. Venters, *Earth Planet. Sci. Lett.* **16**, 147 (1972).

37. B. Weinstock, *Science* **176**, 290 (1972).

38. T. H. Maugh, II, *Science* **177**, 338 (1972).

39. E. Robinson and R. C. Robbins, *Ann. N.Y. Acad. Sci.* **174**, 89 (1970).

40. "Air Quality Criteria for Carbon Monoxide," United States Environmental Protection Agency No. AP-62. Research Triangle Park, North Carolina, 1970.

41. W. L. Faith and A. A. Atkisson, Jr., "Air Pollution," 2nd ed., pp. 163–205. Wiley (Interscience), New York, New York, 1972.

42. E. K. Bigg, A. Ono, and W. J. Thompson, *Tellus* **22**, 550 (1970).

43. C. E. Junge and J. E. Manson, *J. Geophys. Res.* **66**, 2163 (1961).

44. J. Cholak, L. J. Schafer, W. J. Younker, and D. W. Yeager, *Amer. Ind. Hyg. Ass., J.* **19**, 371 (1958).

45. E. A. Tice, *J. Air Pollut. Contr. Ass.* **12**, 553 (1962).

46. "Validation of Improved Chemical Methods for Sulfur Oxides Measurements from Stationary Sources," United States Environmental Protection Agency No. R2-72-105. J. N. Driscoll, Program Manager, Walden Research Corp., Nat. Tech. Inform. Serv., Springfield, Virginia, 1972.

47. H. F. Johnstone and D. R. Coughanowr, *Ind. Eng. Chem.* **50**, 1169 (1958); **52**, 861 (1960).

48. P. Urone, *in* "Proceedings of International Symposium on Air Pollution," pp. 505–520. Union of Japanese Scientists and Engineers, Tokyo, Japan, 1972.

49. P. Urone and W. H. Schroeder, *Environ. Sci. Technol.* **3**, 436 (1969).

50. "Control Techniques for Sulfur Oxide Air Pollutants," United States Environmental Protection Agency No. AP-52. Research Triangle Park, North Carolina, 1969.

51. R. I. Larsen, "A Mathematical Model for Relating Air Quality Measurements to Air Quality Standards," United States Environmental Protection Agency No. AP-89, Research Triangle Park, North Carolina, 1971.

52. A. D. Little, Inc., "Research on Chemical Odors, Part I: Odor Thresholds for 53 Commercial Chemicals." Mfg. Chem. Ass., Washington, D.C., 1968.

53. R. O. Blosser, *Tappi* **55**, 8 (1972).

54. W. L. Faith, D. B. Keyes, and R. L. Clark, "Industrial Chemicals," 3rd ed., p. 741. Wiley, New York, New York, 1965.

55. J. Cholak, L. J. Schafer, and R. F. Hoffer, *Arch. Ind. Hyg. Occup. Med.* **6**, 314 (1952).

56. Southwest Research Institute, "Air Pollution Survey of Houston Area," Tech. Rep. No. 4. Chamber of Commerce, Houston, Texas, 1957.

57. M. Katz, *Air Repair* **4**, 176 (1955).

58. R. A. Schmall, "Atmospheric Quality Protection Literature Review—1972," Tech. Bull. No. 65. National Council of the Paper Industry for Air and Stream Improvement, New York, New York, 1972.

59. R. A. Rasmussen, *Amer. Lab.* **4** (12), 55 (1972).

60. F. A. Cotton and G. Wilkinson, "Advanced Inorganic Chemistry," 2nd ed., pp. 323–357. Wiley (Interscience), New York, New York, 1966.

61. Intersociety Committee, "Methods of Air Sampling and Analysis," pp. 329–336. Amer. Pub. Health Ass., Washington, D.C., 1972.

62. H. Craig and L. I. Gordon, *Geochim. Cosmochim. Acta* **27**, 949 (1963).

63. D. R. Bates and P. B. Hays, *Planet. Space Sci.* **15**, 189 (1967).

64. A. P. Altshuller, *Tellus* **10**, 479 (1958).

65. D. A. Trayser and F. A. Creswick, *Battelle Res. Outlook* **2** (3), 12 (1970).

66. T. C. Hall, Jr. and F. E. Blacet, *J. Chem. Phys.* **20**, 1745 (1952).

67. "Air Quality Criteria for Nitrogen Oxides," United States Environmental Protection Agency No. AP-84. Research Triangle Park, North Carolina, 1971.

68. R. J. Carlson and N. C. Ahlquist, *Atmos. Environ.* **3**, 653 (1969).

69. R. Louw, J. van Ham, and H. Nieboer, *J. Air Pollut. Contr. Ass.* **23**, 716 (1973).

70. B. W. Gay and J. J. Bufalini, *Environ. Sci. Technol.* **5**, 422 (1971).

71. P. L. Hanst, *J. Air Pollut. Contr. Ass.* **21**, 269 (1971).

72. E. A. Schuck, J. N. Pitts, and J. K. Swan, *Int. J. Air Water Pollut.* **10**, 689 (1966).

73. C. E. Junge, "Air Chemistry and Radioactivity," pp. 37–58. Academic Press, New York, New York, 1963.

74. J. P. Lodge, Jr. and J. B. Pate, *Science* **153**, 408 (1966).

75. D. A. Lundgren, *J. Air Pollut. Contr. Ass.* **20**, 603 (1970).

76. D. C. McCune, *Environ. Sci. Technol.* **3**, 720 (1969).

77. J. L. Shupe, *Environ. Sci. Technol.* **3**, 721 (1969).

78. J. D. Tessari and D. L. Spenser, *J. Ass. Offic. Anal. Chem.* **54**, 1376 (1971).

79. J. E. Lovelock, *Nature (London)* **259**, 292 (1974).

80. M. J. Molina and F. S. Rowland, *Nature (London)* **249** (No. 5460), 810 (1974).

81. S. C. Wofsy, M. B. McElroy, and N. D. Siz, *Science* **187**, 535 (1975).

82. J. Cholak, *AMA Arch. Ind. Health* **21**, 312 (1960).

83. R. J. Thompson, T. B. McMullen, and G. B. Morgan, *J. Air Pollut. Contr. Ass.* **21**, 484 (1971).

84. J. D. Stockham, *J. Air Pollut. Contr. Ass.* **21**, 713 (1971).

85. R. C. Specht and R. R. Calaceto, *Chem. Eng. Progr.* **63**, 78 (1967).

86. J. M. Craig, Ph.D. Thesis, University of Florida, Gainesville, Florida, 1970.

87. A. R. Meetham, "Atmospheric Pollution," 2nd ed., p. 215. Pergamon, Oxford, England, 1956.

88. C. D. Leonard and H. B. Graves, Jr., *Proc. Fla. State Hort. Soc.* **79**, 79 (1966).

89. H. C. Hodge and F. A. Smith, *J. Air Pollut. Contr. Ass.* **20**, 226 (1970).

90. "Chlorine and Air Pollution: An Annotated Bibliography," United States Environmental Protection Agency No. AP-99. Research Triangle Park, North Carolina, 1971.

91. "Atmospheric Emissions From Chlor-Alkali Manufacture," United States Environmental Protection Agency No. AP-80. Research Triangle Park, North Carolina, 1971.

92. B. E. Nordlie, *Amer. J. Sci.* **271**, 417 (1971).

93. O. G. Bartels, *Health Phys.* **22**, 387 (1972).

94. J. Cholak, *Proc. Nat. Air Pollut. Symp., 2nd, 1952,* p. 6. [*Chem. Abstr.* **47**, 559c (1953)].

95. "Hydrochloric Acid and Air Pollution: An Annotated Bibliography." Environmental Protection Agency No. AP-100. Research Triangle Park, North Carolina, 1971.

96. P. Buat-Menard and R. Chesselet, *C. R. Acad. Sci., Ser. B* **272**, 1330 (1971); *Chem. Abstr.* **75**, 79245 (1971).

97. C. W. Stanley, J. E. Barney, II, M. R. Helfon, and A. R. Yobs, *Environ. Sci. Technol.* **5**, 430 (1971).

98. J. W. Miles, L. E. Fester, and G. W. Pearce, *Environ. Sci. Technol.* **4**, 420 (1970).

99. G. M. Woodwell, P. P. Craigh, and H. A. Johnson, *Science* **174**, 1101 (1971).

100. J. Cramer, *Atmos. Environ.* **7**, 241 (1973).

101. J. L. Moyers, W. H. Zoller, R. A. Duce, and G. L. Hoffman, *Environ. Sci. Technol.* **6**, 68 (1971).

102. J. L. Moyers, Ph.D. Thesis, University of Hawaii, Honolulu, Hawaii, 1970.

103. H. R. Bowman, J. G. Conway, and F. Asaro, *Environ. Sci. Technol.* **6**, 558 (1972).

104. "Air Quality Criteria for Photochemical Oxidants," United States Environmental Protection Agency No. AP-63. Research Triangle Park, North Carolina, 1970.

105. E. R. Stephens, *in* "Chemical Reactions in the Lower and Upper Atmosphere," Stanford Research Institute (foreword by Richard D. Cadle), pp. 51–69. Wiley (Interscience), New York, New York, 1961.

106. J. J. Bufalini, B. W. Gay, Jr., and K. L. Brubaker, *Environ. Sci. Technol.* **6**, 816 (1972).

107. P. A. Leighton, "Photochemistry of Air Pollution." Academic Press, New York, New York, 1961.

108. R. G. Derwent and H. N. M. Stewart, *Nature (London)* **241**, 342 (1973).

109. R. H. Sabersky, D. A. Sinema, and F. H. Shair, *Environ. Sci. Technol.* **7**, 347 (1973).

110. J. S. Sandberg, R. Thuillier, and M. Feldstein, *J. Air Pollut. Contr. Ass.* **21**, 118 (1971).

111. F. X. Mueller, L. Loeb, and W. H. Mapes, *Environ. Sci. Technol.* **7**, 342 (1973).

112. C. R. Thompson, E. G. Hansel, and G. Kats, *J. Air Pollut. Contr. Ass.* **23**, 881 (1973).

113. United States Environmental Protection Agency, Region IV. Air Pollution Control Office, Atlanta, Georgia (private communication).

114. L. A. Ripperton, H. Jeffries, and J. J. B. Worth, *Environ. Sci. Technol.* **5**, 246 (1971).

115. J. P. Chesick, *J. Chem. Educ.* **49**, 755 (1972).

116. "Effect of Stratospheric Ozone Depletion on the Solar Ultraviolet Radiation Incident on the Surface of the Earth." Institute for Defense Analyses, Sci. Tech. Div., Washington, D.C., 1973.

117. P. F. Woolrich, *Amer. Ind. Hyg. Ass., J.* **35**, 217 (1973).

118. C. E. Knapp, *Environ. Sci. Technol.* **4**, 890 (1970).

119. L. J. Goldwater, *Sci. Amer.* **224**, 15 (May 1971).

120. L. J. Goldwater, "Mercury: A History of Quicksilver." York Press, Baltimore, Maryland, 1972.

121. "Mercury in the Environment," U.S., Geol. Surv., Prof. Pap. No. 713. U.S. Govt. Printing Office, Washington, D.C., 1970.

122. C. E. Billings, A. M. Sacco, W. R. Matson, R. M. Griffin, W. R. Coniglio, and R. A. Harley, *J. Air Pollut. Contr. Ass.* **23**, 773 (1973).

3

Aerosols and the Primary Air Pollutants–Noviable Particles
Their Occurence, Properties, and Effects

Morton Corn

Nomenclature

a	Particle projected area, cm^2
A_1	Constant $= 1.246$, dimensionless [Eq. (20)]
A	Total particle area associated with particles less than size r, cm^2
b	Constant $= 0.87$, dimensionless [Eq. (20)]
B	Constant $= 0.42$, dimensionless [Eq. (20)]
C	Cunningham correction factor, dimensionless
C_1	Constant, dimensionless [Eq. (35]
d_e	Equivalent aerodynamic particle diameter, cm
D	Particle diffusion coefficient, cm^2/second
D_w	Diffusion coefficient of water at temperature T, dyn/cm^2
e	Unit electrical charge, 1.6×10^{-19} C
E	Voltage gradient, statvolt/cm
f	Functional notation, e.g., $f(r)$, $f(r^2)$, etc.
F_r	Resisting force of fluid, dyn
g	Gravitational constant, $cm/second^2$
h	Characteristic linear dimension of collector, cm
i	Electrical current, A
k	Boltzmann constant, 1.38×10^{-16} erg/°C
k_g	Ion of smallest mobility collected with 100% efficiency, cm/second/V/cm
k'	Thermal conductivity of air, cal/second cm °K
k_p	Thermal conductivity of particle, cal/second cm °K
m	Particle mass, gm
M_g	Geometric mean particle radius by count, μm
M_g'	Geometric mean particle radius by weight, μm
n	Particle number, dimensionless
n_t	Particle number per cubic centimeter at time t
n_0	Particle number per cubic centimeter at time $t = 0$
N	Total particle number less than size r, dimensionless
N_1	Number of air molecules per gram mole of air
P	Vapor pressure surrounding a drop, dyn/cm^2
P_∞	Vapor pressure at a plane liquid surface at temperature t, dyn/cm^2
P/P_∞	Supersaturation ratio, dimensionless
P_o	Vapor pressure of water in air at temperature T, dyn/cm^2
P_s	Saturated vapor pressure of water in air at temperature T, dyn/cm^2
q	Electrostatic charge on particle, esu
Q	Airflow rate, cm^3/second
r	Particle radius, cm
r_2	Radius of ion collector inner cylinder, cm
r_e	Equivalent aerodynamic particle radius, cm
r_p	Particle projected area radius, cm
r_s	Particle Stokes' radius, cm
R	Gas constant, 6.23×10^4 cm^3 mm Hg(°K)$^{-1}$ (gm mole)$^{-1}$
R'	Gas constant, 8.3×10^7 gm cm^3 (second)$^{-1}$ (gm mole)$^{-1}$ (°K)$^{-1}$
R_1	Radius of ion collector outer cylinder, cm
R_{e_p}	Particle Reynolds number, dimensionless
t	Time, second

T	Absolute temperature, °K
u	Particle velocity, cm/second
$\bar{u}_t$	Particle terminal settling velocity, cm/second
$\bar{u}_D$	Average "thermal" speed of particle, cm/second
v	Voltage differential between ion collector condenser plates, V
V	Total particle volume associated with particles less than size r, cm³
w	Particle weight, gm
x	Distance, cm
y'	$n/\Delta(\ln r)$ [Eq. (5)]
z	Ions per cubic centimeter
Z	Particle mobility in Stokes' regime, cm/second
Z_D	Particle mobility where diffusion prevails, cm/second
β	Constant, dimensionless [Eq. 34]
γ	Surface tension of water, dyn/cm
$\bar{\Delta}_x$	Mean particle displacement due to diffusion, cm
λ	Mean free path of air molecules (6.53×10^{-6} cm at 20°C and 760 mm Hg)
λ_D	Particle apparent mean free path, cm
τ	Particle relaxation time, second
μ	Viscosity of air, P (1.81×10^{-4} at 20°C and 760 mm Hg)
π	3.1416
ρ	Density of particle, gm/cm³
ρ'	Density of air, gm/cm³
ρ_w	Density of water, gm/cm³
σ_g	Geometric standard deviation, dimensionless
$\mathcal{W}$	Dynamic particle shape factor, dimensionless

I. Introduction

There is, as Turkevich (1) has stated, "a whole world of fine particles—of particles whose size ranges from those of small molecules to those
of ordinary dust and sand visible with an optical microscope." Particles
of fume, dust, soot, fibers, pollen, and droplets can be identified if a sample of air from an urban atmosphere is subjected to examination (Fig.
1). Particles of extraterrestrial origin (2, 3) and from volcanic eruptions
(4, 5) can be found. Each particle is different in shape, size, and composition and has an individual history in the atmosphere if we focus on its
mode of origin, growth, interaction, and decay.

The processes of particle generation and removal in air are continuous
and depend on the specific sources of pollution, both natural and anthropogenic, and the meteorology and topography of the air basin. Once airborne, particles are subject to well-known laws of dynamic behavior, including sedimentation, diffusion, and coagulation. They are also subject
to chemical interactions with components of the gas phase which surrounds them and to the scavenging effects of the various forms of precipitation in the atmosphere. In the past 5 years a flurry of research activity
has been aimed at determining whether a background level of aerosol

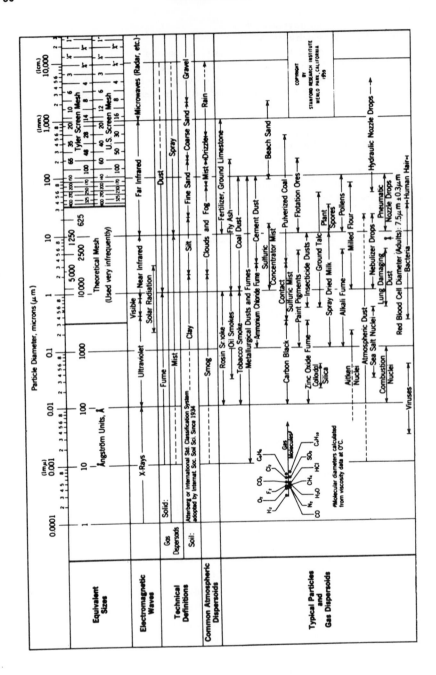

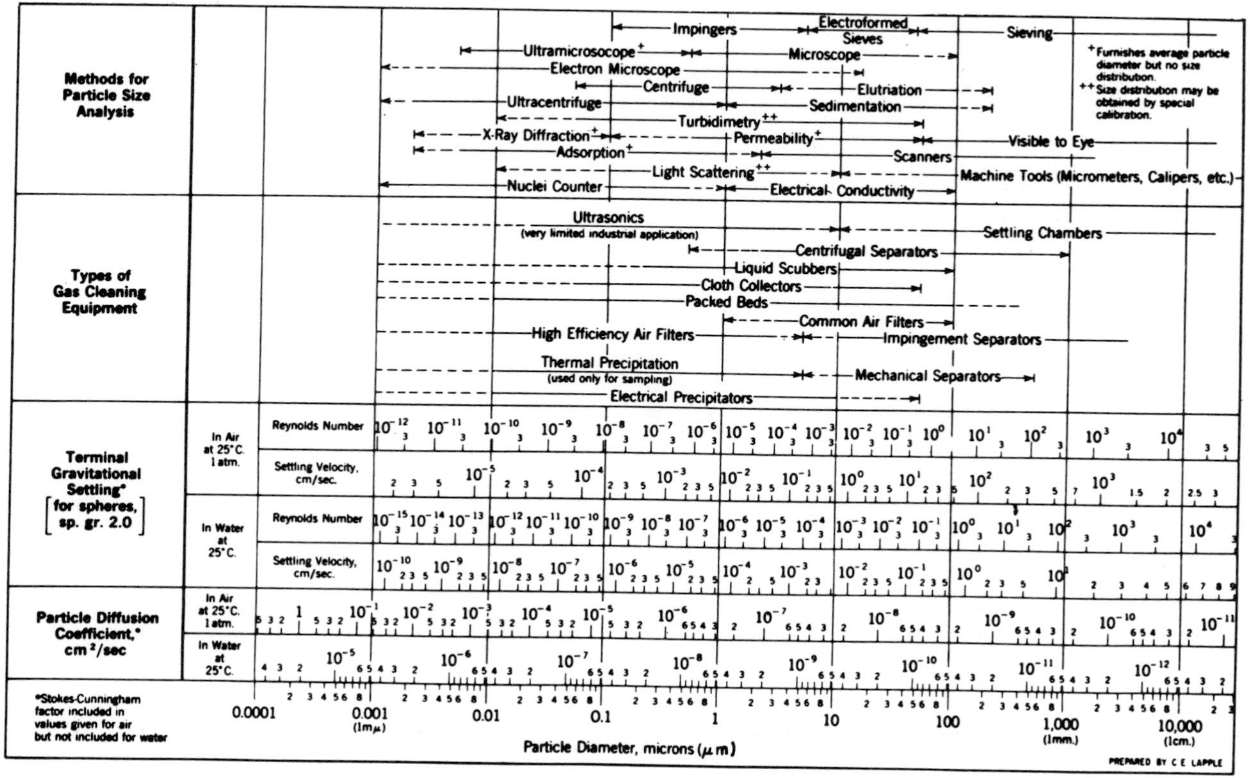

Figure 1. Characteristics of particles and particle dispersoids. Reprinted by permission of *Standford Res. Inst. J.* 5, 95 (1961).

exists (6, 7); if the atmospheric burden of particulate matter is increasing (8); the potential effects of such an increase (9); the quantitation of particulate dynamics in the atmosphere (10); and experimental (11) and theoretical (12) explorations of the regularity of airborne particle size distributions in urban and nonurban areas. Efforts continue on a worldwide basis, to determine the concentration and trends of concentration of suspended particulate matter in the atmosphere in urban and nonurban areas (13, 13a, 13b). The study of the distribution of specific chemical species among particles of different size started in 1964 (14), and continues to be studied (15, 16, 16a, 16b, 16c) because of the toxicological significance of the chemical species, the elucidation of particle sources (16d), and its relationship to meteorological phenomena (16e, 16f). Finally, our understanding of aerosol phenomena in nature has advanced to the presentation of a first effort to quantitate the global sources of tropospheric aerosols (17).

The purpose of this chapter is to present the properties of single particles and dispersions of particles as they relate to the behavior of particles in the atmosphere. Further details of the subjects discussed can be found in monographs devoted to aerosol science (18–21).

II. Characterization of Particles

A. Terminology*

A dispersion aerosol is formed by the grinding or atomization of solids and liquids and by the transfer of powders into a state of suspension through the action of air currents or vibration. A condensation aerosol is formed when supersaturated or saturated vapors are condensed or when gases react chemically to form a volatile or nonvolatile solid product. Condensation and dispersion aerosols with liquid particles are called mists, regardless of particle size. Dispersion aerosols with solid particles are called dusts, regardless of particle size. Condensation aerosols with a solid disperse phase or a solid and liquid disperse phase are called smokes. In general, dusts are heterogeneous in composition and contain a wide spectrum of particle sizes. Traditionally, the term "smoke" has been used to describe the dispersion produced from incomplete combustion of fuel or other combustible material. "Fog" has denoted a high concentration of water droplets in air. The term "fume" has been used to characterize the generation of a condensed aerosol in the company of a gas or vapor—for example, welding fume. Special nomenclature has arisen to describe the air pollution situation, i.e., "haze" which is used

*See Ref. (18).

in association with decreased visibility in the atmosphere and may indicate the presence of dust, mist, and pollutant gases; "smog" (smoke and fog); and "smaze" (smoke and haze). These are popular terms which encompass complex pollutant gas and aerosol mixtures. When referring to an aerodispersion in the atmosphere, it is recommended that the term atmospheric aerosol be used; in an urban area the term urban aerosol is more specific. The term cloud should be reserved for a free aerodisperse system of any type having a definite size and form.

B. Size, Area, Weight, and Density of a Single Particle

Urban and natural aerosols are polydisperse with respect to particle size, encompassing the size range from approximately 6×10^{-4} to 10^3 μm. The properties of particles are dependent on their size and within this range of sizes certain properties undergo transition to greater or lesser dependence on size. In Figure 2, Group 1 properties are associated with the mean free path of the air molecules ($\lambda = 6.53 \times 10^{-2}$ μm at 20°C and 760 mm Hg) while Group 2 properties are associated with the average wavelength of visible light (0.55 μm). A transition between these groups occurs in the particle size range between approximately 0.5×10^{-4}

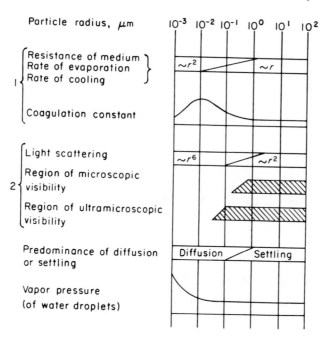

Figure 2. Some properties of aerosols in relation to particle size (18).

and 10^{-4} cm. For the rest of these properties the correspondence is fortuitous.

1. Definition of Particle Size

Examination of particles with the aid of an optical or electron microscope involves the measurement of a linear dimension of a particle silhouette, or comparison of the areas of circles on an eyepiece graticule with the area of the particle silhouette (22). The measured "particle size" is related to the particle perimeter or to the particle projected area diameter, which is the diameter of a circle having the same area as the particle silhoutte. Particle size measurement in this manner cannot account for variation in particle density or in particle shape. It is usual to obtain from microscopic sizing an estimate of particle volume, and, by assuming a density to estimate particle weight. These procedures are discussed in detail in Chapters 1 and 3, Vol. III and Silverman *et al.* (23).

Because of the aggregated nature of atmospheric particles (24, 25), and the attendant large variations in particle density (26), it is necessary to define another quantity as a measure of particle size—a quantity related to the aerodynamic behavior of the particle. The rate of fall of a particle in air depends on the particle weight and the aerodynamic drag. The Stokes' radius r_s is defined as the radius of a sphere having the same falling velocity as the particle, and a density equal to that of the bulk material from which the particle was formed. Because atmospheric particles are formed from materials of many densities, this widely used size parameter is difficult to use. It is more useful to define the equivalent aerodynamic size, r_e or d_e, the radius or diameter, respectively, of a sphere having the same falling velocity as the particle and a density equal to 1 gm/cm³. The ratio between the projected area radius (r_p) and the equivalent (aerodynamic) radius (r_e) for particles collected from a sample of Pittsburgh air (Fig. 3) (27) ranges from 0.25 to 2.4 for most particles, but, in an extreme sample (Fig. 4), is equal to 8.0. In practice, it is not difficult to determine r_e or d_e by utilizing spherical particles of known density to identify the path of deposition of the unknown particles (27–29).

2. Shapes of Airborne Particles

There are great differences in the shapes of atmospheric particles, which, for the purpose of simplification, can be divided into spherical, irregular, cubical, flake, fibrous, and condensation flocs (30). More elaborate classification is possible (25) and a particle atlas with color plates of generic particles is available for particle identification from morpho-

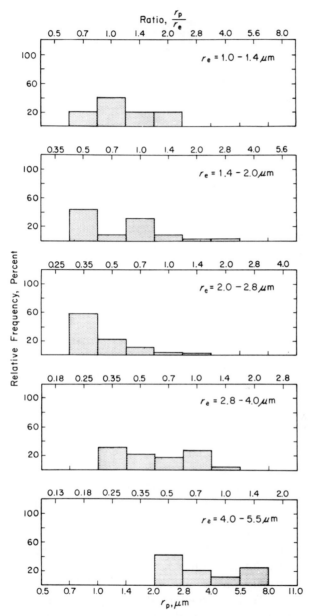

Figure 3. Ratio between projected area diameter and equivalent aerodynamic diameter for particulates in Pittsburgh, Pennsylvania air (27).

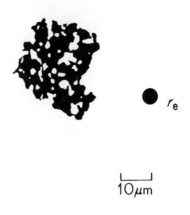

r_e

10μm

Figure 4. An atmospheric particle with projected area diameter to aerodynamic equivalent diameter ratio of approximately 8:1 (27).

logical characteristics (31). Identification is now possible when microscopic techniques are the only ones available (32, 33). This is discussed further in Chapter 3, Vol. III. Spherical particles include pollen and condensation solids, i.e., fly ash. Fibers of wool, cotton, glass, asbestos, and a variety of synthetic materials can be found. Minerals are usually geometrically irregular, regular, or flakelike. Agglomerates can result from particles combining while airborne or can be formed during cooling of hot gases. The latter are characterized by a chainlike appearance and are called flocs; they often are formed during incomplete combustion of fuels and contain a large amount of carbon.

The shape of particles in an atmospheric aerosol sample will depend upon the local emission sources of particles. The shapes of particles larger than 0.5 μm radius in Pittsburgh, Pennsylvania, were reported by Stein, Esmen, and Corn (25). Whitby et al. (30) studied the morphology of particles larger than 0.1 μm radius in the light or electron microscope for samples obtained in Minneapolis, Minnesota; Akron, Ohio; and Louisville, Kentucky (Table I). Stein et al. (25) found large daily variations in airborne particle shapes in Pittsburgh, Pennsylvania. Spurny, Polydorova, and Pixova (34) found three maximum values in the size of airborne particles from Prague, Czechoslovakia, i.e., 0.1, 0.8, and 15.0 μm radii. These sizes correspond to three types of particles, namely, condensation aerosols, ash particles, and coagulated particles. The spherical par-

Table I   Airborne Dust Particle Shapes *(30)*

Shape	Range	Average	Particles[a]
	Percent by weight in sample		
Spherical	0–20	10	Smoke, pollen, fly ash
Irregular ⎱	10–90	40	⎰ Mineral
Cubical ⎰			⎱ Cinder
Flakes	0–10	5	Mineral, epidermis
Fibrous	3–35	10	Lint, plant fiber
Condensation flocs	0–40	15	Carbon, smoke, fume

[a] Larger than 0.1 μm.

ticles were formed by carbon or melted inorganic material; there were also droplets of coal tar.

3. Density of Airborne Particles

The density of individual particles in the air is estimated to vary from 0.5–6.5 gm/cm³ *(25)*. There are few measurements of the density of suspended particulate matter in air. In Pittsburgh, Pennsylvania, the suspended particulate matter collected on membrane filter paper by high volume air sampling during an annual period was characterized by densities (measured by liquid pycnometry) from 1.8–2.3 gm/cm³. The average summer sample density was 2.2 ± 0.1 gm/cm³ (Table II) *(26)*. Whitby *(35)* found that the densities of acetone-insoluble particles in urban air varied from 1.5 to 3.0 gm/cm³; bulk density varied from 0.2 to 1.5 gm/cm³. Durham *et al.* *(35a)* estimated from aerosols measured in Denver, Colorado, that the densities of the submicrometer size background and pollution aerosols were 1.6–1.8 gm/cm³ and 1.1–1.5 gm/cm³, respectively.

4. Specific Surface Area of Airborne Particles

Particles in air have a large specific surface area because of their small size and the presence of surface irregularities and internal pores. It is not possible—or particularly meaningful—to measure this property for individual particles. The only published data for the specific surface area (measured by Brunauer–Emmett–Teller technique with nitrogen or krypton gas) of suspended particulate are for Pittsburgh, Pennsylvania aerosol *(26, 36)*, where some seasonal variation was noted (Table II).

Table II Seasonal Variations in Specific Surface Areas and Densities of Suspended Particulate Matter in Pittsburgh, Pennsylvania[a] (26). [Reprinted with permission from M. Corn, T. L. Montgomery, and N. A. Esmen, *Environ. Sci. Technol.* **5**, 155 (1971). Copyright by the American Chemical Society.]

| Season | Sampling period | Number of samples | Average specific surface area, m^2/gm Arithmetic mean ± S.D.[b] | | Average density, gm/cm^3 Arithmetic mean ± S.D.[b] | | Sampling filter |
			25°C	200°C	Bulk	Specific	
Fall	9/23/67–12/21/67	5	2.66 ± 0.59	4.35 ± 0.87	0.60 ± 0.06	2.1 ± 0.5	Fiberglass
Winter	12/22/67–3/20/68	9	3.05 ± 0.38	4.85 ± 0.75	0.56 ± 0.06	2.1 ± 0.3	Fiberglass
Spring	3/21/68–6/21/68	5	2.36 ± 0.35	3.59 ± 0.54	0.59 ± 0.06	1.8 ± 0.2	Fiberglass
Summer	6/21/68–9/22/68	3	1.90 ± 0.36	2.67 ± 0.35	0.64 ± 0.02	2.1 ± 0.2	Membrane
Fall	9/23/67–12/21/67	4	3.38 ± 0.31	5.36 ± 0.36	0.57 ± 0.07	2.3 ± 0.5	Membrane
Fall	9/23/67–12/21/67	6			0.54	2.3	Membrane
Winter	12/22/67–3/20/68	1	4.36	6.75			Membrane
Summer	6/21/68–9/22/68	6			0.61 ± 0.04	2.2 ± 0.1	Membrane
Summer	6/21/68–9/22/68	2	1.80 ± 0.52	3.22 ± 0.90			Membrane

[a] Samples collected on roof of Graduate School of Public Health with high volume sampling apparatus.
[b] S.D.: standard deviation.

C. Size, Area, and Weight Characterization of the Distribution of Particles

It is essential to consider the dependence of particle properties on particle size (Fig. 2) when setting out to describe the atmospheric aerosol. If the phenomenon of interest is related to the light scattering properties of particles suspended in the atmosphere, it is the distribution of particle size with respect to r^2 that must be established. Figure 2 is a statement of the properties of a single particle. In a polydisperse urban aerosol the distribution of r^2 must be determined, i.e., the frequency distribution curve of the second moment. Thus, the fraction of total number of particles in the size range between r and $r + dr$ is given by

$$dn = f(r) \, dr \tag{1}$$

with the condition that

$$\int_0^\infty f(r) \, dr = 1 \tag{2}$$

The differential distribution of the second moment r^2 which is related to particle projected area a is given by

$$da = f(r^2) \, dr \tag{3}$$

with the condition that

$$\int_0^\infty f(r^2) \, dr = 1 \tag{4}$$

Similarly, if the aerosol property of interest is the weight, then the frequency distribution with respect to the third moment must be determined. These are not trivial distinctions in particle parameters (*37, 38*) since both particle area and weight are related to retention and toxicological significance of dust particles inhaled by man and to deposition of particles on surfaces.

Characterization of an urban aerosol by a distribution function should be related, if at all possible, to the particle parameter most closely associated with the phenomenon of interest, be it light scattering, dustfall, or toxicological potential. The mathematical transformation of a distribution based on number [Eq. (1)] to one based on area [Eq. (3)], or one based on weight, is subject to the introduction of numerous errors (*39, 40*) and should be avoided, if at all possible.

The simplest way to characterize the particle size distribution of a sample of polydisperse particles is to classify the sizes into successive size intervals and to present the data (size interval and frequency of occurrence) in tabular or graphical form (histogram). Equal width rectangles should be used for the size intervals, while the heights of the bars should represent frequency of occurrence. The arithmetic mean size and standard deviation from the mean can then be calculated. Size distributions of particles in air seldom yield symmetrical or normal curves about the mean; they are skewed with the longer tail of the curve characterizing the larger particle sizes. Several mathematical distribution functions have been used to characterize the size distribution of airborne particles, and these will now be briefly discussed.

The two mathematical models which have been frequently used to describe the particle size distribution of the atmospheric aerosol are the logarithmic normal and the log radius particle count, area, or weight distributions. The log-normal particle count distribution is

$$ y' = \frac{1}{\ln \sigma_g \sqrt{2\pi}} \exp - \left[\frac{(\ln x - \ln M_g)^2}{2 \ln^2 \sigma_g} \right] \tag{5} $$

The geometric mean radius M_g is the same as the particle count median radius determined by microscopic methods. Half the number of particles in the distribution are above and below this size, respectively. Smith and Jordan (41) point out that the correct interpretation of y' is that it is equal to $n/\Delta(\ln r)$, where n is the number of particles whose radii have their logarithms lying in the interval $\Delta(\ln r)$. In the practical analysis of particle size data, logarithmic probability paper is often used to plot the percent of particles less than a stated size versus the logarithms of the stated size (Fig. 5). The 50% value of r is taken as M_g and

$$ \sigma_g = 84.3\% \text{ value of } r/50\% \text{ value of } r \tag{6} $$

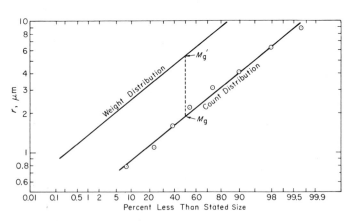

Figure 5. Typical presentation of particle size data on logarithmic probability paper. $\sigma_g = 84.16\%$ value of $r/50\%$ value of r; $\log_{10} M_{g'} = \log_{10} M_g + 6.9 \log_{10}^2 \sigma_g$.

The log-normal distribution is characterized by the same σ_g for moments about the count, area, and weight mean geometric diameters (42). Thus, to calculate the surface median diameter M_a after M_g and σ_g are known use

$$\log_{10} M_a = \log_{10} M_g + 4.60 \log_{10}^2 \sigma_g \qquad (7)$$

and for the mass median diameter M_g' use

$$\log_{10} M_g' = \log_{10} M_g + 6.90 \log_{10}^2 \sigma_g \qquad (8)$$

Figure 6 shows the location of several characterizing statistical parameters on the particle size distribution curve when the log-normal distribution is adhered to (43).

In addition to particle size data, atmospheric dustfall and the concentrations of gaseous pollutants measured at a sampling site have been found to adhere closely to a log-normal distribution.

The log radius number distribution (44) is defined by

$$n(r) = \frac{dN}{d(\log r)} \, \text{cm}^{-3} \qquad (9)$$

where N is the total concentration (number per cubic centimeter) of aerosol particles of radius smaller than r. The number of particles ΔN

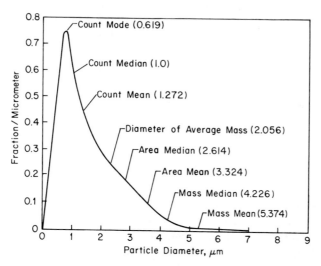

Figure 6. An example of the log-normal distribution function in normalized linear form for $M_g = 1.0$ and $\sigma_g = 2.0$, showing the mode, median, and mean diameters, the mass distribution median, and mean diameters, and the diameter of average mass (43).

between the limits of the interval $\Delta \log r$ can be obtained from a plot of $n(r)$ vs $\log r$ as follows:

$$\Delta N = n(r)\, \Delta(\log r) \tag{10}$$

The log radius–surface area (a) and the log radius–volume (v) distributions are defined by Equations (11) and (12), respectively.

$$a(r) = \frac{dA}{d(\log r)} = 4\pi r^2 \frac{dN}{d(\log r)} \tag{11}$$

$$v(r) = \frac{dV}{d(\log r)} = \tfrac{4}{3}\pi r^3 \frac{dN}{d(\log r)} \tag{12}$$

If the density of the aerosol particles, ρ, is assumed to remain constant for all particles, the log radius–mass distribution is

$$w(r) = \rho v(r) \tag{13}$$

D. Particle Shape Factors

The simplest particle shapes are spheres and are usually associated with condensed aerosols. Coalescence of several liquid particles will result in larger spherical particles, but coalescence of solid particles results in flocculent isometrically shaped or threadlike aggregates. It is often necessary to numerically express the degree of irregularity of primary or aggregated particles, and various shape factors are used to achieve this purpose. The "coefficient of sphericity" is the ratio of the surface area of a sphere with the same volume as the given particle to the surface area of the particle (45). Volume-shape factors are defined by Cartwright (46), and Fuchs defines a dynamic shape factor $\mathcal{W}$ as the ratio between the squares of the equivalent (aerodynamic) radius r_e and the Stokes' radius r_s:

$$\mathcal{W} = r_e^2 / r_s^2 \tag{14}$$

It can be seen that the ratio r_p/r_e cited above is also a form of particle shape factor. If the particle shape factor is accurately specified, it is possible to calculate one particle parameter from another, depending on the factor used. In reality, it is necessary to deal with a variety of particle shapes in the same aerosol, and the factor specified is an average value for the many particles which form the distribution (47). The urban aerosol in Pittsburgh, Pennsylvania was visualized as a sphere with a cylindrically shaped cavity through its center. This solid had the surface, volume, and density characteristics of the entire assemblage of airborne particles sampled from the urban aerosol (*26*).

Dynamic shape factors for nonspherical particles, including clusters of spheres, are presented by Stöber (*48*).

E. Additional Static Particle Properties

The properties of size, area, volume, density, and weight, as discussed above, may be considered "static" properties of particles or assemblages of particles. Other "static" properties relevant to the behavior of nonviable particles in air are the surface electrical charge; strength of adhesion to surfaces; and the proclivity for, and rates associated with, evaporation and condensation of moisture from or onto the particle surface, respectively. The interaction of particles with electromagnetic radiation in the atmosphere is treated elsewhere in these volumes (Chapter 1, Vol. II).

1. Surface Electrical Charge

Particles in air acquire electrical charge by a variety of mechanisms which promote electron transfer to or from the surface, thus producing a negatively or positively charged particle, respectively. The mechanisms include, but are not limited to, diffusion of ions to the particle surface, triboelectrification between solids, and electrolysis (*49*). The maximum surface charge a particle can retain in dry air is about 8 esu/cm² or 1.6×10^{10} electrons/cm², a limit imposed by the electrical breakdown strength of air. Particles larger than 0.1 μm in the atmosphere contain only a small fraction of this charge. Even freshly generated smoke particles contain but a fraction of maximum charge. Table III (*50*) presents data for measured charges on selected particles, expressed in terms of electrons carried. The percentages of maximum charge were calculated by assuming the particles to be smooth spheres and by using the lower and upper values of size and charge, respectively. Because particle mass increases as the cube of particle size, while surface area (and electrical charge) increases with the square of particle size, the influence of charge on particle movement in the field of the earths' atmosphere increases as particle size decreases. Particles with substantial movement or mobility in an electrical field are called either positive or negative ions, according to their charge. Under special conditions of charging, it is possible to charge particles to levels which approach or equal the maximum charge in air (*51*), a procedure used to advantage in industrial electrostatic precipitators to increase particle mobility toward electrically charged collecting plates (*52*).

Table III Electrical Charge Data for Various Aerosols Soon after Generation (50)

Aerosol	Method of formation	Particle diameter, μm	Percent of particles			Particle charge, electrons	Percent maximum charge
			Positive	Negative	Neutral		
Tobacco smoke	Burning	0.1–0.25	40	34	26	1–2	4.0
Magnesium oxide	Burning	0.8–1.5	44	42	14	8–12	3.7
Stearic acid	Condensation	0.2	2	2	96	20–40	19.8
Ammonium chloride	Condensation	0.2	2	2	96	1	1.0
Ammonium chloride	Dispersed from alcohol solution by atomizer	0.8–1.5	40	39	21	12–15	0.05

2. Adhesion

Particles indiscriminately adhere to solid surfaces and to each other, a phenomenon all too familiar to the cleaner of clothes, automobiles, carpets, or furniture surfaces in the home. The basis for many industrial air cleaning methods for particulate-laden gas streams is to precipitate particles onto a surface or onto other particles, where they will be retained while the clean gas exits from the collector.

The properties of the particle and the adhesion surface which influence the strength of the adhesive bond include among others particle and surface chemical composition, the presence or absence of moisture or oil surface films, electrical charge, and surface roughness (53, 54). Forces of adhesion of particles to surfaces or to each other cannot be predicted reliably at present, but simple test methods are available to determine these forces (53). It is generally assumed that airborne particles which contact each other continue to adhere, e.g., the "collision efficiency" is 100%. Particles deposited on the ground can be reentrained by road traffic or by wind (55); those collected on industrial filters are mechanically shaken free or are reentrained by the gas stream (56). The relatively high bulk air velocities required to remove atmospheric dustfall particles from the floor of a small wind tunnel are shown in Table IV (57). Small particles are submerged in the viscous boundary layer adjoining the surface and are difficult to dislodge because the turbulent eddies of the airstream do not effectively penetrate the viscous layer.

3. Condensation and Evaporation Phenomena

Particles in air acquire and lose moisture by molecular transfer to and from the particle surface as the relative humidity of the air increases and decreases. For stationary spherical water drops in air the rate of evaporation is proportional to the surface area of the drop [Eq. (15)] (58):

$$t = [\rho_w r_a{}^2/D_w(p_s - p_0)]\, RT/32M \tag{15}$$

Experiments with freely falling 5–70 μm radius drops in air at temperatures of 0–40°C and 10–100% relative humidity suggest that quasi-stationary steady-state theory and hence Equation (15) can be utilized to approximate evaporation times of pure liquid droplets in air (59). Table V is presented to indicate the rapidity of the evaporation of water drops in air, according to Equation (15).

Particles in the atmosphere are of mixed composition. They combine with water vapor to form solutions containing many solutes; droplets

Table IV Bulk Air Velocity Required to Dislodge Adhering Particles of Atmospheric Dust, Fly Ash, and Glass Beads from a Glass Slide[a] (57)

Bulk air velocity, m/sec	Projected area diameter, μm	Efficiency of removal, percent		
		Glass beads	Fly ash	Atmospheric dust
30	1.3		0.0	0.4
90	1.3		6.0	2.5
150	1.3		21.3	4.9
30	3.8		0.0	0.0
90	3.8		20.1	4.6
150	3.8		57.4	7.9
30	6.4		4.0	5.9
90	6.4		68.0	15.7
150	6.4		100.0	31.6
30	8.9		30.8	6.7
90	8.9		76.9	20.0
150	8.9		100.0	40.0
30	11.5		0.0	0.0
90	11.5		83.3	25.0
150	11.5		100.0	50.0
30	15.9	0		
90	15.9	0		
150	15.9	32		

[a] Relative humidity, 35%; $P = 760$ mm Hg at 20° C.

Table V Evaporation Times of Water Droplets in Still Air at 20°C and $P = 760$ mm Hg

Droplet radius, μm	Evaporation time, seconds	
	0% RH[a]	50% RH[a]
25	7.06×10^{-1}	1.41×10^{0}
10	1.13×10^{-1}	2.26×10^{-1}
5	2.82×10^{-2}	5.64×10^{-2}
1	1.13×10^{-3}	2.26×10^{-3}
0.5	2.82×10^{-4}	5.64×10^{-4}
0.1	1.13×10^{-5}	2.26×10^{-5}

[a] RH: relative humidity.

of pure water are seldom encountered in outdoor air. The introduction of various substances into droplets can cause retardation of the evaporation rate by the mechanisms of solute and surface film interference with surface molecules. Therefore, Equation (15) is only a rough guide to droplet behavior in the atmosphere. Sinclair *et al.* performed *in situ* measurements of the growth of atmospheric aerosol particles with increasing humidity, from 0 to 98% (*59a*). Mason (*58*) reviews many aspects of this complex subject.

III. Particle Dynamics in the Atmosphere

Particles in air extend from those of a size much larger to those smaller than the mean free path of the gas molecules. Particles larger than the mean free path rapidly attain a constant settling velocity which is determined by a balance between the gravitational force and the resisting force of aerodynamic drag. When the particle is comparable to or smaller than the mean free path, it has little weight, and an associated low terminal settling velocity: The random bombardment of air molecules deflects the particle from its terminal settling velocity, and random (Brownian) motion is superimposed on the downward motion due to gravitational force. The major mechanism associated with removal of large particles from the atmosphere is sedimentation, while the smaller particles are removed by diffusion to surfaces and by coagulation with smaller or larger particles, forming even larger particles having appreciable terminal settling velocities.

A. Fluid Resistance and Sedimentation

The fluid resistance experienced by a spherical particle larger than the mean free path settling at terminal velocity in a fluid is given by Stokes' law [Eq. (16)]:

$$F_r = 6\pi\mu ur \tag{16}$$

Because the forces acting on the particle are in equilibrium, the gravitational force acting on the particle can be set equal to the fluid resistance to yield Equation (17), which describes the particle terminal settling velocity:

$$u_t = 2(\rho - \rho')\, gr^2/9\mu \tag{17}$$

$$u_t = \tau g \tag{18}$$

where $\tau = 2r^2(\rho - \rho')/9\mu$. τ has the dimensions of time and is called the relaxation time of the particle. If a particle is released into still air with velocity u, it can be shown that its velocity will decrease to u/e in time τ, where e is Euler's constant, equal to 2.7. The relaxation time can be considered a measure of the time required for a particle to adapt to the changing motion of the surrounding air. Relaxation times of spherical particles of selected size in air are shown in Table VI.

Stokes' law is accurate for the particle size range where the particle is large compared to the mean free path of the gas molecules and where the particle Reynolds number ($R_{e_p} = 2ru\rho/\mu$) is less than approximately 1.0. In air, Stokes' law is reliable to 1% for spherical particles of 1.0 gm/cm³ density and 17 µm radius, and to 5% for particles of 29 µm radius and unit density (60).

The velocity of a particle is often assumed to be proportional to the forces acting on it [Eq. (16)]. The constant of proportionality is called the particle mobility Z. Thus, in Stokes' regime

$$Z = 1/6\pi\mu r \tag{19}$$

The terminal settling velocity of larger spherical particles which do not fall in the Stokes' regime can be calculated from empirical data presented in terms of the particle Reynolds number and the particle drag coefficient (61). The particle drag coefficient is an empirical constant which relates fluid resistance to the particle projected area and settling velocity.

The terminal settling velocities of nonspherical particles have been measured (48, 62); nonspherical particles sediment at slower rates than spherical particles because they have higher ratios of surface area to weight. For instance, a spherical particle will settle in still air approxi-

Table VI Root-Mean-Square Brownian Displacement per Second, Terminal Velocity, and Relaxation Times of Spheres of Unit Density in Air at 760 mm Hg and 20°C

Particle radius, µm	Displacement, cm	Terminal velocity, cm/sec	Relaxation time, sec
0.05	3.70×10^{-3}	8.71×10^{-5}	8.88×10^{-6}
0.1	2.01×10^{-3}	2.27×10^{-4}	2.31×10^{-5}
0.2	1.30×10^{-3}	6.85×10^{-4}	6.99×10^{-5}
0.5	7.43×10^{-4}	3.49×10^{-3}	3.56×10^{-5}
1.0	5.06×10^{-4}	1.29×10^{-2}	1.32×10^{-5}
5.0	2.26×10^{-4}	3.23×10^{-1}	3.30×10^{-4}
10.0	1.60×10^{-4}	1.29×10^{-1}	1.32×10^{-3}

mately 28% faster than an ellipsoid of revolution of the same density having an axial ratio of 4:1.

When the particle size approaches the mean free path of air, the resistance of the air may be thought of as "discontinuous" and particles "slip" between air molecules. It is necessary to apply a correction factor to Equation (17) to account for the increased rate of particle fall. The first correction factor was proposed by Cunningham and although more accurate factors are now available (63), his name is still associated with the term. The factor is

$$C = 1 + A_1(\lambda/r) + B(\lambda/r)\exp(-br/\lambda) \tag{20}$$

The correction factor is 1.164 and 11.554 for 0.5-μm and 0.01-μm radius spheres, respectively. Corrected terminal settling velocities in air of spherical particles of 2.0 gm/cm^3 density are shown in Figure 1.

B. Brownian Motion and Diffusion

The diffusion coefficient is related to particle size as

$$D = RTC/N_16\pi\mu r \tag{21}$$

The diffusion constants for particles in air are shown in Figure 1. The mean square displacement $\bar{\Delta}_x{}^2$ along an axis in a given interval of time t is related to D as originally derived by Einstein:

$$\bar{\Delta}_x{}^2 = 2Dt \tag{22}$$

Table VI demonstrates that in air, particles of unit density smaller than approximately 0.25 μm radius will have larger Brownian displacement than displacement due to gravitational settling.

The mobility of the particle when diffusive movement prevails is

$$Z_D = D/kT \tag{23}$$

where k is Boltzmann's constant (1.38 $\times$ 10^{-16} erg/°C). By using the analogy of the kinetic theory of gases, the average "thermal" speed $\bar{u}_D$ and the mean free path λ_D of particles undergoing Brownian diffusion can be calculated:

$$\bar{u}_D = (8kT/\pi m)^{1/2} \tag{24}$$

$$\lambda_D = 8D/\pi\bar{u}_D \tag{25}$$

C. Coagulation

Aerosol particles adhere together if they come into contact. The process is a continuous one in air and leads to a change in the size and volume

distribution of the aerosol as time progresses. The classical work on coagulation was performed by Smoluchowski (64), who derived Equation
(26) to describe the change in concentration of the number n of monodisperse particles in a given volume of sol. From Equation (26) it can
be seen that the rate at which the particle number concentration is reduced by coagulation depends only on the square of the number present
and the constant K:

$$-\frac{dn_t}{dt} = Kn_t^2 \tag{26}$$

The integrated form of Equation (26), where n_0 is the initial particle
concentration and n_t the concentration at time t is

$$1/n_t - 1/n_0 = Kt \tag{27}$$

Equation (26) has been repeatedly verified in the laboratory, although
the experimental values of the constant K have usually exceeded the
theoretical values. The value of K is dependent on the individual particle
size and its diffusivity, the air temperature, pressure, viscosity, and degree
of turbulence, and the extent of the aerosol polydispersity. It is beyond
the scope of this discussion to review the subject of coagulation here.
The reader is referred to reviews by Green and Lane (20), Hidy and
Brock (12), and Fuchs and Sutugin (65); experimental studies by Devir
(66) and Quon (67); and theoretical studies by Zebel (68, 69). Devir
reported values of K, for a nearly monodisperse 0.5–0.6 μm radius di-
octylphthalate aerosol, of 3.3 to 3.8×10^{-10} cm³/second. Quon reported
values of K which varied from 1.93 to 22.6×10^{-9} cm³/second for nuclei
of radii from 6×10^{-3} to 3×10^{-2} μm.

In the atmosphere the primary mechanism for coagulation is that due
to Brownian motion of the particles. Coagulation phenomena are responsible for the rapid decrease in number and growth in size of particles
smaller than 0.1 μm in the atmosphere. These particles have higher
diffusivities and coagulation rates. Particles larger than approximately
0.3 μm radius have negligible diffusivities and act primarily as acceptors
of the rapidly diffusing smaller particles. Thus, the smaller particles are
removed primarily by the diffusion-dependent process of coagulation; the
larger particles are removed by processes dependent on their inertial
properties, primarily sedimentation. Using a model for the aging atmospheric aerosol, Junge (70) estimated the rate of loss of particles of 10^{-2}
μm radius to be approximately 50% per hour. Particles of 5×10^{-2} μm
radius are lost at the rate of approximately 50% per day.

After analyzing experimental data for the decay of the total number
of airborne nuclei in the smog aerosol of Pasadena, California, Husar,

Whitby, and Liu (*71*) suggested that a value of K equal to 1.8×10^{-8} cm³/second was required to explain the measured decrease in particle numbers during the hours 11:00 p.m. to 8:00 a.m. This value of K exceeds that for a monodisperse aerosol because it includes effects due to polydispersity, inertial effects, and thermal and electrical forces acting on particles. This large value of K, which would not be predicted theoretically, helps to explain anomalies in observations of the changing size spectra of atmospheric aerosols (*72*).

D. Inertial Mechanisms

Curvilinear flow of an aerosol stream can cause the suspended particles to trace paths which differ from that of the surrounding gas; an obstacle in the path of the aerosol stream will collect the particles. If the analysis of particle deposition from a flowing stream onto an obstacle proceeds by considering the particles as points in the gas stream with associated masses larger than the gas, the mechanism of removal is called impaction. If the size of the particle is accounted for and the particle trajectory relative to the obstacle is calculated, the mechanism of particle removal is called interception. The efficiency of collection depends on the velocity of the gas, the size of the obstacle, and the mass and size of the particle (*18*). The efficiency with which particles are removed from an aerosol stream by impaction is a function of the Stokes' number:

$$\text{STK} = 2u\rho r^2/9\mu h \tag{28}$$

The Stokes' number is the ratio of the particle stop distance (that distance required to slow the particle from velocity u to an exceedingly slow velocity) to a characteristic length of the obstacle. In the case of a jet of aerosol impinging normal to a flat plate, h is the jet half-width; for collection by a cylinder transverse to the stream, it is the cylinder diameter. Experimentally measured collection efficiencies by impaction are in good agreement with theoretical predictions (*73*). Particles larger than 15 μm radius in the atmosphere at ground level are undoubtedly removed, to some degree as yet not quantitated, by inertial impaction or interception on buildings, trees, automobiles, etc.

E. Other Influences Affecting Particle Motion in the Atmosphere

1. Thermal Gradients

Particles in the presence of a thermal gradient experience a force which moves them in a direction opposite to that of the gradient, i.e., they are propelled toward the region of lower temperature. For spherical particles which are large compared to the mean free path of the air molecules ($\lambda = 6.53 \times 10^{-6}$ cm at 20°C and 760 mm Hg), the thermal force is given by Equation (29), where $\Delta T/\Delta x$ is the thermal gradient in °K/cm (74).

$$F = -9\pi r \left(\frac{\mu^2}{\rho' T}\right)\left(\frac{k'}{2 + k_{\mathrm{p}}}\right)\left(\frac{\Delta T}{\Delta x}\right) \tag{29}$$

This equation does not adequately describe the behavior of particles of high thermal conductivity, such as metals (75), where an improved equation is necessary (76). The thermal force on particles is utilized in aerosol sampling instruments (77) but its significance for particle deposition from the atmosphere, while certainly minor, remains to be quantitatively assessed.

2. Voltage Gradient

A charged aerosol particle in the presence of an electrical field with a uniform voltage gradient will experience a force which, if the particle is considered to be a point charge, can be calculated:

$$F = Eq \tag{30}$$

These conditions are approximated in the electrical field of the earth's atmosphere ($E \sim 3.3 \times 10^{-3}$ statvolts/cm), where charged particles are subject to an electrical force acting in a direction to or away from the surface of the earth, depending on the polarity of the particle surface

Table VII   Forces Acting on Airborne Spherical Particles of Unit Density in Air at 760 mm Hg and 20°C

Particle radius, μm	Gravitational force, dyn	Electrical force,[a] dyn	Thermal force,[b] dyn	Adhesion force,[c] dyn
0.01	5.13×10^{-16}	3.32×10^{-15}	9.59×10^{-13}	4.5×10^{-4}
0.10	5.13×10^{-13}	3.32×10^{-13}	9.59×10^{-12}	4.5×10^{-3}
0.50	6.41×10^{-11}	8.29×10^{-12}	4.80×10^{-11}	2.25×10^{-2}
1.0	5.13×10^{-10}	3.32×10^{-11}	9.59×10^{-11}	4.5×10^{-2}
5.0	6.41×10^{-8}	8.29×10^{-10}	4.80×10^{-10}	2.25×10^{-1}
10.0	5.13×10^{-7}	3.32×10^{-9}	9.59×10^{-10}	2.25

[a] 1% maximum charge assumed in gradient of 0.33 statvolts/m.
[b] $dT/dx = 10$°K/cm
[c] Glass spheres adhering to flat glass surface in air at 100% relative humidity. Calculated from the equation $F = 2\pi\gamma r$ (53).

charge. The theory of the electrical force acting on particles has been exploited in the fabrication of aerosol sampling instruments (78) and industrial air cleaning devices (52).

Table VII indicates the magnitude of different forces acting on airborne smooth spherical particles of unit density in the atmosphere. The adhesion forces of glass spheres to a flat surface in air of 100% relative humidity are also shown.

IV. Physical and Chemical Properties of Particles in the Air

A. Size Distribution—General

Particles in the air represent a size spectrum that extends from approximately 6×10^{-4} to 2×10^{1} μm, if cloud, fog, and raindrops are disregarded. The largest particles are short-lived because of their large mass and high rate of sedimentation. The smallest particles are electrically charged and have high mobility which leads to their attachment to other particles. Particles representing different regions in the above size range are responsible for different phenomena in the atmosphere. Figure 7 represents the nomenclature used by many meteorologists and cloud physicists to describe atmospheric particles. Superimposed on this figure are the size ranges of particles deposited in the respiratory tract of man and those associated with the two sampling methods accepted in the air pollution field for evaluation of particulate matter in urban atmospheres, i.e., dustfall jar and suspended particulate matter evaluation by high volume sampler (79). It is not unfair to say that both of these measurement techniques were originally adopted and have been continued to be used because of their convenience. It is obvious from Figures 1 and 7 that present routine air pollution particle assessment techniques group into one sample a large spectrum of sizes. While they serve as indicators of the rise or fall of the total concentration of particulates in these groupings, they are not sufficiently specific with respect to size to promote our understanding of alterations in atmospheric phenomena. It has been argued that the total number concentration of particles, as measured by nuclei counter, a measurement substantially independent of particle size in the submicrometer range, is a more effective measure of the degree of particulate pollution of the air (80).

The broad size spectrum of atmospheric particles precludes the use of any one measurement instrument to determine particle concentrations. The optical microscope is not reliable for projected area radii below approximately 0.4 μm (81). The smallest charged particles are measured

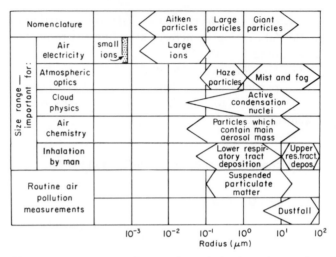

Figure 7. Nomenclature of natural aerosols and the importance of particle size for various atmospheric phenomena and for deposition in upper and lower respiratory tract after inhalation by man.

by their mobility in an electrical field. Condensation nuclei are measured in an Aitken counter. Few complete measurements of the size distribution of the atmospheric aerosol have been made. When they are made, it is necessary to combine the results of measurements by different methods to obtain a single size spectrum.

Whitby and Clark (*82*) described a new instrument which first electrically charges, then counts and sizes particles in the difficult size range from 0.015 to 1.0 μm diameter. This instrument has been used in conjunction with a condensation nucleus counter and an optical particle counter to obtain the particle size distribution of aerosols in the size range 0.0035 to 6.8 μm (*83*).

Because the effects produced by aerosols are associated with the concentrations of particles in selected ranges of the size spectrum, knowledge of the particle size distribution in polluted and nonpolluted atmospheres is essential for understanding aerosol phenomena in these atmospheres. Before dealing with the aerosol spectra in the atmosphere, the properties of particles in subregions of the spectra will be discussed.

B. Air Ions

The air always contains positive and negative carriers of electricity, which are called ions. They may consist of gas molecules, or molecular groups, which are charged due to an excess or deficiency in planetary electrons (small ions); or they may be finely divided particulate matter which has lost or gained electrical charge during or after formation (large ions) (*84–86*). The charged particles, or ions, have an excess of energy due to their charge, and it has been demonstrated (*85, 86*) that in the equilibrium condition the Boltzmann distribution law successfully describes the partition of energy between charged and uncharged particles in the aerosol in the size range greater than 10^{-4} μm radius.

Atmospheric aerosols are surrounded by bipolar ions; the particles come to charge equilibrium because of frequent collisions between particles and ions, thus charging or redistributing charge on the particles. Table VIII (*87*) presents the charge distribution among particles comprising a polydisperse aerosol in charge equilibrium with bipolar ions. The average charge and the ratio of the total concentration of particles to the concentration of uncharged particles are also shown. Below 0.1 μm radius the average charge per ion is assumed to be that of the elementary quantum of electricity, i.e., 4.77×10^{-10} esu. Above 0.1 μm radius, multiple charges on ions occur, and the assumption of a single electrostatic unit particle charge for those particles with charge is no longer valid.

Two main classes of ions are generally recognized—small and large. Small ions are molecules or groups of molecules in air possessing a unit electrical charge either positive or negative. Small ions of each sign are present in air in approximately equal numbers.

The average velocity with which an ion drifts through a gas under the influence of an electrical field is proportional to the strength of the field. The ratio of the velocity to the field strength is called the mobility of the ion and the unit of measurement is centimeters per second per volt per centimeter. Because the average charge per ion for particles less than 0.1 μm radius is 1 esu, differences in particle mobility in air arise due to differences in particle size, density, and shape factor. The mobilities and sizes of the various classes of ions are shown in Table IX (*88*). Between the small and large ions there are also intermediate ions whose mobility in air is a function of the vapor pressure (*89*). Small negative ions appear to have higher average mobilities than small positive ions for reasons which are not well understood (Table X) (*90*).

1. Formation of Air Ions

Small ions are formed from solar and cosmic radiation and from radioactive materials that are always present in small quantities in the atmosphere and in the crust of the earth. At sea it is believed that cosmic radiation is responsible for practically the complete rate of ion formation,

Table VIII Distribution of Charges on Particles in Equilibrium with a Bipolar Ion Atmosphere (87)

Diameter, μm	Number of charges on particle											Average charge	Total particle concentration Concentration of uncharged particles
	0	1	2	3	4	5	6	7	8	9	10		
0.01	0.993	0.007										0.007	1.007
0.015	0.955	0.045										0.045	1.047
0.02	0.900	0.100										0.10	1.111
0.03	0.763	0.236	0.001									0.238	1.316
0.06	0.550	0.430	0.020									0.470	1.818
0.1	0.424	0.48	0.09	0.006								0.677	2.359
0.3	0.241	0.41	0.232	0.093	0.024							1.247	4.150
1.0	0.133	0.253	0.214	0.162	0.109	0.065	0.035	0.017	0.007	0.003	0.001	2.36	7.518

Table IX   Typical Sizes and Mobilities for Atmospheric Ions *(88)*

Type of ion	*Approximate diameter, μm*	*Typical mobility, cm/sec/volt/cm*
Small	0.001–0.005	2.0–0.5
Intermediate	0.005–0.015	0.5–0.1
Large	0.015–0.10	0.01–0.005

resulting in ion pairs, one negatively charged and the other positively charged, at a rate depending upon magnetic latitude, and varying from 1.5 to 2.0 ion pairs/cm³/second. Over most land areas the birth rate of ions in the lower atmosphere is severalfold greater than this because of the additional radiation due to radioactive substances in the air and in the soil. The magnitude of the ionization rate near the earth over land is not readily determined and estimates vary, but 10 ion pairs/cm³/second may be taken as an approximate representative average value *(91, 92)*.

Large ions are formed by combination of small ions with condensed or dispersed aerosol particles, or with drops of water in the air. They can also be formed by combustion processes, by dust transport over land and in dry air, by heated metals, and by the spraying of liquids *(93)*. It has been demonstrated that particles with radii less than 5×10^{-6} cm collect approximately 75–86% of the small ions formed in the atmosphere, while larger particles collect the remainder of the small ions by attachment to their surfaces *(92)*.

Despite the greater rate of production of ions over the land, the electrical conductivity of the air over land generally does not exceed that at

Table X   Average Ion Mobilities *(90)*

Type of ions	*Condition*	*Average mobility, cm/sec/volt/cm*
Small +	In dry air	1.4
Small −	In dry air	1.9
Small +	In moist air	1.1
Small −	In moist air	1.2
Intermediate + and −	—	0.05
Large + and −	—	0.0004
Fog droplets	With single charge	0.5×10^{-6}
Fog droplets	With n charges	$0.5 \times 10^{-6} n$
Electron	In vacuum	100

sea, and at some places, especially near large cities, is much less. For example, in the outskirts of Washington, D.C. it is about one-seventh the value at sea (*91*). Atmospheric ionization is the result of competitive processes, recombination of positive and negative ions, attachment of small ions to airborne particles or droplets to form large ions, and diffusion to and absorption of ions by solid and liquid conductors. The physics and kinetics of small and large ion formation and recombination have been studied by Nolan and Pollak and their colleagues during the past 40 years and have been extensively reported in the literature (*85, 86, 94–96*). The subject is reviewed by Bricard and Pradel (*92*).

In polluted atmospheres, with many solid and liquid particles, small ions are transformed into large ions and the conductivity of the air is thus reduced. The mean life of a small ion varies considerably, depending on the rate of ion formation, particulate matter present in the air, humidity, and air movement. The situation is further complicated by observations that charge equilibrium in the atmosphere is frequently absent (*95*). A small ion may exist for 5–6 minutes in clean air but for only a few seconds where high concentrations of particulates are present. These considerations suggest that large variations in ion concentration should be anticipated in different localities and at the same site, from hour to hour.

2. Measurement of Air Ions

The number of positive or negative ions in a unit volume of air is usually counted by drawing a stream of air through a cylindrical or plate condenser to which an electric field has been applied (Fig. 8). The trajectory of an ion in the condenser is the result of two mutually perpendicular forces, i.e., the velocity of the air and the electric field. Only ions with sufficient mobility to traverse the space between the condenser wall and the ion-collecting electrode will be collected before they are carried away by the airstream.

The "limiting mobility" k_g of the condenser characterizes the device and states that all ions whose mobility is greater than, or at least equal to, this mobility are deposited. The limiting mobility of an ion collector with cylindrical condenser can be calculated by equating the longitudinal travel time of an ion in the unit to the maximum radial travel time:

$$k_g = Q \ln(R_1/r_2)/2\pi v l \tag{31}$$

Ions whose mobility k is smaller than k_g will be collected with efficiency $100(k/k_g)\%$. The concentration z of ions collected from the air sample

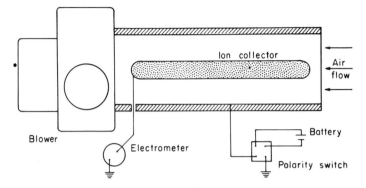

Figure 8. Schematic drawing of a typical instrument to measure air ion concentrations. [B. J. Steigerwald; *in* "Air Pollution" (A. C. Stern, ed.), Vol. I, p. 72. Academic Press, New York, New York, 1962; reprinted with permission from Academic Press, New York, New York.]

is calculated by assuming that each ion was associated with unit charge, i.e., 1.6×10^{-19} C:

$$z = i/eQ \tag{32}$$

Additional methodological details and literature references for ion measurements and instrument design are given by Hoegl (*97*), Mendenhall and Fraser (*98*), Hicks and Beckett (*99*), Israel (*100*), and Hurd and Mullins (*101*).

The method of ion measurement described above should be clearly differentiated from the measurement of net space charge, whereby all ions above a limiting mobility are collected on a single electrode and the net charge or current is recorded. The measurement of net space charge of aerosols and the usefulness of such a measurement is discussed by Masters (*102*).

3. Concentration of Air Ions

Wait and Parkinson (*103*) utilized condensation nuclei measurements of Landsberg (*104*) to estimate large ion concentrations and also reported values for large, intermediate, and small ions in air samples in Washington, D.C. Kornbleuh (*105*) reviewed the results of air ion measurements at nearly 30 locations in the world. These results are summarized in Figure 9.

The small ion content of the air is generally lowest during winter months and highest during the summer, just opposite to the relationship found for the large ion content. Small ion concentrations usually reach

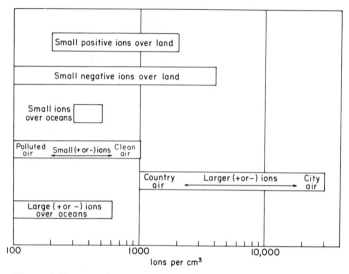

Figure 9. Measured concentrations of large and small ions in air.

maximum values during early morning and lowest values in the early afternoon. As pointed out by Yaglou (*106*), who measured ion concentrations in an apparatus having limiting mobility of 0.2 cm/second/V/cm, ion concentrations were highest on sunny days and lowest on rainy, foggy, or overcast days. During fair weather the concentration of positive small ions exceeds that of negative small ions by 10 or 20% (*103*). This is presumably due to repulsion of negative ions upward from the earth by the negatively charged ground. Also, the higher mobility of small negative ions results in a higher rate of combination with atmospheric particles.

4. Air Pollution and Air Ions*

Almost all industrial and domestic activities which produce fine particle pollution also produce air ions. The electrical conductivity of the atmosphere is largely dependent on the small ion content, and it has been known that the electrical conductivity in and around cities is abnormally low because small ions diffuse to the larger, less mobile particles and thereby are less able to conduct electricity (*107*). Gunn's (*108*) 1962 measurements of electrical conductivity over the North Atlantic, when compared to similar measurements made in 1929, indicated a 5% decrease in electrical conductivity during this period. As stated by Gunn, this result "implies a measurable and parallel increase in the average fine parti-

* See also Chapter 1, Vol. II.

cle pollution of the atmosphere that is doubtless worldwide and represents an increasing hazard to the health of all terrestrial life."

Schilling (109) suggests that Gunn's estimate of fine particle increase is low because the net effect of radioactive particulate matter introduced into the atmosphere during the period 1945–1962 has been to increase the electrical conductivity of air by increasing the rate of production of the conducting ions.

Support for Gunn's hypothesis is suggested by a recent estimate for the increase in the mass of the airborne tropospheric aerosol of approximately 0.4% per year through the year 2000, provided trends in emissions and control of emissions follow those in the United States (17). On the other hand, the evidence for the increase in the global particulate pollution on the basis of decreased atmospheric visibility is not consistent. Peterson and Bryson (110) suggest that an increase in atmospheric turbidity occurred during the past decade in Hawaii, while Flowers et al. (111) and Volz (112) could not discern a long-term increase in turbidity from analysis of United States and European data.

The association of ion concentrations with the occurrence of specific atmospheric pollutants, including gases, and suspended particulate matter, was studied by Steigerwald (113) in the center of Cleveland, in the suburbs, and in a rural area. The only strong positive correlation found to exist was between atmospheric lead concentration (presumed to be indicative of automobile exhaust) and large ion concentration. The fluctuation of large air ions was not related to temperature, relative humidity, or sulfur dioxide concentration of the air.

The physiological effects of air ions after inhalation by man has been a controversial subject for over 40 years (93, 114, 115). It is not within the scope of this discussion to review the evidence for or against the production of effects on man after exposure to elevated concentrations of ions such as those used for therapeutic purposes. However, pioneering investigations by Yaglou and his colleagues in the 1930's at the Harvard School of Public Health failed to uncover any effects, either physiological or psychological, on humans exposed to ions at concentrations normally present in occupied air spaces (115–117). The ions measured in these studies were those with mobilities of at least 0.2 cm/second/V/cm. Yaglou in 1961 stated that the work of other investigators in the intervening years had not altered his original conclusions, namely, that there were not significant changes in metabolic rate, blood pressure, respiration rate, body temperature, or red and white blood cell counts after 1 and 2 hours of exposure to small positive or negative ions at concentrations of 5000 to 10,000/cm^3 (118). These small ion concentrations are higher than those found in polluted atmospheres.

C. Aitken Nuclei

1. Characteristics of Aitken Nuclei

Condensation of vapor is facilitated by the presence of solid foreign particles, the surfaces of which are wetted by the vapor (119, 120). Particles smaller than 0.1 μm radius in the atmosphere are often referred to as condensation nuclei because they have been measured by means of instruments in which water vapor is made to condense on very small particles by supersaturating the vapor. Aitken pioneered these measurements with a nuclei counter which produced supersaturations that resulted in counting of large numbers of particles in the 0.01–0.1 μm range. Airborne particles smaller than 0.1 μm are generally referred to as Aitken nuclei (Fig. 7). The techniques for counting and interpreting results of measurements of these particles are described by Mason (121) and Pollak (122, 123).

The maximum supersaturation achieved, and hence the size of the smallest nuclei detected in an Aitken counter, is not well established, but the value of 4×10^{-3} μm is thought to be correct (124). The term condensation nuclei has been used synonymously with the term Aitken nuclei, which is incorrect. A particle which will undergo transition to form a droplet under conditions of water vapor supersaturation may not undergo transition to a droplet in the outdoor atmosphere. Aitken nuclei have radii less than 0.1 μm and require supersaturation of 0.5–2.0% to carry the water easily through the vapor-to-liquid transition, while large nuclei with radii from 0.1 to 1.0 μm require supersaturations less than that or subsaturation, if soluble. A nucleus of sodium chloride of 10^{-9} gm of dry mass (approximately 5 μm radius) becomes a water drop of about 25 μm radius at a relative humidity of 99% (120).

Thus, all Aitken nuclei are not condensation nuclei nor are they all ions. It is now clearly established that Aitken nuclei are not normally involved in cloud formation and that their primary influence is associated with electrical phenomena in the atmosphere (Fig. 7).

2. Formation, Concentration, and Size Distribution of Aitken Nuclei

Aitken nuclei are formed in nature by evaporation of sea spray (125), dust storms, volcanic activity, forest fires, and air reentrainment of surface dust. They can be produced by α, β, γ, or ultraviolet radiation (126). The yield increases when traces of sulfur dioxide, hydrogen sulfide, or ammonia are present (127). Nuclei can also be produced from gas phase reactions involving oxides of nitrogen and hydrocarbons (*127*). In the presence of solar and ionizing radiations or ozone, nuclei may be formed by oxidation, hydration, acid–base reactions, and addition and recombination reactions involving chlorine, oxides of nitrogen, and sulfur dioxide in the presence of water vapor (*128–130*).

Industrial activities, particularly combustion processes, are an enormous source of nuclei (*131*). The relative importance of natural and man-made sources of Aitken nuclei can be judged from the geographic distribution of Aitken particles shown in Table XI.

Went (*132*) reported a daily rhythm of Aitken nuclei counts in nature. It is lowest in the late night hours, rises suddenly about 20% during the rest of the day, and decreases again after sunset. In cities the nuclei counts were most closely related to automobile traffic, while in rural areas they parallel the production of organic vapors by vegetation. Counts made in St. Louis, Missouri, at or before sunrise, were 15,000–50,000/cm³. This number rose rapidly to 100,000 or over during hours of peak traffic. On a smoggy day counts rose to 200,000/cm³. Went estimated the world-wide production of nuclei by vegetation to be 3×10^6 tons per day (*133*). Exposure of relatively nuclei-free country air to iodine- or turpentine-saturated air produced very large numbers of Aitken nuclei, an effect which could not be duplicated with samples of city air, leading Schaeffer (*134*) to suggest that the clusters of molecules forming nuclei with these vapors are inactivated in polluted air.

The variation of the Aitken nuclei concentration with the nature of the air mass (continental versus maritime), atmospheric precipitation, and the buildup and evaporation of cumulus clouds has been reported

Table XI   Concentration of Aitken Nuclei in Different Locations (*104*)

Location	Number of places	Number of observations	Nuclei number/cm³		
				Average	
			Average	Maximum	Minimum
Large city	28	2500	147,000	379,000	49,100
Town	15	4700	34,300	114,000	5900
Country (inland)	25	3500	9500	66,500	1050
Country (sea shore)	21	7700	9500	33,400	1560
Mountain					
500–1000 m	13	870	6000	36,000	1390
1000–2000 m	16	1000	2130	9830	450
2000 m	25	190	950	5300	160
Ocean	21	600	940	4680	840

(*135*). Maximum Aitken nuclei concentrations of approximately
$1.5 \times 10^6/cm^3$ were measured in Los Angeles, California smog (*71*).

The difficulties associated with chemical identification of Aitken nuclei
can be appreciated from consideration of the weight associated with a
spherical nucleus of 0.05 μm radius of 1 gm/cm^3 density, e.g., 5.24×10^{-16}
gm. These nuclei represent a small fraction of the weight of the atmospheric aerosol and are difficult to separate from the rest of the airborne
particulate matter. Therefore, complete direct chemical analyses of the
composition of Aitken nuclei have not been reported. By using neutron
activation techniques on collected atmospheric nuclei, Megaw and Wiffen
(*126*) detected the presence of copper, manganese, and sodium. Their

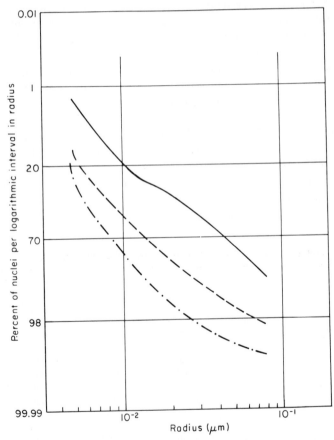

Figure 10. Log-normal plot of typical distributions in clean and polluted air (*137*).
Dot-dashed line: highly polluted air, 50,000 nuclei/cm^3; dashed line: moderately
polluted air, 18,000 nuclei/cm^3; solid line: fairly clean air, 5,700 nuclei/cm^3.

studies suggested that a large proportion of the nuclei were organic in nature, the same conclusion reached by Went (*132, 133*).

Twomey and Severynse employed diffusion decay techniques (*136*) and a condensation nuclei counter to measure the size distribution of atmospheric aerosol particles below 0.1 μm radius (Fig. 10) (*137*). It was concluded that "polluted air samples could be fairly well approximated by a log-normal relationship, while cleaner air samples deviated much more—perhaps a result of aggregation altering the original spectrums." An interesting feature of the measurements was the consistently observed relative absence of particles with a radius of 0.015 μm. These data are presented here because the same measurement technique was utilized to measure particles smaller than 0.1 μm radius in polluted and nonpolluted air. Extensive particle size spectra have been measured in urban air using a combination of instruments to obtain a broader size spectrum (*83*).

3. Measurement of Aitken Nuclei

To count nuclei, an air sample is raised to 100% relative humidity in a humidifier and then passed into a cloud chamber where it expands adiabatically, causing the sample to cool and the relative humidity to rise to a supersaturation up to 400% (Fig. 11) (*138*). The lower limit

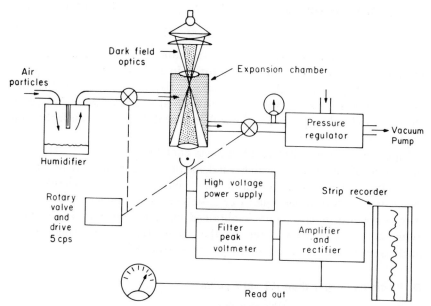

Figure 11. Schematic drawing of an automatic recording condensation nuclei detector (*138*). [Reprinted with permission from F. W. van Luik and R. E. Rippere, Anal. Chem. **34**, 1617 (1962). Copyright by the American Chemical Society.]

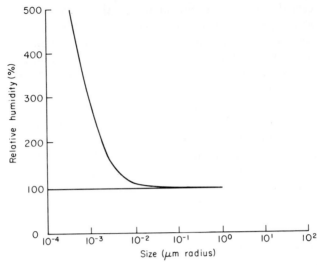

Figure 12. Relative humidity necessary for condensation on aerosol particles of different size.

of particle size which will act as a condensation nucleus is given by the Thompson–Gibbs equation [Eq. (33)] shown graphically in Figure 12:

$$\ln(P/P_\infty) = 2\gamma M/RT\rho r \tag{33}$$

The water droplets formed are all the same size and each contains a nucleus. Their concentration is measured in a dark field optical system by collecting scattered light. The apparatus requires components to establish an overpressure equivalent to the desired supersaturation, to record the scattered light signal on a multiplier phototube, and to flush the sample in a cyclic manner (Fig. 11).

Equation (33) and Figure 12 indicate that by varying P/P_∞ the minimum size of particle on which condensation will occur can be altered. An instrument has been described which has provision for changing the expansion ratio in a continuous or a step-fashion manner (*139*). In this way, a series of nuclei counts is obtained for sizes above a variable minimum size. Nesti (*140*) describes the application of the condensation nucleus counter to a variety of air pollution studies.

D. Particles Larger than 0.1 μm Radius

1. Characteristics and Formation

Particles in this size range are almost entirely responsible for the weight of suspended material in the atmosphere, clouds, and haze (Fig. 7). The nature and origin of the nuclei involved in cloud formation has been a source of conflicting opinion for many years, but the evidence is now strong that the total number of sea-salt particles, formed by the action of the wind and waves, is far too small to be of any importance for the formation of cloud droplets (*141*). The concentration of all chloride particles larger than 0.1 μm radius is of the order of 1–10/cm³ over ocean or land. At present it is believed that the major sources of the large and giant nuclei which are responsible for cloud and rain formation are products of combustion, airborne soil material, and particulate industrial pollution (Fig. 1).

There are indications that water-insoluble particles in the size range larger than 0.1 μm radius acquire, by the mechanism of agglomeration with smaller hygroscopic particles or by adsorption of gases or vapors, an outer water-soluble crust (*142*). They become "mixed nuclei," capable of acting as condensation nuclei. The relative humidity associated with a prevailing visibility of less than one-half mile in different cities in the United States varies from 96 to 100% in Atlanta, Georgia to 85–100% in Denver, Colorado (*143*). Changes in prevailing visibility are related to phase transitions undergone by hygroscopic particles in the air (*71*). The majority of particles larger than approximately 1 μm radius collected from a sample of air changed phase at 73–76% relative humidity (RH) but some particles changed phase at 80–82% RH, 87% RH, and 91–93% RH (*142*). These particles appeared to be mixed nuclei of sodium chloride, sodium carbonate, and ammonium sulfate. Neiburger and Wurtele (*144*) suggest that the number of active hygroscopic nuclei in the industrial area of Los Angeles vary at each relative humidity. Twomey's work (*142*) and that of Husar, Whitby, and Liu (*71*) and Covert, Charlson, and Ahlquist (*143a*) support this suggestion.

Because the deliquescence of aerosol particles is a function of their chemical composition, the variation in atmospheric turbidity can be an indicator of chemical composition (*145*). If one only considers compounds known to be present in atmospheric aerosols (*142, 146*) then deliquescence should occur only above 70% relative humidity, the value adopted by the State of California as the upper limit for application of its visibility standard (*147*). Studies of aerosol size spectra in Pasadena, California suggested a strong correlation of growth with humidity and solar radiation, rather than with either alone, at humidities below 40% (*71*). The effect of relative humidity on the size of atmospheric aerosol particles is addressed by Sinclair, Countess, and Hoopes (*59a*).

2. Measurement

Enormous effort has been expended in recent years to improve measurement techniques in order to obtain size spectra of airborne particles in the size range greater than 0.01 μm radius up to sizes of approximately 20–30 μm radius. The latter was believed to be the upper size limit of the natural aerosol over land but measurements by Jaenicke and Junge (*148*) at several remote sites in Europe did not show an upper limit in the aerosol distribution up to 150 μm radius. In urban areas where local sources predominate, an upper limit is not observed up to 500 μm radius and larger debris is not uncommon (*25, 149*). It can be seen from the category "Methods for Particle Size Analysis" on Figure 1 and from review of the various aerosol sampling and analysis techniques (see Chapters 1–3, Vol. III) that a complete analysis of particulates in the desired size range often necessitates the use of data from more than one measurement technique, with subsequent pooling of the results. The reader is urged to familiarize himself with the limitations of the respective measurement techniques in order to fully appreciate the meaning of the data hereinafter cited.

Accurate description of the aerosol size distribution enables one to deduce the effects produced by airborne particles without directly measuring these effects (Fig. 7). Short-term fluctuations and long-term trends can be followed. Together with information on the particle size distribution and chemical composition of emission sources (*150, 150a, 150b, 150c, 150d*), it may be possible to evaluate the significance of individual source contributions to particles which remain aloft for long times (*16d, 150e, 150f, 150g*). Present practice is to merely prepare an emission inventory on the basis of total particulate emissions by weight (*38*).

3. Size Distribution: Total Particles

Early work by Junge (*70, 151*) on the size distributions of particles in the particle size range from approximately 5×10^{-3} to 2.10×10^{1} μm radius were obtained by combining Aitken nuclei counts with results from cascade impactor sampling. He suggested that measurements of the number density N of the atmospheric aerosol in terms of particle radius r could be approximated for the range $r = 8 \times 10^{-2}$ μm to $r = 10$ μm by a "power law" of the form

$$\frac{dN}{dr} = Cr^{-\beta} \tag{34}$$

where C is a constant independent of r, but which varies with location, and β is approximately 4 for continental surface air and 5 for upper tropospheric air (*152*). The corresponding log radius number distribution is

$$\frac{dN}{d(\log_{10} r)} = C_1 r^{-3} \tag{35}$$

Measurements made at an elevation of 3000 ft above sea level demonstrated the same size distribution characteristics, but concentrations were an order of magnitude lower. The above size distribution yields a constant log radius–volume or log radius–mass distribution, assuming that average particle density does not change in the size range 0.1–10 μm. Thus, the mass of large particles and giant particles (Fig. 7) were approximately equal. Aitken particles contributed only 10–20% of the aerosol mass.

Okita (153) used a sedimentation technique and also sampled particles on web threads to determine the size distribution of particles in the heavily polluted area of Asahigawa, Japan. The particle size distributions approximated the log radius–number distribution, but the exponent β differed for different particle size ranges as follows:

$$0.5 \ \mu\text{m} < r < 10 \ \mu\text{m}; \ \frac{dN}{dr} = (\text{const}) \ r^{-3.6} \tag{36}$$

$$10 \ \mu\text{m} < r < 40 \ \mu\text{m}; \ \frac{dN}{dr} = (\text{const}) \ r^{-4.7} \tag{37}$$

$$r > 40 \ \mu\text{m}; \ \frac{dN}{dr} = (\text{const}) \ r^{-7.5} \tag{38}$$

After proposing the log radius–number distribution for atmospheric particles, Junge (70) stressed the point that the relationship is statistical, and appears only when the number of observations is large enough.

Whitby et al. (30, 35), utilizing membrane filters in Akron, Ohio; Minneapolis, Minnesota; and Louisville, Kentucky, for sampling, and light and electron microscopes and sedimentation cells for sizing, found that individual acetone-insoluble particle size distribution curves between approximately 0.1 and 15 μm radius were bimodal. This characteristic was diminished by sample-averaging processes, with the average distribution approaching the log normal. When all samples were averaged, outside air had a geometric mean size by volume of 3.56 μm (variance = 33%). The Louisville mean size (2.65 μm) was less than the Minneapolis mean size (4.31 μm). Jacobs et al. (154, 155) sampled airborne particulates in New York City with membrane filters and sized by light microscope techniques. In one study 80% of the particles were 1 μm or less in Feret's diameter (22), in the other 74% were 1 μm or less.

Waller, Brooks, and Cartwright (156) used a thermal precipitator to sample airborne particulates in polluted air and sized the particles by means of an electron microscope. The number distribution of particle size generally showed a maximum at about 0.1 μm projected area diameter. The mass median size ranged from 0.5 to 1.0 μm. Sampling and preparation methodology were acknowledged by the authors to be the source of a variety of errors, including loss of soluble and vaporizable particles.

Friedlander and Pasceri (*157, 158*) reported composite atmospheric aerosol size distributions based on sampling by cascade impactor and a new rotating disk device, and sizing by optical and electron microscopes. The distributions were similar to those of Cartwright, Nagelschmidt, and Skidmore (*159*) who sampled the air of Sheffield and Buxton, England with a thermal precipitator and sized by electron microscope. For particle radii between 0.3 and 10 μm, the cumulative size distribution was closely approximated by a dependence on r^{-3} (*159*). For particles below 0.04 μm radius, Friedlander and Pasceri found large variations in the size distribution function at different places and times. The similarity of the size distribution function of widely differing samples for particle sizes greater than 0.04 μm radius suggested that there are stabilizing mechanisms operating in the atmospheric aerosol; the term "self-preserving aerosol" came into usage to describe aerosol spectra which approach a consistent particle size distribution with the passage of time (*160–162*). The reason for the distribution is inherent in the aging process of polydisperse atmospheric aerosols; the principal physical mechanisms are particle diffusion, coagulation, and sedimentation. The most recent reviews in which the similarity theory of aging aerosols is considered in relation to reported aerosol size distributions are those by Takahashi and Kasahara (*163*) and Hidy and Brock (*164*). There is confirmation that data for atmospheric size spectra can be transformed to a form consistent with the theory for certain particle size ranges. However, Brock (*165*) injected the cautionary note that several mathematical distributions can be fitted to distributions of data with characteristically long tails, an attribute of atmospheric aerosol size data in the larger sizes. It is necessary to derive a satisfactory model to explain a distribution of data points in addition to fitting the data to one of a variety of closely fitting mathematical functions. Present distribution functions for the atmospheric aerosol are not convincing because a satisfactory mechanistic model does not exist.

In general, there is a systematic decrease of concentration with size from a peak concentration near 0.01 μm. The decrease is approximated by Equation (34), where β is the slope of the distribution dN/dr vs r on log–log paper [Eq. (35)]. Because of the vital importance of size spectra to effects associated with particulate pollution, certain measured spectra will be presented and discussed.

Figure 13 (*168*) shows particle size distributions measured at ground level in Pasadena, California by Whitby *et al.* (*83*) with a hybrid analyzer system which combines nuclei counting, particle electric mobility

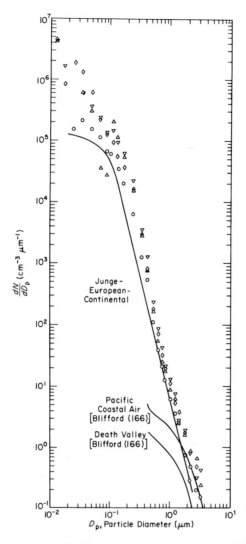

Figure 13. Aerosol size distributions measured in Pasadena, California, September 3, 1969, by Whitby *et al.* (*171*) (light to moderate smog). Results of Blifford for nonurban air are shown for comparison. ○: 0400 PDT; △: 0820 PDT; ▽: 1240 PDT; ◇: 1900 PDT. (PDT, Pacific daylight saving time.)

analysis, and optical particle counting. Also shown are measurements at 15 m altitude in Death Valley and at Pacific offshore sites by Blifford (*166*) in which a cascade impactor and optical microscope were used to collect and analyze samples, respectively (*167*). The results of Junge's measurements on the Zugspitze peak in Germany (*70, 151*) are included

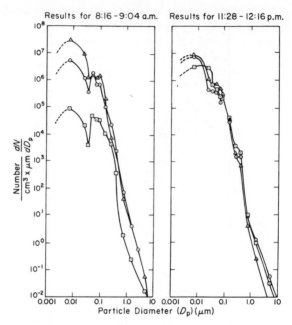

Figure 14. Number-size relative frequency distribution for morning of February 6, 1967 *(169)*, Minneapolis, Minnesota. △: 70-ft level; ☉: 170-ft level; □: 500-ft level.

for comparison. Figure 14 shows similar spectra, collected with the same instrumentation, used by Whitby *et al.* *(83)* at three heights above ground level in Minneapolis, Minnesota before and after dispersal of an atmospheric inversion *(169)*. Figure 13 indicates that polluted atmospheres have many more particles in the submicron range than do nonpolluted atmospheres. Figure 14 clearly shows the stratification of particulate concentration produced by an inversion—in all sizes, but which is most pronounced in the submicron range—and the mixing fostered by inversion breakup.

The most recent and extensive measurements of urban aerosol size spectra were performed in Pasadena, California and led the investigators to propose that the size spectrum be considered in five discrete intervals to fully understand aerosol phenomena occurring in a photochemical smog atmosphere *(83)*. The curves for volume, area, and number distributions of the "average" Los Angeles smog aerosol are presented in Figure 15 in normalized form by dividing the number, area, and volume ordinates by the number total (NT), surface total (ST), or volume total (VT), respectively. The suggested size subranges are scaled by approximately one decade each. For example, subrange 3 extends from 0.05 to 1.05 μm.

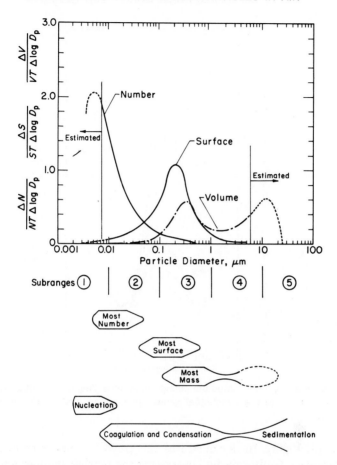

Figure 15. Grand average number, surface area, and volume distributions of Los Angeles smog; the linear ordinate normalized by total number, area, or volume is used so that the apparent area under the curves is proportional to the quantity in that size range. The numbers in each decade of particle size give the name of that subrange. At the bottom the principal mechanism affecting the growth and removal of particles in that size range is indicated. The broken curves were not actually measured but are extrapolations (83).

Particles in this range include 4.7% of the particle number, 50% of the aerosol mass, and are best characterized by a log-normal distribution of $\Delta V/\Delta \log d_p$. Mass in this range is attributed to coagulation of particles smaller than 0.05 μm and from condensation of photochemical reaction products on submicron aerosols. The bimodal nature of the volume distribution curve should be noted because nonphotochemical smog atmospheres are not associated with such a pronounced bimodal effect, if with

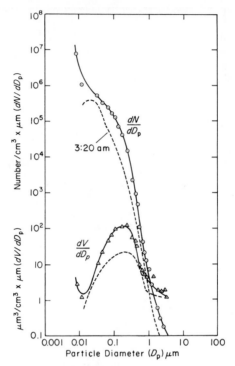

Figure 16. Aerosol under medium smog conditions August 8, 1969, 12:50 p.m., run number 92. Note the substantial growth in numbers and volume in subranges 2 and 3 (83).

any at all. Figure 16 demonstrates the growth of particles in subrange 3 in Pasadena, California by contrasting the particle number and volume distributions in the early hours of the morning and at 12:50 p.m. when solar radiation is intense.

The variation in particle size spectra with height above ground is illustrated by the data of Blifford and Ringer for Scottsbluff, Arizona [Fig. 17; (167)]. The spectra are plotted as the average $dN/d \log_{10} r$ vs r for several collections by cascade impactor at altitudes of 15 m (ground level) up to 9.1 km. The decrease with altitude of the number of particles less than 1.0 μm and those greater than approximately 3 μm is apparent. Variations in the atmospheric light extinction coefficient with height above the ground is reported by Ahlquist and Charlson (170); conversion to concentration profiles is discussed (146).

Table XII summarizes selected investigations of atmospheric aerosol spectra (70, 83, 148, 157, 158, 166, 167, 169, 171–190). It is certain that with the present availability of instruments for particle size measure-

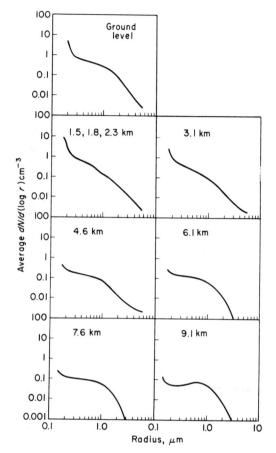

Figure 17. Average $dN/d(\log r)$ (cm^{-3}) versus particle radius at ground level (*167*).

ment, the number of these studies will increase dramatically in the coming years. A clear understanding of instrumental limitations and required assumptions for usage are necessary to obtain meaningful size spectra from raw data (*190e*). Concomitant with spectra measurement there has been, and will continue to be, accelerated interest in the chemical composition of the atmospheric aerosol as a function of particle size.

4. Size Distribution: Selected Chemical Components

The variation in free silica content of industrial dusts as a function of particle size was reported many years ago (*191*). In what was destined to be the first of many studies devoted to variation of chemical composi-

Table XII   Selected Investigations of the Particle Size Distribution of Atmospheric Aerosols

Description of sampling site	Particle sizes measured, μm	Methodology	Comments	Reference(s)
Pasadena, California	0.015–6.8 (diameter)	Hybrid analyzer consisting of condensation nucleus counter, electric mobility analyzer and optical counter (83)	363 size distributions measured during a variety of meteorological conditions and during diurnal cycle	171
Minneapolis, Minnesota	0.015–6.8 (diameter)	Same methodology as above (83)	Spectra variation with height above ground (up to 500 ft) and with diurnal cycle	169, 171–173
Frankfurt and Zugspitze, Germany	0.01–5.0 (diameter)	Nuclei counter, ion counter, and cascade impactor		70
Several sites in Germany	20–300 (diameter)	Impactor	Particles up to 150 μm radius found in continental aerosol	148
Baltimore, Maryland	0.01–50 (radius)	Rotating disc sampler and microscope; impactor with microscope	Laboratory and outdoor air sampling	157, 158
Death Valley, U.S.A.; Barbados; Palmyra (Central Pacific); Scottsbluff, Arizona	~0.13–5.5 (radius)	Cascade impactor and microscope	Spectra obtained at different altitudes from 0.015 to 9.1 km	166, 167
Seattle, Washington	~0.02–3 (radius)	Thermal precipitator and microscope	Maritime and polluted air samples compared	174
Washington State locations	0.01–5; 5–100 (radius)	Thermal precipitator and microscope; cascade impactor	Three distinct subranges in spectra, each described by a power function $dN/d(\log r) = C_1 r^{-\beta}$	175
Urban Vienna Austria, and top of 1600-m peak	0.27–1.1 (diameter)	Goetz Aerosol Spectrometer and microscope	Modal values of concentration vs size curves in range 0.2–0.4 μm; samples collected during different conditions of weather; data expressed as number per cm³ vs size	176

Location	Size range (µm)	Method	Remarks	Reference
Houston, Texas; Santa Fe, New Mexico; Douglas, Arizona; Cedar Key, Florida	0.3–5.0 (diameter)	Sampling from aircraft; sizing by optical counter	Samples collected downwind of pollution sources to contrast polluted and nonpolluted air; data expressed as number per cm³ vs size	177
Pittsburgh, Pennsylvania	0.50–9.2 (diameter)	Cascade impactor	Large daily fluctuations noted and attributed to proximity to local sources	178
Central Pennsylvania near major sources of pollution	~0.2–5.0 (diameter)	Electrostatic precipitator sampling followed by scanning electron microscopy		179
Los Alamos, New Mexico	~0.3–10 (diameter)	Cascade impactor and cascade centripeter	Nonurban samples; concentration of suspended particulate matter 10–25 µg/m³	180
Marine air samples off the Atlantic Coasts of North Africa and the Iberian Peninsula	<2 to 32 (diameter)	Terylene mesh collection and microscopic evaluation		181
Riverside, California	0.5–17 (diameter)	Cascade impactor	Samples also analyzed chemically; variations in size and chemical composition with time presented	182, 183
Budapest, Hungary	<0.14–3.8 (radius)	Cascade impactor and membrane filter	Water-soluble particles also studied. NH_4^+, SO_4^{2-}, and Cl^- reported	184
Moscow, USSR	~0.5–5.0 (diameter)	Impaction traps	Ground-level measurements interpreted re: haze and visibility	185
Over the tropical Atlantic Ocean	~0.4–1.5 (radius)	Size distribution calculated from light extinction measurements	Daily variations indicated	186
Maui, Hawaii	0.1–2.0 (radius)	Parallel impactor and aerosol spectrometer	Aitken nuclei also counted	187
Over the Atlantic Ocean	0.001–100 (radius)	Condensation nucleus counter, optical counter, and various impactors	Background aerosol showed steady decrease in concentration for sizes >0.3 µm	188

Table XII (*Continued*)

Description of sampling site	Particle sizes measured, μm	Methodology	Comments	Reference(s)
Chicago, Illinois; Cincinnati, Ohio; Philadelphia, Pennsylvania; Denver, Colorado; St. Louis, Missouri; London, England and Ankara, Turkey	0.5–3.5 (diameter)	U.S. National Air Sampling Network cascade impactor (Modified Andersen Impactor) (189)	Large variations in mass median diameters and standard deviations for different cities. Larger mass median diameters in Ankara and London	190
Denver, Colorado, at two sites, one downtown and the other near an industrial complex	0.001–17.5 (diameter)	Minnesota Aerosol Analyzing System (MAAS), using nuclei counter, electric mobility analyzer and optical counter	Large variations in size spectra were associated with urban activities, meteorological phenomena and proximity to sources	191a
Eleven sites in Great Britain	~0.5–10 (diameter)	Cascade impactor	During heating season, 71–85% of mass < 1 μm diameter; during nonheating season 67–79% of mass < 1 μm diameter	190b
Mojave Desert, California	0.003–40 (diameter)	MAAS together with cascade impactor and filter samples for chemical analyses	Very low aerosol total volume measured (1.85 μm^3 cm^{-3}). Nuclei concentrations less than 100 cm^{-3}. Considered a "background" aerosol in clean, desert air with range of volume 8–13 μm^3 cm^{-3} and nuclei concentrations from 47 cm^{-3} to 20,000 cm^{-3}	190c
Adjacent to Harbor Freeway, Los Angeles, California	0.003–40 (diameter)	See above Ref. (190c)	Rush hour traffic contribution to aerosol volume and chemical constituents evaluated	190d

tion of atmospheric aerosol with size, Pittsburgh, Pennsylvania air was sampled (*192*) in a particle size selective instrument based on the principle of horizontal elutriation (*193*). The first stage collected 3.55 μm radius spheres of density 1.0 gm/cm³ with 100% efficiency and 2.5 μm radius spheres with 50% efficiency. All particles passing the elutriator were collected on a second-stage filter. Table XIII summarizes the compositional analysis of the filter, i.e, "respirable" fraction and the total sample of particulates collected. These results suggested that there was pronounced segregation, with respect to particle size, of chemical components in the particulate fraction of air pollution. In addition to the obvious size dependency of any toxicological effects of a polluted atmosphere, the particulate pollution potential for corrosion, visibility, and crop damage is also size dependent.

Table XIV is a summary of selected investigations of the size dependency of chemical species in the atmospheric aerosol (*14, 182–184, 194–208*). A typical presentation of data obtained in studies of this type is shown in Figure 18. Because sampling periods of at least 4 hours are currently required to obtain sufficient sample for chemical analyses with impactors, data represent average distributions for the sampling period. A future need is for sampling apparatus which will reduce sampling time but still collect sufficient material for chemical analyses.

a. SULFATES AND SULFURIC ACID DROPLETS. Interest in sulfate aerosols in polluted atmospheres stems primarily from the irritant potential of sulfuric acid droplets and particles of zinc ammonium sulfate, zinc sulfate, and ammonium sulfate, as measured by a significant increase in

Table XIII   Percentage Composition of Elutriated and Total Airborne Particles (*192*)

| Constituent | Elutriated (*i.e., "respirable"*) particles[a] (%) | | Total particles (%) | | Ratio A/B |
	Range	Mean A	Range	Mean B	
Benzene-soluble organics	3.1–11.9	7.7	2.2–5.0	3.5	2.2
Total sulfates	14.1–17.3	15.3	6.7–11.4	9.2	1.7
Total nitrates	0.8–1.8	1.2	0.8–1.2	1.1	1.1
Lead	0.3–1.5	0.9	0.3–0.6	0.4	2.3
Iron	3.3–10.4	5.8	11.8–23.7	17.3	0.3

[a] Caught on filter after passing through horizontal elutriator (*193*).

Table XIV Selected Investigations of Atmospheric Aerosol Composition as a Function of Particle Size

Location	Chemical species	Sampling methodology	Comments	Reference(s)
Pittsburgh, Pennsylvania	Sulfate	Cascade impactor	Mass median diameter approximately 3 μm	14, 194
Los Angeles and San Francisco, California; Chicago, Illinois; Cincinnati, Ohio; Philadelphia, Pennsylvania; and remote sites in Arizona and Oklahoma	Lead	Goetz aerosol spectrometer	Similar size distributions in all areas; Average mass median diameter of 0.25 μm; Upper quartile size of 0.43 μm	195
Downtown and suburban Cincinnati, Ohio	Iron, cadmium, copper, lead, sulfate	Cascade impactor	Each chemical species was associated with a different size spectrum; Spectra also differed at different sampling sites	196
Harwell, England (rural site)	Ammonium sulfate	Membrane filter and electron microscopy	Mass median diameter of 0.6 μm; 80% of the mass of particles in range 0.08–1.0 μm was associated with particles greater than 0.35 μm	197
Chicago, Illinois; Philadelphia, Pennsylvania; and downtown and suburban Cincinnati, Ohio	Phosphate, nitrate, chloride, and ammonium particulate	Cascade impactor	Average mass median diameters decreased in the order phosphate > chloride > ammonium > nitrate	198
Downtown and suburban Cincinnati, Ohio; Philadelphia, Pennsylvania; and Chicago, Illinois	Sulfate	Cascade impactor	Size spectra varied with site and with morning, midday, or evening sampling period; Mass median diameters from 0.42–0.66 μm	199
Los Angeles and San Francisco, California	Sulfur containing compounds	Goetz aerosol spectrometer	75% of mass associated with particles less than approximately 1.0 μm diameter	200
Cincinnati, Ohio and Chicago, Illinois	Sulfates	Cascade impactor		201

Location	Substances	Method	Remarks	Reference
Budapest, Hungary	Ammonium, sulfate, and chloride	Cascade impactor	90+% of SO_4^{2-} and NH_4^+ associated with particles larger than 1.2 μm radius; 67% of chloride associated with same size range	184
Budapest, Hungary	Sulfate	Cascade impactor	Seasonal and diurnal variations discussed	202
Locations in United States; over the Pacific Ocean and Venezuela	Chloride, sulfur, potassium, sodium, silica, calcium, titanium	Ground level and aircraft sampling with cascade impactor	Profiles of concentration and mass fraction ratios presented from 0.015 to 9.1 km above ground; standardized particle size distribution used to calculate mass ratios	203
Riverside, California	Sulfate, nitrate, lead, iron	Cascade impactor	Results of simultaneous measurements of pollutant gases also presented	182, 183
London, England; Ann Arbor, Michigan	Sulfuric acid droplets; Lead, copper, cadmium	Cascade impactor; Cascade impactor	Copper and cadmium showed similar size distributions, which differed from that of the lead; about 60% of the lead estimated to be associated with particles less than 0.2 μm diameter	204; 205
Pittsburgh, Pennsylvania	Polynuclear aromatic hydrocarbons	Horizontal elutriator	"Respirable" and nonrespirable fractions analyzed; more than 75% by weight of PAH associated with "respirable" particulates	206
Osaka, Japan	Iron, lead, manganese, vanadium, copper	Cascade centripeter	Mass median diameters estimated to be 1.0 μm for iron; 0.5 μm for manganese; 0.1 μm for vanadium; and 0.3 μm for copper	207, 208
Willamette Valley, western Oregon	26 trace elements	Cascade impactor	Effects of agricultural field burning on rural aerosol evaluated	208a

Table XIV (Continued)

Location	Chemical species	Sampling methodology	Comments	Reference(s)
San Francisco, California	Lead, bromine	Cascade impactor	Simultaneous sampling at nine sites	208b
Boston, Massachusetts	18 trace elements; polynuclear aromatic hydrocarbons	Cascade impactor	Three sampling sites utilized. Size spectra for each element were distinctive. Several could be related to emission sources where size composition data were known	216a
Toronto, Ontario	Polynuclear aromatic hydrocarbons	Cascade impactor	Five sampling sites. Atmospheric concentrations and size spectra reported for ten polynuclear aromatic hydrocarbons sampled over 24- to 48-hour periods	208c
New York, New York	Organic fraction components	Cascade impactor and hi-vol samplers operated in parallel over periods of 54–96 hours at a single site	Concentrations of organic compounds were strongly dependent on particle size	208d
San Francisco, California	Vanadium; aluminum	Cascade impactor at nine sites	Concluded that atmospheric vanadium at these sites originates primarily from soil (small particles) and combustion sources (large particles)	208e

pulmonary flow resistance in guinea pigs exposed to these aerosols (*209, 210*). The association between high concentrations of ammonium and sulfate (33 and 130 $\mu g/m^3$, respectively) led Eggleton to ascribe most of the visibility loss in Tees-side in the northeast of England to ammonium sulfate in mists (*211*). In rural areas the primary source of sulfate particles is probably the atmospheric oxidation of hydrogen sulfide produced by anaerobic bacteria; sea-spray sulfate nuclei are a minor source (*212*). In urban areas the oxidation of sulfur dioxide is a major source of particulate sulfates. At least three mechanisms may contribute to the conversion depending on atmospheric conditions: (a) the photooxidation of SO_2 in the presence of unsaturated hydrocarbons and NO_2; (b) the oxidation of SO_2 in water droplets catalyzed by metal ions; and (c) the catalytic oxidation of SO_2 adsorbed on solid particles (*213*). There is laboratory evidence to support these mechanisms in the atmosphere. The reaction proceeds rapidly in the presence of fog droplets (*214, 215*) and at a slower, but significant rate at lower humidities in the presence of sunlight

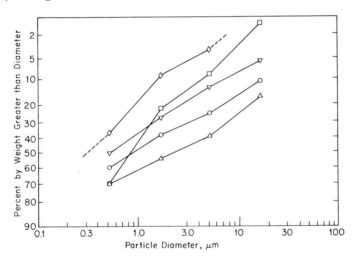

Figure 18. Average size distributions for 10 impactor samples collected in Riverside, California (*182*).

Sample	Symbol	Mass median diameter, μm	$\sigma_g = \dfrac{diameter\ 16\%\ size}{diameter\ 50\%\ size}$
Iron	△	2.2	8
Total particle	○	0.9	11
Nitrate	□	0.8	3
Lead	▽	0.5	7
Sulfate	◇	~0.3	4

(*216*). The conversion of SO_2 to sulfate aerosol in the atmosphere is accelerated in the presence of water-soluble aerosols (*217*) and certain insoluble suspended particulate matter (*218*).

In London, prior to the beneficial effects produced by the Clean Air Act of 1956 and its subsequent amendments, concentrations of acid droplets were especially high at times of fog and reached levels of 678 $\mu g/m^3$. Typical winter and summer daily average concentrations were 18 and 7 $\mu g/m^3$, respectively (*219*). The size distribution of these droplets was estimated by Waller (*204*), who used a cascade impactor for sampling (Table XV). The mass median diameter for this sample was 0.5 μm. Waller states that at lower relative humidities "virtually 100% of the acid is on the filter paper," a conclusion endorsed by Eggleton (*211*) with respect to acid mist at Tees-side, England. Reduction of suspended particulate matter ("smoke") in London from an average concentration of approximately 280 $\mu g/m^3$ in 1959 to approximately 70 $\mu g/m^3$ in 1970 has also reduced the occurrence of acid "smogs." It is reasoned that in "the absence of a pall of morning smoke the sun can warm the earth more rapidly and break up temperature inversions, which might otherwise persist and constitute the old-type 'smog' " (*220*).

The size distribution of airborne particulate sulfates at several locations (Table XIV) indicates that, in general, the bulk of the sulfate mass is associated with submicrometer particles. The association of sulfate with particles larger than 1 μm in Budapest, Hungary (*184, 202*) and in Pittsburgh, Pennsylvania (*14, 194*) may be due to the proximity of sampling sites to local sources of sulfate emission.

The future heavy reliance on coal as the major fossil fuel for energy production in the United States has triggered a major political-scientific

Table XV   Distribution of Sulfuric Acid Droplets in Sample Taken on January 14, 1959 in London under Conditions of High Humidity (204)

	Effective drop size, μm	*Concentration of acid, $\mu g/m^3$ air*	*Cumulative percent of total acid*
Filter	0.5	19	50
Fourth slide	1.7	8	71
Third slide	4.0	5	84
Second slide	13.0	4	95
First slide	23.0	2	100
		38	

investigation of sulfate particulates in the air. Because greatly increased SO_2 emissions are projected and the conversion of SO_2 to SO_4^{2-} particulates is not well understood, the ultimate effects of increased coal consumption on atmospheric particulate sulfate concentrations cannot, at the present time, be reliably predicted. The scientific aspects of the particulate sulfates-in-air controversy were addressed by the National Academy of Engineering. The issue of sulfates-in-air epitomizes all the complexities of atmospheric aerosols, including size dependencies of emissions to the air and removal mechanisms and heterogeneous catalysis and photochemical mechanisms to form sulfate particulates from SO_2 (220a).

b. LEAD. The size distributions of airborne lead particles in a selection of United States urban and remote areas were measured by Robinson and Ludwig (195). Air sampling was performed with a Goetz aerosol spectrometer (221), and the deposited particles were chemically analyzed for lead content. Table XVI presents the results of these studies expressed as mass median equivalent diameter and upper and lower quartiles of the distributions. On the basis of 59 urban lead aerosol size distribution analyses, the average mass median equivalent diameter was 0.25 μm and the associated upper and lower quartile points were 0.43 and 0.16 μm, respectively. All size distributions represented integrated samples taken over periods of 4–9 hours. The uniformity of results was attributed to the dominance of the automobile as the emission source for lead. The combustion of gasoline containing antiknock additives with lead alkyls is the principle source of lead particles in the urban atmosphere (222). The particle size distribution of lead particles in automobile exhaust emissions has been demonstrated to be in the submicrometer range (223). The entire subject of airborne lead was recently extensively reviewed by a special committee of the National Academy of Science (224). Suffice it to say that there is evidence that lead in the atmospheric aerosol is ubiquitous (225, 226); that it is increasing on a global basis (226); and that the estimated loss in visibility attributable to lead particles in the air of one urban area, Los Angeles, is 20% of the total visibility loss (227). It is interesting to note, however, that in an intensive survey of lead in the atmosphere of three urban communities (Cincinnati, Ohio; Los Angeles, California; and Philadelphia, Pennsylvania) a trend could not be discerned toward increasing concentrations during the period 1957–1962; in fact, the suggestion of a decrease in atmospheric contributions of lead was noted in Cincinnati, Ohio (228). The lead background concentration is around 0.5–0.1 μg/m³, while the levels in city streets are often 6–11 μg/m³ (229), thus exceeding the World Health Organization air quality guide for the level I criterion of 2 μg/m³ (230).

Table XVI Lead Aerosol Size Distribution Data; Citywide Averages (195)

		Size frequency distribution, μm^a					
		25% Diameter		MMED[b]		75% Diameter	
Location	Number of samples	Average	Range	Average	Range	Average	Range
Los Angeles, California area	25	0.16 (23)	0.05–0.25	0.24	0.08–0.32	0.44 (23)	0.13–0.74
San Francisco Bay area, California	8	0.13	0.06–0.20	0.24	0.15–0.34	0.43 (7)	0.23–0.89
Chicago, Illinois	12	0.19 (7)	0.10–0.29	0.29	0.16–0.64	0.40 (10)	0.28–0.62
Cincinnati, Ohio	7	0.15 (3)	0.09–0.24	0.24	0.16–0.28	0.45	0.30–0.68
Philadelphia, Pennsylvania	7	0.14 (3)	0.09–0.25	0.24	0.19–0.31	0.41	0.28–0.56
Remote areas	2	0.25 (1)	—	0.29	0.27–0.31	0.53	0.34–0.71
5-city average	59	0.16		0.25		0.43	—

[a] Parentheses denote number of samples analyzed for result tabulated when different from column 2.
[b] Mass median equivalent diameter.

c. OTHER METALS.   Figure 19 shows the particle size distributions of selected metals associated with airborne particles collected with cascade impactors in Fairfax, Ohio, a rural suburb of Cincinnati, Ohio (*196*). The data indicate that each metallic component is characteristically distributed with respect to particle size (*230a*). Indeed, data collected at other sites confirm this finding (Table XIV). The size distribution for any component of airborne particulate matter is dependent on production and emission sources, transformation reactions in the air, and removal mechanisms. The lead and sulfate components of the atmospheric aerosol were initially stressed by professional public health workers because of their possible health effects following inhalation by the public. However, data are now being collected for other components of the atmospheric aerosol, including metals, chloride, nitrate, bromide, and asbestos.

5. Concentration and Chemical Composition

From the discussion up to this point, the reader will realize that particulate pollution of an urban atmosphere is a very complicated process, dependent on numerous properties of the atmosphere, the particles, and the emission sources. A complete description of particle pollution of the

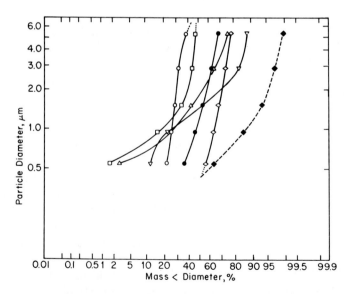

Figure 19. Cumulative particle size distributions of various metals in Fairfax air. Curves represent average distributions from data for three 4-day samples taken during February 1967. ○: Mg; □: Cd; ●: gross particulate; △: Cr; ◇: Pb; ▽: Fe; ◆: SO_4^{2-} (*198*).

air at a particular site encompasses the physical characterization of the aerosol, including the total size distribution and those with respect to chemical composition, as well as the variation of these quantities with respect to time. Less difficult, time-consuming, and costly, albeit less informative, measurements of particulate pollution can be obtained by the measurement of dustfall, weight of suspended particulate matter, and stain or soiling index. Before discussing each of these measurements in detail, the availability to the public of these air quality data will be considered.

Sources of air quality data available for general utilization are extremely heterogeneous throughout the world. In general, measurements of air pollution are made by one or more agencies at all levels of government, national, provincial, city, town, and borough, as well as by public and private laboratories, institutes, and universities in many instances. Primary departmental responsibility at the national level usually resides in the federal environmental agency (as in the United States), in the health service (as in the USSR), or in a science and technology ministry (as in Japan and the United Kingdom). Detailed air quality monitoring and surveys on the local level are carried out by municipal hygiene laboratories (as in Paris), by a town planning commission (as in Liège), by an independent, government related agency (as in São Paulo), and sometimes by a complex of agencies (as in Milan).

In the United Kingdom, the coordinated National Survey of Air Pollution (231) has consisted of 24-hour measurements of "smoke" and SO_2 at about 1000 sites in over 200 cities and towns, with hourly data available at a few locations. These data have been regularly made available in a bulletin of the monthly summary of results as well as complete annual tables.

In the United States the national program of air quality measurement has been the National Air Sampling Network (NASN) of the Environmental Protection Agency's National Environmental Research Center, Research Triangle Park, North Carolina. It has been operated in cooperation with state and local health departments, air pollution agencies, and other organizations. Samples of suspended particulates at over 200 sites (urban and rural) (232) and several common gaseous pollutants, at over 50 of those stations have been collected over a 24-hour period once every 2 weeks.

The NASN has since 1970 included a cascade impactor network in selected United States cities to characterize the particulate size distribution of ambient aerosols in major United States urban areas (233).

Annual summaries of analyzed NASN data (e.g., frequency distributions, means, and variations) are published and periodic general summaries of several years of data are also prepared (232). The data have also been analyzed to determine trends (13a, 13b, 234, 235). Some state and local air pollution control agencies (236) issue monthly air quality data, but the practice is far from universal.

V. Traditional Measures of Particulate Pollution*

A. Dustfall

Dustfall refers to a particulate sample collected after airborne particles settle by sedimentation into dustfall cans or jars. The dustfall measurement is an indication of the quantity of larger particles having appreciable settling velocities and relatively short atmospheric residence times. Dustfall is usually expressed in tons per square mile per 30-day period (Table XVII) (237). It is a useful indicator of the amount of material deposited from stack emissions. Fairweather, Sidlow, and Faith measured by a sedimentation technique the size distribution of dustfall at sites dòwnwind from industrial sources (149). Of the 20 samples studied, all but two showed 93.5–99.9% by weight of the sample to be associated with particles larger than 15 μm projected area radius. The two exceptions were cement dust samples taken on the roofs of buildings almost directly beneath a kiln stack, because they were not considered to be representative of community dustfall. The largest particles observed in the study were approximately 500 μm radius. Dustfall particles, because of their large size, are offensive to the visual sense and constitute a nui-

Table XVII Guidelines for Dustfall in Allegheny County, Pennsylvania[a,b]

	Dustfall	
Classification	tons/mile²/30 days	mg/cm²/30 days
Slight	0–20	0–0.7
Moderate	20–40	0.7–1.4
Heavy	40–100	1.4–3.5
Very heavy	>100	>3.5

[a] Referred to as "cleanliness index."
[b] Collection according to ASTM D 1739-62 for dustfall evaluation (237).

* See also Chapter 1, Vol. III.

sance to housewives and other members of the community. The establishment of the relationship between the nuisance value of deposited dust and the area of a clean surface covered by the dust was the subject of a fascinating paper by Carey (238). He estimated that the rate of road surface coverage by dustfall in "clean" rural areas in Britain is 0.05% per day. In highly industrialized areas the rate is 1% per day. Carey estimated that people recognize a pavement to be "dirty" when 0.4% of the surface is covered by particles. No correlation between particle projected area and particle weight has yet been developed. A method for evaluation of the projected area of dustfall samples collected on glass slides has been reported. Sampling times as short as 1 hour and coverage of 0.01% of the surface can be assessed (239).

The major value of the present method of assessment of dustfall is that fluctuations will be detected at a given station over a period of time, thus revealing improvement or deterioration of this aspect of air quality. Monthly variations in dustfall at a single station in the Monongahela Valley, Pennsylvania, for instance, have been as high as 600% (240). Therefore, only trends of improvement or deterioration over a long period of time should be used as indicators of altering conditions.

The adoption by pollution agencies throughout the world of dustfall collectors with different shapes and physical dimensions makes it difficult to compare dustfall values obtained in different cities (241). Sensenbaugh and Hemeon (242), Fisher (243), Nader (244), and Sanderson, Bradt, and Katz (245) tried to overcome this difficulty by comparing the collection efficiencies of different types of collectors. Fisher found differences of up to 15% by weight in the material collected by identical collectors and up to 100% for different collectors. Sanderson, Bradt, and Katz reported differences of up to 140% for different collectors. A collector similar to the British Standard Gage (241) was the most consistent, showing 8% variation in weight of dustfall in duplicate collectors at the same site. Katz (246) applied correction factors suggested by Fisher (243) to dustfall data from different cities and countries. The correction factors used were 23 and 85%. These considerations suggest that it is unwise to compare dustfall values for different cities where different collectors are employed.

The particulates collected in dustfall cans or jars are usually analyzed for water-soluble and -insoluble components, tars, carbons, and ash content (79, 237). In addition, specific constituents can be evaluated. Tables XVIII and XIX show the average results of analysis of dustfall samples collected in industrial and residential areas of Yokohama, Japan, during an extensive air pollution survey (247). The water-insoluble component of dustfall was higher in industrial areas than in rural areas, as was the

Table XVIII Composition of Dustfall Collected in the City of Yokohama, Japan[a]

Area	Water-insoluble component percent by weight				Loss on ignition	Soluble component percent by weight		Total
	Tar	Carbon	Ash	Subtotal		Ash	Subtotal	
Industrial	0.4	10.3	54.7	65.4	11.3	23.3	34.6	100
Semi-industrial	4.6	7.5	58.6	66.7	10.9	23	33.9	100
Commercial and residential	0.8	8.4	49.7	58.9	15.9	25.2	41.1	100
Rural	0.7	9.2	50.3	60.2	17.3	22.5	39.8	100

[a] Average values, 1956–1961 (247).

Table XIX   Constituents Identified in Dustfall Samples Collected in Yokohama and Kawasaki, Japan[a]

	Location		
Constituent	Industrial area	Semi-industrial area	Residential area
Loss on ignition	20.81	25.86	28.74
SiO_2	17.21	25.32	27.05
Fe_2O_3	14.56	9.54	7.02
Al_2O_3	6.87	6.95	9.99
CaO	21.70	11.56	9.18
MgO	1.91	2.58	2.08
MnO	0.15	1.13	0.08
SO_4^{2-}	13.48	12.36	10.42
Cl^-	1.66	3.06	3.10

[a] Average values, 1956–1961 (247).

tar, carbon, and ash content of the water-soluble portion. Water-soluble organic material was higher in rural area samples. Sulfates were higher in industrial samples, while silicon dioxide and chlorides were lower. A seasonal variation also occurs, i.e., dustfall increases during the heating season. The figures in Table XVIII for ash and water-soluble components do not differ greatly from those presented by Katz for the Greater Windsor area (246), which may be fortuitous since it is to be expected that the chemical composition of dustfall will vary with the nature of fuels used and the character of industries associated with a particular air basin. Gould (248) suggests that dustfall data adhere to a log normal frequency distribution.

B. Suspended Particulate Matter

Suspended particulate matter consists of particles which are airborne in the vicinity of the sampling apparatus used for collection. The particle size selection characteristics of the sampling apparatus routinely used for this measurement by the National Air Sampling Network (NASN) in the United States (249) are not defined (232). Particles with a terminal settling velocity in air greater than approximately 30 cm/second (equivalent to a sphere of density 1 gm/cm^3 and about 50 μm radius) cannot possibly enter the standard NASN shelter. The size distribution of the material on the collection filter has not been measured; indeed, it is impossible to redisperse the sample so as to reproduce its original state of dispersion in the atmosphere. In Pittsburgh, Pennsylvania,

O'Donnell, Montgomery, and Corn (*178*) designed an entry cover for a six-stage cascade impactor in order to simulate the aerosol entry conditions of the high volume sampler used in the National Air Sampling Network (NASN). It was found that approximately 85% by weight of particles collected in this manner had less than 9.2 μm effective cutoff diameter. The size of particles collected by the NASN sampler can also be deduced from analogous measurements by Whitby *et al.* (*30, 35*) and the measurements of the size distribution of dustfall by Fairweather *et al.* (*149*). Ninety percent of the weight of suspended particulate matter reported by the NASN can be associated with particles having terminal settling velocities less than that of a sphere of approximately 8 μm radius and 1 gm/cm³ density. Thus, suspended particulate matter consists of particles which travel with air currents and remain aloft for appreciable times; they do not settle in the vicinity of the source of emission.

Studies of the vertical distribution of suspended particulate matter in the atmosphere suggest that below 400 m there can be layers of different particulate concentration, especially during atmospheric inversion conditions (*170, 172, 250*). After breakup of the inversion layer by solar heating, the vertical distribution of suspended particulate matter is more uniform.

McCormick and Kurfis (*251*) measured the variation in atmospheric turbidity in 61-m intervals from the surface up to 610 m. They concluded that buoyancy forces, rather than mechanical factors of wind speed and surface roughness, are the major contributors to the dispersal of aerosols over urban areas.

These studies were performed with an optical device which measured atmospheric transmissivity, and it was necessary to assume particle size distribution in order to infer particulate concentrations. The assumptions relative to aerosol particle size distribution and particle characteristics relative to light scattering, as well as the confidence limits for an aerosol light scattering–mass concentration correlation, are reviewed by Charlson (*146*).

Measurements by the National Air Sampling Network indicated that there was a trend toward lower average concentrations of suspended particulate pollution during 1957 through 1963 (*235*). A closer look at the data reveals that the patterns of individual areas reveal increases, decreases, or lack of a trend in suspended particulate matter concentrations (*236*). The concentration of suspended particulate matter in urban air basins in the United States is highly associated with the population of the city (Table XX). These data give support to an equation derived by Sheleikhovskii (*252*) for the concentration of fine aerosol pollution in towns. Sheleikhovskii found the pollution to be directly proportional

Table XX Summary of NASN Suspended Particulate Samples for Urban Stations by Population Class, 1957–1963, United States (232)

Population class	Number of samples	Number of stations[a]	Minimum ($\mu g/m^3$)	Maximum ($\mu g/m^3$)	Arithmetic mean ($\mu g/m^3$)	Geometric mean ($\mu g/m^3$)
3 million and over	316	2	57	714	182	167
1–3 million	519	3	34	594	161	146
0.7–1.0 million	1191	7	14	658	129	113
0.4–0.7 million	3053	19	18	977	128	112
0.1–0.4 million	9531	92	10	1706	113	100
50,000–100,000	5806	81	6	982	111	93
25,000–50,000	1606	23	5	679	85	71
10,000–25,000	484	6	11	539	80	63
<10,000	150	5	22	396	100	84

[a] 64 Stations participate every year; the remaining stations participated 1 or more years during the 7-year period.

to the density of population per unit area and the average aerosol discharged per capita per second. It was inversely proportional to average wind velocity and gustiness.

There are yearly, weekly, and diurnal cycles associated with suspended particulate matter concentrations in the atmosphere. The weekly cycle of suspended particulate matter in the United States is pronounced in urban areas, where weekend concentrations are about 15% lower than the weekday levels. At nonurban stations differences between weekend and weekday concentrations of suspended particulate matter have not been observed (*244*).

The diurnal cycle of suspended particulate matter concentrations in the atmosphere has been observed in different cities and shows two maxima, but the times of occurrence of the maxima differ in different cities (*83*). In Los Angeles, California, the aerosol volume was synchronized with solar radiation and peaked at about noon at a value of about 60 μm^3/cm^3 while the condensation nuclei concentration peaked at about 2:30 p.m. at a value of about 2×10^5/cm^3. In Minneapolis, Minnesota, the condensation nuclei count peaked at a value about an order of magnitude greater, possibly due to combustion sources; the effect of solar radiation on the volume distribution was not pronounced (*83*). Meetham (*253*) attributed the daily cycle in England to the following causes: (a) the fuel consumption cycle, (b) increased domestic heating in the morning, (c) the diurnal cycle in atmospheric turbulence, and (d) the effect of suspended particulates blown from outlying industrial districts.

It is difficult to compare the concentrations of suspended particulate matter in different areas of the world because, as in the case of dustfall, methods of evaluation differ in different nations. The poor correlations at a single site in Denmark between simultaneous measurement methods used by different nations has led to a plea for adoption of an international standard method (*253a*). Some data are presented which indicate the extent of differences to be expected when measuring suspended particulate matter using United States and United Kingdom methodology (*254*). Lobner has tabulated results for the United States, France, Russia, England, and Germany and briefly described the sampling and evaluation techniques used (*255*).

The distribution of the concentrations of 24-hour samples of suspended particulate matter at individual sampling stations exhibit log normal characteristics (*256*).

The chemical composition of suspended particulate matter in urban and nonurban locations in the United States, expressed as arithmetic mean and maximal values, is known from the analysis of 14,494 urban and 3114 nonurban samples, each collected over a 24-hour period at an

MORTON CORN

Table XXI Mean and Maximum Concentrations of Selected Particulate Contaminants in United States Atmospheres,[a] 1957–1964

Pollutant	Pollutant concentration ($\mu g/m^3$)	
	Geometric mean	Maximum
Suspended particulates	98.	1706.
Benzene-soluble organics	7.4[b]	128.3
Nitrates	1.68	24.8
Sulfates	9.35	95.3
Antimony	c	—
Bismuth	c	—
Cadmium	c	—
Chromium	0.020	0.710
Cobalt	c	—
Copper	0.063	10.00
Iron	1.99	74.00
Lead	0.54	17.00
Manganese	0.064	4.70
Molybdenum	c	—
Nickel	0.028	0.830
Tin	0.024	1.00
Titanium	0.042	1.14
Vanadium	c	—
Zinc	0.09	58.00
Radioactivity (in Ci/m^3)	4.7[d]	5435.00

[a] Based on samples collected during a 24-hour period (249).
[b] 1957–1963 Mean, 1964 samples have been composited by quarters.
[c] Concentrations in most samples are below minimum detectable quantity.
[d] Arithmetic average of national monthly averages.

average airflow rate of 40–50 ft^3/minute (1.41–1.76 m^3/minute) (Table XXI). Each city is characterized by a unique chemical profile of suspended particulates which results from the nature of the fuels and industrial activity, as well as local agricultural and natural sources of pollution (Table XXII).

New York City offers an example of a study to determine the ingredients of airborne particulates, their sources, and the factors which influence their variation in space and time (257). Table XXIII has been compiled from the reported work of a diversity of investigators who have studied the airborne concentrations of selected anions and cations (196,

Table XXII   Particulate Analyses from Selected Urban Locations: Arithmetic Mean Values for 1966 Expressed as Micrograms per Cubic Meter (249) (Copyright by the American Association for the Advancement of Science, 1970)

	Atlanta, Georgia	Birmingham, Alabama	Baltimore, Maryland	Albuquerque, New Mexico
Suspended particulates	97.	142.	146.	120.
Benzene-soluble organics	7.4	10.0	9.2	7.7
Benzo[a]pyrene	0.0014	0.018	2.76	2.02
Ammonium	0.1	0.1	0.8	a
Nitrates	2.7	2.8	2.9	a
Sulfates	8.2	12.2	16.0	a
Antimony	0.000	0.000	0.000	0.000
Beryllium	0.000	0.000	0.000	0.000
Bismuth	0.000	0.000	0.000	0.000
Cadmium	0.017	0.008	0.003	0.000
Chromium	0.002	0.005	0.018	0.001
Cobalt	0.000	0.000	0.000	0.000
Copper	0.04	0.06	0.06	0.07
Iron	1.2	1.7	0.8	a
Manganese	0.06	0.15	0.08	0.03
Nickel	0.007	0.004	0.034	0.000
Tin	0.02	0.01	0.01	0.01
Titanium	0.03	0.03	0.01	0.01
Vanadium	0.001	0.003	0.071	0.001
Zinc	0.52	1.09	0.34	0.00

[a] Not analyzed.

225, 257–270). The methodologies of sampling and analyses varied widely in these studies.

The organic constituents of polluted air have undergone scrutiny mainly because of the carcinogenicity of many of the compounds to both laboratory animals and man. A more detailed analysis of benzene-soluble organics in samples from cities in the National Air Sampling Network was presented by Tabor and Fair (256). The introduction of organic aerosols into urban air through the use of insecticides was discussed by Tabor (271). Talc dust, a dry solid carrier and diluent for insecticides, has been extensively found in atmospheric dusts, but is still a local, rather than a global reflection of agricultural activity (272). A model of the circulation of DDT on earth has been proposed (273). The national average suspended particulate sample contained 8.2% benzene-soluble material. However, the range of this percentage was large; 80% of the values were between 3.8 and 14.0% with individual cities exhibiting great variation. The concentrations of selected constituents of particulate matter at urban and nonurban stations in the United States are shown in Table

Table XXIII Reported Average Concentrations ($\mu g/m^3$) of Selected Anions and

	Fe	Pb	Zn	Cu	Ni	Mn	V	Ca	Al
Chicago, Illinois	1.1	3.2	0.5	0.1	0.06	0.03	0.06		
Cincinnati, Ohio	1.8	1.8	1.7	0.2	0.06	0.17			
Denver, Colorado	0.8	1.8	0.1	0.4	0.06	0.02			
Philadelphia, Pennsylvania	0.7	1.6	0.4	0.1	0.06	0.05	0.14		
St. Louis, Missouri	1.1	1.8	0.3	0.1	0.06	0.03			
Washington, D.C.	0.6	1.3	0.3	0.2	0.06	0.02	0.09		
New York (metropolitan)	2.98	1.37		0.29	0.18		0.17	1.17	2.04
Bronx, New York		3.82		0.133	0.15	0.054	1.46		
Lower Manhattan, New York		2.99		0.212		0.071	1.19		
Tuxedo, New York		0.409	0.21	0.044	0.068	0.033	0.115		
NASN New York (urban)	3.4	1.9	0.4	0.37	0.187	0.05	0.905		
NASN New York (nonurban)	0.37	0.12	0.05	0.266	0.004	0.019	0.009		
Cambridge, Massachusetts						0.02	0.6		0.7
Niles, Michigan	1.9					0.062	0.005	1.0	1.2
East Chicago, Illinois	13.8					0.255	0.0181	7.0	2.175
Detroit, Michigan		4.8							
New York, New York		4.1							
Los Angeles, California		7.6							
Chilton, Berkshire, England						0.033	0.015		0.25
Warwick, England		0.81							
San Diego, California		1.8							
Laguna Mountain, California		0.004							
Frankfurt, Germany		0.95							
Berlin, Germany (residential)		1.0							
Berlin, Germany (heavy traffic)		3.8							
Cincinnati, Ohio	3.12	2.78							
Fairfax, Ohio	1.15	0.69							
Osaka, Japan		0.30							
University of Alaska		3.7							
Downtown Fairbanks, Alaska		2.0							
Hawaii									
Cincinnati, Ohio									
Fairfax, Ohio									
Chicago, Illinois									
Philadelphia, Pennsylvania									

^{a} ng/m^3.

XXIV (274). "Remote" stations are farthest from large population centers and are the best available indicators of geophysical background concentrations. "Intermediate" stations are relatively closer to populated areas and usually have agricultural activity in the neighborhood. "Proximate" stations are technically in nonurban locations, but are conspicuously influenced by proximity to large urban centers (274). At "urban" stations the average benzene-soluble content of suspended particulate matter was 6.6%. It decreased to 5.6, 5.4, and 5.1% at proximate, intermediate, and remote stations, respectively.

The sources of polycyclic hydrocarbons in the atmosphere all involve the combustion or pyrolysis of carbonaceous material at high temperature

Cations in Suspended Particulate Matter in Air

Si	Na	Mg	Cd	Cr	Cl^-	Br^-	I^{-a}	Phosphate	Nitrate	Chloride	NH_4^+	SO_4^{2-}	References
													258
													258
													258
													258
													258
													258
4.84	1.08	0.49											259
			0.014	0.049									257
			0.023	0.063									257
			0.003	0.009									257
			0.02	0.030									260
			0.000	0.002									260
	0.8												261
	0.17	0.50		0.0095									262
	0.455	2.40		0.113									262
													263
													263
													263
	0.85					2.2	0.040						264
													265
													225, 266
													267
													267
													267
													267
		7.21		0.31				0.22	2.96				196
		0.42		0.28				0.31	2.83	2.50	5.75	7.2	196
										3.54	4.00	8.7	260
					0.50	0.035	3.0			3.02	9.45	12.4	268, 269
					1.44	0.50	3.9						268, 269
						0.017							268
								0.22	2.96	2.50			270
								0.31	2.83	3.54			270
										3.02			270

(275, 276). Although great interest centers on the polynuclear aromatic hydrocarbons in air (276), the average weight ratio of known aliphatic hydrocarbons to aromatic hydrocarbons associated with suspended particulates is about 13 (277). This figure was derived from analysis of a particulate sample composited from samples collected during 1963 in downtown areas of approximately 100 communities. The concentrations of compounds detected in this sample are shown in Table XXV. Only 4.6% of the benzene-soluble fraction of this sample could be quantitatively determined (277).

In the United States the analysis of the benzene-soluble portion of suspended particulate samples has been more extensive and detailed than anywhere else in the world, where attention has centered mainly on

Table XXIV Selected Particulate Constituents as Percentages of Gross Suspended Particulates, United States (1966–1967)

| | Urban stations (217) | | Nonurban | | | | | | |
| | | | Proximate (5) | | Intermediate (15) | | Remote (10) | |
	µg/m³	Percent	µg/m³	Percent	µg/m³	Percent	µg/m³	Percent
Suspended particulates	102.0		45.0		40.0		21.0	
Benzene-soluble organics	6.7	6.6	2.5	5.6	2.2	5.4	1.1	5.1
Ammonium ion	0.9	0.9	1.22	2.7	0.28	0.7	0.15	0.7
Nitrate ion	2.4	2.4	1.40	3.1	0.85	2.1	0.46	2.2
Sulfate ion	10.1	9.9	10.0	22.2	5.29	13.1	2.51	11.8
Copper	0.16	0.15	0.16	0.36	0.078	0.19	0.060	0.28
Iron	1.43	1.38	0.56	1.24	0.27	0.67	0.15	0.71
Manganese	0.073	0.07	0.026	0.06	0.012	0.03	0.005	0.02
Nickel	0.017	0.02	0.008	0.02	0.004	0.01	0.002	0.01
Lead	1.11	1.07	0.21	0.47	0.096	0.24	0.022	0.10

TABLE XXV   Concentrations of Large Organic Compounds in the Average American Urban Atmosphere (277)

Compound	Airborne particulate ($\mu g/gm$)	Amount associated with airborne suspended particulate matter — $\dfrac{\mu g}{1000\ m^3\ air}$
Benzo[f]quinoline	2	0.2
Benzo[h]quinoline	3	0.3
Benzo[a]acridine	2	0.2
Benzo[c]acridine	4	0.6
11H-Indeno[1,2-b]quinoline	1	0.1
Dibenz[a,h]acridine	0.6	0.08
Dibenz[a,j]acridine	0.3	0.04
Benz[a]anthracene	30.	4.
Fluoranthene	30.	4.
Pyrene	42.	5.
Benzo[a]pyrene	46.	5.7
Benzo[e]pyrene	42.	5.
Perylene	5.5	0.7
Benzo[g,h,i]perylene	63.	8.
Anthanthrene	2.3	0.26
Coronene	15.	2.
n-Heptadecane	20.	2.5
n-Octadecane	110.	14.
n-Nonadecane	160.	20.
n-Eicosane	180.	23.
n-Heneicosane	320.	40.
n-Docosane	480.	60.
n-Tricosane	620.	77.
n-Tetracosane	480.	60.
n-Tentacosane	480.	60.
n-Hexacosane	85.	11.
n-Heptacosane	260.	32.
n-Octacosane	340.	43.
	3800	480.

benzo[a]pyrene, one of the most measured of all air pollutants. Data are now on record for the concentrations of this compound associated with particles in the air of South Africa, England, Italy, Denmark, Sweden, Norway, and Germany (278) as well as Vienna, Leningrad, Tokyo (279), and Budapest (280). Benzo[a]pyrene is routinely measured in the air of a variety of United States urban and nonurban areas; data are reported by the National Air Sampling Network (232). Concentrations throughout the world are comparable in urban and nonurban loca-

tions (*281*). The fate of benzo[a]pyrene, once airborne, is not well understood, but it does disappear only to be continuously replaced (*282*).

DeMaio and Corn (*206*) measured six polynuclear aromatic hydrocarbons associated with "respirable" and "nonrespirable" particulate matter
in Pittsburgh air. The "respirable" particulates, although low in weight,
concentrate the condensed organic compounds, presumably due to the increased surface area associated with particles as their state of subdivision
is increased. Similar studies with *p,p*-DDT associated with the same particulate samples confirmed the finding of condensed liquid concentration
with the smaller size fraction (*283*). These studies again highlight the
need to obtain simultaneous data for particle size, chemical composition,
and concentration of airborne particulates.

Finally, it should be noted that the national primary and secondary
ambient air quality standards for particulate matter in the United States
are based on the high volume sampling method of the NASN (*232*). The
primary standard, which is adopted to provide a margin of safety to protect the public health, is 75 $\mu g/m^3$ annual geometric mean and 260 $\mu g/m^3$
maximum 24-hour concentration not to be exceeded more than once per
year. The secondary standard, which is adopted to protect the public
from any adverse effect of a pollutant, not a health effect, is 60 $\mu g/m^3$
annual geometric mean, and 150 $\mu g/m^3$ maximum 24-hour concentration
not to be exceeded more than once per year (*284, 285*).

C. Staining or Soiling Index

The darkness of stains produced by drawing polluted air through filter
paper has been used to estimate suspended particulate matter concentrations for more than 50 years. Waller (*286*) reviews the history of this
type of measurement and points out that these measurements are being
made on an ever-increasing scale throughout the world. In some cases
the results are expressed directly in terms of optical density of the stain,
and in others a calibration curve serves to transform light transmission
or reflectance measurements of the stain to weight concentrations. The
shape of the curve relating reflectance or transmission of light to particulate weight depends on the nature of the pollution. Waller cites work
aimed at development of calibration curves in Canada, Australia, Belgium, and Great Britain. A tentative calibration curve, together with
standard sampling and measurement procedures was proposed and subsequently adopted by the Working Party on Methods of Measuring Air
Pollution and Survey Techniques of the Organization for Economic Cooperation and Development (*287*). In the United Kingdom the National
Survey of Air Pollution utilizes reflectance measurements of samples collected with tape samplers at approximately 1100 town sites (*288*). A stan-

dard "smoke" calibration curve is used to convert from reflectance measurements to concentration per unit volume (μg/m^3). The technique is simple, and the measurement of soiling index at one sampling site over an extended period of time has great practical value. Particulate concentration is also expressed as COH units per 1000 linear feet of air drawn through filter paper. The COH is an acronym for "coefficient of haze." One COH unit is defined as that quantity of particulate matter which produces an optical density of 0.01 on filter paper. Optical density (O.D.) is defined as O.D. = $\log_{10}(100/\%$ transmission). Optical density is measured by the transmission of light through the particulate-laden filter. In Allegheny County, Pennsylvania, the following optical density classifications are used for soiling index; slight pollution, 0–1.0 COHS/1000 linear feet; moderate pollution, 1.0–2.0 COHS/1000 linear feet; heavy pollution, 2.0–3.0 COHS/1000 linear feet; very heavy pollution, 3.0–4.0 COHS/1000 linear feet.

If reflectance of light from the filter stain is measured, the reference standard is the reflectance of the clean filter tape. The percentage reflectance of the sample is expressed in RUDS per 100 linear feet of air; RUDS is an acronym for "reflectance unit of dirt shade." Reflectance measurement is replacing transmission measurement at most United States sites.

Meetham (253) indicates that prior to the Clean Air Act improvements in the 1960's, average concentrations of "smoke" in England during winter were two to three times the summer value. He also cites weekend decreases in smoke pollution of 20–40% in urban areas. The range of seasonal variations has been reduced in a spectacular manner; large variations still exist with respect to location of site and with height above ground (288). A season of weekly and diurnal cycles of atmospheric smoke in Montreal, Quebec were analyzed in detail by Summers (289) on the basis of data collected during 1960–1963 at three sampling stations with American Iron and Steel Institute (A.I.S.I.) smoke samplers. Summers found two types of diurnal cycles: a winter type with small amplitude and morning and evening maxima of similar magnitude; and a summer type with large amplitude and a marked morning peak. Weekend reduction in smoke averaged 20% and variations with altitude (up to 600 ft) were also noted.

The dependence of optical measurements on the nature of the dust collected on the tape sampler was documented by Stalker et al. (290) for transmittance of samples collected at different sites in Nashville, Tennessee. Attempts to calibrate tape samplers in order to convert from optical measurements to concentration units (μg/m^3) include those by Kemeny (291), Ellison (292), Sullivan (293), and Sanderson and Katz (294). Clarification of the extent of the dependence of RUDS and COHS on

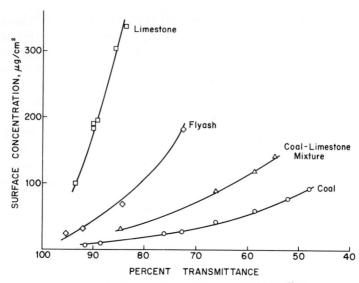

Figure 20. Effect of type of dust on COH reading.

the nature of the dust was greatly needed and was offered in the studies of Pedace and Sansone (295) and Saucier and Sansone (296). The refractive index of the particles on the filter is of overwhelming importance for converting from optical units of evaluation to weight of particles on the filter. Figure 20 shows the calibration curves obtained for test dusts which varied in refractive index from absorbing (coal) to nonabsorbing (silica). The conclusion is clear—there is no standard calibration curve for airborne particles in the atmosphere if optical evaluation of tape samples is the measurement technique. Removal of black particles or "smoke" from urban air could have an enormous effect on the improvement of air quality if optical evaluation of tape samples is the criterion. However, this improvement would certainly not be the same if weight of airborne particles per unit volume of air is the criterion. A calibration curve is required for each site to relate COHS or RUDS to $\mu g/m^3$. Furthermore, periodic updating of the calibration is necessary as the filter stain indicates trends in removal or addition of blackness.

VI. Indoor versus Outdoor Particulate Pollution*

There is a vast literature of air pollution measurements outdoors, but relatively few studies have been devoted to pollution which enters or

* See also Chapter 3, Vol. II.

originates in buildings (*297*), notwithstanding the extensive reports of exposure of workers to airborne toxic agents on the job. This is surprising because most urban residents spend the greater part of their day indoors. It seems appropriate at this time to briefly summarize our knowledge of particulate pollution inside buildings.

A theoretical model has been developed which relates indoor pollutant concentrations to those outside (*297a*). It has been confirmed for ozone (*297b*) but the applicability of the model to the disperse phase of pollutants in air remains to be tested. The particle size distributions of particles indoors and outdoors have been found to be similar, but show a larger proportion of submicrometer particles indoors (*154, 155, 298*). Jacobs *et al.* (*154, 155*) used optical microscopic techniques and determined that indoor samples contained a larger fraction of fibers, presumably from carpets, curtains, etc. The relationships between concentrations of six air pollutants, including particulates, outside and inside eleven buildings during summer and fall have been reported (*298a*). Where foot traffic was light or ventilation rates were low, reduced particulate concentrations were measured. An intensive study of indoor/outdoor air pollution involved three pairs of buildings in and around Hartford, Connecticut (*298*). Particulate samples were evaluated by the weight increase of filter papers containing samples and by optical evaluation of filter tape stains. Results were expressed as the ratio of indoor to outdoor concentrations. Three types of structures were used in the study, i.e., public building, office building, and private homes. Table XXVI presents results for concentrations inside and outside of these structures and for the enrichment ratio. The designations of sampling sites are explained as follows:

Far outdoors: representing ambient air quality in the area of the building

Near outdoors: located next to a window separating the indoor from the outdoor environment

Near indoors: just inside the near outdoor sample

Far indoors: located well inside the structure and representing levels of pollutants from outdoors or generated indoors to which the occupants are exposed

In general, daytime concentrations were higher, both indoors and outdoors, than nighttime concentrations. The indoor/outdoor ratio shows substantial loss of particulate matter in the penetration process. In contrast to total particulate matter concentrations, the organic content (benzene-soluble fraction) of particulate matter was enriched in the indoor samples as contrasted to outdoor samples. This finding was attributed to smoking and cooking activities indoors.

Table XXVI Suspended Particulate Matter at Indoor and Outdoor Sampling Sites^a (298)

Location	Sampling point	Summer		Fall		Winter	
		Day	Night	Day	Night	Day	Night
Public building (library; airtight structure over four lane highway)	Far outdoor	132	82	150	100	425	189
	Near outdoor	98	66	115	77	293	130
	Near indoor	70	45	61	46	74	51
	Far indoor	66	43	57	44	67	45
	Indoor/outdoor	0.50	0.52	0.38	0.44	0.16	0.26
Office building (16-story structure with air conditioning)	Far outdoor	104	93	48	38	124	81
	Near outdoor	118	98	47	40	137	89
	Near indoor	49	49	34	24	39	41
	Far indoor	50	46	36	27	38	39
	Indoor/outdoor	0.48	0.49	0.75	0.71	0.31	0.48
Private home (in the midst of a freeway interchange network)	Far outdoor	79	65	96	74	114	86
	Near outdoor	87	65	93	70	109	79
	Near indoor	67	51	52	42	45	32
	Far indoor	70	56	54	45	49	35
	Indoor/outdoor	0.87	0.86	0.56	0.61	0.43	0.41

^a Concentration, grams per cubic meter; indoor/outdoor ratio, dimensionless.

The half-life of smoke indoors was demonstrated to be 145–300 minutes, considerably longer than that of sulfur dioxide (54 to 71 minutes) (*299*). A half-life of smoke four or five times longer than that of sulfur dioxide is reported elsewhere (*300, 301*).

An investigation aimed at quantitating indoor pollution from cigarettes and cigars was performed by Bridge and Corn (*302*). Knowledge of the room volume and ventilation rate in room-air changes per hour, together with the occupancy level and number of cigarettes and cigars smoked, permitted these investigators to calculate reasonably accurately the subsequently measured concentrations of carbon monoxide and suspended particulate matter at social gatherings. It was concluded that in rooms with 7–10 air changes per hour and areal (ft^2/smoker) and volumetric (ft^3/smoker) densities of approximately 10–25 and 100–200, respectively, suspended particulate matter concentrations in occupied spaces could exceed 5 mg/m^3 when steady-state conditions are reached. These conditions are typical of those encountered at a beer or cocktail party. Hinds and First (*302a*) measured nicotine-in-air concentrations in a variety of public places and found average concentrations varied from 1.0 to 10.3 μg/m^3. Tobacco smoke concentrations were calculated from nicotine concentrations to average 40 to 400 μg/m^3. The characteristics of tobacco sidestream smoke and factors influencing its concentration and distribution in occupied spaces are summarized by Corn (*302b*).

In summary, it is apparent that, at least from the point of view of potential effects on the health of nonsmokers in occupied spaces (*303*), indoor particulate and gaseous pollution is a relatively neglected area of study when compared to our efforts with outdoor air pollution. This discrepancy requires rectification as soon as possible.

VII. Concluding Remarks

Investigations of the past 5–6 years have added enormously to our understanding of the atmospheric aerosol. They have also revealed the complexity of a complete description and the volume of work remaining. Perhaps the most important tasks for the future are to quantitate the physical and chemical removal mechanisms for airborne particles of different sizes and to demonstrate the feasibility of a material balance between sources of emission and suspended particulate matter in a relatively simple air basin. The accomplishment of this balance on a global basis will be a monumental feat, but the first "pencil and paper" attempts at a global inventory of suspended particulate matter is already on record.

The formation of aerosols from the gas phase and chemical alterations of particles by reaction with the gas phase continue to be elusive subjects for research. The particles are difficult to deal with experimentally and the amount of material available for analysis is measured in micro- or nanograms. Progress has been made with describing the fate of atmospheric sulfur dioxide and hydrogen sulfide and their conversion to sulfate. The aerosol interactions of nitrogen dioxide and organic gases in the atmosphere, complex as they are, must be unravelled. Inextricably woven into the fabric of these studies is delineation of aerosol changes as they undergo "aging" in the atmosphere.

It is still appropriate, unfortunately, to highlight the inadequacies of our present methods of aerosol assessment. Successful techniques for particle size measurement utilize "hybrid" units composed of different instruments operating on different principles of measurement, each requiring a host of assumptions to translate the raw data into a particle size spectrum. In addition, the instruments are costly and technically complex. Methods for determination of chemical composition of atmospheric aerosols require hours for sample collection; valuable information on transient behavior cannot be resolved. The instrumentation barrier remains a reality to those who would study the properties of the atmospheric aerosol. This barrier is undoubtedly responsible for the still limited number of studies of the spatial and temporal variations of the atmospheric aerosol at different locations on the surface of the earth.

Finally, attention has been focused here on the relatively neglected subject of indoor particulate pollution. Preliminary evidence suggests that there are significant differences between indoor and outdoor particulate pollution. Because most urban dwellers spend the bulk of their day indoors, the subject of indoor air pollution deserves more investigatory attention.

REFERENCES

1. J. Turkevich, *Amer. Sci.* **17**, 97 (1959).
2. F. W. Wright and P. W. Hodge, *J. Geophys. Res.* **68**, 5575 (1963).
3. N. Bhandari, J. R. Arnold, and D. Parkin, *J. Geophys. Res.* **73**, 1837 (1968).
4. A. J. Dyer and B. B. Hicks. *Quart. J. Roy. Meteorol. Soc.* **94**, 545 (1968).
5. F. E. Volz, *J. Geophys. Res.* **75**, 5185 (1970).
6. W. H. Porch, R. J. Charlson, and L. F. Roake, *Science* **170**, 317 (1970).
7. W. H. Fischer, *Science* **171**, 828 (1971).
8. J. T. Peterson and R. A. Bryson, *Science* **158**, 120 (1968).
9. R. A. McCormick and J. H. Ludwig, *Science* **156**, 1358 (1967).
10. G. M. Hidy, *in* "Assessment of Airborne Particles" (T. T. Mercer and P. E. Morrow, eds.), pp. 81–115. Thomas, Springfield, Illinois, 1972.

11. W. Clark and K. T. Whitby. *J. Atmos. Sci.* **24**, 677 (1967).

12. G. M. Hidy and J. R. Brock, "The Dynamics of Aerocolloidal Systems," pp. 350–358. Pergamon, Oxford, England, 1970.

13. R. Spirtas and H. J. Levin, "Characteristics of Particulate Patterns, 1957–1966," Publ. No. AP-61. U.S. Dept. of Health, Education and Welfare, Nat. Air Pollut. Contr. Admin., Raleigh, North Carolina, 1970.

13a. "Monitoring and Air Quality Trends Report, 1973," Rep. No. EPA-450/1-74-007. U.S. Environmental Protection Agency, Research Triangle Park, North Carolina, 1974.

13b. "Special Report: Trends in Concentrations of Benzene-Soluble Suspended Particulate Fraction and Benzo(a)Pyrene, 1960–1972," Rep. No. EPA-450/2-74-022. U.S. Environmental Protection Agency, Research Triangle Park, North Carolina, 1974.

14. M. Corn and L. DeMaio, *Science* **143**, 803 (1964).

15. R. E. Lee, R. K. Patterson, and J. Wagman, *Environ. Sci. Technol.* **2**, 288 (1968).

16. R. E. Lee and R. K. Patterson, *Atmos. Environ.* **3**, 249 (1969).

16a. E. S. Gladney, W. H. Zoller, A. G. Jones, and G. E. Gordon, *Environ. Sci. Technol.* **8**, 551 (1974).

16b. R. C. Pierce and M. Katz, *Environ. Sci. Technol.* **9**, 347 (1975).

16c. P. T. Cunningham, S. A. Johnson, and R. T. Yang, *Environ. Sci. Technol.* **8**, 131 (1974).

16d. S. K. Friedlander, *Environ. Sci. Technol.* **7**, 235 (1973).

16e. R. A. Reck, *Atmos. Environ.* **9**, 89 (1975).

16f. B. J. Palmer, *Environ. Lett.* **5**, 249 (1973).

17. G. M. Hidy and J. R. Brock, *Proc. Int. Clean Air Congr., 2nd, 1970* p. 1088 (1971).

18. N. A. Fuchs, "The Mechanics of Aerosols." Pergamon, Oxford, England, 1964.

19. C. N. Davies, ed., "Aerosol Science." Academic Press, New York, New York, 1966.

20. H. L. Green and W. R. Lane, "Particulate Clouds: Dusts, Smokes and Mists," 2nd ed. Van Nostrand-Reinhold, Princeton, New Jersey, 1964.

21. G. M. Hidy and J. R. Brock, "The Dynamics of Aerocolloidal Systems." Pergamon, Oxford, England, 1970.

22. C. N. Davies, *Nature (London)* **195**, 768 (1962).

23. L. Silverman, C. E. Billings, and M. W. First, "Particle Size Analysis in Industrial Hygiene," p. 87. Academic Press, New York, New York, 1971.

24. H. F. Johnstone, *Ill. Univ., Eng. Exp. Sta., Bull.* No. 16 (1959).

25. F. Stein, N. A. Esmen, and M. Corn, *Atmos. Environ.* **3**, 443 (1969).

26. M. Corn, T. L. Montgomery, and N. A. Esmen, *Environ. Sci. Technol.* **5**, 155 (1971).

27. F. Stein, R. Quinlan, and M. Corn, *Amer. Ind. Hyg. Ass. J.* **27**, 39 (1966).

28. V. Timbrell, *Brit. J. Appl. Phys., Suppl.* **3**, 586 (1954).

29. W. Walkenhorst and E. Burckman, *Staub* **36**, 45 (1966).

30. K. T. Whitby, A. B. Algren, R. C. Jordan, and J. C. Annis, *Heat., Piping, Air Cond.* **29**, 185 (1957).

31. W. C. McCrone and J. G. Delly, "The Particle Atlas," 2nd ed., 4 vols. Ann Arbor Sci. Publ., Ann Arbor, Michigan, 1973.

32. J. S. Ferguson and E. G. Sheridan, *J. Air Pollut. Contr. Ass.* **16**, 669 (1966).

33. E. M. Hamilton and W. D. Jarvis, "The Identification of Atmospheric Dust

by Use of the Microscope," Monograph. Central Electricity Generating Board, London, England, 1963.

34. K. Spurny, M. Polydorova, and J. Pixova, *Proc. Int. Clean Air Conf., 1959* p. 181 (1960).
35. K. T. Whitby, A. B. Algren, and R. C. Jordan, *Trans. Amer. Soc. Heat. Vent. Eng.* **61**, 463 (1955).
35a. J. L. Durham, W. E. Wilson, T. G. Ellestad, K. Willeke, and K. T. Whitby, *Atmos. Environ.* **9**, 717 (1975).
36. M. Corn, T. L. Montgomery, and R. J. Reitz, *Science* **159**, 1350 (1968).
37. C. N. Davies, *Ann. Occup. Hyg.* **3**, 219 (1961).
38. M. Corn, *Proc. Int. Clean Air Congr., 2nd, 1970* p. 965 (1971).
39. M. Corn, *Amer. Ind. Hyg. Ass. J.* **26**, 8 (1965).
40. L. Silverman, C. E. Billings, and M. W. First, "Particle Size Analysis in Industrial Hygiene," Chapter 6. Academic Press, New York, New York, 1971.
41. J. E. Smith and M. L. Jordan, *J. Colloid Sci.* **19**, 549 (1964).
42. T. F. Hatch and S. Choate, *J. Franklin Inst.* **207**, 369 (1933).
43. O. G. Raabe, *in* "Inhalation Carcinogenesis" (M. G. Hanna, P. Nettlesheim, and J. R. Gilbert, eds.), CONF-691001, p. 123. Clearinghouse Fed. Sci. Tech. Inform., Nat. Bur. Stand., U.S. Dept. of Commerce, Springfield, Virginia, 1970.
44. C. E. Junge, "Air Chemistry and Radioactivity," p. 115. Academic Press, New York, New York, 1963.
45. H. Wadell, *J. Franklin Inst.* **217**, 459 (1934).
46. J. Cartwright, *Ann. Occup. Hyg.* **5**, 163 (1962).
47. J. R. Hodkinson, *Amer. Ind. Hyg. Ass. J.* **26**, 64 (1965).
48. W. Stöber, *in* "Assessment of Airborne Particles" (T. T. Mercer and P. E. Morrow, eds.), pp. 249–288. Thomas, Springfield, Illinois, 1972.
49. L. B. Loeb, "Static Electrification." Springer-Verlag, Berlin, Germany and New York, New York, 1958.
50. H. L. Green and W. R. Lane, "Particulate Clouds: Dusts, Smokes and Mists," 2nd ed., p. 13. Van Nostrand-Reinhold, Princeton, New Jersey, 1964.
51. K. T. Whitby and B. Y. H. Liu, *in* "Aerosol Science" (C. N. Davies, ed.), Chapter 11. Academic Press, New York, New York, 1966.
52. H. J. White, "Industrial Electrostatic Precipitation." Addison-Wesley, Reading, Massachusetts, 1963.
53. M. Corn, *in* "Aerosol Science" (C. N. Davies, ed.), Chapter 11. Academic Press, New York, New York, 1966.
54. A. D. Zimon, "Adhesion of Dusts and Powders." Plenum, New York, New York, 1969.
55. G. Sehmel, *Atmos. Environ.* **7**, 291 (1973).
56. M. Corn and F. Stein, in "Surface Contamination" (B. Fish, ed.), pp. 45–54. Pergamon, Oxford, England, 1966.
57. M. Corn and F. Stein, *Nature (London)* **211**, 60 (1966).
58. B. J. Mason, "The Physics of Clouds," 2nd ed. Oxford Univ. Press (Clarendon), London, England and New York, New York, 1971.
59. G. Kinzer and R. Gunn, *J. Meteorol.* **81**, 71 (1951).
59a. D. Sinclair, R. J. Countess, and G. S. Hoopes, *Atmos. Environ.* **8**, 1111 (1974).
60. C. N. Davies, *Jrans. Inst. Chem. Eng.* **25**, 25 (1947).
61. R. H. Perry, C. H. Chilton, and S. D. Kirkpatrick, eds., "Chemical Engineers' Handbook," 4th ed., pp. 5–60. McGraw-Hill, New York, New York, 1963.
62. J. F. Heiss and J. Coull, *Chem. Eng. Progr.* **48**, 133 (1952).

63. N. A. Fuchs, "The Mechanics of Aerosols," p. 27. Pergamon, Oxford, England, 1964.

64. M. V. Smoluchowski, *Phys. Z.* 17, 557 (1916).

65. N. A. Fuchs and A. G. Sutugin, "Visokoldyspersny Aerosoly" (Highly Dispersed Aerosols). Sov. Acad. Sci. Publ House, Moscow, USSR, 1969.

66. S. E. Devir, *J. Colloid Sci.* 18, 744 (1963) ; 23, 80 (1967).

67. J. E. Quon, *Int. J. Air Water Pollut.* 8, 355 (1964).

68. G. Zebel, *Kolloid-Z.* 156, 102 (1958).

69. G. Zebel, *in* "Aerosol Science" (C. N. Davies, ed.), Chapter 2. Academic Press, New York, New York, 1966.

70. C. E. Junge, *J. Meteorol.* 12, 13 (1955).

71. R. B. Husar, K. T. Whitby, and B. Y. H. Liu, *J. Colloid Interface Sci.* 39, 211 (1972).

72. C. E. Junge, *J. Atmos. Sci.* 26, 603 (1969).

73. G. M. Hidy and J. R. Brock, "The Dynamics of Aerocolloidal Systems," pp. 68–80. Pergamon, Oxford, England, 1970.

74. L. Waldmann and K. H. Schmitt, *in* "Aerosol Science" (C. N. Davies, ed.), Chapter VI. Academic Press, New York, New York, 1966.

75. C. F. Schadt and R. D. Cadle, *J. Phys. Chem.* 65, 1689 (1961).

76. J. R. Brock, *J. Colloid Sci.* 17, 768 (1972).

77. D. L. Swift, *in* "Air Sampling Instruments" (M. Lippman, ed.), pp. Q1–Q7. Amer. Conf. Govt. Ind. Hyg., Cincinnati, Ohio, 1972.

78. M. Lippman, *in* "Air Sampling Instruments," pp. P1–P14. Amer. Conf. Govt. Ind. Hyg., Cincinnati, Ohio, 1972.

79. Intersociety Committee, "Methods of Air Sampling and Analysis," pp. 365–375. Amer. Pub. Health Ass., Washington, D.C., 1972.

80. T. A. Rich, *Aerosol Sci.* 2, 185 (1971).

81. W. N. Charman, *J. Opt. Soc. Amer.* 55, 415 (1963).

82. K. T. Whitby and W. E. Clark, *Tellus* 18, 2 (1966).

83. K. T. Whitby, B. Y. H. Liu, R. B. Husar, and N. J. Barsic, *J. Colloid Interface Sci.* 39, 136 (1972).

84. R. Gunn, *J. Colloid Sci.* 10, 107 (1955).

85. J. Keefe, P. J. Nolan, and T. A. Rich, *Proc. Roy. Irish Acad., Sect. A* 60, 6 (1959).

86. T. A. Rich, *Int. J. Air Pollut.* 1, 288 (1959).

87. K. T. Whitby and B. Y. H. Liu, *in* "Aerosol Science" (C. N. Davies, ed.), p. 65. Academic Press, New York, New York, 1966.

88. V. A. Gordieyeff, *AMA Arch. Ind. Health* 14, 471 (1956).

89. G. R. Wait, *Phys. Rev.* 48, 383 (1935).

90. H. Neuberger, "Introduction to Physical Meteorology." Penn State Univ. Press, University Park, Pennsylvania, 1951.

91. O. H. Gish, *in* "Compendium of Meteorology" (T. F. Malone, ed.), p. 103. Amer. Meteorol. Soc., Boston, Massachusetts, 1951.

92. J. Bricard and J. Pradel, *in* "Aerosol Science" (C. N. Davies, ed.), Chapter IV. Academic Press, New York, New York, 1966.

93. L. B. Loeb, *Heat., Piping, Air Cond.* 6, 437 (1934).

94. D. Keefe and P. J. Nolan, *Geofis. Pura Appl.* 50, 155 (1961).

95. P. J. Nolan and D. J. Doherty, *Proc. Roy. Irish Acad., Sect. A* 53, 163 (1950).

96. T. A. Rich, L. W. Pollak, and A. L. Metneiks, *Geofis. Pura Appl.* 44, 233 (1959).

97. A. Hoegl, *Z. Angew. Phys.* 16, 252 (1962).

98. A. H. Mendenhall and D. A. Fraser, *Amer. Ind. Hyg. Ass. J.* **24**, 555 (1963).
99. W. W. Hicks and J. C. Beckett, *Trans. Amer. Inst. Elec. Eng., Part 1* **76**, 108 (1957).
100. H. Israel, *in* "Compendium of Meteorology" (T. F. Malone, ed.), p. 144. Amer. Meteorol. Soc., Boston, Massachusetts, 1951.
101. F. K. Hurd and J. C. Mullins, *J. Colloid Sci.* **17**, 91 (1960).
102. J. E. Masters, *Rev. Sci. Instrum.* **24**, 586 (1953).
103. C. R. Wait and W. D. Parkinson, *in* "Compendium of Meteorology" (T. F. Malone, ed.), p. 124. Amer. Meteorol. Soc., Boston, Massachusetts, 1951.
104. H. Landsberg, *Gerlands Beitr. Geophys., Suppl.* **3**, 155 (1938).
105. J. H. Kornbleuh, *Bull. Amer. Meteorol. Soc.* **41**, 361 (1960).
106. C. P. Yaglou, L. C. Benjamin, and S. P. Choate, *Heat., Piping, Air Cond.* **3**, 865 (1931).
107. G. R. Wait, *J. Wash. Acad. Sci.* **36**, 321 (1946).
108. R. Gunn, *J. Atmos. Sci.* **21**, 168 (1964).
109. C. F. Schilling, Comments on "The Secular Increase of the World-Wide Fine Particle Pollution." Rand Corp., Santa Monica, California, 1964 (available as AD-600371 from Clearinghouse Fed. Sci. Tech. Inform., Sills Bldg., Springfield, Virginia, 1964).
110. J. T. Peterson and R. A. Bryson, *Science* **158**, 120 (1968).
111. E. C. Flowers, R. A. McCormick, and K. R. Kurfis, *J. Appl. Meteorol.* **8**, 955 (1969).
112. F. E. Volz, *Tellus* **21**, 625 (1969).
113. B. J. Steigerwald, *55th Annu. Meet. Amer. Inst. Chem. Eng., 1962.*
114. W. W. Hicks, *J. Franklin Inst.* **261**, 209 (1956).
115. C. P. Yaglou, A. D. Brandt, and L. C. Benjamin, *Trans. Amer. Soc. Heat. Vent. Eng.* **39**, 357 (1933).
116. C. P. Yaglou, A. D. Brandt, and L. C. Benjamin, *J. Ind. Hyg.* **5**, 341 (1933).
117. C. P. Yaglou, L. C. Benjamin, and A. D. Brandt, *J. Ind. Hyg.* **5**, 8 (1933).
118. C. P. Yaglou, *in* "The Air We Breathe" (S. M. Farber and R. H. L. Wilson, eds.), p. 269. Thomas, Springfield, Illinois, 1961.
119. J. Hirth and G. Pound, "Condensation and Evaporation." Pergamon, Oxford, England, 1963.
120. H. R. Byers, *Ind. Eng. Chem.* **57**, 32 (1965).
121. B. J. Mason, "The Physics of Clouds," 2nd ed. Oxford Univ. Press (Clarendon), London, England and New York, New York, 1971.
122. L. W. Pollak, *Int. J. Air Pollut.* **1**, 293 (1959).
123. L. W. Pollak, *Geofis. Pura Appl.* **22**, 75 (1952).
124. C. E. Junge, "Air Chemistry and Radioactivity," p. 115. Academic Press, New York, New York, 1963.
125. A. H. Woodcock, *J. Meteorol.* **9**, 200 (1952).
126. W. J. Megaw and R. D. Wiffen, *Geofis. Pura Appl.* **50**, 118 (1961).
127. F. Verzar and D. Evans, *Geofis. Pura Appl.* **43**, 259 (1959).
128. A. P. Altshuller, D. L. Klosterman, P. W. Leach, I. J. Hindawi, and J. E. Sigsby, Jr., *Int. J. Air Water Pollut.* **10**, 81 (1966).
129. K. G. Vohra and P. V. N. Nair, *Aerosol Sci.* **1**, 127 (1970).
130. K. G. Vohra, K. N. Vasudevan, and P. V. N. Nair, *J. Geophys. Res.* **75**, 2951 (1970).
131. J. E. Quon, *Arch. Environ. Health* **7**, 600 (1963).
132. F. W. Went, *Proc. Nat. Acad. Sci. U.S.* **51**, 1259 (1964).
133. F. W. Went, *Tellus* **18**, 549 (1968).

134. V. J. Schaeffer, *Science* **170,** 851 (1970).
135. L. F. Radke and P. V. Hobbs, *J. Atmos. Sci.* **26,** 281 (1969).
136. S. Twomey and G. T. Severynse, *J. Atmos. Sci.* **20,** 392 (1963).
137. S. Twomey and G. T. Severynse, *J. Atmos. Sci.* **21,** 558 (1964).
138. F. W. Van Luik and R. E. Rippere, *Anal. Chem.* **34,** 1617 (1962).
139. G. Riediger, *Staub* **31,** 11 (1971).
140. A. J. Nesti, Jr., *Opt. Spectra* p. 76 (1970).
141. C. E. Junge, "Air Chemistry and Radioactivity," p. 160. Academic Press, New York, New York, 1963.
142. S. Twomey, *J. Meteorol.* **11,** 334 (1954).
143. R. E. Kerr, Jr., J. R. Thompson, and R. D. Elliot, "Quantitative Assessment of the Performance Characteristics of the Airways Terminal Forecasting System," Final Rep., Contract Cwb. 10077. Aerometric Res. Inc., Goleta, California, 1962.
143a. D. S. Covert, R. J. Charlson, and N. C. Ahlquist, *J. Appl. Meteorol.* **11,** 968 (1972).
144. M. Neiburger and M. G. Wurtele, *Chem. Rev.* **44,** 321 (1949).
145. D. A. Lundgren and D. W. Cooper, *J. Air Pollut. Contr. Ass.* **19,** 243 (1969).
146. R. J. Charlson, *Environ. Sci. Technol.* **3,** 913 (1969).
147. State of California, "California Standards for Ambient Air Quality and Motor Vehicle Exhaust," pp. 74–75. California Department of Public Health, Berkeley, California, 1960.
148. R. Jaenicke and C. Junge, *Beitr. Phys. Frei. Atmos.* **40,** 129 (1947).
149. J. H. Fairweather, A. F. Sidlow, and W. C. Faith, *J. Air Pollut. Contr. Ass.* **15,** 345 (1965).
150. M. J. Pilat, D. S. Ensor, and J. C. Bosch, *Amer. Ind. Hyg. Ass. J.* **32,** 508 (1971).
150a. R. L. Davison, D. F. S. Nautsch, J. R. Wallace, and C. A. Evans, Jr., *Environ. Sci. Technol.* **8,** 1107 (1974).
150b. R. S. Lee, Jr., H. L. Crist, A. E. Riley, and K. E. MacLeod, *Environ. Sci. Technol.* **9,** 643 (1975).
150c. J. T. Ganley and G. S. Springer, *Environ. Sci. Technol.* **8,** 340 (1974).
150d. E. J. Schulz, R. B. Engdahl, and T. T. Frankenberg, *Atmos. Environ.* **9,** 111 (1975).
150e. S. K. Friedlander, *Environ. Sci. Technol.* **7,** 235 (1973).
150f. S. L. Heisler, S. K. Friedlander, and R. B. Husar, *Atmos. Environ.* **7,** 633 (1973).
150g. G. Gartrell and S. K. Friedlander, *Atmos. Environ.* **9,** 279 (1975).
151. C. E. Junge, *Tellus* **5,** 1 (1953).
152. C. E. Junge, *J. Atmos. Sci.* **26,** 603 (1969).
153. T. Okita, *J. Meteorol. Soc. Jap.* [2] **33,** 291 (1955).
154. M. B. Jacobs, A. Monoharan, and L. J. Goldwater, *Int. J. Air Water Pollut.* **6,** 205 (1962).
155. M. B. Jacobs, M. M. Braverman, C. Theophil, and S. Hochheiser, *Amer. J. Pub. Health* **47,** 1430 (1957).
156. R. E. Waller, A. G. F. Brooks, and J. Cartwright, *Int. J. Air Water Pollut.* **7,** 779 (1963).
157. S. K. Friedlander and R. E. Pasceri, *J. Atmos. Sci.* **22,** 571 (1965).
158. R. E. Pasceri and S. K. Friedlander, *J. Atmos. Sci.* **22,** 577 (1965).
159. J. Cartwright, G. Nagelschmidt, and J. W. Skidmore, *Quart. J. Roy. Meteorol. Soc.* **82,** 82 (1956).
160. S. K. Friedlander, *J. Meteorol.* **17,** 479 (1960).

161. S. K. Friedlander, *J. Meteorol.* **18**, 753 (1961).
162. S. K. Friedlander, *Proc. Nat. Conf. Aerosols, 1st, 1962* p. 115 (1965).
163. K. Takahashi and M. Kasahara, *Atmos. Environ.* **2**, 441 (1968).
164. G. M. Hidy and J. R. Brock, "The Dynamics of Aerocolloidal Systems," pp. 337–358. Pergamon, Oxford, England, 1970.
165. J. R. Brock, *Atmos. Environ.* **5**, 833 (1971).
166. I. H. Blifford, *J. Geophys. Res.* **75**, 3099 (1970).
167. I. H. Blifford and L. D. Ringer, *J. Atmos. Sci.* **26**, 716 (1969).
168. G. M. Hidy and S. K. Friedlander, *Proc. Int. Clean Air Congr., 2nd, 1970* p. 391. (1971).
169. C. M. Peterson and H. J. Paulus, *Amer. Ind. Hyg. Ass. J.* **29**, 111 (1968).
170. N. C. Ahlquist and R. J. Charlson, *Environ. Sci. Technol.* **5**, 363 (1968).
171. K. T. Whitby, R. B. Husar, and B. Y. H. Liu, *J. Colloid Interface Sci.* **39**, 177 (1972).
172. C. M. Peterson, H. J. Paulus, and G. H. Foley, *J. Air Pollut. Contr. Ass.* **19**, 795 (1969).
173. W. E. Clark and K. T. Whitby, *J. Atmos. Sci.* **24**, 677 (1967).
174. K. T. Noll, *Trend Eng.* **19**, 21 (1967).
175. K. T. Noll and M. J. Pilat, *Atmos. Environ.* **5**, 527 (1971).
176. H. Horvath, *Environ. Sci. Technol.* **1**, 651 (1967).
177. R. O. McCaldin, W. Johnson, and N. T. Stephens, *Science* **166**, 381 (1969).
178. H. O'Donnell, T. L. Montgomery, and M. Corn, *Atmos. Environ.* **4**, 1 (1970).
179. R. L. Byers, J. W. Davis, E. W. White, and R. E. McMillan, *Environ. Sci. Technol.* **5**, 517 (1971).
180. H. J. Ettinger and G. W. Royer, *J. Air Pollut. Contr. Ass.* **22**, 108 (1972).
181. R. Chester and L. R. Johnson, *Mar. Geol.* **11**, 251 (1971).
182. D. A. Lundgren, *J. Air Pollut. Contr. Ass.* **20**, 603 (1970).
183. D. A. Lundgren, *Atmos. Environ.* **5**, 645 (1971).
184. E. Meszaros, *Tellus* **20**, 443 (1968).
185. G. V. Rosenberg, *Atmos. Oceanic Phys.* **3**, 936 (1967) (transl. from Russian by J. Findlay).
186. H. Quenzel, *J. Geophys. Res.* **75**, 2915 (1970).
187. K. Bullrich, R. Eiden, R. Jaenicke, and W. Nowak, *Pure Appl. Geophys.* **69**, 280 (1968).
188. C. Junge and R. J. Jaenicke, *Aerosol Sci.* **2**, 305 (1971).
189. R. E. Lee, Jr. and S. Goranson, *Environ. Sci. Technol.* **6**, 1019 (1972).
190. R. E. Lee, Jr., *Science* **178**, 567 (1972).
190a. K. Willeke, K. T. Whitby, W. E. Clark, and V. A. Marple, *Atmos. Environ.* **8**, 609 (1974).
190b. R. E. Lee, Jr., J. Caldwell, G. G. Akland, and R. Fankhauser, *Atmos. Environ.* **8**, 1109 (1974).
190c. G. M. Sverdrup, K. T. Whitby, and W. E. Clark, *Atmos. Environ.* **9**, 493 (1975).
190d. K. T. Whitby, W. E. Clark, V. A. Marple, G. M. Sverdrup, G. J. Sem, K. Willeke, B. Y. H. Liu, and D. Y. H. Pui, *Atmos. Environ.* **9**, 463 (1975).
190e. R. B. Husar, "Recent Developments in In Situ Size Spectrum Measurement of Submicron Aerosols." *Instrumentation for Monitoring Air Quality,* pp. 151–192. ASTM ST P555, Amer. Soc. for Testing and Materials, Philadelphia, Pennsylvania, 1974.
191. T. F. Hatch and C. Moke, *J. Ind. Hyg. Toxicol.* **18**, 91 (1936).
192. F. Shanty and W. C. L. Hemeon, *J. Air Pollut. Contr. Ass.* **13**, 211 (1963).

193. R. J. Hamilton, *Brit. J. Appl. Phys., Suppl.* 3, S29 (1954).
194. M. Corn and L. DeMaio, *J. Air Pollut. Contr. Ass.* 15, 1 (1965).
195. E. Robinson and F. L. Ludwig, *J. Air Pollut. Contr. Ass.* 17, 664 (1967).
196. R. E. Lee, Jr., R. K. Patterson, and J. Wagman, *Environ. Sci. Technol.* 2, 288 (1968).
197. M. J. Heard and R. D. Wiffen, *Atmos. Environ.* 3, 337 (1969).
198. R. E. Lee, Jr. and R. K. Patterson, *Atmos. Environ.* 3, 249 (1969).
199. J. Wagman, R. E. Lee, Jr., and J. C. Axt, *Atmos. Environ.* 1, 479 (1967).
200. F. L. Ludwig and E. Robinson, *Atmos. Environ.* 2, 13 (1968).
201. J. F. Roesler, H. J. R. Stevenson, and J. S. Nader, *J. Air Pollut. Contr. Ass.* 15, 576 (1965).
202. E. Meszaros, *Tellus* 22, 235 (1970).
203. D. A. Gillette and I. H. Blifford, Jr., *J. Atmos. Sci.* 28, 1199 (1971).
204. R. E. Waller, *Int. J. Air Water Pollut.* 7, 779 (1963).
205. P. R. Harrison, W. R. Matson, and J. W. Winchester, *Atmos. Environ.* 5, 613 (1971).
206. L. DeMaio and M. Corn, *J. Air Pollut. Contr. Ass.* 16, 67 (1966).
207. T. Hasegawa and A. Sugimae, *Clean Air J. Jap. Air Cleaning Ass.* 9(1), 1 (1971); cited in Air Pollution Abstracts No. 13654, APTIC No. 31325. U.S. Environmental Protection Agency, Durham, North Carolina, 1970.
208. T. Hasegawa, A. Sugimae, J. Fujii, Y. Matsuo, and Y. Okuyama, *J. Jap. Soc. Air Pollut.* 5, 210 (1970); cited in Air Pollution Abstracts, APTIC No. 28650. U.S. Environmental Protection Agency, Durham, North Carolina, 1970.
208a. Y. S. Shum and W. D. Loveland, *Atmos. Environ.* 8, 645 (1974).
208b. C. S. Martens, J. J. Wesolowski, R. Kaifer, and W. John, *Atmos. Environ.* 7, 905 (1973).
208c. R. C. Pierce and M. Katz, *Environ. Sci. Technol.* 9, 349 (1975).
208d. L. L. Ciaccio, R. L. Rubino, and J. Flores, *Environ. Sci. Technol.* 10, 935 (1974).
208e. C. S. Martens, J. J. Wesolowski, R. Kaifer, W. John, and R. C. Harris, *Environ. Sci. Technol.* 7, 817 (1973).
209. M. O. Amdur, *in* "Inhaled Particles and Vapors" (C. N. Davies, ed.), p. 281. Pergamon, Oxford, England, 1961.
210. M. O. Amdur and M. Corn, *Amer. Ind. Hyg. Ass. J.* 24, 336 (1963).
211. A. E. J. Eggleton, *Atmos Environ.* 3, 355 (1969).
212. M. V. Jensen and N. Nakai, *Science* 134, 2102 (1961).
213. A. P. Altshuller, *Bull. W. H. O.* 40, 616 (1969).
214. A. P. Van Den Heuvel and B. J. Mason, *Quart. J. Roy. Meteorol. Soc.* 89, 271 (1963).
215. H. F. Johnstone and A. J. Moll, *Ind. Eng. Chem.* 52, 861 (1960).
216. P. A. Leighton, "Photochemistry of Air Pollution," p. 246. Academic Press, New York, New York, 1961.
217. R. T. Cheng, M. Corn, and J. O. Frohliger, *Atmos. Environ.* 5, 987 (1971).
218. M. Corn and R. T. Cheng, *J. Air Pollut. Contr. Ass.* 22, 8 (1972).
219. B. T. Commins, *Analyst* 88, 364 (1963).
220. P. J. Lawther, *Proc. Roy. Inst. Gt. Brit.* 44, 714 (1971).
220a. National Academy of Engineering-National Research Council, "Air Quality and Stationary Source Emissions Control." Report prepared for U.S. Senate, Serial No. 94-4. U.S. Government Printing Office, Washington, D.C., 1975.
221. A. Goetz and T. Kallai, *J. Air Pollut. Contr. Ass.* 12, 479 (1962).

222. T. J. Chow and M. S. Johnstone, *Science* **147**, 502 (1965).
223. P. K. Mueller, H. L. Helwig, A. E. Alcocer, W. K. Gong, and E. E. Jones, *Amer. Soc. Test. Mater., Spec. Tech. Publ.* **32**, 60 (1964).
224. Committee on Biological Effects of Atmospheric Pollutants, "Lead: Airborne Lead in Perspective." Nat. Acad. Sci., Washington, D.C., 1972.
225. T. J. Chow, J. L. Earl, and C. F. Bennett, *Environ. Sci. Technol.* **3**, 737 (1969).
226. T. J. Chow and J. L. Earl, *Science* **169**, 557 (1970).
227. R. J. Charlson and J. M. Pierrard, *Atmos. Environ.* **3**, 479 (1969).
228. Working Group on Lead Contamination, "Survey of Lead in the Atmosphere of Three Urban Communities," Pub. Health Serv. Publ. No. 999-AP-12. U.S. Dept. of Health, Education and Welfare, Raleigh, North Carolina, 1970.
229. H. W. George and D. Jost, *Atmos. Environ.* **5**, 725 (1971).
230. J. R. Goldsmith, *J. Air Pollut. Contr. Ass.* **19**, 714 (1969).
230a. R. E. Lee, Jr. and D. J. von Lehmden, *J. Air Pollut. Contr. Ass.* **23**, 853 (1973).
231. "Atmospheric Pollution Bulletins, National Survey of Air Pollution." Dept. Sci. Ind. Res., Warren Springs Lab., Stevenage, Herts, England.
232. "Air Pollution Measurements of the National Air Sampling Network, Analysis of Suspended Particulates, 1957–1961" (Supplements for 1962, 1963, 1964, 1965, 1966), Pub. Health Serv. Publ. No. 978. U.S. Dept. of Health, Education and Welfare, Washington, D.C., 1962.
233. R. E. Lee, Jr. and S. Goranson, "The NASN Cascade Impactor Network; First Year Operation," AP-108. U.S. Environmental Protection Agency, Research Triangle Park, North Carolina, 1972.
234. T. B. McMullen and R. Smith, "The Trend of Suspended Particulates in Urban Air: 1957–1964," Pub. Health Serv. Publ. No. 999-AP-19. U.S. Dept. of Health, Education and Welfare, Div. Air Pollut., Cincinnati, Ohio, 1965.
235. R. Spirtas and H. J. Levin, "Characteristics of Particulate Patterns, 1957–1966," Nat. Air Pollut. Contr. Admin. Publ. No. AP-61. U.S. Dept. of Health, Education and Welfare, Raleigh, North Carolina, 1970.
236. For example, see Monthly Activities Report, Bureau of Air Pollution Control, Allegheny County Health Department, Pennsylvania.
237. American Society for Testing and Materials, *Book ASTM Stand.* p. 97 (1962).
238. W. F. Carey, *Int. J. Air Pollut.* **2**, 1 (1959).
239. M. Corn, R. Quinlan, and J. Katz, *Atmos. Environ.* **1**, 227 (1967).
240. Monthly Activities Report, Bureau of Air Pollution Control, Allegheny County Health Dept., Pittsburgh, Pennsylvania, 1965.
241. M. B. Jacobs, "The Chemical Analysis of Air Pollutants," Chapter 5. Wiley (Interscience), New York, New York, 1960.
242. J. D. Sensenbaugh and W. C. L. Hemeon, *Amer. Soc. Test. Mater., Proc.* **53**, 1160 (1953).
243. A. F. Fisher, *J. Air Pollut. Contr. Ass.* **7**, 47 (1957).
244. J. S. Nader, *J. Air Pollut. Contr. Ass.* **8**, 35 (1958).
245. H. P. Sanderson, P. Bradt, and M. Katz, *J. Air Pollut. Contr. Ass.* **13**, 461 (1963).
246. M. Katz, *in* "Air Pollution," p. 97. World Health Organ., Geneva, Switzerland, 1961.
247. "Technical Report on Air Pollution in Yokohama-Kawasaki Industrial Area, 1957–1962," p. 25. Kanagawa Prefecture Govt., Japan, 1963.
248. G. Gould, *Int. J. Air Water Pollut.* **8**, 657 (1964).

249. G. B. Morgan, G. Ozolins, and E. C. Tabor, *Science* **170**, 289 (1970).
250. R. A. McCormick and D. M. Baulch, *J. Air Pollut. Contr. Ass.* **12**, 492 (1962).
251. R. A. McCormick and K. R. Kurfis, *Quart. J. Roy. Meteorol. Soc.* **92**, 392 (1966).
252. G. V. Sheleikovskii, *in* "Smoke Pollution of Towns" (R. A. Babayants, ed.), p. 97. Nat. Sci. Found., Washington, D.C., 1961 (transl. from Russian).
253. A. R. Meetham, "Atmospheric Pollution, its Origins and Preventions," 3rd ed., pp. 213–245. Macmillan, New York, New York, 1964.
253a. S. Dalager, *Atmos. Environ.* **9**, 690 (1975).
254. R. E. Lee, Jr., J. S. Caldwell, and G. B. Morgan, *Atmos. Environ.* **6**, 593 (1972).
255. A. Lobner, *Staub* **25**, 221 (1965).
256. E. C. Tabor and D. H. Fair, *J. Air Pollut. Contr. Ass.* **11**, 403 (1961).
257. T. J. Kneip, M. Eisenbud, C. D. Strehlow, and P. C. Freudenthal, *J. Air Pollut. Contr. Ass.* **20**, 144 (1970).
258. R. E. Lee, Jr., S. S. Goranson, R. E. Enrione, and G. B. Morgan, *Environ. Sci. Technol.* **6**, 1025 (1972).
259. N. L. Morrow and R. S. Brief, *Environ. Sci. Technol.* **5**, 786 (1971).
260. T. Hasegawa, A. Sugimae, J. Fujii, Y. Matsuo, and Y. Okuyama, *J. Jap. Soc. Air Pollut.* **5**, 210 (1970).
261. W. H. Zoller and G. E. Gordon, *Anal. Chem.* **42**, 257 (1970).
262. R. Dams, J. A. Robbins, K. A. Rahn, and J. W. Winchester, *Anal. Chem.* **42**, 861 (1970).
263. J. M. Colucci, C. R. Begeman, and K. Kumler, *J. Air Pollut. Contr. Ass.* **19**, 255 (1969).
264. J. R. Keane and E. M. R. Fisher, *Atmos. Environ.* **2**, 603 (1968).
265. J. Bullock and W. M. Lewis, *Atmos. Environ.* **2**, 517 (1968).
266. R. A. Duce, J. W. Winchester, and T. W. V. Nahl, *Tellus* **18**, 238 (1966).
267. H. W. Georgii and D. Jost, *Atmos. Environ.* **5**, 725 (1971).
268. D. Randerson, *Atmos. Environ.* **4**, 249 (1970).
269. J. W. Winchester, W. H. Zoller, R. A. Duce, and C. S. Benson, *Atmos. Environ.* **1**, 105 (1967).
270. R. E. Lee, Jr. and R. K. Patterson, *Atmos. Environ.* **3**, 249 (1969).
271. E. C. Tabor, *Trans. N.Y. Acad. Sci.* [2] **28**, 569 (1966).
272. H. Windom, J. Griffin, and E. D. Goldberg, *Environ. Sci. Technol.* **1**, 923 (1967).
273. J. Cramer, *Atmos. Environ.* **7**, 241 (1973).
274. T. B. McMullen, R. B. Faoro, and G. B. Morgan, *J. Air Pollut. Contr. Ass.* **20**, 369 (1970).
275. E. Sawicki, T. R. Hauser, W. C. Elbert, F. T. Fox, and J. E. Meeker, *Amer. Ind. Hyg. Ass. J.* **23**, 137 (1962).
276. Committee on Biological Effects of Air Pollutants, "Polynuclear Aromatic Hydrocarbons." Nat. Acad. Sci., Washington, D.C., 1972.
277. E. Sawicki, S. P. McPherson, T. W. Stanley, J. Meeker, and W. C. Elbert, *Int. J. Air Water Pollut.* **9**, 515 (1965).
278. C. W. Louw, *Amer. Ind. Hyg. Ass. J.* **26**, 520 (1965).
279. A. P. Altshuller, *Anal. Chem.* **39**, 10R (1967).
280. M. Kertesz-Saringer, J. Morik, and Z. Morlin, *Atmos. Environ.* **3**, 417 (1969).
281. E. Sawicki, W. C. Elbert, T. R. Hauser, F. T. Fox, and T. W. Stanley, *Amer. Ind. Hyg. Ass. J.* **21**, 443 (1960).
282. J. F. Thomas, M. Mukai, and B. D. Tebbens, *Environ. Sci. Technol.* **2**, 33 (1968).

283. P. Antommaria, M. Corn, and L. DeMaio, *Science* **150**, 1476 (1965).
284. *Federal Register* **36**, No. 21, 1503 (1971).
285. "Air Quality Criteria for Particulate Matter." U.S. Dept. Health, Education and Welfare, Washington, D.C., 1969.
286. R. E. Waller, *J. Air Pollut. Contr. Ass.* **14**, 323 (1962).
287. "Methods of Measuring Air Pollution. Report of the Working Party on Methods of Measuring Air Pollution and Survey Techniques." Organization for Economic Cooperation and Development, Paris, France, 1964.
288. "National Survey of Air Pollution, 1961–1971," Vol. I. Warren Spring Laboratory, HM Stationery Office, London, England, 1972.
289. P. W. Summers, *J. Air Pollut. Contr. Ass.* **16**, 432 (1966).
290. W. W. Stalker, R. C. Dickerson, and G. D. Kramer, *Amer. Ind. Hyg. Ass. J.* **24**, 68 (1963).
291. E. Kemeny, *J. Air Pollut. Contr. Ass.* **12**, 278 (1962).
292. J. M. Ellison, *Staub* **28**, 240 (1968).
293. J. L. Sullivan, *J. Air Pollut. Contr. Ass.* **12**, 474 (1962).
294. H. P. Sanderson and M. Katz, *J. Air Pollut. Contr. Ass.* **13**, 476 (1963).
295. E. Pedace and E. B. Sansone, *J. Air Pollut. Contr. Ass.* **22**, 348 (1972).
296. J. Y. Saucier and E. B. Sansone, *Atmos. Environ.* **6**, 37 (1972).
297. F. B. Benson, J. J. Henderson, and D. E. Caldwell, "Indoor-Outdoor Air Pollution Relationships: A Literature Review," Publ. No. AP-112. U.S. Environmental Protection Agency, Research Triangle Park, North Carolina, 1972.
297a. F. H. Shair and K. L. Heitner, *Environ. Sci. Technol.* **8**, 444 (1974).
297b. C. H. Hales, A. M. Rollinson and F. H. Shair, *Environ. Sci. Technol.* **8**, 452 (1974).
298. J. E. Yocum, W. L. Clink, and W. C. Cote, *J. Air Pollut. Contr. Ass.* **21**, 251 (1971).
298a. C. R. Thompson, E. G. Hensel, and G. Kats, *J. Air Pollut. Contr. Ass.* **23**, 881 (1973).
299. M. J. G. Wilson, *Proc. Roy. Soc., Ser. A* **300**, 215 (1968).
300. R. J. Shepherd, G. C. Carey, and J. J. Phair, *AMA Arch. Ind. Health* **17**, 236 (1958).
301. K. Biersteker, H. de Graaf, and C. A. G. Nass, *Int. J. Air Water Pollut.* **9**, 343 (1965).
302. D. P. Bridge and M. Corn, *Environ. Res.* **5**, 192 (1972).
302a. W. C. Hinds and M. W. First, *N. Engl. J. Med.* **292**, 844 (1975).
303. M. Corn, *Environ. Lett.* **1**, 29 (1971).

4

The Primary Air Pollutants—Viable Particulates
Their Occurrence, Sources, and Effects

Alvin R. Jacobson and Samuel C. Morris

I. Viable Particulates in the Air

The viable particulates found in air are primarily of three broad types—
pollen, microorganisms, and insects. Table I indicates the size range of
viable particulates (1). Although the pollen of most seed plants is not

Table I   Size Range of Viable Particulates (1)

Particulate	Stokes' diameter (μm)
Viruses	0.015–0.45
Bacteria	0.3–15
Fungi	3–100
Algae	0.5
Protoza	2–10,000
Moss spores	6–30
Fern spores	20–60
Pollen grains (wind-borne)	10–100
Plant fragments, seeds, insects, other microfauna	100+

dispersed in the air, airborne pollen grains are a vital link in the life cycle of many plants. Pollinosis (hay fever) is an unwelcome side effect which severely affects 5–10% of the population of the United States.

The microorganisms in the air include algae, protozoa, fungi, yeasts, molds, rusts, spores, bacteria, and viruses. Although many of these have purposes in the ecological cycle useful to man, others are associated with disease in man, animals, and plants. Insects and insect parts are found in varying numbers in the atmosphere. While insects perform vital functions, they are a nuisance to man at times; they destroy vast quantities of crops, and frequently are the vectors of disease.

While most viable particulates are of natural origin, man provides artificial sources such as sewage treatment and rendering plants. Man's influence is even stronger, however, when he modifies natural sources by his use of the land.

A. Pollen

Pollen grains are the male gametophytes of gymnosperms and angiosperms. Pollen grains are discharged into the atmosphere from weeds, grasses, and trees. Table II lists a number of common plant sources (12). In most localities from 20 to 40 genera will emit pollen that becomes airborne in appreciable quantities (3).

Because wind pollination is very wasteful of pollen, thousands of pollen grains must be liberated. During the ragweed season, daily pollen concentrations over much of the eastern and central United States commonly reach 250–1000 grains/m³ of air (4). In Ann Arbor, Michigan, peak concentrations as high as 4400 grains/m³ have been reported (5).

Table II Common Wind-Pollinated Plants (2)

Common name	Botanical name	Pollen diameter (μm)	Pollen specific gravity
Giant ragweed	*Ambrosia trifida*	19.25	0.52
Burweed marsh elder	*Iva xanthifolia*	19.3	0.79
Short ragweed	*Ambrosia elatior*	20.0	0.55
False ragweed	*Franseria acanthicarpa*	22.0	0.75
Marsh elder	*Iva ciliata*	23.0	0.58
Southern ragweed	*Ambrosia bidentata*	23.0	0.50
Western ragweed	*Ambrosia psilostachya*	26.4	0.57
Cocklebur	*Xanthium commune*	27.0	0.45
Russian thistle	*Salsola pestifer*	23.6	0.90
Palmer's amaranth	*Amaranthus palmeri*	25.8	1.02
Western water hemp	*Acnida tamariscina*	27.5	1.01
Mexican fireweed	*Kochia scoparia*	32.7	0.97
Annual sage	*Artemisia annua*	20.4	1.02
Tall wormwood	*Artemisia caudata*	21.0	1.04
Sagebrush	*Artemisia tridentata*	25.85	1.03
Nettle	*Urtica gracilis*	14.0	0.77
Red sorrel	*Rumex acetosella*	21.45	0.78
Hemp	*Cannabis sativa*	25.0	0.82
English plantain	*Plantago lanceolata*	27.5	0.97
Bluegrass	*Poa pratensis*	28.0	0.90
Bluegrass	*Poa pratensis*	30.0	0.90
Bermuda grass	*Capriola dactylon*	28.5	1.01
Orchard grass	*Dactylis glomerata*	34.0	0.91
Timothy	*Phleum pratense*	34.0	0.90
Rye	*Secale cereale*	49.5	0.98
Corn	*Zea mays*	90.0	1.00
Sycamore	*Platanus occidentalis*	22.22	0.92
Mountain cedar	*Juniperus sabinoides*	22.8	1.08
Hazlenut	*Corylus americana*	23.6	1.09
Birch	*Betula nigra*	24.6	0.94
Alder	*Alnus glutinosa*	26.0	0.97
Ash	*Fraxinus americana*	27.1	0.90
Cottonwood	*Populus virginiana*	30.0	0.79
Elm	*Ulmus americana*	31.2	1.00
Bur oak	*Quercus macrocarpa*	32.3	1.04
Shingle oak	*Quercus imbricaria*	33.1	1.04
Walnut	*Juglans nigra*	35.75	0.93
Beech	*Fagus grandifolia*	44.0	0.94
Hickory	*Carya ovata*	45.0	0.79
Scotch pine	*Pinus sylvestris*	52.0	0.45
Bull pine	*Pinus ponderosa*	60.0	0.45

Airborne pollen is of interest in regard to pollination, geochronology, and the allergic responses it produces in sensitized individuals. Because of the latter effect, airborne pollen fall in the class of substances known as aeroallergens.

While air-transported pollen grains range chiefly between 10 and 50 μm in size, some have been measured as small as 5 μm and as large as 100 μm in diameter. Most grains are quite hygroscopic and therefore vary in weight with humidity. Most airborne tree pollen are shed during spring and early summer, grass pollen during midsummer, and weed pollen during the late summer and fall.

In the United States, more than 90% of pollinosis is caused by pollen from species of *Ambrosia* (ragweed) which belong to the family of flowering plants known as the Compositae, the sunflower family. Within this family only the ragweeds and a few relatives such as the cockleburs and marsh-elders are wind pollinated. Approximately 50 species of ragweed are known, most occurring only in North America. Payne (5) describes the ragweed pollen grain as a lightly flattened sphere, bearing numerous short spines scattered over the surface, and three small germination pores equidistantly spaced about the equator. Within the wall of the pollen grain three air spaces develop between the germination pores.

Various ragweeds are found from Canada to Mexico. The principal species of ragweed found in the United States are the common ragweed, *Ambrosia artemisiifolia*, also called short ragweed, dwarf ragweed, hogweed, bitterweed and Roman wormwood. It grows from 1 to 4 ft tall and has finely divided, lacy, fernlike leaves that wilt during hot, dry afternoons. The next most common is the giant ragweed, *Ambrosia trifida*, which grows 3–12 ft tall and has rather large, rough, three- or five-lobed leaves. Another variety found less extensively is the perennial ragweed, *Ambrosia coronopifolia*, which produces pollen and seeds; once it has established growth, however, it propagates mainly by shoots from narrow underground roots to form large clumps; a diameter of 9 to 12 m or more is not uncommon. Finally, when common ragweed and perennial ragweed grow near each other, a hybrid, *Ambrosia intergradiens*, tends to form. Its fruit is sterile, but since it is perennial it can nevertheless form large populations; and it does produce some allergenic pollen.

The pollen grains of the two principal varieties of ragweed, *A. artemisiifolia* and *A. trifida*, are similar in size and form, ranging from 16.5 to 20 μm in diameter. One ragweed plant can release billions of pollen grains in a season; it may produce several million granules per day.

Most ragweed pollen never become airborne, but rather fall within 1 m of the plant and remain there. The airborne grains may be carried upward by convection currents to heights as great as 12,000 m. Not infrequently there is a higher concentration at an altitude of 1200–1800 m than nearer the surface of the ground. Pollen showers, similar to rainfall, have occurred infrequently in various parts of the world. In absolutely still air, ragweed pollen settles at a rate of about 3 m/minute, but with a slight wind the rate of settling is decreased considerably, owing to the buoyancy of the pollen grains.

Outside North America, except for a few small areas, *Ambrosia* is not found in sufficient amounts to cause concern, and the incidence of hay fever is one-tenth that of the United States (*3*). The major sources of pollinosis in Europe are grass pollen, particularly bluegrass, timothy, orchard grass, and rye (*6*).

Ragweed pollen preserved in peat deposits indicate that the plant has occupied much of the eastern United States since the early Pleistocene (*4*). Ragweeds establish themselves quickly in freshly overturned soil, but tend to be crowded out again after a period of years by other vegetation if the soil is not disturbed. In Michigan, cereal grain fields have been reported to be the primary refuge of ragweed with up to 172,000 ragweed plants per acre. Urban areas support little ragweed, except in new subdivisions where over 50,000 plants per acre have been reported (*4*). Table III indicates the abundance of ragweed found in different land uses.

Table III   Abundance of Ragweed According to Land Use Category (*58*)

Ragweed plants per square mile	Percent of areas in category					
	0	*<0.5*	*0.5–1*	*1–10*	*>10*	*Total*
Cropland						
Corn	8.5	50	33	8.5	0	100
Wheat	0	0	0	33	67	100
Oats	0	0	$33\frac{1}{3}$	$33\frac{1}{3}$	$33\frac{1}{3}$	100
Alfalfa meadow	75	25	0	0	0	100
Pasture	81	0	9.5	0	9.5	100
Grass meadow	91	0	0	9	0	100
Parklands	90	10	0	0	0	100
Woods	100	0	0	0	0	100
Marshes	100	0	0	0	0	100
Roadsides	87	10	1	2	0	100
Residence property	100	0	0	0	0	100
Soybeans	0	50	50	0	0	100
Clover and mixtures	40	0	40	20	0	100
Timothy and mixtures	100	0	0	0	0	100
1- to 3-year abandonment	0	44	14	28	14	100
Summer fallowed fields	50	50	0	0	0	100

Ragweed pollen counts reported by news media have limited usefulness since they indicate only the pollen levels of the previous day. Raynor and Hayes have done extensive work on the development of methods for prediction of pollen levels (7).

Ragweeds can be controlled by cutting or pulling but these methods are laborious, impractical, and expensive. The plants must be pulled or cut immediately prior to the flowering season; otherwise they will pollinate despite the control efforts. Chemical treatment is the recommended method of control. The herbicide should be selective, readily absorbed by the plants, nontoxic or as slightly toxic to humans as possible, and effective in minute concentrations. There are three general classes of these, namely, soil-sterilizing agents, nonselective contact herbicides, and selective herbicides. The selective herbicides may be sprayed on an area to destroy unwanted growth, yet leave desirable vegetation. Several chemicals are available, but the most effective for ragweed control is 2,4-D (2,4-dichlorophenoxyacetic acid). It kills many broad-leaved or dicotyledenous plants, is relatively inexpensive, and is harmless to grasses and more resistant plants in the concentration used. From 200 to 300 gal of a 1000 ppm solution are sufficient to treat an acre of ragweed. Spraying should be done only when there is relatively little or no wind to prevent any damage to adjacent less resistant vegetation.

The eradication of ragweed is not a simple and foolproof procedure. Many difficulties are obvious but most of them can be overcome through a centrally directed and administered program. A great many states, counties, and municipalities have shown considerable interest in promoting ragweed control programs. Several voluntary agencies have established educational and research programs to promote a better understanding of this problem and to encourage the establishment of control programs for the relief of hay fever victims. These programs have met with varied degrees of success dependent on the administrative, scientific, and financial support of the community.

However, there is no assurance that ragweed eradication in one community or govermnental jurisdiction will necessarily lower hay fever incidence, because of sporadic exposure to pollen from uncontrolled areas. Pollen, like any other form of air pollution, are no respecter of governmental boundaries and unless ragweed control is carried out on a regional basis there can be no assurance that the incidence of hay fever will be lower.

Persons sensitive to pollen may obtain a measure of relief by (a) receiving a series of preseasonal injections of pollen extracts in increasing doses to build up tolerance to the offending ragweed pollen, (b) receiving symptomatic relief by the use of antihistamines, (c) avoiding the pollen by effecting a change of climate, i.e., escaping to the mountains or sea-shore, or by taking an ocean voyage, or (d) eliminating the offending substance from the air in the office or home by means of air filtration.

B. Microorganisms—Algae, Protozoa, Fungi, Yeasts, Molds, Rusts, Spores, Bacteria, and Viruses

Microorganisms of almost every type are found in the air, either as individual organisms or attached to some other particulate substances—usually dust. Microorganisms in the atmosphere are the same as those found in or on humans, animals, birds, plants, soil, manure, or decaying vegetation. Bruch has reported the presence of microorganisms at heights up to 28,000 m (8).

1. Algae and Protozoa

Algae are chlorophyll-bearing plants which can be dispersed in the air as cysts or in a vegetative state. Protozoa are primitive animals with no tissues, measuring from 2 μm to several centimeters, usually transported in the air as cysts. Puschkarew (9) estimates that there are about 2.5 viable protozoan cysts per cubic meter of air. Schlichting (10, 12) found 7 viable species of algae and protozoa in the atmosphere in Michigan and 37 species of algae and 7 species of protozoa in Texas. Brown and co-workers (13) cultured over 60 genera of algae found in the air over 14 states (Table IV). Concentrations of from 1 to 200 algae colonies/m³ were reported at a stationary sampling point 25 miles above ground level in Austin, Texas, while up to 3000 algae/m³ were reported in dust clouds. Exposures of petri dishes for 1 minute from a plane at 1100 m over central Texas indicated a heterogeneity of airborne algae at that altitude often equivalent to those similarly exposed in automobiles at ground level. Brown and co-workers concluded that airborne algae primarily originate from the soil, and that cultivated soils, when blown as dusts, yield a greater quantity and diversity of algae than do undisturbed soils. The specific composition of the airborne algae flora is dependent upon proximity to various soil algae populations and meterological conditions. The algae population in blowing dust frequently exceeds that of fungi, and may be a cause of allergy through inhalation of particles.

Besides their possible role as aeroallergens, airborne algae and protozoa are of importance because they tend to concentrate radioisotopes and pesticides and potentially could be carriers of pathogenic bacteria and viruses (14). Algae and protozoa are of economic importance in the fouling of air filters.

Table IV   Algae Recovered from Air and Cultivated in or on "Bold's Basal Medium" (13)

Chlorophyta

Borodinella	*Cylindrocystis*	*Pamellococcus*	*Scenedesmus*
Bracteacoccus	*Dictyochloris*	*Planktosphaeria*	*Spongiochloris*
Chlamydomonas[a]	*Friedmannia*	*Pleurastrum*	*Spongiococcum*
Chlorella[a,b]	*Hormidium*[a]	*Protococcus*-like[a]	*Stichococcus*
Chlorococcum[a,b]	*Hormotilopsis*	*Protosiphon*	*Tetracystis*[a,b]
Chlorosarcina	*Nannochloris*[a,b]	*Psuedoulvella*-like	*Tetraspora*
Chlorosarcinopsis[a,b]	*Neochloris*[b]	*Radiococcus*	*Trebouxia*[b]
Chlorosphaeropsis	*Oocystis*	*Radiosphaera*	*Ulothrix*
Coelastrum	*Ourococcus*	*Roya*	*Westella*
Cosmarium	*Palmella*		

Cyanophyta

Anabaena[b]	*Gloeocapsa*[b]	*Myxosarcina*	*Schizothrix*
Anacystis[b]	*Lyngbya*	*Nostoc*[b]	*Synechococcus*
Arthrospira	*Merismopedia*	*Oscillatoria*[b]	*Scytonema*[b]
Chroococcus-like	*Microcoleus*	*Phormidium*[a,b]	*Tolypothrix*
Frenivella			

Chrysophyta

Hantschia	*Navicula*[a,b]	*Heterococcus*	*Tribonema*
Melovira-like	*Botrydiopsis*[a,b]	*Monocilia*	

[a] Indicates those genera most frequently encountered.
[b] New genus, unpublished.

In addition to their occurrence in outdoor air, algae are common constituents of house dust. Berstein and Safferman (15) reported the presence of viable algae in samples of dust from 41 homes and from commercially derived dust samples. This work has demonstrated that algae cells in the resting stage can survive the dessication caused by the low humidity conditions of the home. Viability was maintained even after samples were stored up to a year (16).

2. Fungi, Yeasts, Molds, Rusts, and Spores

Fungi are practically ubiquitous in soil and dust, and many species possess spores that are adapted to aerial dispersion. Certain fungi, including some pathogens, are known to be very common in the air at certain times. As early as 1904, Klebahn (17) suggested that the outbreaks of rust of wheat and other cereal grains in the different parts of the world might be caused by spores that were carried long distances by the wind. Savulescu (18) found that the principal sources of rust inocula of *Puccinia graminis*, *P. triticina*, and *P. glumorum* (stripe rust) were airborne urediospores from distant surrounding wheat-producing areas. Stakman and his co-workers (*19*, *21*) obtained evidence that vast numbers of urediospores of *P. graminis* are picked up by the wind in southern regions of the United States and dropped in countless billions on remote grain fields. In addition to spores, detached fragments of fungus mycelium are frequently found. Such fragments have been observed in the air over Canada, the Canadian Arctic, England, the Atlantic Ocean, the Pacific Ocean, and the Mediterranean Sea by various investigators. Pady and Gregory (*22*) observed that many of these fragments were viable, suggesting that they may at times be important components of the airborne fungus flora and potentially able to act as allergens.

Some fungi may produce as many as four distinct spore types, and the properties of individual spores vary with external conditions, particularly humidity (*1*).

The largest percentage of airborne fungi occurs below 1500 m altitude and decrease above that (*2*). Bruch (*8*) reports consistent recovery of fungi, primarily *Alternaria* and *Cladosporium*, at altitudes of from 18,000 to 28,000 m, but at concentrations as low as one organism per 280 m^3. Pady (*23*) found concentrations at an elevation of 45 m to be 1800 to 25,000 fungi/m^3 in summer and 170 to 700/m^3 in winter. Pady, Kramer, and Wiley (*24–26*), using silicone slides and nutrient plates exposed in slit samplers, found a great variation in number of airborne fungi from day to day and hour to hour in Kansas. The highest spore count was 95,000/m^3 in July and the highest number of colonies 38,000/m^3 in October. Commonest spores were a one-celled hyaline type. Basidiospores were next, and *Cladosporium* were present in all series, with highest numbers in late summer and fall. Yeast colonies were generally present, but with few exceptions, numbers were low. In most series, spore viability ranged from 30 to 60% but occasionally reached as high as 90%.

Basidiospores, one-celled hyaline-type spores, and yeast displayed diurnal periodicity with distinct daytime peaks. *Cladosporium*, *Alternaria*, and hyphal fragments were very irregular but tended to have daytime peaks with occasional twin peaks. Morrow and co-workers (*27*) made an extensive survey of airborne molds throughout the United States (Table V). Ogden and Lewis (*28*) found that the total number of fungus spores from 35 sampling stations in New York State usually closely approached, and often exceeded, the total number of pollen grains collected. *Cladosporium* (including *Homodendrum*) gave the highest counts, often being more than half the total count. *Alternaria* averaged second in abundance.

Fungi spores may become airborne either through violent or passive liberation. Violent discharge, which may involve turgid living cells, can take place only under reasonably damp conditions, or during the process

Table V   Molds Most Frequently Isolated from 41 Sampling Stations in the United States (27)

Universal dominant genera

Alternaria	*Trichoderma*
Homodendrum	*Fusarium*
Aspergillus	*Helminthosporium*
Penicillium	*Cryptococcus*
Pullularia	*Rhodotorula*
Phoma	

Other dominant genera which occur more or less consistently

Rhizopus	*Gliocladium*
Mucor	*Monilia*
Stemphylium	*Mycogone*
Nigrospora	*Paecilomyces*
Botrytis	*Chaetomium*

of drying. Discharge by controlled bursting of turgid living cells is found especially in Ascomycetes, the largest class of fungi (*29*).

Although spores may be "shot" up to 50 cm, the usual range is 0.5 to 2.0 cm. Even this short distance is often sufficient to project the spores out of the laminar boundary layer of air into the turbulent layers above. Passive liberation of spores involves wind, rain, or insects.

Gregory *et al.* (*30*) found that a large drop of water striking the already wet stomata of *Nectria cinnabarina* growing on a twig broke into 2000 or more droplets, varying in size from 12 to 800 μm, all containing spores. Jarvis (*31*) suggested that the percussion waves associated with the impact of large falling raindrops may blast spores from moldy fruit. Vibration and leaf flutter produced by heavy rain must also be a significant factor in passive liberation of spores (*29*). The action of rain in the liberation of fungi spores into the air is supported by repeated observations of steep increases in air concentrations at the onset of heavy rain. However the effect of rain in washing spores out of the atmosphere must also be considered (*29*).

Smith (*32*) has demonstrated a direct relationship between wind and spore liberation. Airborne spores do not, however, originate to any great extent from soil fungus florax that is blown into the air with dust. Most spores seem to have been propelled directly into the air from their sporophores (*29*). Once liberated into the air, they are carried upward by wind eddies which have high velocities compared with the terminal settling velocities of spores, i.e., 0.5 to 2.0 cm/second. At least 90% of spores from near ground sources are usually deposited onto the ground within 100 m of their source, and spores emitted from weak sources are undetectable after traveling a short distance. Hirst and Hurst (*33*) detected movement extending more than 100 miles with, against, and across the direction of the prevailing westerly winds in Great Britain.

Recently there have been reports of airborne dispersion of microorganisms in indoor air from contaminated air conditioning systems. Banazak *et al.* (*34*) have reported four cases of office workers developing a condition similar to farmer's lung.

3. Bacteria

Bacteria are found almost everywhere. They are present in air, as well as in water, soil, most food and drink, and in the bodies of animals and plants. The soil is a natural habitat for saprophytic bacteria, which live on organic remains, as well as a reservoir for parasitic bacteria, which infect living plants and animals. Saprophytic and parasitic pathogenic and nonpathogenic bacteria may become airborne. While soil and water bodies are the primary source of bacteria in outdoor air, rendering plants (*35*) and sewage treatment plants (*36, 37*) have also been shown to be important localized sources of airborne bacteria. Adams and Spendlove (*37*) demonstrated substantial increases in *Escherichia coli* up to 800 m downwind from trickling filter sewage treatment units as compared with upwind controls. Additional work (*38*) has shown that aerosols generated by trickling filters are generally less than 5 μm in diameter. Wind speeds of 3–5 m/second produced greater aerosol emission than higher or lower wind speeds. Sorber and Guter (*39*) report preliminary work by Goff suggesting that spray irrigation of chlorinated sewage effluent produces biological aerosols at about the same rate as seen from trickling filters. They point out, however, that *E. coli* is a poor indicator organism for airborne measurements since it is killed much more rapidly on dessiccation than some pathogens, specifically *Klebsiella*. While potentially pathogenic bacterial aerosols are formed in sewage treatment processes, no evidence exists to date of disease outbreaks from this source. Nonetheless, the potential aerosol problem has contributed to a possible shift in future practice from spray irrigation to overland flow in which waste water is allowed to trickle over gently sloping ground preventing the formation of aerosols (*40*).

Direct dissemination of bacteria from people is of particular importance in indoor air. Bacteria-laden particles are liberated in the air from human beings by two principal means. Primarily, they arise from activities involving the respiratory tract, such as sneezing and coughing; and from movements which shed bacteria-bearing particles from the skin or wound dressings where they may have accumulated from contact with the skin, orifices, or wounds. Secondarily, they arise from the redissemination of organism-bearing particulates which have accumulated in the dust of rooms, streets, sidewalks, etc. The bacterial content of air comparatively free of dust is exceedingly small.

Darlow and Bale (41) showed that the flushing of commonly used water closets can produce a bacterial aerosol of considerable concentration locally, of which a large number of particles are within the respirable size range. Air collected on the summit of Mt. Blanc contained only from 4 to 11 bacteria per cubic meter. Proctor (42) showed that bacteria, particularly of the spore-forming variety, could be recovered from the air up to an elevation of 6000 m. Pincus and Stern (43) examined over 2000 samples of indoor and outdoor air at various locations in New York City and found 100 to 1000 bacteria per m³. While marine air usually contains fewer microorganisms than that of continental or terrestrial origin, it is by no means germfree. Pure air collected in midocean yielded only four or five bacteria in 10 m³. Bacteria are transferred from the sea water into the air along with droplets of water.

Blanchard and Syzdek (44) demonstrated that bacteria can be carried into the air by drops from bursting bubbles and that concentration of bacteria in the drops can far exceed that in the water from which the bubbles broke.

Unlike the air over terrestrial areas where molds predominate, preponderance of bacteria over other kinds of microorganisms is usually characteristic of marine air. Zobell and Matthews (45) found appreciable numbers of marine bacteria in the air 6 m above the water level at all stations occupied by the Scripps Institution of Oceanography at distances ranging from 5 to 130 miles from land. Pady and Kelly (46) sampled the atmosphere for bacteria throughout two flights from Montreal to London, using quantitative and qualitative methods. Bacteria varied from 0.0 to 18/m³ in polar air and 0.0 to 32/m³ in tropical air.

4. Viruses

Viruses have not been recovered in significant numbers from outdoor air. They have been isolated from indoor air, and there is no doubt that indoor airborne transmission of virus particles is an important factor in the spread of respiratory viral infection. There is some epidemiological evidence which indicates that viruses can be carried in outdoor air over distances of several miles.

Gerone and co-workers (47) conducted extensive studies with Coxsackievirus A, type 21. They recovered viruses from both artificially produced aerosols and from sneezes and coughs of infected volunteers, demonstrating that virus can be aerosolized in the process of sneezing or coughing, in some instances in sufficient quantities to account for infection of susceptible individuals in the environment. Using large volume samplers, they then recovered viruses from rooms inhabited by infected volunteers. Up to 185 $TCID_{50}$ (tissue culture infectious dose-50) were recovered from 12-minute samples of 120 m^3 of air.

The mode of transmission of viruses in air appears to be droplets (over short distances), droplet nuclei, or dust particles infected by contact with virus-bearing droplets. Smallpox virus can survive desiccation in dried crusts of skin. Although most smallpox transmission occurs by fairly close contact, instances have occurred which suggested airborne spread over some distance (48).

Investigations of the spread of foot-and-mouth disease during the 1967–1968 epidemic in Great Britain provide strong epidemiological evidence of airborne spread of the virus (49, 50). Other investigators have reported similar evidence of the airborne spread of Q fever virus (51, 52). Sewage works are potentially important sources of airborne enteric viruses. There is no evidence of disease outbreaks from this source, however, nor is there evidence on the survival of enteric virus in air.

The human population supports about 100 antigenically different varieties of respiratory virus which appear to be in a comparatively rapid state of evolution (53). As the population grows and becomes more concentrated in urban areas, the problem of airborne spread of virus infection is becoming increasingly important.

C. Insects

Approximately 82,500 species of insects are known to exist in the United States. Entomologists have estimated that approximately 10,000 species are injurious to man in one way or another, i.e., harmfully affect human beings, crops, domestic animals, buildings, furniture, clothing, and other possessions of man. Insects are carriers of diseases of man, plants, and animals in many complex ways, but also can be beneficial. Many are indispensable as pollenizers of plants; predators and parasites of other insects; and producers of foodstuffs, e.g., bees. The number of individual insects in any given area is very difficult to estimate. The population is not only dependent upon such factors as the soil and the plants, but will vary from season to season, and from one minute to the next.

Dispersal of insects in the air falls into two classes: active and passive. Actively flying insects may swarm or migrate in large numbers. Tremendous swarms of locust, such as those sent as a plague on the children of Egypt as described in the Old Testament of the Bible, are reported in many parts of the world. The most spectacular examples in the United States were the migrating swarms of Rocky Mountain grasshoppers (*Melanoplus sprelus*) in the Great Plains in the 1870's. The locusts left fields as barren as if they had been burned over. One observer in Nebraska recorded that one of the invading swarms of locust averaged a half mile in height and was 100 miles wide and 300 miles long. With an estimate of 27 locusts/yd^3, he estimated that there were almost 28 million/mile^3. Based on the fact that they were migrating at a speed of 5 mph and the migration lasted for at least 6 hours, he calculated that more than 124 billion locusts were on the move in the one direction. The senior author has been a witness to a similar, though not as spectacular, experience in the early 1930's during the drought and dust storms of the Dakotas. Other insects which may be present in the atmosphere in considerable numbers are mayflies, flying ants, termites, and butterflies. Gable and Baker (*54*) recorded a migration of snout butterflies, *Libytheana bachmanii*, which were so numerous that an average of about 1,250,000 of them per minute flew in a front 250 miles wide. At the main observation point the migration continued at the same level of intensity for 18 days.

Insects that are passively wind-borne and do not fly actively include aphids, which appear to be constantly encircling the globe (*55*), and gypsy moth larvae. The latter can be carried by wind up to 50 km and have been recovered at altitudes near 600 m (*56*). Use of high resolution radar has shown that the vertical distribution of insects is influenced by atmospheric structure. Peak concentrations were observed by Richter and co-workers above and below an inversion layer (*57*).

Insect parts may also be collected in air samples. Ogden and Lewis (*28*) found that scales of insects were common and often abundant in samples throughout New York State. This category was comprised of tiny roundish scales, long narrow scales (hairs), portions of wings, legs, and even whole insects (usually mites). Five sampling stations averaged several hundred. Only 7 of the 35 sampling stations averaged less than 100 scales. Incidentally, hairs of animals, such as dogs, cats, horses, and bats have been also found occasionally, but were never common.

II. Factors Influencing the Numbers of Viable Particulates in the Air

A. Pollen

Botanists, allergists, meteorologists, and others have made careful observations of the various climatic and meteorological factors involved in the emission, dispersion, and deposition of ragweed pollen. These observations have been made on plants in their natural state and plants grown artificially in greenhouses with the pollen released during the prepollen season.

Studies at the University of Michigan (5, 58) and at Brookhaven National Laboratory (59) have identified the role of factors such as temperature, humidity, wind speed, and turbulence in the release of airborne pollen. Pollen emission is extremely low during the night, begins to increase near sunrise, and reaches a peak a few hours later (Fig. 1). At this time of day, several meteorological factors are changing rapidly;

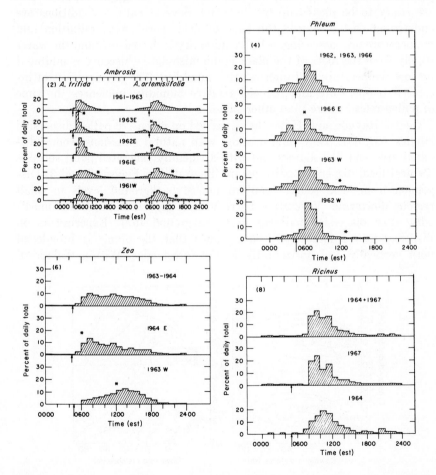

Figure 1. Diurnal emission patterns from *Ambrosia, Phleum, Zea,* and *Ricinus* pollen sources. Arrows mark mean time of sunrise and asterisks time of sample change. Values are seasonal means. (Courtesy Brookhaven National Laboratory.)

solar radiation, ambient air temperature, relative humidity, and wind speed. Further observations of pollen emission on cloudy, humid days, and on days when there were showers followed by rapid drying, confirm the hypothesis that the dehiscence of florets ready to discharge their pollen is triggered by a drop in relative humidity. Early in the morning between 4:00 and 6:00 am., a floret which is ready to release its pollen, i.e., a dehiscent floret, swells slightly and forcibly pushes out the pollen granules. The grains fall in clusters, held together by a substance called tapetal fluid, which makes some of them adhere to the leaves of the parent or neighboring plant. As this fluid dries, the grains separate and are ready to be picked up by the wind. Several natural conditions are undergoing change during the time of dehiscence: solar radiation and temperature are increasing, relative humidity is decreasing, and the water supply from the roots of the plant is diminishing. Dehiscence is inhibited by low temperature and high relative humidity. The drying effect of increased temperature and reduced relative humidity influences the time of pollen release more than other factors.

Ogden, Raynor, and Hayes have carried on extensive pollen dispersion studies (*60, 68*). They have worked with natural emissions from cultivated area and line sources and with artificially generated point sources. Typical local dispersion patterns are illustrated in Figures 2 and 3. Dispersion from local sources can far exceed background levels in small regions downwind. Vegetative barriers such as arborvitae hedge markedly alter dispersion patterns near the ground (*67*). Experiments on dispersal into a forest (*68*) have shown that the plume is broadened both vertically and horizontally at the forest edge. Within the forest,

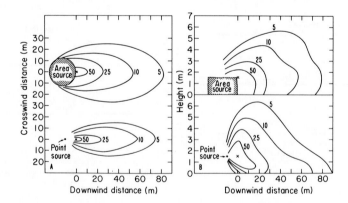

Figure 2. Typical dispersion patterns from area and point sources of ragweed pollen. Concentrations are expressed in percentage of maximum concentration at point "×." (Courtesy Brookhaven National Laboratory.)

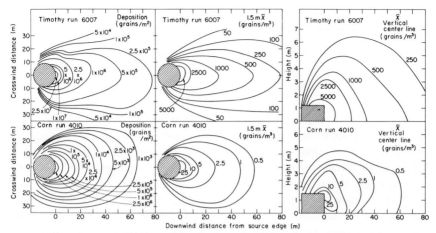

Figure 3. Typical dispersion patterns from timothy and corn pollen. (Courtesy Brookhaven National Laboratory.)

concentration decreases at a faster rate than in the open. Most pollen loss is to the forest foliage rather than to the ground. Transport and dispersion of pollen over distances of about 100 km were studied from aircraft (66). Figure 4 shows results of measurements made in the New York City region. Ground-level concentrations due to local sources which often exceed 100 grains per m³ are not shown in the figure. Large quantities of pollen are transported over these distances, often in large discrete clouds.

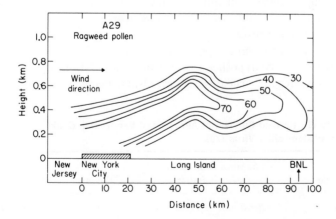

Figure 4. Aircraft measurements showing transport of ragweed pollen over Long Island from source areas west of New York City. Regions of high ground-level concentration not shown. (Courtesy Brookhaven National Laboratory.)

B. Microorganisms

The numbers of microorganisms in the air vary greatly under the influence of the complex dynamics of weather conditions and other factors which influence both their introduction to and retention in the atmosphere.

Wright et al. (69) reported that the concentration of microorganisms at an urban location was independent of wind direction or speed and showed little relation to temperature or relative humidity, all factors that, based on laboratory studies, should affect concentration (70–72). Nonetheless, the effect of temperature and humidity must still be considered of importance, particularly in indoor air. The well-known bactericidal and viricidal effects of ultraviolet light (73, 74) make solar radiation another important consideration in the maintenance of viability in airborne microorganisms (75). Tests on survival of Escherichia coli in aerosols exposed to sunlight indicate that survival was longer in large particles than in small ones (76).

Environmental conditions, particularly sunlight and relative humidity, have been shown by several experimenters to affect the viability of viruses in the air (77–80). While sunlight is definitely viricidal, humidity seems to have mixed effects. Berendt and Dorsey (79) exposed aerosolized Venezuelan equine encephalomyelitis virus to simulated solar radiation and to two levels of relative humidity for 1-hour test runs. The results are tabulated in Table VI. Working with adenovirus 12, Davis and co-workers (72) obtained the results shown in Table VII. Hemmes (80) reports that influenza and influenza like viruses appear to be less stable at higher relative humidities, while the converse seems to hold for rhinoviruses and adenoviruses.

The presence of chemical pollutants in the air has an effect on viable organisms. An early hypothesis concerning the relationship of man-made air pollution with airborne disease organisms (made even before the germ

Table VI   Percent Survival of Aerosolized Venezuelan Equine Encephalomyelitis Virus (79)

Relative humidity, percent	Exposed to simulated solar radiation		Not exposed to radiation	
	After 5 min	After 1 hr	After 5 min	After 1 hr
30	25.5	0.02	113.5	90.
60	5.8	0.006	55.5	78.

Table VII   Percent Recovery of Aerosolized Adenovirus 12 (72)

Relative humidity, percent	Virus recovered from aerosol, percent	Virus recovered from lungs of exposed newborn Syrian hamsters,[a] percent
89 ± 4	4.6	12.6
51 ± 2	1.05	8.0
32 ± 1	0.02	0.82

[a] Percent of maximum theoretical inhaled dose after 20 minute exposure.

theory of disease) was presented by Dr. William H. Denny. In the same 1826 volume of the Pittsburgh *Directory* where one observer commented "The atmosphere is darkened with a sulphureous canopy which nearly conceals the place from view," Dr. Denny speaks of the "antimiasmatic" quality of the smoke. Because of the smoke's "sulphurous and antiseptic qualities, no putrid disease has ever been known to spread in the place," he wrote (*81*). The bactericidal qualities of Pittsburgh's air were later demonstrated through more scientific investigations (*82*).

Hood (*83*) and Drvett (*84*) demonstrated the toxic effect of chemical air pollutants on viable aerosols. The bactericidal agent is believed to be a product of photochemical reactions. Length of viability is increased with larger particles, but humidity does not seem to be a factor.

Speculating on the possible implications of interactions between chemical air pollutants and viable particulates, Lightart *et al.* (*85*) suggest damage to necessary saprophytic bacteria, undesirable genetic effects, and altered allergenic potential of airborne bacteria.

III. Special Considerations in Sampling for Viable Particulates

In addition to all of the physical considerations involved in sampling nonviable particulates, the sampling of living particles often imposes the additional requirement that viability be maintained. In the case of microorganisms it is only after growth in a culture that identification and measurement can be made. The requirement for culturing is thus important. Provision must often be made to split the sample for growth by multiple culturing techniques.

The original pollen samplers were of the sedimentation type. Pollen and large spores (over 6 μm) can best be sampled with rotating impactors (Fig. 5), while smaller particles are best sampled with suction devices

Figure 5. Swing-shield intermittent rotoslide sampler. Curved plates prevent collection on slides when not sampling. On rotation, centrifugal force draws plates away exposing slides. (Courtesy Brookhaven National Laboratory.)

such as the Hirst trap (Fig. 6). Ogden, Raynor, and their co-workers have done considerable work in the development and testing of pollen samplers (*86–88*). Microorganisms are best sampled with Anderson-type

Figure 6. Burkard recording volumetric spore trap. Similar to the Hirst Trap in design. Particles are impacted on an adhesive coated rotating drum. Vacuum pump and drive motor are contained in the device. (Courtesy Burkard Manufacturing Co. Ltd.)

cascade impactors or membrane filters. The duration and rate of sampling is important, since too great a buildup of organisms on a filter may result in a loss of viability. The airflow past the collected particles can have a dessicating effect.

Sampling of airborne insects is divided into two categories: random trapping and selective trapping. Random techniques include moving nets, sticky barriers, and suction traps. Selective trapping involves luring insects to the traps, usually by olfactory or visual stimulation (*89*). Location and height of sampling stations are very important. Consideration must be given to the source of the insects and to their manner of dispersal. For example, an insect trap near a swamp might yield biased results.

Lack of quantification of numbers and concentrations of viable particulates has been due to inadequacies in sampling devices and techniques. Improvements in sampling and in the manner of handling the collected samples should result in considerable advances in the overall understanding of airborne viable particulates.

IV. Effects on Man, Animals, and Plants

A. Aeroallergens

Plants physiologically cannot be affected by allergens, although the presence of unwanted pollens in the air may present difficulties in agriculture due to undesired cross pollination. Animal response to aeroallergens has been induced in experimental studies (90).

The National Health Survey reported over 14 million asthma or hay fever sufferers in the United States in 1964 (91). Ogden estimates that as much as 10% of the United States population is severely affected by hay fever (3). While most victims have an uncomplicated type of hay fever in which the symptoms disappear at the termination of the pollen season, some develop sinusitis, catarrh, bronchitis, bronchial asthma, and forms of dermatitis. It is believed that 5–10% of cases of untreated hay fever develop into bronchial asthma (92).

While there are many indications that asthma and hay fever have a more complex etiology than mere exposure to airborne pollen, demonstration that foreign students with no history of the disease develop allergies after 3–5 years in ragweed areas in the United States, and with the same frequency as natives of the United States shows the important role of airborne pollen (93).

B. Microorganisms

Airborne infection and contact infection are the two primary modes of transmission of respiratory diseases. The original sources of human infections are mainly other human beings, and less commonly, animals. Water, soil, insects, and animals may act as temporary reservoirs of the infection.

While the original demonstration of airborne microorganisms goes back to Pasteur, Wells (73, 74) sprayed a variety of bacteria and viruses into a closed chamber and demonstrated that they remain in viable and infectious form for hours and even days. The ability of airborne microorganisms to cause respiratory disease depends largely on particle size. Classic studies were carried out by Wells *et al.* (*94*) and Hatch (*95*).

Airborne spread of bacterial infection has been well established in hospitals, laboratories, ships, and barracks. Dilution in outdoor air seems adequate to make the spread of bacterial disease by that medium generally inconsequential, although Larson (*96*) suggests that much of the disease in developing countries is transmitted via the airborne route. The airborne transmission of virus disease has been more difficult to demonstrate epidemiologically. The airborne route, particularly indoor air, is believed to be of importance in the transmission of influenza and the common cold.

Because of the increasing numbers of people who spend an increasing amount of time in indoor environments with no direct ventilation with outdoor air, the airborne transmission of respiratory disease is of increasing interest. In discussing the need for disinfection of indoor air, Dowling (*97*) states, "Air in a crowded room may be as contaminated as water in the average farm pond. But just as our ancestors paid no attention to the water so long as it looked clear, so the public pays no attention to the air when it seems pure."

Unlike pathogenic bacteria and virus, disease-producing fungi are found in outdoor air in great numbers in certain areas. Coccidioidomycosis, for example, was known as San Joaquin Valley Fever because of outbreaks in that part of California. The fungus is generally endemic throughout Arizona, California, New Mexico, Nevada, Texas, and Utah. In the parts of the endemic area with the highest prevalence nearly 100% of the population is infected, half of them severely enough to cause temporary incapacity (*98*).

Coccidioidomycosis is a systemic mycosis beginning as a respiratory infection, but it may become a progressive and highly fatal granulamatous disease characterized by lung lesions and single or aggregated abscesses throughout the body. The fungus apparently propagates in soil, especially in and around rodent burrows, in regions where temperature and moisture requirements are satisfactory, and infects man, cattle, dogs, horses, burros, sheep, swine, and wild rodents. The source of infection is soil and spore-laden dust, and the mode of transmission is through inhalation of these spores from dust, soil, and dry vegetation. It has been estimated that 35,000 new cases of coccidioidomycosis occur annually in California alone (*98*).

The soil supports numerous fungus organisms, mostly harmless saprophytes, but some of them are pathogenic to man. While apparently occurring naturally in soils, some are closely associated with soil contaminated with droppings from birds and bats. These include *Histoplasma capsulatum* and *Cryptococcus neoformans*.

Histoplasmosis is a systemic mycosis of varying severity, with the primary lesions usually, but not always, in the lungs. Infection is fairly common, sometimes to the extent of 80% of a population, in wide areas of the Americas. Loosli (*93*) estimated that as of 1955, 30 million people in the United States had experienced some form of histoplasmosis infection. Furcolon (*100*) estimated that 500,000 new infections of histoplasmosis occur in the United States each year. The clinical disease, however, is uncommon; and the chronic progressive cavitary form most uncommon. The reservoir and source of infection is the soil around old chicken houses, caves, starling roosts, houses sheltering the common brown bat, and other soils high in the organic content in which the fungus grows. The inhalation of airborne spores on dust is the principal mode of transmission of the pulmonary infection. Dodge *et al.* (*101*) found that the existence of hyperendemic sensitivity to histoplasmin in Milan, Michigan school children was one result of the fortuitous sharing of a schoolyard by large numbers of children and roosting starlings.

In urban areas, outbreaks of histoplasmosis have been associated with the razing of old buildings contaminated with bird or bat droppings and with open spaces within urban areas. In Jacksonville, Texas, where 53 cases of histoplasmosis were reported in a 3-year period, 87% of the school children tested positive for a histoplasmin skin test as opposed to 13% in a nearby community. Although several of the clinical cases were linked directly to a blackbird starling roost, the striking prevalence of positive skin tests in children suggests more widespread sources of infection (*102*).

Cryptococcosis is a mycosis most frequently recognized in the United States as a chronic, usually fatal meningitis (*Torula* meningitis). The reservoir is saprophytic growth in the external environment. The mode of transmission is presumably by inhalation of spore-laden dust. Emmons (*103, 106*) found virulent strains of *Cryptococcus neoformans* in 63 of 91 specimens of dried and weathered pigeon manure from old nests or under roosting sites, and from fresh pigeon droppings and soil from a city park in Washington, D.C., where pigeons are fed by visitors. While cryptococcus is a rare disease, the fungus is one of the most frequently found molds in the atmosphere (*107*). Recent introduction of specific serological tests may uncover a broader disease prevalence (*108*).

Aspergillosis is a chronic pulmonary mycosis in which compost piles undergoing fermentation and decay are prominent reservoirs and sources of infection. Fungi are found also in hay stored when damp, decaying vegetation, and in cereal grains stored under conditions which permit "heating." Pulmonary infection results from inhalation of airborne spores.

Systemic blastomycosis is a chronic granulomatous mycosis, primarily of the lungs. The soil is the reservoir of infection, and the immediate source of infection is probably spore-laden dust.

Nocardiosis is a chronic mycotic disease often initiated in the lungs, with hematogenous spread to produce peritonitis, meningitis, brain abscess, and other pyogenic lesions leading to a high fatality rate. The reservoir of the fungus is the soil, and the mode of transmission of the pulmonary infections presumably occur through inhalation of organisms suspended in dust. Nocardiosis may also be transmitted by direct contact with contaminated soil through minor traumatic wounds and abrasions.

REFERENCES

1. R. L. Edmonds, *in* "Ecological Systems Approaches to Aerobiology" (W.S. Benninghoff and R. L. Edmonds, eds.), pp. 6–11. Univ. of Michigan, Ann Arbor, Michigan, 1972.
2. H. Finkelstein, "Air Pollution Aspects of Aeroallergens." Litton Systems, Inc., Bethesda, Maryland, 1969.
3. E. C. Ogden, *in* "Aerobiology Objectives in Atmospheric Monitoring" (W. S. Benninghoff and R. L. Edmonds, eds.), p. 21. Univ. of Michigan, Ann Arbor, Michigan, 1971.
4. R. L. Edmonds, *in* "Ecological Systems Approaches to Aerobiology" (W. S. Benninghoff and R. L. Edmonds, eds.), pp. 44–50. Univ. of Michigan, Ann Arbor, Michigan, 1972.
5. E. W. Hewson, W. W. Payne, A. L. Cole, J. B. Harrington, and W. R. Solomon, *J. Air Pollut. Contr. Ass.* **17,** 651 (1967).
6. V. Pitila, L. Noro, and A. Laamanen, *Acta Allergol.* **18,** 113 (1963).
7. G. S. Raynor and J. V. Hayes, *Ann. Allergy* **28,** 580 (1970).
8. E. C. Bruch, *in* "Airborne Microbes" (P. H. Gregory and J. L. Monteith, eds.), p. 345. Soc. Gen. Microbiol., Cambridge, England, 1967.
9. B. Puschkarew, *Arch. Protistenk.* **28,** 323 (1913).
10. H. E. Schlichting, Jr., Ph.D. thesis. Michigan State University, East Lansing, Michigan, 1958.
11. H. E. Schlichting, Jr., *Lloydia* **27,** 2 (1961).
12. H. E. Schlichting, Jr., *Lloydia* **27,** 1 (1964).
13. R. M. Brown, D. Larson, and H. C. Bold, *Science* **143,** 583 (1964).
14. H. E. Schlichting, Jr., *in* "Aerobiology Objectives in Atmospheric Monitoring" (W. S. Benninghoff and R. L. Edmonds, eds.), p. 16. Univ. of Michigan, Ann Arbor, Michigan, 1971.
15. I. L. Bernstein and R. S. Safferman, *Nature (London)* **227,** 851 (1970).
16. I. L. Bernstein and R. S. Safferman, *in* "Ecological Systems Approaches to Aerobiology" (W. S. Benninghoff and R. L. Edmonds, eds.), p. 106. Univ. of Michigan, Ann Arbor, Michigan, 1972.

17. H. Klebahn, "Die Wirtswechselnden Rpstpolze." Borntraeger, Berlin, Germany, 1904.
18. T. Savulescu, *Vestn. Cesk. Akad. Zemed.* **14,** 328 (1938); *Rev. Appl. Mycol.* **17,** 510 (1938).
19. E. C. Stakman, *Proc. Pac. Sci. Congr., 5th, 1933* p. 3177 (1934).
20. E. C. Stakman and R. C. Russell, *Phytopathology* **28,** 20 (1938).
21. E. C. Stakman, W. L. Popham, and A. C. Cassell, *Amer. J. Bot.* **27,** 90 (1940).
22. S. M. Pady and P. H. Gregory, *Trans. Brit. Mycol. Soc.* **46,** 4 (1963).
23. S. M. Pady, *Mycologia* **49,** 399 (1957).
24. S. M. Pady, C. L. Kramer, and B. J. Wiley, *Mycologia* **54,** 168 (1962).
25. C. L. Kramer, S. M. Pady, and B. J. Wiley, *Mycologia* **55,** 380 (1963).
26. C. L. Kramer, S. M. Pady, and B. J. Wiley, *Trans. Kans. Acad. Sci.* **67,** 3 (1964).
27. M. B. Morrow, G. M. Myer, and H. E. Prince, *Ann. Allergy* **22,** 575 (1964).
28. E. C. Ogden and D. M. Lewis, "Airborne Pollen and Fungus Spores of New York State," Bull. No. 378. New York State Museum and Science Service, Albany, New York, 1960.
29. C. T. Ingold, *in* "Airborne Microbes," p. 102. Soc. Gen. Microbiol., Cambridge, England, 1967.
30. P. H. Gregory, E. J. Guthrie, and M. E. Bunce, *J. Gen. Microbiol.* **20,** 328 (1959).
31. W. T. Jarvis, *Trans. Brit. Mycol. Soc.* **45,** 549 (1962).
32. R. S. Smith, *Trans. Brit. Mycol. Soc.* **49,** 33 (1962).
33. J. M. Hirst and G. W. Hurst, *in* "Airborne Microbes" (P. H. Gregory and J. L. Monteith, eds.), p. 307. Soc. Gen. Microbiol., Cambridge, England, 1967.
34. E. F. Banazak, J. N. Fink, and W. H. Theiede, *N. Engl. J. Med.* **283,** 271 (1970).
35. J. C. Spendlove, *Publ. Health Rep.* **72,** 176 (1957).
36. C. R. Albrecht, unpublished thesis. University of Florida, Gainesville, Florida, 1958.
37. A. P. Adams and J. C. Spendlove, *Science* **169,** 1218 (1970).
38. G. D. Goff, J. C. Spendlove, A. P. Adams, and P. S. Nicholes, *HSMHA Health Rep.* **88,** 640 (1973).
39. C. A. Sorber and K. J. Guter, *Amer. J. Pub. Health* **65,** 47 (1975).
40. R. E. Thomas, K. Jackson, and L. Penrod, "Feasibility of Overland Flow for Treatment of Raw Domestic Wastewater," United States Environmental Protection Agency, No. 660/2-74-087. USEPA, Corvallis, Oregon, 1974.
41. H. M. Darlow and W. R. Bale, *Lancet* **1,** 1196 (1959).
42. B. E. Proctor, *Proc. Amer. Acad. Arts Sci.* **69,** 315 (1943).
43. S. Pincus and A. C. Stern, *Amer. J. Pub. Health* **27,** 321 (1937).
44. D. C. Blanchard and L. Syzdek, *Science* **170,** 626 (1970).
45. C. E. Zobell and H. M. Matthews, *Proc. Nat. Acad Sci. U.S.* **22,** 567 (1936).
46. S. M. Pady and C. D. Kelley, *Can. J. Bot.* **32,** 202 (1954).
47. P. J. Gerone, R. B. Couch, and G. V. Keefer, *Bacteriol. Rev.* **30,** 576 (1966).
48. C. H. Stuart-Harris, *in* "Virus and Rickettsial Diseases of Man" (S. Bedson, ed.), p. 55. Arnold, London, 1967.
49. R. J. Henderson, *J. Hyg.* **67,** 21 (1969).
50. M. E. Hugh-Jones and P. B. Wright, *J. Hyg.* **68,** 253 (1970).
51. A. D. Langmuir, *Bacteriol. Rev.* **25,** 173 (1961).
52. H. H. Welsh, E. H. Lennette, R. R. Abinanti, and J. R. Winn, *Ann. N.Y. Acad. Sci.* **70,** 528 (1958).

53. C. A. Mims, *in* "Aerobiology" (I. H. Silver, ed.), p. 241. Academic Press, New York, New York, 1970.
54. C. H. Gable and W. A. Baker, *Can. Entomol.* **54**, 265 (1922).
55. D. O. Wolfenberger, *in* "Aerobiology Objectives in Atmospheric Monitoring" (W. S. Benninghoff and R. L. Edwards, eds.), p. 253. Univ. of Michigan Press, Ann Arbor, Michigan, 1971.
56. M. L. McManus, *in* "Ecological Systems Approaches to Aerobiology" (W. S. Benninghoff and R. L. Edwards, eds.), p. 76. Univ. of Michigan, Ann Arbor, Michigan, 1972.
57. J. H. Richter, D. R. Jensen, V. R. Noonkester, J. B., Kreasky, M. W. Stimmann, and W. W. Wolf, *Science* **180**, 1176 (1973).
58. J. M. Sheldon and E. W. Hewson, "Atmospheric Pollution by Aeroallergens," Progr. Rep. 1-5. University of Michigan Engineering Research Institute, Ann Arbor, Michigan, 1957–1967.
59. E. C. Ogden, J. V. Hayes, and G. S. Raynor, *Amer. J. Bot.* **56**, 16 (1969).
60. G. S. Raynor, J. V. Hayes, and E. C. Ogden, *Arch. Environ. Health* **19**, 92 (1969).
61. G. S. Raynor, E. C. Ogden, and J. V. Hayes, *J. Allergy* **41**, 217 (1968).
62. G. S. Raynor, E. C. Ogden, and J. V. Hayes, *J. Appl. Meteorol.* **9**, 885 (1970).
63. G. S. Raynor, E. C. Ogden, and J. V. Hayes, *Agr. Meteorol.* **9**, 347 (1971/1972).
64. G. S. Raynor, E. C. Ogden, and J. V. Hayes, *Agron. J.* **64**, 420 (1972).
65. G. S. Raynor, E. C. Ogden, and J. V. Hayes, *Agr. Meteorol.* **11**, 177 (1973).
66. G. S. Raynor, J. V. Hayes, and E. C. Ogden, *J. Appl. Meteorol.* **13**, 87 (1974).
67. G. S. Raynor, E. C. Ogden, and J. V. Hayes, *Agr. Meteorol.* **13**, 181 (1974).
68. G. S. Raynor, J. V. Hayes, and E. C. Ogden, *Boundary Layer Meteorol.* **7**, 429 (1974).
69. T. J. Wright, V. W. Greene, and H. J. Paulus, *J. Air Pollut. Contr. Ass.* **19**, 337 (1969).
70. C. S. Cox, *Appl. Microbiol.* **21**, 482 (1971).
71. R. Ehrlich, S. Miller, and R. L. Walker, *Appl. Microbiol.* **20**, 884 (1970).
72. G. W. Davis, R. A. Griesemer, J. A. Shadduck, and R. L. Farrell, *Appl. Microbiol.* **21**, 676 (1971).
73. W. F. Wells, *Amer. J. Hyg.* **20**, 611 (1934).
74. W. F. Wells, H. L. Radcliffe, and C. Crumb, *Amer. J. Hyg.* **47**, 11 (1948).
75. R. F. Bereualt and E. L. Dorsey, *Appl. Microbiol.* **21**, 447 (1971).
76. A. M. Hood, *J. Hyg.* **69**, 607 (1971).
77. K. H. Kindgon, *Amer. Rev. Resp. Dis.* **81**, 504 (1960).
78. J. C. DeJong, *in* "Aerobiology" (I. H. Silver, ed.), p. 210. Academic Press, New York, New York, 1970.
79. R. F. Berendt and E. L. Dorsey, *Appl. Microbiol.* **21**, 447 (1971).
80. J. H. Hemmes, K. C. Winkler, and S. M. Kool, *Nature (London)* **188**, 430 (1960).
81. J. Duffy, *Allegheny County Pub. Health Bull.* **7**, 2 (1963).
82. W. I. Holman, "The Bacteriology of Soot," Smoke Invest. Bull. No. 9. Mellon Institute of Industrial Research, Pittsburgh, Pennsylvania, 1914.
83. A. M. Hood, *J. Hyg.* **69**, 607 (1971).
84. H. A. Drvett, *in* "Aerobiology" (I. H. Silver, ed.), p. 212. Academic Press, New York, New York, 1970.
85. B. Lighthart, V. E. Hiatt, and A. T. Rossano, *J. Air Pollution Control Ass.* **21**, 639 (1971).

86. E. C. Ogden and G. S. Raynor, *J. Allergy* **40**, 1 (1967).
87. G. S. Raynor, *Atmos. Environ.* **6**, 191 (1972).
88. J. V. Hayes, *Ann. Allergy* **27**, 575 (1969).
89. R. I. Gara, *in* "Ecological Systems Approaches to Aerobiology" (W. S. Benninghoff and R. L. Edmonds, eds.), p. 109. Univ. of Michigan, Ann Arbor, Michigan, 1972.
90. R. Patterson, J. J. Pruzansky, and W. W. Y. Chang, *J. Immunol.* **90**, 35 (1963).
91. U.S. Public Health Service, "Vital and Health Statistics from the National Health Survey," Publ. No. 1000, Ser. 10, No. 39. U.S. Department of Health, Education, and Welfare, Washington, D.C., 1967.
92. L. Brodler, P. B. Barlow, and R. J. Horton, *J. Allergy* **33**, 524 (1962).
93. C. J. Maternonski and K. P. Mathews, *J. Allergy* **33**, 130 (1962).
94. W. F. Wells, H. L. Ratcliffe, and C. Crumb, *Amer. J. Hyg.* **47**, 11, (1948).
95. T. F. Hatch, *Bacteriol. Rev.* **25**, 237 (1961).
96. E. Larson, *in* "Ecological Systems Approaches to Aerobiology" (W. S. Benninghoff and R. L. Edmonds, eds.), p. 35. Univ. of Michigan, Ann Arbor, Michigan, 1972.
97. H. F. Dowling, *Bacteriol. Rev.* **30**, 485 (1966).
98. L. Ajello, *in* "Proceedings on the International Symposium on Mycoses," p. 197. Pan Amer. Health Organ., Washington, D.C., 1970.
99. C. G. Loosli, *J. Chronic Dis.* **5**, 473 (1957).
100. M. L. Furcolon, *Arch. Environ. Health* **10**, 4 (1955).
101. H. J. Dodge, L. Ajello, and O.K. Engelke, *Amer. J. Pub Health* **55**, 8 (1965).
102. M. S. Dickerson, J. Goldfeder, and H. Davenport, *Morbidity Mortality Weekly Rep.* **21**, 231 (1972).
103. C. W. Emmons, *Pub. Health Rep.* **75**, 4 (1960).
104. C. W. Emmons, *Pub. Health Rep.* **72**, 11 (1957).
105. C. W. Emmons, *Pub. Health Rep.* **73**, 7 (1958).
106. C. W. Emmons, *Pub. Health Rep.* **76**, 7 (1961).
107. M. B. Morrow, G. H. Mayer, and H. E. Prince, *J. Infec. Dis.* **125**, 412 (1972).
108. D. J. Spencer, *Annu. Rev. Microbiol.* **25**, 465 (1971).

5

The Primary Air Pollutants—Radioactive
Their Occurrence, Sources, and Effects

Merril Eisenbud

Atmospheric radioactivity originates from natural as well as artificial sources (1). Although the natural radioactivity of the atmosphere was first reported at the turn of the century, systematic inquiry has been undertaken only during the past two decades, originally as a by-product of studies of the atmospheric radioactivity produced by nuclear and thermonuclear explosions. More recently, interest in atmospheric radioactivity has been sustained by the need to understand the behavior of gaseous emissions from nuclear reactors and fuel reprocessing plants.

In addition to its public health implications, the subject of atmospheric radioactivity has attracted the attention of geophysicists who find both the natural and artificial radionuclides in the atmosphere to be useful as tracers in the study of various atmospheric transport mechanisms.

I. Natural Radioactivity

A. Radioactivity of Terrestrial Origin

The natural radioactivity of the atmosphere results from the presence of radionuclides which originate either from radioactive minerals in the earth's crust or from the interaction of cosmic (often called galactic) radiations with the gases of the atmosphere.

The soils and rocks contain naturally radioactive minerals in variable amounts, depending on local geological factors. Table I (2) gives the average content of various naturally occurring radionuclides in rocks. Although soil dusts of these substances undoubtedly find their way into the atmosphere, their contribution to its natural radioactivity is insignificant. Certain of these radionuclides, however, contribute greatly to atmospheric radioactivity by the formation of radioactive noble gases which emanate from the earth's crust. These gases, radon (^{222}Rn) and thoron (^{220}Rn), are the radioactive progeny of two nuclides of radium, ^{226}Ra and ^{228}Ra. Their respective decay schemes are as given in Table II.

Each of the gaseous isotopes diffuses into the atmosphere to some extent. Radon, with a half-life of 3.8 days, has a reasonably high probability of escaping into the atmosphere before it decays, whereas thoron, with a half-life of only 54 seconds, has a correspondingly smaller probability

Table I   Average Radium, Uranium, Thorium, and Potassium Contents in Various Rocks (2)

Type of rock	^{226}Ra (pCi/gm)	^{238}Pu (pCi/gm)	^{232}Th (pCi/gm)	^{40}K (pCi/gm)
Igneous	1.3	1.3	1.3	22.0
Sedimentary				
Sandstones	0.71	0.4	0.65	8.8
Shales	1.08	0.4	1.1	22.0
Limestones	0.42	0.4	0.14	2.2

Table II   Decay Schemes of ^{226}Ra and ^{228}Ra

Nuclide	Half-life	Radiation	Nuclide	Half-life	Radiation
^{226}Ra	1622 years	α	^{228}Ra	5.7 years	β
^{222}Rn	3.8 days	α	^{228}Ac	6.1 hr	β
^{218}Po	30.5 min	α	^{228}Th	1.9 years	α
^{214}Pb	26.8 min	β	^{224}Ra	3.6 days	α
^{214}Bi	19.7 min	β	^{220}Rn	54.5 sec	α
^{214}Po	1.6×10^{-4} sec	α	^{216}Po	0.16 sec	α
^{210}Pb	22 years	β	^{212}Pb	10.6 hr	β
^{210}Bi	5.0 days	β	^{216}At	3×10^{-4} sec	α
^{210}Po	138 days	α	^{212}Bi	60 min	β
^{206}Tl	4.2 min	β	^{212}Po	3×10^{-7} sec	α
^{206}Pb	Stable		^{208}Tl	3.1 min	β
			^{208}Pb	Stable	

of diffusing from its place of birth to the atmosphere. It is estimated that radon diffuses from soil at an average rate of 1.4 ± 0.73 pCi/m² second (3).

The atmospheric concentration of these noble gases and their daughter products depends on many geologic and meterological factors, some of which have not been thoroughly studied.

It should be noted from Table II that the radon decay scheme permits rapid secular equilibrium to be achieved between the radon and the first four of its progeny, ^{218}Po, ^{214}Pb, ^{214}Bi, and ^{214}Po. This follows from the fact that the longest half-life of the nuclides in this series is only 30.5 minutes. The ^{214}Po then decays to ^{210}Pb, a β emitter having a half-life of about 22 years. We will see that the relatively long half-life of ^{210}Pb makes this isotope particularly significant from several points of view. The thoron series has no long-lived members, the time for secular equilibrium for the full series being determined largely by the 10.6-hour ^{212}Pb.

When radon and thoron decay, their daughter products tend to attach themselves to the inert dusts of the atmosphere. If the radioactive gases coexist with dust in the same air mass for a sufficiently long time, the parents, and their various daughters will achieve radioactive equilibrium. In the case of radon, this process is almost complete after about 2 hours for the isotopes preceding ^{210}Pb. In the case of thoron, equilibrium is reached in about 2 days. Adsorbed radioactive daughters thus have the effect of endowing ordinary dusts of the atmosphere with apparent radioactivity.

Wilkening (4) has observed that radon daughters tend to distribute themselves on atmospheric dust in a manner that depends on the particle size of the dust, and that the bulk of the activity is contained on particles having diameters less than 0.035 μm. Kawasaki and Kato (5) have also investigated the particle-size spectrum of the dust on which radon daughters are adsorbed, and have concluded that the radioactivity is associated with dust particles in the range 0.084–0.2 μm.

Another source of natural atmospheric radioactivity is the combustion of fossil fuels. In coal ash, a number of radionuclides, including the radium isotopes, originate from traces of uranium-238 and thorium-232. On the basis of the measured ^{226}Ra and the ^{228}Ra content of coal ash it has been estimated that ^{228}U and ^{232}Th are present in coal in concentrations of 1.1 and 2.0 ppm, respectively. A 1000-MW (electrical) coal-burning power plant without dust collection equipment would discharge annually about 350 mCi of ^{228}Ra and 550 mCi of ^{226}Ra. These figures would be reduced by a factor of 50 to 100 with well-maintained modern dust collection equipment (6, 7).

Oil-burning plants normally discharge nearly all of the combustion products into the atmosphere: a 1000-MW station which consumes 460 million gallons of oil per year will discharge about 0.5 mCi of ^{226}Ra and ^{228}Ra. There is also evidence that ^{210}Pb, the 22-year product of ^{222}Rn decay, is discharged in measurable amounts when residual oil is burned. This might be explained by the relatively high solubility of radon in oils and fats.

Lockhart (8) has undertaken measurements of ground-level atmospheric radioactivity at a number of places throughout the world and his summary of several years of data is given in Table III. The radon concentra-

Table III Summary of Measurements of Natural Radioactivity in Ground-Level Air (8) (Copyright 1964. Reprinted by permission of Univ. of Chicago Press, Chicago, Illinois)

Site	Period of observation	Radioactivity $(pCi/m^3)^a$		Activity ratio $^{214}Pb/^{212}Pb$
		^{214}Pb	^{212}Pb	
Wales, Alaska	1953–1959	20	0.16	125
Kodiak, Alaska	1950–1960	9.9	0.04	250
Washington, D.C.	1950–1961	122	1.34	91
Yokosuka, Japan	1954–1958	56	0.48	117
Lima, Peru	1959–1962	42	1.33	28
Chacaltaya, Bolivia	1958–1962	40	0.53	72
Rio de Janeiro, Brazil	1958–1962	51	2.54	20
Little America, Antarctica	1956–1958	2.5	<0.01	>250
South Pole	1959–1962	0.47	<0.01	>50

^{a} One picocurie equals 2.2 disintegrations per minute.

tions were inferred from measurements of ^{214}Pb, a reasonable procedure since secular equilibrium is reached relatively quickly and is nearly complete when the radon-laden air and dust coexist for about 2 hours. It is seen in Table III that Washington, D.C., which is some distance from the ocean, had the highest concentration of ^{214}Pb, followed by seaports, midocean islands, and finally Antarctica. This is consistent with our knowledge that the concentration of ^{226}Ra in ocean water is low relative to the concentration observed in rocks and soils.

Variations from day to day are dependent on meteorological factors which influence the rate of emanation of the gases from the earth. The rate of emanation from soil may increase during periods of diminishing atmospheric pressure. The history of an air mass for several days prior to observation also influences its radon and thoron content (9). Passage of the air over oceans and precipitation tend to reduce the concentration of these gases, whereas periods of temperature inversion cause them to increase.

Prospero and Carlson (10) have shown the ^{222}Rn content of North Atlantic air to be influenced by the amount of dust carried from North Africa by the trade winds. Blifford and his associates (11) investigated the relationships between the concentrations of radon and its various decay products in the normal atmosphere and found, as expected, that the atmosphere is markedly depleted in ^{210}Pb, relative to the precursors of this isotope. This is because the inert dust of the atmosphere, the radon, and the radon daughters coexist long enough under normal circumstances for equilibria to be reached between radon and the more short-lived daughters. Since the longest half-life prior to ^{210}Pb is 26.8-minute ^{214}Pb, equilibrium is reached in about 2 hours. The ^{210}Pb, which has a 22-year half-life, would take about 100 years to reach equilibrium. However, various mechanisms exist for removing dust from the atmosphere, and the ratio of ^{210}Pb to its shorter-lived ancestors was shown by Blifford to be indicative of the length of time the dust resides in the atmosphere. He concluded by this method of analysis that the mean life of the atmospheric dust to which the radon daughters are attached is 15 days.

Wilkening (12) found that the atmospheric content of ^{222}Rn daughters is depleted during passage of a thunderstorm, which he attributes to the action of electric fields that, in his measurements, changed from a normal value of about 1.8 to -340 V/cm during the storms.

The natural radioactivity of atmospheric dust, due primarily to the adsorbed daughters of radon, can be readily demonstrated. When air is drawn through a filter, the radon daughters attached to the filtered atmospheric dust cause both the α and β activity of the filter to rise. Curve A of Figure 1 illustrates the manner in which this increase of α radioac-

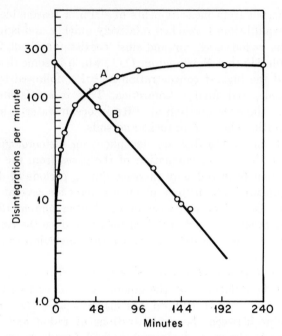

**Figure 1. The buildup and decay of radioactivity on filter paper through which
is passed 1 ft³/minute of air containing 5 × 10 Ci/liter of radon in equilibrium
with short-lived daughter products. A: Buildup of radioactivity. B: Decay of radioactivity after cessation of flow.**

tivity occurs in the case of normal air containing 5×10^{-4} Ci/liter of
radon in equilibrium with its daughter products. The rise in α activity
increases for about 2 hours, at the end of which time equilibrium is approached, and the accumulated daughters decay at a rate which is compensated for by the decay rate of newly deposited daughters. The radioactivity of the filter will not increase beyond this equilibrium unless
either the rate of airflow or the concentration of radon is increased. When
airflow ceases, the α radioactivity of the filter will diminish, as shown
in curve B, with an apparent half-life of about 40 minutes.

B. Radioactivity Induced by Cosmic Rays

Interactions of cosmic rays with atmospheric gases produce a number
of radioactive species of which the most important are tritium (^{3}H), ^{14}C,
and ^{7}Be; of lesser importance are ^{10}Be, ^{22}Na, ^{32}P, ^{33}P, ^{35}S, and ^{39}Cl. The
properties of these isotopes and the extent to which they have been re-
ported to be present in the atmosphere are given in Table IV which is

Table IV Singly Occurring Natural Radionuclides Produced by Cosmic Rays (13) [Reproduced from *Health Physics* **11**, 1297 (1965) by permission of the Health Physics Society.]

Radio-nuclide	Half-life	Average atmospheric production rate (atoms/cm² sec)	Tropospheric concentration (pCi/kg air)	Principal radiations and energies (MeV)	Observed average concentrations in rainwater (pCi/l)
^{3}H	12.3 years	0.25	3.2×10^{-2}	$\beta^-0.0186$	—
^{7}Be	53.6 days	8.1×10^{-3}	0.28	$\gamma\,0.477$	18.0
^{10}Be	2.5×10^6 years	3.6×10^{-2} a	3.2×10^{-8}	$\beta^-0.555$	—
^{14}C	5730 years	~ 2 b	3.4	$\beta^-0.156$	—
^{22}Na	2.6 years	5.6×10^{-5}	3.0×10^{-5}	$\beta^+0.545,\ \gamma\,1.28$	7.6×10^{-3}
^{24}Na	15.0 hours	—	—	$\beta^-1.4,\ \gamma\,1.37,\ 2.75$	0.08–0.16
^{32}Si	$\approx$650 years	1.6×10^{-4}	5.4×10^{-7}	$\beta^-0.210$	—
^{32}P	14.3 days	8.1×10^{-4}	6.3×10^{-3}	$\beta^-1.71$	} "a few"
^{33}P	24.4 days	6.8×10^{-4}	3.4×10^{-3}	$\beta^-0.246$	
^{35}S	88 days	1.4×10^{-3}	3.5×10^{-3}	$\beta^-0.167$	0.2–2.9
^{36}Cl	3.1×10^5 years	1.1×10^{-3}	6.8×10^{-9}	$\beta^-0.714$	—
^{38}S	2.87 hours	—	—	$\beta^-1.1,\ \gamma\,1.88$	1.8–5.9
^{38}Cl	37.3 minutes	—	—	$\beta^-4.91,\ \gamma\,1.60,\ 2.17$	4.1–67.6
^{39}Cl	55.5 minutes	1.6×10^{-3}	—	$\beta^-1.91,\ \gamma\,0.25,\ 1.27,\ 1.52$	4.5–22.5

a From Ref. (14).
b From Ref. (15).

summarized from data reported by Perkins and Nielsen (*13*) and Korff (*14*).

Carbon-14 is produced by the ^{14}N capture of cosmic ray neutrons. The incident cosmic ray neutron flux is approximately 2 neutrons/second/cm^2 of the earth's surface, and essentially all of these neutrons disappear by ^{14}N capture. The incident neutron flux, integrated over the surface of the earth, yields the natural rate of production of ^{14}C atoms, which is believed to have been unchanged for at least 10,000–15,000 years prior to 1954, during which time the ^{14}C existed in an equilibrium concentration in the carbon of living biological substances in a constant amount of 7.5 ± 2.7 pCi/gm of carbon. After living things die, the ^{14}C content of their remains diminishes at a rate of 50% every 5730 years, thus making it possible to use ^{14}C concentration to measure the age of archeological specimens.

Since 1954, the atmospheric content of ^{14}C has increased, owing to the production of this isotope by neutrons produced in thermonuclear explosions. The concentration of tropospheric ^{14}C increased in the northern hemisphere to about double its normal value. By 1969, the ^{14}C concentration in the troposphere of both the northern and southern hemispheres was about 60% above normal (*15*).

Food residues, textiles, paper, and other organic constituents of refuse are not likely to contain material so old that a significant fraction of the ^{14}C has decayed. On the other hand, because there has been essentially complete decay of the ^{14}C originally present in the substances from which fossil fuels evolved, ^{14}C analyses make it possible for one to estimate the relative contribution of the two sources of particulate carbon. The introduction into the atmosphere of fossil-fuel carbon tends to reduce the specific ^{14}C activity of atmospheric carbon, and the concentration of ^{14}C in atmospheric carbon tends to be lower in urban and industrial areas. Clayton, Arnold, and Patty (*16*) and Lodge, Bien, and Suess (*17*) have used the ^{14}C content of particulate atmospheric carbon to estimate the fraction of dust originating from refuse incineration and from combustion of fossil fuels.

Tritium, a radioactive isotope of hydrogen (^3H), is formed from several interactions of cosmic rays with gases of the upper atmosphere. Existing in the atmosphere principally in the form of water vapor, tritium precipitates in rain and snow. Tritium, like ^{14}C, is produced in thermonuclear detonations, and the atmospheric concentration of the isotope has increased since 1954 (*15*).

Other isotopes formed from cosmic ray interactions with the atmosphere may be potentially useful as tracers for studying atmospheric transport mechanisms, but relatively few observations have been reported at this time. A radionuclide of particular interest is 53-day half-life [7]Be, which is produced in the upper atmosphere by the interaction of cosmic ray-produced neutrons. Concentrations of the order of 0.1 pCi/minute/m[3] have been reported in air at ground level.

II. The Fission and Activation Products

The principal artificial radioactive nuclides encountered in the atmosphere are frequently the same regardless of whether the source of atmospheric contamination is a nuclear reactor, a nuclear or thermonuclear bomb, or a plant reprocessing spent reactor fuel.

The individual artificially produced radionuclides originate either as fission products or activation products. The former are the radioactive fragments from the fissioning of uranium or plutonium, from which nuclides are produced in yields ranging from 10^{-5} to about 10^{-1} of the atoms produced, depending on the mass number of the nuclide. The fission product spectra of [235]U, [233]U, and [239]Pu are distributed in bimodal curves in which the locations of the peaks are shifted only slightly, depending on the fissile material. The two modes occur at about mass numbers 95 and 135.

The significance of the fission products as atmospheric contaminants depends on the type of energy they emit, their radioactive half-lives (Table V), and the manner in which they are metabolized when absorbed into the body (1).

Three of the most important fission products are ^{90}Sr, ^{137}Cs, and [131]I. Strontium-90 is regarded as potentially the most dangerous from the standpoint of its long-range public health significance because: (a) it is formed abundantly in the fission process (about 5% of the total atoms produced in fission); (b) it is long lived (28-year half-life); and (c) it is chemically similar to calcium. Because of its chemical properties, it is readily absorbed by living things and thus may pass through the food chains to man. With calcium, it is deposited in the skeleton, thus presenting the potential hazard of bone cancer or, possibly, injury to the blood-forming tissue in the bone marrow. While it is one of the most dangerous of the fission products, it is far less labile than the cesium or iodine and in many situations may be relatively unavailable in the ecological sense.

Cesium-137 is formed in the fission process in slightly greater amounts than [90]Sr. Its half-life is about 30 years, and it is a relatively soluble substance, somewhat similar to potassium in chemical properties. Unlike ^{90}Sr, it does not become fixed in the skeleton, but distributes throughout

Table V Some of the More Important Radionuclides Produced by Nuclear Reactions

Nuclide	Half-life
Fission products	
^{85}Kr	10.4 years
^{89}Sr	50 days
^{90}Sr	28 years
^{95}Zr	65 days
^{131}I	8.05 days
^{133}I	21 hours
^{135}I	6.7 hours
^{137}Cs	30 years
^{140}Ba	12.8 days
^{144}Ce	285 days
Activation products	
^{3}H[a]	12 years
^{14}C	5730 years
^{54}Mn	314 days
^{55}Fe	2.7 years
^{59}Fe	45.6 days
^{60}Co	5.3 years
^{65}Zr	245 days
^{238}Pu	86.4 years
^{239}Pu	24000 years

[a] Hydrogen-3 (tritium) is also produced in fission.

the body, from which it is eliminated with a biological half-life of about 100 days. The principal significance of ^{137}Cs for man is that it is a γ emitter and as such is a major contributor of long-lived γ activity. The potential hazard from ^{137}Cs is thought to be primarily a genetic one from irradiation of gonadal tissue.

Several radioactive isotopes of iodine are produced in fission, but ^{131}I is of principal concern because of its relatively long half-life (8.1 days). When absorbed into the body, the radioiodines tend to concentrate in the thyroid which, in the case of a child, can weigh less than a gram. The potential hazard of the radioiodines results from the fact that they are capable of such a high degree of concentration and can thus deliver a relatively high dose to a very small volume of tissue.

Because of their volatility, the radioiodines tend to escape to the atmosphere to a greater degree than most other radionuclides. The radioiodines may be inhaled if an individual is immersed in a passing cloud, but most extensive exposure is apt to result from deposition of ^{131}I on pasture land. The cow grazes large areas of grass surface each day, from which the radioiodine is absorbed and transmitted to humans via milk. When this happens, the public health risk can be minimized by diverting the contaminated milk to powdered or condensed milk production, thereby permitting time for the 8-day iodine to decay. Another precautionary measure would be to remove the cows out of pasture and feed them silage until the iodine in the fields has decayed. Although the radiological half-life of ^{131}I is 8.1 days, it has been reported from several sources that the effective half-life on dairy farms is about 5 days due to the combined effects of radiological decay and weathering.

As noted earlier, the intense production of neutrons in thermonuclear detonations produces ^{14}C in relatively significant amounts as compared with the ^{14}C inventory normally present in nature. Similarly, the inventory of tritium has also increased measurably. Tritium, like ^{14}C, is formed by neutron bombardment of light nuclei in nuclear explosions, but it is also a fission product, being produced in 1 out of 10,000 fissions.

The activation products are radioactive nuclides that are produced by the interaction of neutrons on substances in the vicinity of fission or fusion reactions. We have seen previously that cosmic ray-produced neutrons interact with nonradioactive atmospheric constituents to produce radioactive activation products such as carbon-14 and tritium. In the case of exploding nuclear weapons and in the operation of nuclear reactors, a number of activation products are produced, of which the principal ones are listed in Table V.

III. Mining, Milling, and Fuel Fabrication

The industrial, research, and military uses of radioactive materials require a vast network of mines, plants, and laboratories, schematically illustrated in the flow diagram of Figure 2, which traces uranium ore from the mines to its ultimate uses.

The air pollution problems of mines, mills, and refineries are primarily those of the chemical industry. Such dusts and fumes as are produced are only mildly radioactive, and their noxious properties, if any, would be due to the presence of inert radicals associated with the radioactive cations.

When uranium mining is conducted underground, inadequate ventilation may result in hazardous conditions for the miners, and cases of lung cancer have resulted from such exposures. However, the problem is basically one of occupational hygiene, and not of community air pollution.

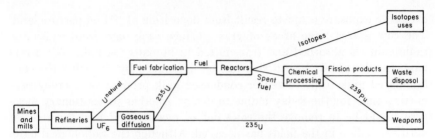

Figure 2. Processes in the atomic energy industry.

Radon and its daughter products are discharged to the atmosphere in the course of mine ventilation, but the amounts are not significant so far as the communities are concerned. The required ventilation rates for the mines vary from 1000 ft^3/minute to over 200,000 ft^3/minute, and the discharged air contains radon in concentrations that range from 0.5 to 20 μCi/minute/1000 ft^3 of air (18). This is a relatively small amount of radon, the maximum emission rate of which is equivalent to the normal radon emissions from about 2 $\times$ 10^5 m^2 of earth's surface.

The tailings piles have caused additional problems because of wind-blown dust and evolution of radon. Snelling (19, 20) showed the external radiation levels to range from 0.5 to 0.7 mR/hour, 3 ft above the piles, which covered 65 acres at one location.

Breslin and Glauberman (21) measured the airborne dust downwind from tailings piles associated with the uranium mills and demonstrated clear relationships between the distance from the tailings piles and the concentrations of uranium and ^{210}Pb. Of the three tailings piles sampled, the air concentrations were well below permissible levels in two cases, but approached the upper limits recommended in 10CFR20 (21a) at a distance of about 1000 ft from the tailings piles.

The piles of tailings are now stabilized with topsoil and plantings. Radon evolution (22), although higher than background, does not exceed the recommended limit for the public. However, the concentration of radon in several cities in which mills are located has been found to be elevated over normal values. A notable example is Grand Junction, Colorado, where the average concentration is about 10 pCi/liter (23).

The radiation exposure of Grand Junction residents has also been increased by the past practice of using mill tailings as a material of construction of homes and public buildings. The extent of exposure from this source has been under study by the United States Environmental Protection Agency and Public Health Service.

Klevin et al. (24) have reported on the contamination of soils in the vicinity of uranium refineries which were the major producers of uranium for several years, beginning early in World War II. The dust collection devices initially provided were inadequate, and over 50 tons of uranium were estimated to have been discharged from each of two plants during their period of operation. The soil concentrations within 2000 ft of the plants were somewhat elevated above the normal uranium background of the region, about 5 μg/gm of soil, but the amounts present were not excessive from the standpoint of either hazard or nuisance. The relatively high costs of uranium and thorium, and the relatively modest rates at which they are processed, in comparison with other metals of comparable hazard, such as lead or arsenic, limit the extent to which these metals could occur as atmospheric contaminants.

The refineries produce uranium salts of natural isotopic composition which may be (a) reduced to metal for fabrication into fuel for plutonium-producing reactors, or (b) converted to uranium hexafluoride and processed by gaseous diffusion plants to increase the $^{235}U/^{238}U$ ratio. The "enriched" uranium from the diffusion plants may then be converted either to UO_2 or metal, for fabrication into weapon components or reactor fuel elements. In one process, natural uranium is converted to a fissionable form by isotopic enrichment in the gaseous diffusion process. In another process, the uranium is transmuted to plutonium in reactors designed to operate with natural uranium. The air pollution problems of the uranium industry up to the point of reactor operation are minimal. They are of concern primarily from the standpoint of occupational exposure, and require attention as off-site problems only because the uranium is associated with nonradioactive acid radicals.

IV. Nuclear Reactors

The type and quantity of radioactive emissions released to the atmosphere by nuclear reactors depends on the kind of reactor.

A. Types of Reactors

Reactors can be constructed to serve primarily as sources of radiation or heat. Included among those constructed as radiation sources are (a) "production" reactors in which products of neutron irradiation such as plutonium, other transuranic elements, or ^{60}Co are produced; (b) research reactors such as those located on university campuses and other research centers; and (c) industrial-type test reactors that are used to study the effects of radioactivity on materials of construction and equipment components. The reactors used as sources of heat are used primarily as sources of power for electric generators but may also be used to generate steam for space or process heat. Production and test reactors are not included in this discussion.

B. Research Reactors

Research reactors function basically as sources of neutrons or other radiations used to activate samples, study the structure of matter, or irradiate living things. At the end of 1970, there were 72 operating research reactors in the United States, and nearly 400 operating in 47 countries throughout the world (25). The only type of research reactor that is a significant source of atmospheric radioactivity is the air-cooled reactor. Two such reactors, one at Oak Ridge and one at Brookhaven National Laboratory, were operated in the United States for many years, but were eventually shut down. The principal source of gaseous radioactivity was argon-41, produced by neutron activation of the argon-40 normally present in the cooling air. One of the more popular reactors during the late 1950's and early 1960's was the so-called "pool reactor," which ranges in power up to about 5 MW. Moderate amounts of gaseous activation products are produced by the pool reactors when the power levels are greater than 1 MW. More modern reactors such as the Argonaut or Triga are so designed as to preclude production of significant quantities of gaseous radioactivity (1).

C. Power Reactors

Contemporary nuclear power reactors are largely of two types, boiling water or pressurized water reactors. Both types are usually fueled with uranium oxide pellets assembled in tubes of stainless steel or zirconium. Water serves as both the moderator and coolant.

1. Boiling Water Reactors

The water is heated during passage through the core of the reactor (BWR) and is converted to steam. The steam passes through a dryer to the turbines and is condensed in condensers for return to the reactor.

Atmospheric radioactivity from boiling water reactors originates in a number of ways. Neutron irradiation of nitrogen, oxygen, and argon dissolved in the water produces short-lived radioactive gases that appear in the steam. In addition, noble gases, including isotopes of krypton and xenon, are able to diffuse through the fuel-element cladding and be carried by the steam to the turbines. Other fission products may escape into the water through defects in the fuel cladding or may originate from small amounts of uranium contaminating the surface of the reactor components.

It has been found that the noble gases boil off with the steam but are not condensable under the conditions that prevail in the system. They escape to the atmosphere via the condenser air ejector, which is ventilated to a stack. Most of the radioactivity that escapes via the air ejector is short lived, so that by designing for a 30-minute delay between the air ejector and the top of the stack, the bulk of the radioactivity can be made to disappear.

The noble gases discharged by boiling water reactors have short half-lives, and because of their chemical inertness do not concentrate in biological systems. Their maximum permissible atmospheric concentrations are based on the dose delivered externally to individuals immersed in a cloud of the gases. The only long-lived radioactive noble gas is ^{85}Kr, which has a half-life of 10.3 years.

Fission and activation products other than the noble gases tend not to carry over with the steam and have not constituted a significant fraction of the gaseous radioactive discharges from boiling water reactors. Iodine-131 has not been reported in the environs of operating boiling water reactors. Other fission products such as ^{90}Sr and ^{137}Cs would be expected to be even less of a problem because of their lower volatility. They diffuse relatively slowly from the fuel and have less tendency for the radioactive nuclides to carry over from the coolant.

For many years, the United States Atomic Energy Commission (AEC) limited the permissible discharge from reactors to a rate that controlled the exposure of members of the public to less than 0.5 rem/year. In calculating the permissible discharge rate, the physical and biological characteristics of the radioactive gases and meteorological characteristics of the site are taken into consideration. In actual practice, it was found that contemporary plants could operate so as to discharge a very small fraction of the permissible atmospheric emissions. Accordingly, the AEC in 1972 proposed that, consistent with the concept that all radiation exposure should be reduced to the lowest practicable dose, emissions from light water reactors should be reduced so that the dose to the maximum exposed individual is less than 5–10 mrem/year. To assure that this objective is accomplished, the more modern boiling water reactors are equipped to provide additional hold-up time to allow for more decay of the relatively short-lived noble gases.

2. Pressurized Water Reactors

The development of the pressurized water reactor (PWR) in the United States received great impetus from the decision of the United States Navy to standardize on this type of reactor to provide power for submarine and surface vessels. The success of the navy nuclear submarine program is well known. It is less well known that one vessel, the aircraft carrier U.S.S. *Enterprise*, is powered by eight nuclear reactors. There are now more than 100 power reactors on vessels of the United States Navy.

In pressurized water reactors water serves as both a moderator and coolant, but is retained within a closed loop. The pressure is adjusted so that the boiling of the water does not take place within the reactor. The temperature of the water is increased in passing through the reactor, after which the pressurized water is allowed to transfer its heat to a second loop in which steam is produced. The two-loop system is used to prevent fission products from entering the turbines and thereby complicating maintenance problems and adding to the complexity of radiation protection. In the event of a fuel element failure in a two-loop reactor, the fission products remain in the primary system and contaminate neither the secondary system nor the turbines.

The composition of the gaseous discharges from a nuclear reactor vary not only depending on the type of reactor, but on its operating history and, in particular, the condition of its fuel. The principal radionuclides released in the gaseous emissions from a boiling water reactor are given in Table VI (*26*), in which it is seen that except for minor amounts of tritium and krypton-85, the emissions are primarily short-lived noble gases. In the event of defects in the fuel cladding, it is possible for nuclides such as cesium and radioiodine among others to be emitted, but this has not been observed to be a significant problem at any of the operating reactors to date. For purposes of monitoring and regulation, the gaseous emissions are customarily divided into the noble and activation

Table VI   Principle Radionuclides Released in Gaseous Emission from Boiling Water Reactors (*26*)

Radionuclides	Half-life	Release rate, $\mu Ci/sec$
^{m}Krypton-85	4.4 hours	3×10^2
Krypton-85	10.7 years	1×10^{-1}
Krypton-87	76 minutes	7×10^2
Krypton-88	2.8 hours	5×10^2
^{m}Xenon-133	2.3 days	1×10^1
Xenon-133	5.3 days	3×10^2
Xenon-135	9.1 hours	8×10^2
Xenon-138	17 minutes	2×10^3
Hydrogen-3 (tritium)	12 years	5×10^{-2}

gases, and halogens and particulates with half-lives greater than 8 days. The latter would include iodine-131, and possibly cesium-137, cesium-134, and barium-140.

The releases of gaseous effluents from commercial power reactors operating during calendar year 1971 are given in Table VII. It is seen that the reactors all operated at a fraction of the emission rate permitted by AEC regulations. The "permissible" rates of emission shown in Table VII (*27*) are those prescribed in the license issued by the AEC, and are generally rates that have been calculated to meet the original AEC regulation that the dose to the maximum exposed individual should be less than 500 mrems/year. The rate of emission can be calculated so as to meet this requirement if the meteorological characteristics of the site and the radiological and biological characteristics of the gaseous effluents are known. As noted earlier, the AEC has recently reduced this maximum permissible exposure by approximately a factor of 100 on the basis that it is practicable for light-water reactors of contemporary design to meet this limit.

The principal way in which the general population can be exposed to external radiation from reactors is from a cloud of passing radioactive gases being discharged from the plant. The AEC has assumed in the past that the average exposure to the nearby population will be one-third of the dose received by the maximum exposed individual (500 mrem/year). In other words, the average dose to the population must be kept under 170 mrem/year. There is a very substantial inherent safety factor in this assumption, as can be seen from examination of a hypothetical case in which a boiling water reactor stack is located 100 m from a 360° fence at which the dose is assumed to be 500 mrem/year. This dose would be received by people living at the property fence. It can be calculated that if one million people are uniformly distributed around the fence at a density of 100 people/km², the annual per capita dose for such a population would be somewhat less than 0.3 mrems/year. The average dose to the population would thus be 1/1800 of the maximum, not one-third as is assumed in the AEC regulations.

All factors being taken into consideration, it can be concluded that the gaseous emissions from commercial power reactors under normal operating conditions do not constitute an important source of public exposure.

D. Reactor Accidents

Power reactors accumulate enormous inventories of fission products and extensive precautions are taken to prevent release of the radioactive

Table VII Releases of Radioactivity in Gaseous Effluents from Nuclear Power Plants, 1971 (27)

Facility	Noble and activation gases			Halogens and particulates with half-lives >9 days		
	Released (Ci)	Permissible (Ci/year)	Percent of permissible	Released (Ci)	Permissible (Ci/year)	Percent of permissible
Oyster Creek	516,100	8,300,000	6.2	2.14	126	1.7
Millstone Point	275,700	25,000,000	1.1	4.0	94.6	4.23
Indian Point-1	360	5,300,000	0.007	0.21	7.6	2.8
Dresden-1	753,000	18,000,000	4.3	<0.67	76	0.9
Dresden-2/3	580,000	28,000,000	2.1	8.68	73	11.9
H. B. Robinson	0.018	160,000	<0.0001	Not detected	~0.16	
Humboldt Bay	514,300	1,600,000	32	0.3	5.6	5.4
Fermi-1	<180	900	<20	<0.001	2.5	<0.04
Big Rock Point	284,000	30,000,000	0.91	0.61	38	1.6
Nine Mile Point	253,000	26,000,000	0.97	<0.80	47.3	1.65
R. E. Ginna	31,850	3,160,000	~1	0.17	1.7	10.1
Saxton	437	3,800	11.7	0.007	10	0.07
Peach Bottom-1	122	190,000	0.06	<0.0003	0.09	<0.33
La Crosse	529	310,000	0.17	<0.001	1.6	<0.06
San Onofre	7,667	1,700,000	0.45	<0.0001	0.8	<0.012
Point Beach-1	838	2,400,000	0.035	<0.0001	a	a
Monticello	75,800	8,500,000	0.89	0.052	~10	0.5
Yankee-Rowe	13	22,000	0.058	<0.0001	22.4	<0.008
Connecticut Yankee	3,250	280,000	1.12	0.031	0.2	15

a Included in noble and activation gases.

substances to the environment. It is commonly accepted that for light-water reactors such as described above, the maximum credible mishap would be the sudden structural failure of the coolant system. This is the so-called "loss-of-coolant accident" in which the water coolant and moderator would flash to steam, rendering the reactor subcritical, but causing a rapid rise in core temperature due to the heat produced by radioactive decay of the massive quantities of fission products contained within the fuel.

A fundamental question is whether the massive failure of the piping is, in fact, credible. There have been no such accidents in 30 years of experience with nuclear systems and, more important, massive failures are unknown in high-pressure central station steam boilers which provide considerably more experience than is available in the nuclear industry. A study of about 500 boiler steam drums designed for pressures greater than those used in the nuclear industry, and representing 4000 boiler years of operating experience, showed no failures of the steam drums themselves. Failures did occur in other parts of the high-pressure steam system, but they were not of the massive type that could cause a sudden release of coolant from a water reactor (*28*).

The basic requirement is that the reactor system be designed in such a way that a loss-of-coolant accident or less severe accidents cannot occur. This is accomplished by applying very conservative design criteria and very strict standards of quality assurance. However, having completed the reactor design, the engineer must then make the assumption that the accident, in fact, does occur and then design safeguards to reduce the environmental impact to a minimum.

An essential safeguard is the provision of a containment building designed to confine the radioactive steam and any radionuclides that might be volatilized from the reactor core. The first containment vessels were simple spheres about 125 ft in diameter and fabricated from 1-in. welded steel plate (Fig. 3). Subsequent containment vessels have been constructed of massive concrete, with linings fabricated from steel plate. Some of the contemporary designs include provision for venting the steam into a water-filled torus that serves as a heat sink (Fig. 4). Figure 5 shows another type of vapor suppression system, designed by Westinghouse, in which the steam is condensed by venting through an ice-filled structure (*29*). Additional protection is provided by a system of sprays that wash the radioiodine and other fission products from the containment atmosphere into a sump. Addition of sodium thiosulfate to the spray adds to the efficiency of the radioiodine scavenging process. Alternatively or concurrently, the atmosphere within containment can be recirculated through activated charcoal to remove the radioiodines. One of the most

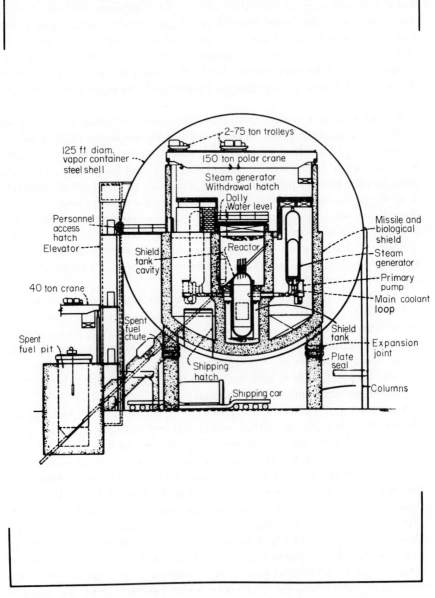

Figure 3. The spherical steel containment for the pressurized water reactor at Rowe, Massachusetts.

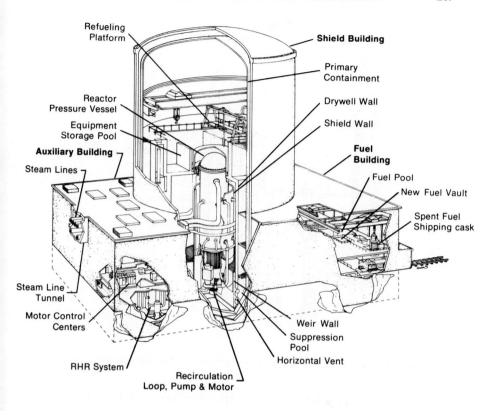

Figure 4. Boiling water reactor conical concrete pressure suppression containment (General Electric Company).

important of the engineered safeguards is the emergency core cooling system, which would flood the exposed reactor core and cool it sufficiently to prevent it from melting. Considerable controversy has arisen over the question of whether the emergency core cooling systems as currently designed would be capable of injecting the water into the core against the steam pressure that would be produced within the core in the event of a loss-of-coolant accident. One of the underlying difficulties is that full-scale tests of the system are impractical, and scaling factors are not fully understood.

Reactor licenses are issued by the AEC following lengthy and complex inquiry into the safety of the reactor. Figure 6 illustrates the mass of

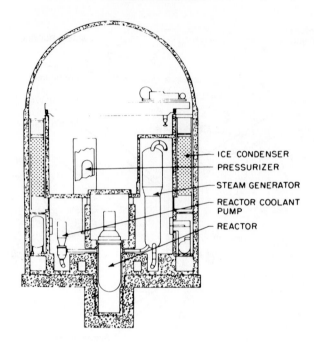

Figure 5. Elevation section of ice-condenser containment system (29).

technical reports and hearing transcripts that accumulate in the course of such an application. The reactor designs and estimates of potential risk from failure of a reactor component are subject to continuing inquiry that takes the form of theoretical investigations and laboratory tests. It is commonly accepted that the principal risk in the event of a massive reactor failure would be due to the release of the radioiodines. The emergency problem would be due to inhalation of the iodines rather than contamination of milk supplies. If the latter happened, there would be ample times in which to take remedial actions such as removing the cows from pasture, converting the milk to a powdered form or, as a last resort, confiscation of the milk.

There has been only one major reactor accident in the United States, in which a nuclear excursion resulted in the death of three operating personnel at Idaho Falls in January, 1961 (30). This accident occurred in connection with the operation of an experimental military reactor, the Stationary Low Power Reactor (SL-1). Although this accident was a severe one, the radioactivity released from the core was substantially confined within the building in which the reactor was housed.

Figure 6. Mass of technical reports and hearing transcripts accumulated in the course of application for an AEC power reactor operating license. Shown are the Preliminary Safety Analysis Report, Final Safety Analysis Report, Environmental Reports, and transcripts of AEC construction license hearings (Consolidated Edison Company).

The only reactor accident that has resulted in significant environmental contamination up to the present time occurred in October, 1957, in England at the Adlermaston Weapons Establishment (Windscale) reactor. The primary cause of the accident was the release of stored energy within the graphite used as moderator. It was originally predicted by Wigner that irradiation might result in storage of energy within the graphite, owing to alterations in its crystalline structure. Dimensional changes due to this effect were first observed in the Clinton pile at Oak Ridge, Tennessee, during World War II.

The Windscale accident (*31*) was caused by the sudden release of this stored energy, resulting in a sudden rise of the temperature of the reactor core. The accident actually occurred when the reactor was shut down for the purpose of annealing the graphite and thereby achieving a controlled release of the stored energy. A too rapid rise in the temperature of the metallic uranium fuel cartridges during this procedure caused fuel failure and ignition. The reactor instrumentation was inadequate to warn of the rise in temperature, and the high temperatures caused oxidation of the defective uranium cartridges, thereby producing more heat, and more rapid oxidation by reactions between the metallic uranium, the air, and the graphite. The mishap was not appreciated until the fourth day after the original overheating had occurred, when the filter at the top of the stack showed a sharp rise in radioactivity. Emergency procedures were instituted and the fire in the core was extinguished by the end of the fifth day.

The fission products, mainly ^{131}I, from this accident were carried across the surrounding countryside to the English Channel and thence across much of Europe. The radioactivity was detectable for great distances, and the levels of contamination were high enough to require precautionary procedures downwind in a coastal strip about 30 miles long and 6–10 miles wide. In this area, airborne radioactive iodine deposited on foliage and was consumed by dairy cows. The resulting iodine content of the milk necessitated its being withheld from public consumption for several weeks following the accident.

The maximum observed thyroid dose among inhabitants of the area was 19 rads in one child. This is not large in comparison with the 30-rad dose suggested by the United States Federal Radiation Council as the dose above which protective action should be taken in the event of a contaminating event.

The Windscale mishap was a major reactor mishap. It is significant that only the gaseous iodine escaped in appreciable amounts and that ^{90}Sr and other isotopes of the more toxic elements remained, for the most part, within the core, or were trapped in the air filter on top of the stack.

The Windscale accident is sometimes cited to illustrate the potential hazards of nuclear power. This is not a valid comparison, since the Windscale reactor was designed for the production of plutonium, and had a number of features, deliberately provided, which were desirable for plutonium production but which would have been contraindicated for a power reactor. For example, the fuel (metallic uranium), the moderator (graphite), and the coolant (air) were a pyrophoric combination. In contrast, in both pressurized and boiling water reactors the fuel is uranium oxide, which is a relatively inert material. Water is used as both the coolant and moderator. Thus, the exothermic reactions that took place between the uranium, graphite, and air could not take place in a boiling water or pressurized water reactor. Moreover, the fact that the reactor was air cooled necessitated direct discharge of contaminants to the atmosphere with reliance only on mechanical filters to remove particulates.

Water-cooled and moderated nuclear power plants are contained in sealed structures capable of confining any releases to the atmosphere.

However, the Windscale accident highlighted the potential danger of radioiodine contamination following reactor accidents, and much of the research undertaken since 1957 has been concerned with the chemical and physical properties of radioiodine released to the atmosphere and the manner in which these nuclides can be removed from gaseous and liquid wastes.

It is clear that not every mishap results in atmospheric pollution, because the records of many reactors in this country and abroad contain many instances of mishaps which resulted in no release of fission products. In the Windscale accident, the only example of extensive atmospheric contamination, the effects were by no means disastrous and not unusually costly by standards of other industrial episodes with which we are all familiar. There were no injuries to people. The principal cost of the Windscale reactor accident was due to the damage to the reactor itself.

In summary, we can say that after 30 years of reactor experience, their operation has not resulted in problems of atmospheric contamination under normal operating conditions. This has been particularly true of contemporary reactors built for the production of electrical power.

The ultimate consequences of uncontrolled release of fission products from an overheated reactor core are unquestionably a risk with which we must be concerned continuously, but modern engineering has made it possible to minimize the possibility of such core meltdowns while at the same time providing engineered safeguards that will minimize the consequences of such an event if it did occur. While it cannot be said that there is no risk from atmospheric pollution in the event of accidents having very low probabilities of occurrence, it should be recognized that installation of a power plant deriving its power from the nuclear process makes it possible to avoid the discharge of copious amounts of sulfur dioxide, fly ash, and other noxious effluents. In the years to come the benefits of nuclear power to air conservation will surely be a major factor in public acceptance of this relatively new form of energy.

V. Reprocessing of Spent Fuel

When spent fuel is removed from a reactor, it is taken to a chemical processing plant where the fuel is dissolved and the solutions processed in order to separate the unfissioned uranium and plutonium from the radioactive waste products. Because of the enormous amount of radioactivity produced in the fuel, the processes of chemical separation are fraught with great potential hazards. The fuel is transported in massive shielded casks, and the chemical processes are carried out by remote control in "canyons" consisting of thickly shielded concrete cells. These installations are so expensive that very few exist in the world at the present time. In the United States, spent fuel processing has been undertaken for many years at government facilities at Hanford, Idaho Falls, Savannah River, and Oak Ridge. The first privately owned fuel reprocessing plant to be constructed in the United States began operation in New York State several years ago, and at least two others are nearing completion.

A highly specialized technology has been developed to make it possible for these separation plants to operate efficiently and safely. In the years to come, as the civilian reactor industry grows, it will be necessary to erect many such plants in various parts of the world to accommodate the need for conveniently located reprocessing facilities.

The types of wastes generated in the fuel reprocessing plant depend on the type of fuel and the particular process being used. As previously noted, most contemporary nuclear power plants use uranium oxide fuel. In addition, it is necessary for the reprocessing plants to be prepared eventually to handle other chemical forms of uranium as well as thorium and plutonium. Some variation in technique is also required, depending on the type of cladding, which may consist of aluminum, zirconium, or stainless steel.

The amount of radioactivity contained in the spent fuel is very large but the potential hazards are mainly those associated with the management of liquid and solid wastes, which will not be discussed here. The two most important gaseous effluents are ^{85}Kr, which has a 10.4 year half-life, and ^{131}I. The amount of 8.1-day ^{131}I present in the core is dependent on the length of time since the fuel was removed from the reactor. The amount initially present is sufficient so that a considerable inventory may remain even after 3–6 months. The permissible amount of iodine that may be released from the stack is governed by the local climatology, and the location of dairy herds and human habitation. Iodine-131 is sufficiently reactive so that chemical means are available for removing it from stack effluents. Krypton-85, being a noble gas, cannot be removed by ordinary chemical or mechanical means. However, it can be removed by cryogenic techniques, as well as by the use of clathrates. Removal of ^{85}Kr from the waste streams of the fuel reprocessing plants will probably be necessary to prevent worldwide accumulations of this nuclide. It is estimated that if ^{85}Kr recovery is not practical, the skin dose to the world's population could equal the dose received from nature by the middle of the next century. Exposure of the public would be limited to the skin because krypton is not absorbed in the body in significant amounts. Exposure would be to the β particles emitted from krypton contained in the atmosphere.

VI. Nuclear Weapons

Shortly after a nuclear detonation larger than a few kilotons of trinitrotoluene (TNT) in explosive equivalence, fission products are easily detected in the atmosphere everywhere in the world. Until 1954, the devices were relatively small and the contamination they produced was confined to the troposphere, from which the radioactive dust would disappear within a few weeks after cessation of tests. However, beginning in 1954, numerous detonations of nuclear devices took place which had explosive yields equivalent to many millions of tons of TNT, and the upper atmosphere became a reservoir of fission products from which continuous deposition has occurred ever since. As a result of this pollution of the earth's atmosphere and the resulting deposition on soils, some of the fission products have been present in trace, but detectable, quantities in all biological material formed during the past decade.

The intense heat that accompanies the detonation of a nuclear device results in volatilization of the fission products produced in the explosion. A mass of luminescent gas is formed and cools as it rises, sucking into itself varying amounts of dust raised from the ground by convection. Some of this convected dust will vaporize when it reaches the fireball and become admixed with the radioactive materials originally involved in the detonation. Dust particles sucked into the fireball in more advanced stages of cooling are likely to act as nuclei on which fireball vapors condense as a radioactive coating. Other fireball vapors may condense on ionic nuclei and form discrete fume particles, which may or may not aggregate with larger particles of inactive dust. The particles of fumes and dust diffuse throughout the atmosphere and ultimately deposit on the earth as "fallout" (*1, 32*).

The amounts of radioactivity produced in nuclear detonations are enormous. A weapon having a fission yield equivalent to 1 megaton of TNT produces, 1 minute after detonation, about 4×10^{12} Ci of γ activity. This radioactivity decays rapidly and the residual activity may be estimated at any future time by the relationship $A = A_1 t^{-1.2}$, where A_1 is the activity at one unit of time (t) after the detonation. As a rule of thumb, the radioactivity will decrease tenfold for every sevenfold increase in time.

Within a matter of hours after an explosion, the radioactive dust has been partitioned into three portions. Radioactivity in relatively large particles (for the most part larger than 50 μm) will have deposited within 10 or more miles downwind of the detonation. This fraction is intensely radioactive because its fallout occurs soon after detonation and before appreciable diffusion has occurred. This close-in fallout is particularly significant for those detonations in which the fireball actually touches the ground, and is the component of the radioactivity that in time of war might cover large areas with lethal radioactive deposits.

A second portion of the initial radioactive cloud is distributed within the troposphere. This fraction may pass several times around the world before it finally deposits completely on the earth's surface, but complete deposition is likely to occur within a few weeks after the detonation.

For weapons having yields greater than 0.1 to 1 megaton (of TNT), a third portion punctures the tropopause and injects radioactive material into the stratosphere, within which the cloud slowly diffuses. Residence time in the stratosphere is a function of the height at which the cloud stabilizes and of the latitude of injection. In general, transfer from the stratosphere to the troposphere proceeds with a half-time of about 1 year.

The manner in which the debris partitions itself among the three types of fallout depends on the size of detonation, the height of burst above ground, and meteorological factors.

It is estimated that, from World War II until the end of 1962, the total explosive yield of all nuclear detonations by the United States, United Kingdom, and Soviet Union was equivalent to 511 megatons of TNT, as shown in Table VIII (*33*).

Table VIII Approximate Fission and Total Yields of Nuclear Weapons Tests Conducted in the Atmosphere by All Nations (Yield in Megatons) (*33*)

Inclusive years	Fission yield		Total yield	
	Air	Surface	Air	Surface
1945–1951	0.19	0.52	0.19	0.57
1952–1954	1	37	1	59
1955–1956	5.6	7.5	11	17
1957–1958	31	9	57	28
Subtotal	37.8	54	69.2	104.6
1961	25[a]	—	120	—
Subtotal	63	54	189	105
1962	76[a]	—	217	—
Total	139	54	406	105

[a] The small yield tests conducted in Nevada do not contribute significantly to the worldwide distribution of ^{90}Sr to which this summary is related.

Testing was first conducted on a large scale in 1952, prior to which the total yield is estimated to have been less than the equivalent of 1 megaton of TNT. From 1952 until the temporary suspension of weapons tests in 1958, the nuclear and thermonuclear explosions amounted to the equivalent of about 174 megatons of TNT. Following resumption of testing in 1961, an additional 337 megatons were exploded until the second moratorium, this time on open-air testing, was adopted by the United States, United Kingdom, and Soviet Union in 1963. Since then there has been a small amount of venting from underground tests conducted by the United States and the Soviet Union, but this has not represented a significant addition to the total atmospheric inventory. The Chinese and French have in the meantime begun to test weapons, but the yields of these explosions have not added appreciably to the totals prior to 1962. During tests prior to 1963, it is estimated that about 30% of the radioactivity was deposited in the immediate vicinity of the test sites.

Measurements of the atmospheric radioactivity resulting from nuclear weapons tests have been made at hundreds of locations throughout the world and at elevations ranging from ground level to more than 100,000 ft. The basic measurements are contained in reports submitted from many nations to the United Nations Scientific Committee on the Effects of Atomic Radiation which was established in 1955 to serve as the clearinghouse for the collection of both physical and biological data on a global scale. As of this writing, the last and most comprehensive report of this committee was published in 1972 (15).

It is widely accepted that the hazard to man can best be evaluated on the basis of the deposited fallout rather than the concentration of radioactive debris in the atmosphere. This arises primarily from the fact that most of the debris is carried to the earth's surface in rainfall and the greatest source of human exposure is due to radionuclides absorbed by man via the food chain. Thus, in the case of ^{131}I, the dose received from inhalation of atmospheric iodine has been shown to be insignificant compared to the dose due to ingestion of radioiodine in milk.

The principal dose received from inhaled radioactive debris from weapons tests is to the lung and originates from ^{95}Zr and ^{144}Ce. Wrenn *et al.* (34) have shown that during periods of weapons testing, ^{95}Zr and its radioactive daughter, ^{95}Nb, are present in human lungs and can be measured *in vivo* using a sodium iodide crystal positioned directly in contact with the chest. In New York City, where their measurements were made, Wrenn and his associates found that the dose to the lungs was about 3 μrem/day from this pair of isotopes. In additional work, Wrenn demonstrated that the dose from inhaled ^{144}Ce was about 21 μrem/day (35). These doses are very small compared to the dose received by the lung from the radon normally present in the atmosphere. During the period of heavy weapons testing in 1962 and 1963, it is estimated that the dose to the lungs of all inhaled radionuclides was of the order of a few millirads per year.

Food chain contamination may occur either by direct deposition on plants, or by contamination of the soil and subsequent absorption by plants. This, of course, is different than the case of most pollutants where the potential hazard is primarily from inhalation and where the risk is likely to be dissipated once deposition on surfaces has occurred. However, radioactive debris does not constitute a significant problem until it has settled out of the atmosphere. As we have seen, the amount present at any given time in the atmosphere as a result of weapons testing is so slight that the inhalation hazard is relatively minor. On the other hand, when the material has precipitated from the atmosphere, it remains in the soil and may become available to the biosphere in proportion to its presence in soil. Deposition on forage can be ingested by cows and concentrated in milk. For this reason a useful parameter to judge the risk from fallout is the number of millicuries per square mile on the ground at any given time.

The fallout in any interval of time can be estimated in a variety of ways, using pots, pans, soil samples, and adhesive surfaces. Measurement from atmospheric weapons tests has been undertaken on a very extensive scale because knowledge of the global distribution of the debris makes it possible to forecast the dose that will ultimately ensue to human beings in various parts of the world. In addition, the radioactive dusts make excellent tracers by which global circulation can be studied. Since much of the debris is injected into the stratosphere, the dust is also an excellent tracer for studying transfer of air from the stratosphere to the troposphere and the movement of stratospheric air across the equator.

Most of the worldwide fallout measurements have been of ^{90}Sr because this isotope is so much more toxic than other long-lived constituents of the radioactive debris. Figure 7 gives the cumulative deposition of ^{90}Sr as of the end of 1970 (36). The data are given for the northern and southern hemispheres, and it is seen that the deposition in northern latitudes is about three times that below the equator. This reflects the fact that testing has been performed mainly in the northern hemisphere.

Dairy products are the main dietary source of ^{90}Sr in western countries. The ^{90}Sr content of fresh milk in New York City is shown in Figure 8.

^{137}Cs is another long-lived nuclide of importance, which is widely present in the staples of diet. When absorbed by humans, it deposits in muscle from which it is eliminated with a half-time that ranges from 50 to 150 days.

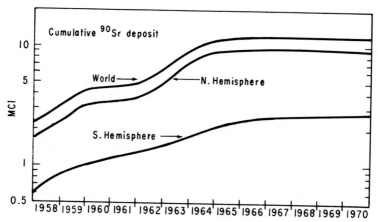

Figure 7. Accumulation of ^{90}Sr on the earth's surface from 1958 through 1970 (*36*).

Of the short-lived nuclides, 8-day radioiodine is of particular importance because of its tendency to pass via cows' milk to the human thyroid (*37*).

VII. Aerospace Applications of Nuclear Energy

A potentially important application of nuclear technology is the thermoelectric conversion of decay heat as a source of power for the operation of satellites and space probes (*1*).

With the advent of miniaturized, transistorized electronic circuits, there has developed a need for small power sources to provide the power necessary to operate the instrumentation and controls on space vehicles. In order for an isotope to be useful as a source of thermoelectricity, its half-life must be sufficiently long to offer the prospect of a reasonably long period of service, and yet sufficiently short so that its specific activity is not too low. An additional requirement is that the particle emitted

Figure 8. ^{90}Sr in whole milk in New York City, 1954–1971. One liter of milk contains ~1 gm of calcium (United States Atomic Energy Commission, Health and Safety Laboratory).

by the isotope be sufficiently energetic to yield a reasonable amount of power per curie.

Isotopic power generators were originally conceived as satellite nuclear auxiliary power (SNAP) units, but several terrestrial applications have also developed. In general, strontium-90 has been popular as a heat source for land-based devices, and plutonium-238 and polonium-210 are utilized in the space program. The first space unit was launched in 1961 to provide 5 W of power for a navigational satellite that continues in operation at this writing. More recently, in 1972, the astronauts on Apollos 16 and 17 placed two additional units on the lunar surface to provide about 75 W to supply power for experimental apparatus left on the moon. Several kilowatts of power are visualized for the late 1970's and even higher power requirements for space exploration can perhaps be met by reactors of special design (1).

The SNAP units have been designed to meet conflicting criteria. On the one hand, they must be so fabricated as to withstand launch pad aborts without compromising the integrity of the capsule holding the radioactive substance. On the other hand, having achieved orbit, they have been designed to volatize upon reentry to the earth's surface.

A navigational satellite launched on April 21, 1964 (SNAP 9-A) failed to reach orbital velocity and reentered the atmosphere at about 150,000 ft over the Indian Ocean. The isotopic power unit contained about 17,000 Ci of plutonium-238. The stratospheric distribution of the debris during the next several years followed predictions that had been made on the basis of transport models of the stratosphere developed through studies of the debris from weapons testing. However, the concentration of plutonium-238 in ground-level air was somewhat lower than had been predicted. About 95% of the amount originally injected was estimated to have been deposited on the earth's surface by the end of 1970.

VIII. Peaceful Uses of Nuclear Explosives

The enormous amount of energy available from nuclear devices of compact design and the relative ease with which the energy equivalent of millions of tons of TNT can be emplaced have resulted in widespread interest in the United States and the Soviet Union concerning the use of nuclear explosives for civilian purposes. In the United States, the program is known allegorically as "Plowshare," and since its inception in 1957, a considerable amount of laboratory and field work has been undertaken. The principal interest is focused on natural gas stimulation, because of the projected acute shortages in natural gas that are expected to develop in the next 15–20 years (*38*). Nuclear explosives can be used to fracture rocks containing tightly bound natural gas that would otherwise be unavailable to wells. Three test explosions called Gasbuggy, Rulison, and Rio Branco have been conducted in New Mexico and Colorado. However, other applications of nuclear explosives have fared less well. Extensive studies were made of the feasibility of constructing a second canal across the Panama Isthmus, using a series of nuclear cratering explosions, but the plan was abandoned in 1970. Other applications that have been considered include production of transuranic elements, power production by exploitation of the heat deposited by deep underground nuclear explosions, underground fracturing for the recovery of oil and low-grade copper ores, and construction of artificial harbors (*1*).

A large number of test explosions have been conducted underground in Nevada and these have occasionally vented small amounts of radioactive debris to the atmosphere. However, the total amounts have been small compared to those associated with nuclear weapons testing.

The amounts of radioactivity that will remain essentially *in situ* following a Plowshare explosion, and the quantities that will be dispersed to the environment, will depend on the depth and size of the explosion, the underground hydrology, and the amount of radioactivity that vents to the atmosphere. There are three types of Plowshare explosions: (a) the energy and debris are contained totally below ground, in which case there is no gross venting to the atmosphere, although some gradual seepage of the more volatile constituents takes place; (b) the explosion in relation to its size is so close to the surface that debris is ejected to the atmosphere with partial fallback into the crater produced; and (c) the explosion vents to the atmosphere but it is arranged so that there is almost total fallback of debris, producing what is called a "retarc" in Plowshare nomenclature.

The ideal Plowshare device would be totally thermonuclear, deriving its energy from the fusion of light elements such as hydrogen or lithium. In discussing the characteristics of radioactivity produced by nuclear explosives, Miskel (*39*) assumed, for purposes of calculation, that the explosive yield is derived 99% from fusion and 1% from fission. This minimizes the production of potentially noxious fission products such as radioiodines, strontium-90, or cesium-137. However, substantial quantities of tritium are produced, as well as activation products caused by neutron irradiation of the soil and rock. Generally speaking, the activation products are less noxious than the fission products.

Studies have been undertaken of the potential public health risks due to radioactive contamination of natural gas obtained by Plowshare stimulation. Three nuclides, tritium, carbon-14, and krypton-85, are the principal isotopes observed in the gas a few months following detonation. Tritium has been identified as the most important of the three nuclides, and calculations have been undertaken (40) of the population dose in a metropolitan area that uses such gas. Present estimates are that the maximum exposure from tritium would be about 2.5 mrem/year, with an average exposure of 0.5 mrem/year. Future studies of the gas produced in the experimental programs will no doubt make it possible to refine projections of this kind (41).

REFERENCES

1. M. Eisenbud, "Environmental Radioactivity," 2nd ed. Academic Press, New York, New York, 1973.
2. "Report of the United Nations Scientific Committee on the Effects of Atomic Radiation." United Nations, New York, New York, 1958.
3. J. E. Pearson and G. E. Jones, *J. Geophys. Res.* 70, 5279 (1965).
4. M. H. Wilkening, *Rev. Sci. Instrum.* 23, 13 (1952).
5. K. Kawasaki and K. Kato, *J. Phys. Soc. Jap.* 14, 234 (1959).
6. M. Eisenbud and H. G. Petrow, *Science* 144, 3616 (1964).
7. J. E. Martin, E. D. Harward, D. T. Oakley, J. M. Smith, and P. H. Bedrosian, *Proc. Environ. Aspects Nucl. Power Sta.,* Rep. No. SM-146/19. International Atomic Energy Agency, Vienna, Austria (1971).
8. L. B. Lockhart, Jr., *in* "The Natural Radiation Environment" (J. A. S. Adams and W. M. Lowder, eds.), pp. 331–344. Univ. of Chicago Press, Chicago, Illinois, 1964.
9. F. Barreira, *Nature (London)* 190, 1092 (1961).
10. J. M. Prospero and T. N. Carlson, *Science* 167, 974 (1970).
11. I. H. Blifford and L. B. Lockhart, *U.S. Nav. Res. Lab. Rep.* 4036 (1952).
12. M. H. Wilkening, *in* "The Natural Radiation Environment" (J. A. S. Adams and W. M. Lowder, eds.), p. 359. Univ. of Chicago Press, Chicago, Illinois, 1964.
13. R. W. Perkins and J. M. Nielsen, *Health Phys.* 11, 12 (1965).
14. S. A. Korff, personal communication (1971).
15. "Report of the United Nations Scientific Committee on the Effects of Atomic Radiation." United Nations, New York, New York, 1972.
16. G. D. Clayton, J. R. Arnold, and F. A. Patty, *Science* 122, 751 (1955).
17. J. P. Lodge, Jr., G. S. Bien, and H. E. Suess, *Int. J. Air. Pollut.* 2, 309 (1960).
18. D. A. Holaday, *in* "Industrial Radioactive Waste Disposal," Hearings before Joint Committee on Atomic Energy. U.S. Govt. Printing Office, Washington, D.C., 1959.
19. R. N. Snelling, *Radiol. Health Data Rep.* 10, 475 (1969).
20. R. N. Snelling, *Radiol. Health Data Rep.* 12, 17 (1971).
21. A. J. Breslin and H. Glauberman, *in* "Environmental Surveillance in the Vicinity of Nuclear Facilities" (W. C. Reinig, ed.), p. 535. Thomas, Springfield, Illinois, 1970.
21a. U.S. Atomic Energy Commission, "Code of Federal Regulations," Title 10, Part 20. Washington, D.C., 1969.
22. S. D. Shearer, Jr. and C. W. Sill, *Health Phys.* 17, 77 (1969).

23. U.S. Public Health Service, "Evaluation of Radon-222 near Uranium Failings Piles," DER 69–1. U.S. Department of Health, Education and Welfare, Washington, D.C., 1969.
24. P. B. Klevin, M. S. Weinstein, and W. B. Harris, *Amer. Ind. Hyg. Ass., Quart.* **17,** 189 (1956).
25. International Atomic Energy Agency, "Power and Research Reactors in Member States." Vienna, Austria, 1970.
26. B. Kahn, R. L. Blanchard, H. L. Krieger, H. E. Kolde, D. B. Smith, A. Martin, S. Gold, W. J. Averett, W. L. Brinck, and G. J. Karches, *Proc. Environ. Aspects Nucl. Power Sta.,* International Atomic Energy Agency, Vienna, Austria.
27. U.S. Atomic Energy Commission, "Report on Releases of Radioactivity in Effluents from Nuclear Power Plants for 1971." Directorate of Regulatory Operations, Washington, D.C., 1972.
28. E. C. Miller, "The Integrity of Reactor Pressure Vessels," Rep. No. 15. Nucl. Safety Inform., Oak Ridge Nat. Lab., Oak Ridge, Tennessee, 1966.
29. S. J. Weems, W. G. Lyman, and P. B. Haga, *Nucl. Safety* **11,** 215 (1970).
30. U.S. Atomic Energy Commission Report by the SL-1 Incident Investigation Board, Washington, D.C., 1961.
31. "Accident at Windscale No. 1 Pile on 10th October, 1957." HM Stationery Office, London, England, 1957.
32. S. Glasstone, "The Effects of Nuclear Weapons." U.S. Atomic Energy Commission, Washington, D.C., 1962.
33. Federal Radiation Council, Rep. No. 4. Washington, D.C., 1963.
34. M. E. Wrenn, R. Mowafy, and G. R. Laurer, *Health Phys.* **10,** 12 (1964).
35. M. E. Wrenn, Ph.D. thesis, New York University, New York, New York, 1966.
36. H. L. Volchok and M. T. Kleinman, "Worldwide Deposition of Sr⁹⁰ thru 1970," Rep. HASL-243. U.S. Atomic Energy Commission, Washington, D.C. 1971.
37. M. Eisenbud and M. E. Wrenn, *Health Phys.* **9,** 12 (1963).
38. S. Smith, *Nucl. Technol.* **11,** 331 (1971).
39. J. A. Miskel, *Proc. Plowshare Symp. Eng. Nucl. Explosives, 3rd, 1964* (1964).
40. C. J. Barton, D. G. Jacobs, M. J. Kelly, and E. G. Struxness, *Nucl. Technol.* **11,** 335 (1971).
41. D. G. Jacobs, C. J. Barton, M. J. Kelly, C. R. Bowman, S. R. Hanna, F. A. Gifford, Jr., and W. M. Culkowski, "Theoretical Evaluation of Consumer Products from Project Gasbuggy," Final Rep., Phase II, ORNL-4748. Oak Ridge Nat. Lab., Oak Ridge, Tennessee, 1972.

Part **B**

THE
TRANSFORMATION
OF
AIR
POLLUTANTS

6

Atmospheric Reactions and Scavenging Processes

A. J. Haagen-Smit and Lowell G. Wayne

I. Principles of Atmospheric Reactions

A. Primary and Derived Air Pollutants

Substances emitted to the atmosphere are subjected to a variety of physical and chemical influences that may lead to the formation of objectionable products or, on the other hand, objectionable products may be converted into harmless ones. Investigations of these reactions are helpful in establishing the nature of the precursors of the substances found in the air; consequently, a great deal of research has been carried out to explore in detail the mechanism of these reactions and the effect of the resulting products on physiological and toxicological phenomena in plants and animals.

The emissions undergoing these changes are products of human activity, from industrial as well as individual sources, and represent a variety of inorganic and organic materials. Evaporation of gasoline alone accounts for a mixture of hundreds of species of organic molecules. Even more complicated are the secondary transformation products found in automobile exhaust and in the emissions from incineration, which comprise a mixture of dry distilled, steam distilled, and destructively distilled products from essentially every material used by people. These thousands of compounds are then mixed with the natural components of the air and remain in the atmosphere for a considerable time. During this residence time unstable molecules will rearrange or couple with other substances. Both primary and secondary products are exposed to further changes through oxidation and photochemical reactions.

The overall tendencies toward the formation of atmospherically stable products are seen in a practical example of the effect of atmospheric reaction on concentrations of pollutants emitted in the Los Angeles area (1). Using known emission of nonreactive carbon monoxide as a reference, the expected atmospheric concentrations of other pollutants were calculated from source-testing data and then compared with actual measurements of the atmosphere. This study showed that oxides of nitrogen, sulfur dioxide, and hydrocarbons found are less than the amounts calculated from the emission of known sources. On the other hand, aldehydes and organic acids, both oxidation products of hydrocarbons, had increased above their calculated concentrations.

An attempted air pollution balance comparison of automobile emissions before and after exposure to atmospheric influences is found in the work of Weaver et al. (2). From their spectrographic data analysis of smog condensates, a gradual oxidation of hydrocarbons and disappearance of the more fragile olefins with formation of oxygen-containing carbonyl compounds and acids can be deduced.

B. Atmospheric Assimilation and Removal of Pollutants

Speculations about the ultimate result of the reactions of inorganic and organic materials in the atmosphere can be made with reasonable certainty. Through turbulence and diffusion, substances released at ground level will eventually be exposed to conditions in the upper atmosphere and become accessible to high energy photons which break up even the most stable molecules. It is in these regions that molecular oxygen is dissociated, and at heights above 50 miles oxygen exists almost exclusively in monatomic form. At lower levels, where concentrations are higher, the monatomic, quite reactive atoms partially combine to form diatomic molecular oxygen and triatomic ozone. The region of greatest ozone concentration is located between 10 and 20 miles above the earth (*3*).The reactions of oxygen in the upper atmosphere are summarized in the following equations (M represents an energy-accepting third body):

$$O_2 + h\nu(\lambda 2000) \rightarrow O + O \tag{1}$$
$$O + O + M \rightarrow O_2 + M \tag{2}$$
$$O + O_2 + M \rightarrow O_3 + M \tag{3}$$
$$O_3 + h\nu(\lambda 2900-2000) \rightarrow O_2 + O \tag{4}$$

It is generally assumed that, through the continuous mixing taking place in the atmosphere, some of the ozone is carried downward to establish the concentration of 0.03 ppm of ozone normally found in the lower atmosphere. In higher atmosphere regions several oxides of nitrogen—N_2O, NO, and NO_2—occur. Their reactions with ozone have received considerable attention because of their relation to the phenomenon of airglow. In order to reach ground level the oxides must pass the ozonosphere, which results in a complete oxidation of nitric acid. It is therefore not expected that nitrogen oxides formed in the upper atmosphere will materially increase the ground-level concentration due to industrial and man-made pollution (*4*). Conversely, these man-made oxides of nitrogen, when reaching the upper atmosphere, will adjust themselves to the equilibrium conditions prevalent at the higher regions.

At lower altitudes photochemical effects are less drastic, owing to the filtering effect of the ozone layer on ultraviolet radiation and the sharp termination of the solar spectrum at 2900 Å. For these lower regions we can predict that under the milder conditions inorganic materials will be converted to stable salts—mostly sulfates, nitrates, and chlorides—which will eventually be added to the soil. Organic compounds will end up as carbon dioxide and water. Their nitrogen will be found in ammonium salts and nitrates.

In this assimilation process the composition of the earth and its atmosphere is not changed qualitatively. Noticeable quantitative changes in the atmosphere do occur, however, through the years, by the release of billions of tons of carbon dioxide in the burning of fossil fuels. The concentration level of carbon dioxide is rising slowly, and has been estimated to increase by almost 1 ppm/year. Its effect on living conditions on our planet has been a matter of speculation. The problem of the effects of continued increases in atmospheric carbon dioxide on weather and plant growth has been discussed by Peterson ($4a$).

C. Formation of Intermediate Reaction Products

For air pollution control purposes we are usually more interested in the chain of events that precedes the formation of stable end products. The investigation of the intermediates formed from an exceedingly complex mixture of substances present in minute quantities is very difficult. Fortunately, we can establish some restrictions or simplifications in the discussion of atmospheric reactions. One of the first requirements is that the reaction rate be reasonably fast, so that the reactions will have significantly advanced during the time that the pollutants are still in the area being studied, and are present in concentrations sufficiently high to affect us in some way.

The large preponderance of some atmospheric constituents over others is another factor which guides us in predicting the probability of certain reactions. For example, the relatively high concentration of oxygen (209,400 ppm, v/v) makes it one of the most important participants in various reactions with air pollutants. Since rates of reactions are dependent on the concentrations of the participating reactants, reactions with oxygen, present at concentrations often a million times greater than that of the pollutant, are more likely to occur than when both substances are present in concentrations of only a fraction of a part per million. For example, when a bimolecular reaction with participants in the concentration range of 10% (100,000 ppm) requires 0.0036 second to go halfway to completion, it will take 10^6 times as long, or a whole hour, to reach the same point with concentrations of 0.1 ppm. If such reactions are to be at all significant, their rate must be quite high as compared with those with oxygen. The same holds true, although to a far lesser degree, for reactions with water and carbon dioxide, which are present, respectively, in concentrations of 1000–50,000 and 300–1000 ppm v/v.

D. Photochemical and Photosensitized Reactions

1. Energy Requirements

For chemical reactions to take place it is necessary to add a certain amount of energy derived from molecular collisions or from radiation. In atmospheric reactions the temperature is usually limited to a narrow range which does not supply enough energy for most reactions. The light-activated processes, on the other hand, take a most important place in atmospheric reactions, and may be considered initiators of most dark reactions by endowing atoms and molecules with the necessary activation energy. The limitations set to the photochemical reactions are contained in the two fundamental laws of photochemistry.

The first law requires that light must be absorbed by the reacting atoms or molecules. For example, when visible light shows some photochemical activity there is a colored substance involved. In plants it is the green chlorophyll which makes the energy of the sun available for photosynthesis. In air pollution problems it is often the orange-colored nitrogen dioxide that plays the role of light energy acceptor.

The second law of photochemistry states that one molecule of a reacting substance may be activated by the absorption of one light quantum. A light quantum is the smallest amount of energy that can be removed from a beam of light by any material system. A molecule therefore can absorb several of these quanta, but not less than one. The size of this energy unit is directly proportional to the frequency of the light, and is usually expressed as $h\nu$, where ν is the frequency of the light, and h a constant (Planck's constant) having the value of 6.62×10^{27} erg seconds. As seen in Figure 1, the energy contained in quanta in the long wavelength area of infrared is relatively small as compared with that needed for the breaking of bonds between atoms such as the carbon–carbon or carbon–hydrogen bonds, respectively, 84 and 100 kcal per mole

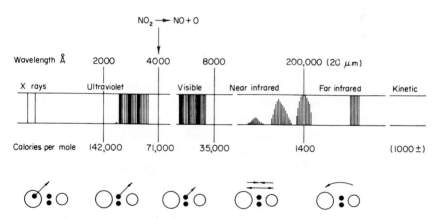

Figure 1. Schematic representation of different types of spectra, showing wavelength ranges and energy ranges in which they occur, and the corresponding electronic, vibrational, and rotational motions (5). [Reprinted with permission from F. Daniels, "Outlines of Physical Chemistry," 1st ed. Wiley New York, New York, 1951. Copyright John Wiley & Sons.]

(5). In this region, quanta will influence vibration and rotation and will heat the molecule, but they cannot supply enough energy to an individual molecule to overcome the forces which hold its atoms together. In the visible and especially the ultraviolet range of the spectrum, chemical bonds can be broken, and the only limit set is due to the presence of the ozone layer in the upper atmosphere which does not permit radiation of wavelengths below 2900 Å to reach the earth's surface.

A typical example of a photochemical primary reaction is given by the photodissociation of the common air pollutant, nitrogen dioxide. The bond strength between the nitrogen and oxygen is of the order of the energy corresponding to wavelengths smaller than 4500 Å. The photodissociation actually takes place at wavelengths smaller than 3800 Å; above this region, the "excited" molecule retains the energy without dissociating until it loses the extra energy by emitting light in fluorescence or phosphorescence, or dissociates as a result of secondary reaction. In other cases atoms and molecules endowed with the extra energy may transfer this to other molecules.

The energy required to cause dissociation of a molecule is usually recoverable when the fragments—free atoms or radicals*—are allowed to recombine. However, in the gas phase, these highly reactive species seldom recombine immediately. Instead, for some finite time, they retain their individual identities while colliding with other molecules in the gas phase. They are very likely to react with ordinary molecular constituents of the atmosphere whenever such reactions are possible, rather than to return to their original partners. These reactions thus utilize the energy originally furnished by the absorption of the light quantum.

The energy of excitation of nondissociating molecules may also promote chemical changes. Although absorbing only faintly in the red end of the visible spectrum, the high concentration at which oxygen occurs will supply a relatively large number of excited oxygen molecules. The excited molecules may possibly react with aldehydes or with hydrocarbons to form free radicals. Both reactions are of special importance in air pollution studies. The maximum rate of reaction between excited oxygen molecules and a pollutant present at concentrations of 50 pphm has been estimated at 1–7 pphm/hour (6). In these reactions the excited oxygen returns to its normal state after it has functioned as a temporary carrier of light energy.

---

* In the remainder of this chapter, free radicals or atoms possessing odd numbers of electrons will be denoted by a dot associated with the chemical formula; e.g., $CH_3 \cdot$, methyl radical. This convention is not used for nitric oxide (NO) and nitrogen dioxide (NO_2), even though these are odd-election molecules.

2. Light Source

In addition to the limitations set by the absorbing materials, there is an important factor in the variable intensity of the light. This factor has been discussed in extensive theoretical reviews by Leighton and Perkins (6–8). In their calculations the cyclic effects of direct and scattered light at ground level have been considered. The sunlight passing through the atmosphere is subject to molecular and particulate scattering, reflection, refraction, diffraction, and absorption. The intensity depends further on the length of the path of the direct solar radiation and is a function of the zenith angle and surface reflection. At conditions favorable to scattering, the absorption at short wavelengths may be greater than that of direct sunlight. The scattering within the polluted layer may cause the absorption rate at the top of the layer to be several times greater than at the bottom, and at noon the average absorption rate within the layer may be 20–30% greater than it would be in the absence of scattering. At ground level, Renzetti (9) determined that over the period August–November, 1954, the average total daily radiation (sun + sky at all wavelengths) in downtown Los Angeles was some 10% below what it should have been in the absence of pollution. Spectroradiometer measurements conducted in Pasadena by Stair (10) showed that at times of intense pollution solar radiation was reduced in the ultraviolet region near 3200 Å by more than 80%.

3. Primary Photochemical Reactions in Air Pollution

Absorption rates and estimated upper limits for the rates of primary photochemical processes in urban air under certain radiation conditions and absorber concentration of 10 pphm have been calculated by Leighton and Perkins (6:95–96). By comparing these rates with those actually observed in stationary or moving air pollution clouds, we obtain some information regarding the major contributing reactions. For the Los Angeles area the photodissociation of nitrogen dioxide into nitric oxide and atomic oxygen seems to be the most important primary photochemical process [Eq. (5)]. Next in importance is the photodissociation of aldehydes into free radicals [Eq. (6)]:

$$NO_2 + h\nu \rightarrow NO + O \tag{5}$$

$$R-\overset{\displaystyle O}{\underset{\displaystyle H}{\overset{\|}{C}}} + h\nu \rightarrow \dot{R} + H\dot{C}O \tag{6}$$

While this is probably true for most modern urban areas, there is a distinct possibility that under different circumstances and different patterns of emissions, the dominant role of nitrogen dioxide may be taken over by other primary absorbers—organic compounds of many types, halogens, and other inorganic compounds, as, for example, particulate metal oxides.

A case was reported in Midland, Michigan (11), in which an eye irritant was formed by the chemical combination of two effluents—styrene and halogens—both occurring at concentrations of a few parts per million or less. The formation of the irritant in laboratory experiments was catalyzed by ultraviolet radiation, and in practical circumstances seemed to be dependent upon the presence of sunlight. The role of ultraviolet radiation in these tests suggests that some reaction product other than the direct substitution product may be the actual irritant.

4. Free Radical Formation

An important feature of atmospheric reactions is the formation of free radicals. The great dilution of the radicals—less than 1 pphm—results in half-lives of minutes, and even hours, as Johnston (12) has calculated. A possibility that these reactive intermediates may play a role in eye irritation reactions has been suggested (7:195). Free radicals are produced in the photodissociation of aldehydes, as shown in Equation (6), but they are also formed by a great number of other compounds which absorb solar radiation. These compounds include aldehydes; ketones; alkyl, acyl and peroxyacyl nitrates; hydrogen peroxide and organic peroxides; nitrous and nitric acid [Eqs. (7)–(12)] (7:119). Free radical formation may be indicated as follows:

$$\begin{array}{c} R_1 \\ \diagdown \\ C{=}O + h\nu \to \dot{R} + R\dot{C}O \\ \diagup \\ R_2 \end{array} \qquad (7)$$

$$H_2O_2 + h\nu \to 2\dot{O}H \qquad (8)$$

$$RONO + h\nu \to R\dot{O} + NO \qquad (9)$$
$$\searrow \dot{R} + NO_2 \qquad (10)$$

$$HNO_2 + h\nu \to \dot{O}H + NO \qquad (11)$$
$$\searrow \dot{H} + NO_2 \qquad (12)$$

Some reactions which are not photochemical primary processes may also generate free radicals. One possibility is the reaction of ozone with olefins, for example,

$$O_3 + RCH=CHR \rightarrow RCHO + R\dot{O} + H\dot{C}O \tag{13}$$

The reactive free radicals would find themselves surrounded by oxygen molecules, and peroxy radicals would be therefore readily formed. Equation (14) shows this reaction, in which R could be a hydrogen atom as well as an alkyl or acyl group, forming the corresponding peroxy radicals.

$$\dot{R} + O_2 \rightarrow ROO\cdot \tag{14}$$

The peroxy radicals readily react further with nitrogen oxides and other primary and derived air pollutants, to yield a variety of products including alkyl nitrates, peroxyacyl nitrates, alcohols, ethers, acids, and peroxyacids. A selection of these postulated secondary reactions is presented as follows:

$$ROO\cdot + NO \xrightarrow{h\nu} ROONO \rightarrow RO\cdot + NO_2 \tag{15}$$
$$ROO\cdot + NO_2 \rightarrow ROONO_2 \tag{16}$$

$$ROO\cdot + C_2H_3R \rightarrow ROOCHR - \dot{C}H_2 \tag{17}$$

$$ROO\cdot + SO_2 \rightarrow ROO\dot{S}O_2 \tag{18}$$
$$ROO\cdot + O_3 \rightarrow RO\cdot + 2O_2 \tag{19}$$

5. Chain Reactions

The secondary reaction products are again subject to chemical and photochemical attack. For example, photochemical decomposition of a peroxyacyl nitrite may result in the formation of acylate radical and nitrogen dioxide [Eq. (20)], thereby accomplishing a transfer of an oxygen atom from molecular oxygen to nitrogen dioxide

$$\overset{O}{\underset{\|}{R\overset{}{C}}}-OONO \xrightarrow{h\nu} \overset{O}{\underset{\|}{R\overset{}{C}}}O\cdot + NO_2 \tag{20}$$

via the peroxy radical.

Many reactions of free radicals, like the reactions shown in Equations (14), (15), (17), (18), and (19), yield other free radicals as products. Since the product radicals are also highly reactive, they enter into similar reactions yielding still other radicals. The sequence of steps involving these species resembles a chain of reactions linked by the radical species produced in one step and consumed in the next. Such a chain might be endless except for the possibility of chain-terminating reactions from which no radical species emerge, such as (for example) Equations (16) and (21).

$$ROO \cdot + RO \cdot \rightarrow ROR + O_2 \tag{21}$$

If the chain-terminating steps occur only rarely by comparison with the steps in the chain, the chains will be long; conversely, chains are short if chain-terminating steps are plentiful.

Chain-terminating steps which may be expected in the atmosphere are of three main types: (a) those in which two free radicals interact, such as Equation (21); (b) those in which a free radical reacts with nitric oxide or nitrogen dioxide to form an adduct, as in Equation (16) or (22);

$$RO \cdot + NO_2 \rightarrow RONO_2 \tag{22}$$

and (c) those in which a radical impinges on the surface of a particle and adheres, contributing to the formation of a polymeric droplet or adsorbed layer.

Reaction chains of as many as 10^6 steps are not uncommon in gas phase photochemistry. These long chains usually involve extended repetition of a short set of reactions which starts and ends with a particular active radical species. Equations (23) and (24) schematically illustrate a pair of reactions of this type:

$$A + R \cdot \rightarrow RA \cdot \tag{23}$$
$$RA \cdot + B = AB + R \cdot \tag{24}$$

The net result of this pair of reactions (taken in sequence) is the conversion of a molecule of A and a molecule of B to a molecule of their addition product. Since the radical R emerges at the end ready to enter another cycle, the yield of product after a large number of cycles may be much greater than the amount of the radical species present in the system. In the chemistry of air pollution, such mechanisms may account for the rapid conversion of nitric oxide to nitrogen dioxide and the accumulation of ozone in smog in sunlight.

Besides the types of reactions discussed above, free radicals may also enter into chain-branching reactions, in which more radicals emerge from the reaction than enter it. Such a reaction is shown schematically by Equation (25), where the interaction of the radical with an excited molecule

$$R \cdot + AB^* = R \cdot + A \cdot + B \cdot \tag{25}$$

results in the production of two new radicals. Frequently a similar result is reached following a sequence of two or more elementary steps, the last involving dissociation of an energetic molecule into two radical fragments. If the number of chain-branching steps is not too small compared to the number of chain-terminating steps, a very rapid accumulation of free radicals can occur in such systems. This is a typical state of affairs in explosive reactions, but branching chains may also be involved in atmospheric systems.

E. Heterogeneous Reactions

The presence of solid and liquid particles suspended in the atmosphere necessitates the consideration of reactions which may occur within the particles or at their surfaces. In aqueous droplets, for example, neutralization reactions proceed more rapidly than in the gas phase. Further, such reactions as the oxidation of sulfur dioxide to sulfates (or sulfuric acid) and the hydrolysis of some of the oxides of nitrogen to nitrate or nitric acid may be promoted by catalysts dissolved in water droplets.

Again, the decomposition of intermediate organic oxidation products to more stable acids and polymers continues in the liquid state, although it started as a gas reaction. Droplets collected from air polluted with organic material show the presence of a film of polymeric material enclosing an aqueous solution containing organic and inorganic constituents. The catalytic effect of charcoal and metal oxide dusts on oxidation is well known, and labile substances such as ozone are decomposed. On the other hand, the oxidation of organic material is accelerated.

A special field of heterogeneous reactions is the one where light plays an important role (6:62; 13). For example, it has been shown that zinc oxide surfaces and water form hydrogen peroxide upon irradiation. Zinc oxide is therefore photosensitized for peroxide formation. In this process photoconduction electrons are formed, which results in electron transfer and subsequent oxidation–reduction phenomena at the crystal surface. The photosensitized formation of hydrogen peroxide and the oxidation of organic compounds by zinc oxide are not limited to liquid water medium, but have also been established for nonaqueous media and gases (14, 15).

F. Prediction of Atmospheric Concentrations

In principle, the concentrations of pollutants in urban atmospheres could be predicted in advance of their occurrence if the places and rates of contaminant emissions were known, together with the necessary meteorological parameters to specify the motion of the air masses. In practice, available information on both these points is usually exceedingly sketchy.

Nevertheless, computer models to facilitate estimation of the effects of emissions control on air quality have proliferated in recent years. Their potential value in planning control strategy is large and obvious, and substantial efforts to develop a sophisticated prediction capability are in progress, as described in Chapter 1, Vol. V.

Until recently, this problem has been approached principally in terms of local distribution of effluents assumed to be chemically stable. If the focus is shifted to atmospheric contaminants which undergo further reactions, new complications are introduced. For example, sulfur dioxide may be emitted from boiler stacks, together with fly ash containing catalytic metal oxides; under certain conditions of humidity, appreciable conversion to sulfate may take place within the plume, downwind of the stack. Knowledge of the kinetics of the catalytic conversion would be essential, in such a case, to permit prediction of either sulfate or sulfur dioxide concentration within the plume or in the general atmosphere affected by such emissions.*

The production of photochemical secondary pollutants in the atmosphere depends upon energy from sunlight and proceeds most effectively when the sun is brightest. In appropriate weather, the daily cycle of the sun therefore leads to daily cycles in the concentrations both of the reacting primary pollutants and of the photochemical products. With such a substance as ozone, which is simultaneously produced and consumed by atmospheric reactions, the problem of prediction is particularly difficult. Nevertheless, Frenkiel (16) has offered a very interesting treatment, by which he estimated the relative contributions of motor vehicles, industry, and other sources to ozone as a function of time of day at an urban location. The calculations were based on a hypothetical model of the distribution of pollution sources and on certain assumptions as to the kinetics of ozone formation. Figure 2 shows typical results, illustrating the very practical implications of such studies with respect to policy in controlling various types of emission sources.

A point of particular significance, emphasized by Frenkiel, is that owing to the chemical reactions in the atmosphere, the contributions of the various sources to ozone concentrations are not simply additive. Figure 3 shows that the ozone expected (according to the model) when all assumed sources are operating could be nearly twice as much as the sum of the values expected for industry alone and for other sources (without industry). These results illustrate that in predicting atmospheric con-

---

* However, the sum of these two concentrations, being unaffected by the conversion, can be treated in the same manner as the concentrations of chemically stable effluents.

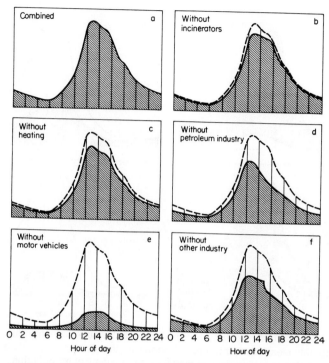

Figure 2. Relative contributions of various pollution sources to the mean concentration of ozone (ordinate) at the California Institute of Technology. These results refer to a hypothetical model of photochemical air pollution in Los Angeles County, California (16).

centrations, it is dangerous to rely heavily on unstudied assumptions regarding the kinetics of atmospheric reactions.

In view of United States federal regulations which require long-range planning with the goal of reducing photochemical oxidant levels to a smaller fraction of the levels prevalent in some cities at the time of the promulgation of these regulations, the development of appropriate air quality models became urgent. The subject of modeling of air quality is addressed in Chapter 10, this volume. We note here, however, that air quality models are validated by comparison of their output with data on contemporary urban atmospheres, while their purpose is to predict air quality under hypothetical future conditions. Therefore, models designed to deal with concentrations of oxidant, nitrogen dioxide, and other photochemically generated pollutants must provide for a realistic simulation of the chemical dynamics of the atmosphere, in addition to simulating the dispersion of pollutants by air mass transport and turbulence.

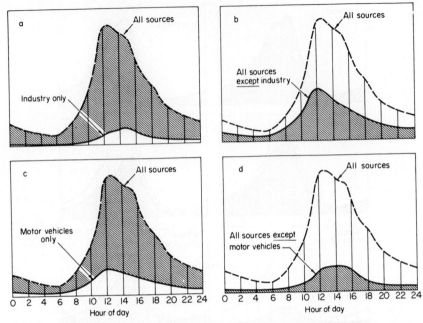

Figure 3. Contributions of industry (including petroleum industry) and of automobiles to the mean concentration of ozone at the California Institute of Technology, according to a hypothetical model of photochemical air pollution.

In the development of models for photochemically generated pollutants, one issue is the degree of validity required in the representation of the chemical kinetics within the models. A rather sophisticated capability for computer simulation of the chemical kinetics, according to chemically valid principles, is already in existence (Section II,B); it depends upon the identification of individual elementary reactions among the numerous long-lived and transitory components of the irradiated atmosphere. However, to utilize this capability and to preserve the degree of validity it can furnish, requires a major share of the computer capacity which can be made available for any air quality model, given the state of the computer art. Determining the appropriate balance between chemical validity (and therefore the overall validity of predictions of photochemical air quality), and cost and convenience of computation, is surely one of the most significant problems in the modeling field.

There have been several attempts to incorporate photochemical kinetics into air quality models. They utilize various degrees of empiricism to reduce the number of individual reaction steps. As a result, these models are deficient theoretically in that the hydrocarbons are dealt with as comprising only one or two chemical species. The system used by Wayne *et al.* (*17*), with this simplification, accommodates 30 or more postulated elementary reactions. Models developed by Roth (*18*) and Eschenroeder and Martinez (*19*) employ additional empirical artifices to reduce the reaction set to below 20, but permit more complicated simulation of other aspects of the atmospheric system.

On the whole, good progress has been made in developing photochemical air quality models, but there is a paucity of detailed atmospheric data suitable for validating them, and there is a hazard that the long-range predictions made by these models may be significantly in error due to an inadequate understanding of the chemical kinetics of the photo-oxidation system.

It is, therefore, essential to develop a mechanism based on elementary reactions, which are the basic units of chemical kinetics. The principles to be applied are well established and have been reviewed in substantial detail by Leighton (*8*).

In the following section, the most probable reactions of various pollutants, common to most urban areas, are discussed, with an emphasis on these mechanistic aspects.

II. Atmospheric Reactions in Air Pollution

A. Inorganic Pollutants

1. Oxides of Sulfur

One of the most common types of pollution, the emission of sulfur dioxide, provides a typical example of a set of reactions leading toward atmospheric stability. When sulfur, present in reduced form in metal sulfides or in complex organic molecules in oil or coal, is roasted or burned, most of it escapes as sulfur dioxide. In a hot plume, such as may be emitted from a smelter or from a fossil fuel-fired power plant, some of the sulfur dioxide is quickly oxidized to sulfur trioxide. This process may be catalyzed by metallic constituents of the suspended ash. Water vapor, also abundant in such plumes, rapidly hydrates the sulfur trioxide to form sulfuric acid, which is largely responsible for the whitish and bluish smokes typical of "sulfur dioxide" emissions.

After the contaminants have been diffused from the hot plume into the general atmosphere, the oxidation continues at a greatly reduced rate, for turbulent diffusion lowers the SO_2 concentration quite rapidly to only a few parts per million at a few hundred feet from the source. Under these conditions the gas-phase dark oxidation is immeasurably slow. Although light absorption by SO_2 extends to the near ultraviolet, the photochemical oxidation is also slow; experimental measurements of the rate in artificial light intended to simulate the intensity of natural sunlight in the ultraviolet yielded values as divergent as 0.01% per minute (20) and 0.4% per minute (21). A few tests in natural sunlight showed an apparent order of magnitude of 0.1% per hour (0.006% per minute) (20), while considerations based on observed quenching rates for excited (triplet) SO_2 suggested a theoretical maximum of 1.9% per hour (0.03% per minute) (22). This rate is only slightly increased by the presence of nitrogen dioxide (23). Sulfuric acid droplets, mainly less than 0.5 μm in diameter, are formed, contributing to the general haziness of inhabited areas.

As soon as some oxidation of SO_2 has taken place, droplets are formed, and the stage is set for reactions in the liquid phase. This phenomenon has been studied in an ingenious manner on single droplets by the University of Illinois workers (24). They found that the rate of oxidation of SO_2 was materially increased by the presence of iron and manganese salts. This accelerated oxidation may perhaps be realized in fogs and stack plumes. The end result of the combined effect of these different types of reaction is the formation of sulfuric acid, which is soon converted to sulfates such as ammonium and, especially, calcium sulfate.

The rate of oxidation of sulfur dioxide is also enhanced by the presence of reactive intermediate species in the photooxidation of hydrocarbons sensitized by oxides of nitrogen; i.e., by photochemical smog; cf. Section II,B. Automobile exhaust and also olefins in the presence of oxides of nitrogen have been shown to cause a more rapid oxidation of SO_2 by the photochemically produced oxidants (23, $25-27$). This phenomenon may also be responsible for some of the higher estimates of the SO_2 photooxidation rate, discussed above (28).

Evidence from experiments with flash photolysis of sulfur dioxide (29) points to the possible existence of a metastable dimer, S_2O_4, formed by the collision of excited molecules of SO_2. This substance would have a natural lifetime of about 1 second in air at atmospheric pressure, and its investigators suggest that it may be an important component of sulfur-bearing stack plumes.

An article entitled "The Sulfur Cycle" describes the production and conversions of man-made as well as natural sources and gives an excellent view of how our actions affect the normal balance of nature ($29a$).

2. Oxides of Nitrogen

The oxides of nitrogen present in the atmosphere are N_2O, NO, and NO_2. In addition, nitrous and nitric acids have been found free or in the form of their salts. Nitrous oxide (N_2O) is present as a regular atmospheric component at concentrations on the order of 0.5 ppm (4, 30). Its stability makes it appear unlikely that it plays an important role in low-level atmospheric reactions. The higher oxides of nitrogen are formed in chemical processes such as nitrations, but by far the largest contributor is combustion at high temperature, whereby nitric oxide is formed.

Nitric oxide (NO) reacts with oxygen to form nitrogen dioxide (NO_2), according to Equation (26):

$$2NO + O_2 \rightarrow 2NO_2 \tag{26}$$

At equilibrium conditions most of the NO is oxidized to NO_2. At a concentration of 1000 ppm colorless NO is seen to turn brown with the formation of NO_2, in a matter of seconds. At the low concentrations occurring in the atmosphere the oxidation rate is much slower. For example, at 1 ppm 100 hours are needed for a 50% conversion of NO to NO_2. At 0.1 ppm the half-life is 1000 hours. Ozone, however, oxidizes nitric oxide much more rapidly [Eq. (27)] (31, 32) and it is calculated that at concentrations of 1 ppm the half-life of NO is 1.8 seconds in this reaction.

$$NO + O_3 \rightarrow NO_2 + O_2 \tag{27}$$

If both reactants are present at 0.1 ppm, only 18 seconds are needed for virtually total oxidation.

Photochemically, nitrogen dioxide is strongly active and absorbs light over the entire visible and ultraviolet range of the solar spectrum available in the lower atmosphere (33, 34). From 6000 to about 3800 Å, the spectrum indicates the formation of excited molecules. Below 3800 Å, NO_2 dissociates to produce NO and oxygen atoms, according to Equation (5). In the atmosphere, atomic oxygen combines with molecular oxygen [Eq. (3)], forming ozone. In this two-step process equal amounts of NO and O_3 would be produced and, as we have seen previously. the rapid reaction between NO and O_3 will greatly decrease the total amount of ozone found. When nitrogen dioxide is present in a concentration of 1 ppm, the ozone formed in this process is 0.1 ppm; at 0.1 ppm of NO_2 it is 0.03 ppm, indicating that the high levels of ozone found in Los Angeles, California, and some other places are not explained by this process (35). In the presence of ozone, NO_2 is readily oxidized to NO_3,

$$NO_2 + O_3 \rightarrow NO_3 + O_2 \tag{28}$$

then to N_2O_5,

$$NO_2 + NO_3 \rightarrow N_2O_5 \tag{29}$$

which can be hydrated to nitric acid,

$$N_2O_5 + H_2O \rightarrow 2HNO_3 \qquad (30)$$

At a nitrogen dioxide concentration of 1 ppm, we estimate the half-life of ozone to be only 8 minutes. The same end results can be obtained in fog droplets, where hydration and catalytic oxidation lead to complete conversion of NO_2 to nitric acid:

$$4NO_2 + 2H_2O + O_2 \rightarrow 4HNO_3 \qquad (31)$$

A similar fate awaits any nitrosyl chloride which could possibly have been formed from sodium chloride.

Both nitric and nitrous acid may be photochemically decomposed, reforming the lower oxides according to Equations (11), (12), and (32).

$$HNO_3 + h\nu \rightarrow HO + \overset{.}{N}O_2 \qquad (32)$$

Although nitric acid vapor has never been detected in the general urban atmosphere, both inorganic and organic nitrates have been found. It is not unlikely that the failure to detect HNO_3 is due to insufficient sensitivity of available methods of measurement and analysis.

In general, we observe that the atmospheric processes tend to bring the oxides of nitrogen to the nitric acid stage. These oxidations may take place in hours or days, and during this time the nitrogen oxides participate in a number of complicated reactions in which they are switched back and forth between the various oxidation stages (36). Eventually, however, they end up largely as nitrates, which are removed from the nitrogen oxide atmospheric pool by rain (4). A smaller percentage is found in aerosols as nitro derivatives of large-molecule organic polymers.

B. Organic Pollutants: Photochemical Smog

1. Oxidation of Organic Materials

Most organic substances oxidize slowly when exposed to atmospheric influences. On a large scale, these oxidations take place in the drying of paints, deterioration of foods, disintegration of rubber, gum formation in petroleum products, and bleaching of pigments. The mechanism common to these reactions involves reaction chains, initiated by the removal of a hydrogen atom from the organic molecule, leaving a radical. This is followed by the addition of an oxygen molecule to the radical; the peroxide formed may remove a hydrogen atom from a previously intact organic molecule, as discussed above.

An example is the reaction of aldehydes with oxygen and light, in which peroxyacids are formed, as shown by the evidence of infrared spectra (*37*). The presumed mechanism, involving acyl and peroxyacyl radicals as chain carriers, is as follows:

$$CH_3-\overset{O}{\overset{\|}{C}}-H + O_2 \overset{h\nu}{\to} CH_3-\overset{O}{\overset{\|}{C}}\cdot + HOO\cdot \tag{33}$$

$$CH_3-\overset{O}{\overset{\|}{C}}\cdot + O_2 \to CH_3-\overset{O}{\overset{\|}{C}}-O-O\cdot \tag{34}$$

$$CH_3-\overset{O}{\overset{\|}{C}}-O-\dot{O} + CH_3-\overset{O}{\overset{\|}{C}}-H \to CH_3-\overset{O}{\overset{\|}{C}}-OOH + CH_3-\overset{O}{\overset{\|}{C}}\cdot \tag{35}$$

Fortunately for the continued existence of the organic world, the initiation step, the formation of radicals, is, for most molecules, a difficult one, requiring activation energy from extrinsic sources. This energy may be provided directly by light or heat, or it may be bypassed by the introduction of easily activated molecules called initiators, which react to generate free radicals at ordinary temperatures. Such materials are found in the metal–organic complexes used as catalysts in the drying of paints and oils, and the peroxides used in the polymerization of unsaturated hydrocarbons to produce plastics and synthetic rubber.

In the atmosphere, organic molecules are subject to the same process, with ozone, nitrogen dioxide, and peroxides serving as light sensitizers or initiators. The many different organic radicals thus generated are presumably consumed either by further degradation of the same type, eventually yielding carbon dioxide or carbon monoxide and water, or by polymerization and adsorption, which incorporate these fragments into the suspended particulate matter, later to be washed out in rain.

2. Symptoms of Photochemical Smog

An interesting example of a large-scale oxidation process, with nitrogen dioxide serving as a light sensitizer, was first observed in Los Angeles, California. The result is known as "photochemical smog," a phenomenon accompanied by eye irritation, plant damage, haze, ozone formation, and a characteristic odor. The typical smog episode occurs in warm, sunny weather and produces an atmosphere with a high oxidant value, due largely to ozone but in part to organic peroxides. Oxidant measurements show a definite daily rhythm, with a maximum during the daytime, and values near zero at night.

Followed in more detail, the following sequence is commonly observed (Fig. 4). During the predawn hours of the morning, when urban activity is at a minimum, concentrations of primary contaminants—carbon mon-

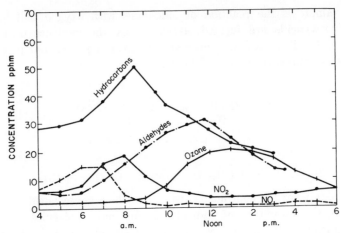

Figure 4. Average concentration during days of eye irritation in downtown Los Angeles, California. Hydrocarbons, aldehydes, and ozone for 1953–1954. Nitric oxide and nitrogen dioxide for 1958. From data of the Los Angeles County Air Pollution Control District (8).

oxide, nitric oxide, hydrocarbons—increase slowly, in the absence of wind. The secondary contaminants, ozone and nitrogen dioxide, remain at negligible levels in the absence of the photochemical reactions. The rising of the sun and the burgeoning of morning traffic speed the accumulation of the raw materials without, however, much effect on the levels of the products until the sun is fairly high, perhaps 2 hours after dawn. By this time the daily injection of nitric oxide and hydrocarbons from motor cars is well underway and nitrogen dioxide begins to be generated at a substantial rate. Usually within 1 or 2 hours more, the nitric oxide is reduced to low values because it has been converted to nitrogen dioxide, and the nitrogen dioxide has therefore reached a peak.

The disappearance of nitric oxide in the late morning hours coincides with the first appearance of ozone in the atmosphere. Ozone now accumulates until, sometime after noon, it reaches a maximum, then gradually declines during the next several hours. The concentration of nitrogen dioxide declines from its peak as the ozone builds up, and is usually negligible by late afternoon. The afternoon traffic injects an additional burden of nitric oxide into the atmosphere, which scavenges the remaining traces of ozone by early evening; then the primary pollutants reaccumulate at a decreasing rate for the remainder of the night.

Although this picture omits the effect of the movement of the air over the city, it is fairly closely followed on many days of the year, and can be accurately reproduced in the laboratory. There is, therefore, no reason to doubt its essential validity.

The effects of photochemical smog can be simulated in fumigations wherein pure hydrocarbons or gasoline fractions are exposed to solar radiation in the presence of oxides of nitrogen. Similar results are obtained in a fumigation with olefins and ozone, without irradiation, and early research established that ozone is, in fact, formed during the photochemical oxidation of hydrocarbons in the presence of nitrogen dioxide (*38–42*). The formation of ozone accounts for the intense rubber cracking noted in the Los Angeles, California area. The high ozone concentration is also responsible for plant damage in areas with photochemical smog (*43*).

However, a more prevalent type of plant damage is caused by peroxyacyl nitrates, as was first demonstrated by Stephens and his colleagues (*44–49*). Other types of plant damage have been attributed to peroxides and other products of the reaction between ozone and olefins (*44*).

The eye-irritating material should also be regarded as a mixture of several irritants. This complex is more stable than the plant toxicant and its half-life has been estimated to be of the order of half a day (*50*:32). Since neither starting products nor end products such as acids show eye-irritating effects at the concentrations used in the fumigations, intermediate oxidation products are again suspect as eye irritants. A part of the irritating effect, but not all, can be explained by the presence of formaldehyde and acrolein, as well as peroxyacyl nitrates (*27*). Other lachyrmators mentioned are peroxides of various types, free radicals, diketene, and nitroolefins. The presence of aerosols may enhance the physiological reaction or, since some of the polymers formed are strong oxidants, aerosols themselves may contribute to the total eye-irritation response.

Aerosol formation, usually accompanying fumigations with photochemically oxidized hydrocarbons, has been studied in connection with automobile exhaust (*51*). These studies confirm the photochemical aerosol formation observed with cyclic olefins and with olefins in the presence of sulfur dioxide (*38, 52*). The chemical composition, as well as the infrared spectra, of these materials obtained by irradiation of gasoline or olefins in the presence of oxides of nitrogen is similar to that obtained from ether-soluble aerosols collected from Los Angeles air. The composition of these aerosols is given in Table I. The empirical formula roughly corresponds to $(CH_2O)_n$, indicating a far-advanced state of oxidation of the original hydrocarbon material.

3. Chemical Nature of Smog Components

As a result of the work of many investigators, the major products of the chemical reactions taking place in urban atmospheres under the in-

Table I   Comparison of Average Elemental Analyses of Air Samples, Synthetic Aerosols, and Gasoline Gums, Percent by Weight

	Air samples		*Synthetic aerosols*[a] *(NO_2 + cracked gasoline)*	*Gasoline*[c] *gums*
	A[a]	*B*[b]		
Carbon	67.9	34.0	69.1	69.8
Hydrogen	9.2	5.3	9.0	8.7
Nitrogen	1.2	1.4	—	—
Oxygen	20.7	59	21.9	21.1
Sulfur	0.6	—	—	0.4

[a] Mader *et al.*, Ref. (*52*).
[b] Faith *et al.*, Ref. (*50*), p. 43.
[c] Ellis, Ref. (*53*), p. 908.

fluence of sunlight have been well identified. Long path infrared studies contributed to the early finding of various oxidation products. Figure 5 shows a spectrogram obtained after irradiation of a synthetic atmosphere containing 3-methylheptane and nitrogen dioxide.

This spectrogram shows the development of formic acid, carbonyl compounds (aldehydes or ketones), and carbon dioxide as degradation prod-

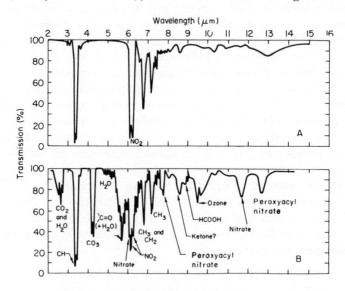

Figure 5. Infrared spectra—10 ppm 3-methylheptane and 5 ppm nitrogen dioxide in 1 atm of oxygen. A: Before irradiation by AH-6 mercury arc; path length 240 m. B: After irradiation (*39*). (Copyright by the American Chemical Society.)

ucts of the hydrocarbon, as well as organic nitrates and peroxyacyl nitrates. The existence of these nitrates can be taken as an indication of the participation of free radicals in the photooxidation mechanism.

The major products found in irradiation of synthetic atmospheres containing nitric oxide, nitrogen dioxide, and various hydrocarbons (either singly or in combinations) are aldehydes, ketones, organic nitrates, carbon dioxide, carbon monoxide, organic peroxides and hydroperoxides, and ozone. When an olefin is the sole initial hydrocarbon constituent in such an irradiation system, the principal carbon-containing products are usually the carbonyl compounds which correspond to fission of the olefin at the double bond; e.g., for 2-methylbutene-2, they would be acetaldehyde and acetone, as shown by the schematic diagram

$$
\underset{\substack{| \\ CH_3}}{CH_3} \underset{\substack{| \\ }}{\overset{CH_3\ H}{\underset{| \quad |}{C=C}}}-CH_3 : CH_3-\overset{CH_3}{\underset{|}{C}}=O + O=\overset{H}{\underset{|}{C}}-HC_3 \tag{36}
$$

This conversion is predominant, but not quantitative. One reason for the lack of stoichiometric agreement is that the carbonylic products are subject to photolysis as well as further reactions with free radicals and possibly other reactive intermediate species in the system. Since the more reactive initial hydrocarbon constituents in urban atmospheres are predominantly olefinic, it is not surprising to find that aldehydes are the quantitatively predominant oxygenates in photochemical smog. Among the aldehydes in smog, formaldehyde is consistently the most abundant, being the sole carbonylic product in the photooxidation of ethylene, which in turn is the most abundant olefinic constituent of automobile exhaust.

The photooxidation of the higher aldehydes by ultraviolet light was shown (54) to yield hydroperoxides with, in each case, one carbon atom less than the reacting aldehyde. These hydroperoxides are undoubtedly present in photochemical smog, although at concentrations lower than the aldehydes. They undoubtedly contribute to the oxidizing capacity of an aged photochemical mixture, although when that oxidizing capacity is large, its major source is usually ozone. Hydrogen peroxide was also shown (55) to be a product in the photolysis of formaldehyde.

A small proportion of the carbon in the photolyzed hydrocarbons also turns up in the form of organic nitrates, principally alkyl nitrates and peroxyacyl nitrates. Since the latter compounds have been found to have lachrymatory properties and to cause damage to vegetation, they have been especially well studied. They appear to form the predominant fraction of the organic nitrate product in smog as well as in irradiation chambers.

Of the peroxyacyl nitrates, the most abundant is the acetyl homologue.

Its concentration has been observed to exceed 0.010 ppm in photochemical smog in the Los Angeles, California, basin, with a value of 0.058 recorded on one occasion in Riverside, California (56). Visible symptoms of injury have been observed on sensitive plants when concentrations of peroxyacetyl nitrate did not exceed 0.014 ppm. However, since higher concentrations of the synthetic product were required to produce injury in the laboratory, it is possible that more potent higher homologues were also present in the atmosphere to account for the observed damage (56). Peroxybenzoyl nitrate has been detected (57) among the smog chamber photooxidation products of some aromatic hydrocarbons. It is a much more potent lachrymator than the peroxyacyl nitrates (at least, those which have been tested). It is probably present in photochemical smog, although its presence has not been demonstrated as yet.

In a study of the photolysis of isobutene, Schuck, Doyle, and Endow (58) have measured the rates of formation of the numerous degradation products. Found in addition to the expected major oxidation products, acetone and formaldehyde, were carbon monoxide, peracetylnitrate, isobutene oxide, acetaldehyde, isobutyraldehyde, propionaldehyde, ethyl nitrate, methyl nitrate, ethyl nitrate, propylene oxide, and methanol.

Accumulation of ozone in the photochemical system is limited to a well-defined region of low concentrations of oxides of nitrogen (59, 60) as shown in Figure 6. In general, the ozone measured during irradiation is the resultant of a light reaction leading to the formation of ozone [Eq. (3)] and dark reactions which destroy it [Eqs. (13) and (19)]. Among

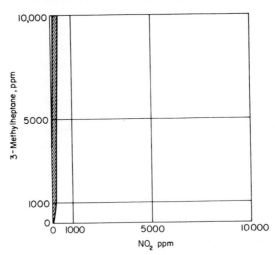

Figure 6. Area of ozone formation with 3-methylheptane and NO₂. Linear chart (60).

these are oxidation of nitric oxide to nitrogen dioxide and nitric acid, the formation of nitropolymers, and the reactions with olefins and other hydrocarbons. The competition between formation and destruction of ozone is readily seen in experiments in which any ozone formed is immediately removed from the reaction mixture by an ozone acceptor such as rubber. Under these conditions it is found that the capacity for producing ozone is several times greater than would be suspected from the concentration of the reactants (Fig. 7). It has been calculated that during a severe smog day, reaching concentrations of 0.5 ppm O_3, 500 tons of ozone are present in the air over the Los Angeles, California, basin at any one moment, below the inversion layer. To maintain this concentration over a period of hours, several thousand tons of ozone must have been formed.

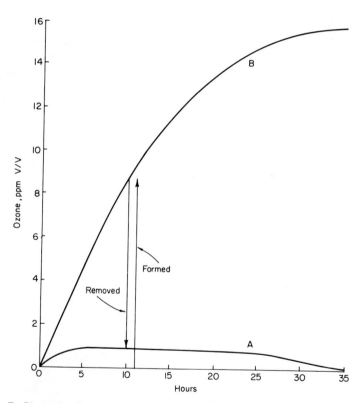

Figure 7. Photochemical formation of ozone with 3-methylheptane and NO_2. A: Observed concentrations of ozone after irradiation. Net result of ozone formation and simultaneous destruction. B: Cumulative formation of ozone measured by continuous removal, using bent strips of rubber as ozone acceptors (59). (Copyright by the American Chemical Society.)

Figure 8 shows the small difference between the rates of formation and decomposition of ozone in the case of irradiation of 3-methylheptane and nitrogen dioxide. The initial rate difference of about 0.2 ppm per hour establishes a concentration of 0.9 ppm in 5 hours. By that time both rates have become practically equal, and a steady level near 1 ppm is maintained until, after 25 hours, the rate of formation diminishes and the ozone gradually disappears. This combination of light and dark reactions explains the limited area of concentrations at which ozone accumulation may occur, as shown in Figure 6. It is also a clue to the controversial statements regarding correlations between oxidant and irritation. The slightly faster rate of the light reaction decreases markedly when less light is available, whereas the dark reaction leading to the irritating products is not affected.

The excess ozone may be rapidly consumed by dark reactions, especially in atmospheres where appreciable amounts of nitric oxide are generated by motor vehicle traffic, while the eye irritants may be less susceptible to destruction. As a result, irritation is often noticed many hours after sundown, when the ozone, and consequently the oxidant, is practically absent.

The relation of eye-irritation severity to initial concentrations of hy-

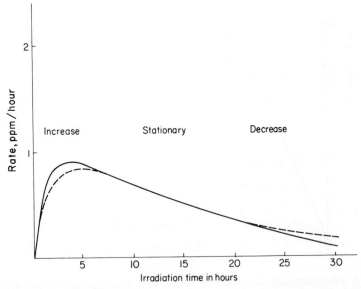

Figure 8. Rates of formation and decomposition of ozone during irradiation of 3-methylheptane (3 ppm) and NO₂ (1 ppm). Note shift in equilibrium with time. Solid line: formation rate; dashed line: decomposition rate (*61*). (Copyright by the American Chemical Society.)

drocarbons and oxides of nitrogen, in irradiated atmospheres containing motor vehicle exhaust gases, is shown in Figure 9 for "static" experiments and in Figure 10 for "dynamic" experiments (tests in which exhaust gases were continuously added to the irradiated atmosphere). Both figures show that, for a given hydrocarbon level, there exists a maximum eye-irritation index, corresponding to some oxides of nitrogen concentration within the experimental range.

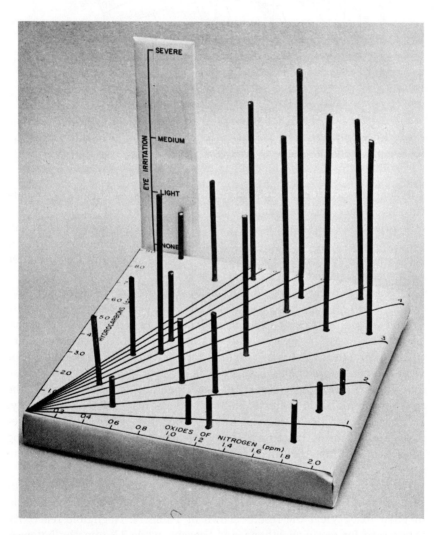

Figure 9. Eye irritation as a function of hydrocarbons and oxides of nitrogen concentrations after irradiation in static experiments (50).

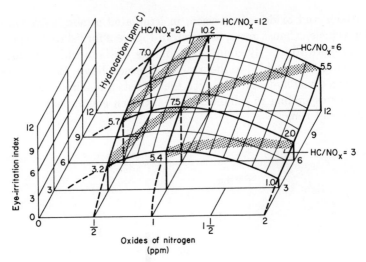

Figure 10. Eye irritation as a function of hydrocarbons and oxides of nitrogen concentrations after irradiation in dynamic experiments (62).

Systematic study of the photooxidation reactions of various hydrocarbons and other organic compounds has yielded several schemes for the comparison of compounds with regard to their smog-producing potential, or "reactivity." Reactivity scales may be based on various measures of chemical reaction rates or on the severity of smog manifestations after some specified period of irradiation.

An element of arbitrary definition is unavoidable in any reactivity scale. Indeed, for most such scales, the relative reactivity values must depend on the particular experimental details used in the comparisons. The values may be expected to change if initial concentrations are changed, or the intensity of the light, or the temperature, humidity, or vigor of stirring are changed; in some cases, such factors may even cause differences in ranking of particular compounds. Needless to say, the rankings obtained on scales of different effects may differ substantially. In general, from the standpoint of reproducibility and ease of measurement, the chemical reactivity scales are more suitable than the smog effect scales, while the effect scales seem to yield an intuitively more satisfying measure of smog potential.

Extensive investigations (54, 63, 72) of the reactivity of various hydrocarbons relative to their effectiveness in promoting the photooxidation of nitric oxide yield the results shown in Table II. Negligible reactivity is observed for methane, ethane, and benzene, while 2,3-dimethylbutene-2 is highly reactive. With the reactivity of ethylene arbitrarily set as unity,

Table II Reactivities of Hydrocarbons Based on Ability to Participate in Photooxidation of Nitric Oxide to Nitrogen Dioxide[a]

Hydrocarbon	Ranking	
	Altshuller and Cohen (72)	Glasson and Tuesday (63)
2,3-Dimethylbutene-2		10
2-Methyl-2-butene		3
trans-2-Butene	2[a]	2
Isobutene	1	
Propylene	1	0.5
Ethylene	0.4	0.3
1,3,5-Trimethylbenzene	1.2	1.2
m-Xylene	1	0.9
1,2,3,5-Tetramethylbenzene	0.9	0.7
1,2,4-Trimethylbenzene	0.6	0.7
o- and p-Xylene	0.4	0.4
o- and p-Diethylbenzene	0.4	0.4
Propylbenzenes	0.3	0.2
Toluene	0.2	0.2
Benzene	0.15	0.04
n-Nonane	0.15	
3-Methylheptane	0.15	
n-Heptane		0.2
Methylpentanes	<0.1	0.2
Pentanes		0.2
2,2,4-Trimethylpentane	0.15	0.15
Butanes		0.1
Ethane		0.03
Methane		<0.01
Acetylene	0.1	

[a] Arbitrarily adjusted to the same ranking as trans-2-butene on Glasson and Tuesday's scale to permit comparison of other hydrocarbons.

the largest value found for an alkane is 1.1; for a 1-alkene, 2.3; for an alkylbenzene, 4.5; for a 2-alkene, 29. It is interesting to note that the relative reactivity increases with each increase in the number of methyl groups adjacent to the double bond, in olefins, and with each increase in the number of methyl groups attached to the benzene nucleus, in aromatics.

Supplementary studies (64) have shown that the lower aliphatic aldehydes also promote photooxidation of nitric oxide, and that their reactivities in this respect resemble those of ethylene, propylene, toluene, and xylene. Ozone is formed in these photooxidations, and since aldehydes are among the products of photooxidation of the hydrocarbons, it is evident that the production of ozone by irradiation of a hydrocarbon–nitric oxide system may continue even after the hydrocarbon initially present has been used up.

Studies of reactivity in terms of the relative rate of consumption of the organic compound in photooxidation have yielded results generally consonant with those just discussed, although some details of ranking differ. Interestingly, the ratio of reactivities for 2,3-dimethylbutene-2 vs 2-methylbutene-2 appears to be substantially less by this measure than in the results shown in Table II. In spite of such differences in detail, however, scales of both types agree in indicating the following reactivity classes:

High reactivity: internally double-bonded olefins, especially tri- and tetraalkyl ethylenes

Moderate reactivity: polyalkyl benzenes, monoalkyl ethylenes, 1,3-alkadienes

Low reactivity: ethylene, monoalkyl benzenes, alkanes (C_4 and higher)

Negligible reactivity: benzene, acetylene, methane, ethane, propane

On the other hand, when reactivity is defined in terms of smog manifestations, such as oxidant level or eye-irritation potential, rankings are often substantially different from those obtained by chemical reactivity studies. Discordant rankings (65) are obtained with various oxidant scales, and eye-irritation potency rankings disagree seriously with hydrocarbon consumption rankings.

A method of predicting effect rankings for some of the compounds of each class, proposed by Yeung and Phillips (66), requires the estimation of "biological effect" factors to be applied (as multipliers) to the reactivity values of a chemical scale. These effect factors are expressed in terms of parameters which can be determined, for each class of compounds, from experiments on a few members of the class. Thus, to determine parameters applicable to eye-irritation reactivity, the organic compounds are assigned to categories according to the expected lachrymatory potential of their photooxidation products.

As illustrated by the authors (Table III), this procedure assigns one parameter to benzylic hydrocarbons; another, lower, value to nonbenzylic alkylbenzenes and ethylene; another, still lower, to internally double-bonded olefins, and so on. Figures 11 and 12 illustrate, respectively, the lack of relation between eye-irritation reactivity and chemical reac-

Table III   Measured and Predicted Eye Irritation Reactivities (66)

Structural groups	Hydrocarbon	Measured eye-irritation reactivity (E.I.R.)		Biological effect factor (f)	Calculated E.I.R. = (R.C.R.) (f)
Benzyls	n-Butylbenzene	6.4	β		6.12
	Isobutylbenzene	5.7	β		5.11
	n-Propylbenzene	5.4	β	$\Big\}$ 5.94	5.82
	Toluene	5.3	β		5.82
Nonbenzyls	sec-Butylbenzene	1.8	α		1.70
	Isopropylbenzene	1.6	α		1.48
	tert-Butylbenzene	0.9	α		0.85
Multialkyl benzenes	m-Xylene	2.9	α	1.30	3.47
	p-Xylene	2.5	α		1.73
	o-Xylene	2.3	α		3.04
Ethylene	Ethylene	1.0	α		1.30
Terminal olefins	Propylene	3.9	$\alpha + \gamma = 1.64$		3.43[a]
	1-Hexene	3.5	$\alpha + \gamma = 1.64$		2.72[a]
Internal olefins	trans-2-Butene	2.3	γ	$\Big\}$ 0.341	2.23
	cis-2-Butene	1.6	γ		1.65
Fully substituted olefin	Tetramethylethylene	1.4	σ	$= 0.048$	1.40
Diene	1,3-Butadiene	6.9	θ	$= 1.60$	6.90
n-Alkanes	n-Butane	0.0			0
	n-Hexane	0.0			0
Branched alkane	Isooctane	0.9	$\frac{1}{2}(\alpha + \gamma) = 0.821$		0.67[a]
Others	Benzene	1.0	$\frac{1}{2}(\beta + \alpha) = 3.62$		1.00[a]
	2-Methyl-2-butene	1.9	$\frac{1}{2}(\gamma + \sigma) = 0.195$		1.68[a]
	Styrene	8.9	$\beta + \alpha = 7.24$		9.27[a]
	Allylbenzene	8.4	$\beta + \alpha = 7.24$		8.76[a]
	β-Methylstyrene	8.9	$\beta + \gamma = 6.28$		8.41[a]

[a] Predicted E.I.R. These are used in estimating correlation coefficient.

tivity, and the degree of correlation achieved by application of a consistent set of biological effect factors.

In the case of eye irritation, it is satisfying to observe that the necessary biological effect factors, as evaluated from an appropriately restricted set of experimental data, are in qualitative accord with expectations based on prior information as to the products of photooxidation and their lachrymatory potential. Thus the largest parameter value is the one which corresponds to the benzylic hydrocarbons; and these, on photooxidation, yield peroxybenzoyl nitrate, the most potent irritant yet

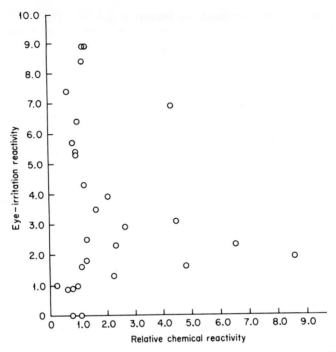

Figure 11. Eye-irritation reactivity vs relative chemical reactivity (66).

identified in these photooxidation systems. Successively smaller values are assigned for classes expected to yield chiefly acrolein, formaldehyde, other aliphatic aldehydes, and ketones.

The same investigators (66) also suggest another parameter scheme, focusing on structural features of the hydrocarbons, rather than on the predicted products of photooxidation.

The rates of formation and decay of the objectionable reaction products of the organic emissions determine to a large extent the type and severity of photochemical smog attacks. Details of the behavior of pure organic substances in photochemical oxidation systems of this sort have been provided for quite a few selected compounds of various types, especially by Tuesday and his co-workers (68–70) and by Altshuller et al. (71–77). A typical chart for the reaction of ethylene in such a system is shown in Figure 13 (78).

The times at which maximum ozone concentrations are produced with different organics vary widely (Fig. 14). After emission, a mixture of these compounds is selectively oxidized, with the most reactive components being used up most rapidly, as shown in Figure 15 (78). This figure

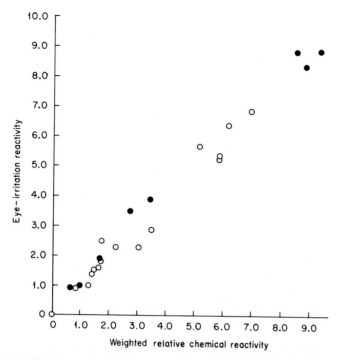

Figure 12. Measured eye-irritation reactivity vs weighted chemical reactivity (*66*). ○: Hydrocarbons used for evaluation of parameters; ●: hydrocarbons used for prediction and correlation.

shows concentration profiles in an irradiated mixture of ethylene, propylene, nitric oxide, and nitrogen dioxide; it is clear that a high level of oxidant is maintained for at least 3 hours after the more reactive component, propylene, is depleted.

Pollution clouds therefore gradually lose their most rapid ozone formers, whose function is then taken over by aldehydes, ketones, slow-reacting olefins (such as ethylene), aromatics, and saturated hydrocarbons (*43*). Thus a high ozone level is established and maintained over a long period of time.

4. Chain Reactions in Photochemical Smog

One of the perplexing puzzles of the photochemistry of urban atmospheres is to understand how the oxidizing power of the irradiated atmosphere is developed. The paradox is that, starting with contaminants which are recognized reducing agents such as nitric oxide and sulfur

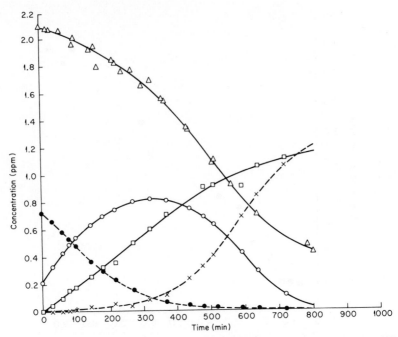

Figure 13. Concentration changes on irradiation of a mixture of $NO + NO_2 + C_2H_4$ (*78*). $\triangle$: C_2H_4; $\square$: RCHO; $\bigcirc$: NO_2; $\times$: O_3; $\bullet$: NO. (Copyright by the American Chemical Society.)

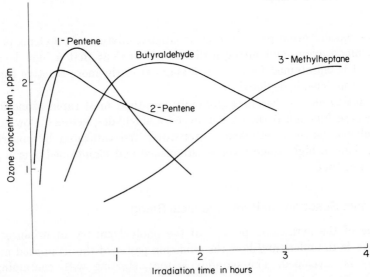

Figure 14. Ozone formation with 5 ppm NO_2 and 10 ppm organic compounds (*39*). (Copyright by the American Chemical Society.)

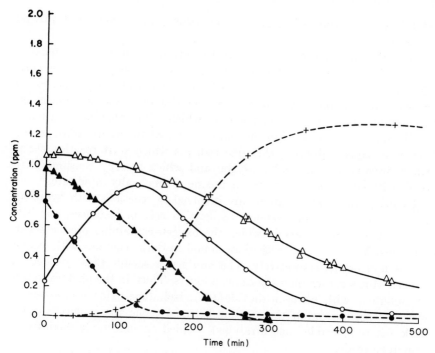

Figure 15. Concentration changes on irradiation of a mixture of $NO + NO_2 + C_2H_4 + C_3H_6$ (78). $\triangle$: C_2H_4; $\blacktriangle$: C_3H_6; $\bigcirc$: NO_2; $\bullet$: NO; $+$: O_3. (Copyright by the American Chemical Society.)

dioxide, and with hydrocarbons which have no oxidizing power, the end effect of sunlight is the production of oxidizing agents as potent as ozone. Clearly, the key is in the production of oxygen atoms by the photodissociation of nitrogen dioxide; but, even given the presence of free oxygen atoms, there is difficulty in accounting for the course of events. Several alternative reactions are available to consume the oxygen atoms, even in the absence of hydrocarbons. They may react with molecular oxygen to form ozone, with nitrogen dioxide to give molecular oxygen and nitric oxide, or with nitric oxide to form nitrogen dioxide. Since molecular oxygen is the most abundant of these possible reagents, ozone is the most likely product; but the ozone, also, is reactive. It can pair with nitric oxide to regenerate nitrogen dioxide, or with nitrogen dioxide to give nitrogen trioxide, leaving an oxygen molecule behind in each case. Nitrogen trioxide is also a powerful oxidizing agent, but its most likely fate is reaction with nitric oxide to regenerate nitrogen dioxide. The net result of these reactions, all proceeding simultaneously, is that nitrogen dioxide is regenerated almost as rapidly as it is photolyzed. However, as long as the irradiation is maintained, the system must contain small "steady state" concentrations of atomic oxygen, ozone, nitric oxide, and nitrogen trioxide.

If nitric oxide and organic vapors are injected into such a system, the nitric oxide is soon converted to nitrogen dioxide. This presents the paradox of a photoreaction system in which the primary absorbing species—in a sense, one of the raw materials of the system—does not vanish, but accumulates as a result of the reactions which follow photodissociation.

The most likely answer to this paradox is that the hydrocarbon, reacting with oxygen atoms, yields free radicals which start reaction chains which involve the molecular oxygen and which thereby consume nitric oxide without dissociating nitrogen dioxide at an equally rapid rate.

Evidence that oxygen atoms are involved in reactions with hydrocarbons in photooxidation systems, at least in certain instances, has been provided by experiments in which the oxygen-containing products were identified and compared with those found in dark reactions with oxygen atoms and ozone, respectively. The products resemble those yielded by oxygen atoms rather than the ozone adducts. Thus, in the photooxidation of isobutene with nitrogen dioxide (58), isobutene oxide was a readily observed product, while the reaction between ozone and isobutene in the dark (at comparable concentrations) yielded no detectable quantity of isobutene oxide.

Evidence that free radicals are involved in the photooxidation systems may be seen, as mentioned above, in the production of organic nitrates and nitrites, which probably arise from the combination of free radicals with nitrogen dioxide and nitric oxide, respectively.

Evidence that reaction chains are involved in photooxidation systems arises especially from consideration of the detailed kinetics of such systems. One item is the so-called "excess rate" of hydrocarbon disappearance. When the rate at which an olefinic hydrocarbon is consumed is compared with the expected rates of reaction with atomic oxygen and ozone, there is often an unexplained excess; that is, the expected rates do not add to the observed rate. The discrepancy for olefins is in some cases a factor of 2 or 3 (79) and for aromatics a factor of 20 or 30 (77).

To account for this excess rate, it is necessary to assume that the hydrocarbon reacts with "intermediates," i.e., reactive chemical species formed by the reactions which initiate the photooxidation process. These intermediates, therefore, propagate reaction chains, as discussed above (Section I,D); they are almost certainly free radicals.

The identity of the intermediate species which propagate the chains is a subject of much interest as well as theoretical importance; these intermediates may be viewed as providing the "driving force" for the photooxidation and thus for the development of photochemical smog. A growing body of opinion (*80*) assigns this role to the hydroxyl radical, HO·; which has been shown to react very readily with olefinic hydrocarbons by addition to the double bond as given by

$$\text{RCH}{=}\text{CH}_2 + \text{HO·} \rightarrow \text{R}-\overset{\overset{\displaystyle \text{OH}}{|}}{\underset{\underset{\displaystyle \text{H}}{|}}{\text{C}}}-\overset{\overset{\displaystyle \cdot}{}}{\underset{\underset{\displaystyle \text{H}}{|}}{\text{C}}}-\text{H} \tag{37}$$

and with other hydrocarbons or oxygenated compounds by abstraction of hydrogen, although this is less rapid.

$$\text{RH} + \text{HO·} \rightarrow \text{R·} + \text{H}_2\text{O} \tag{38}$$

Another item suggesting the importance of reaction chains in these systems is found when the rate of consumption of hydrocarbon is compared with the rate of oxidation of nitric oxide promoted by the hydrocarbon. Ratios of nitrogen dioxide produced to ethylene consumed have been observed (*74*) to be as high as 16 to 1. Although such ratios are very unusual in nonchain stoichiometry, they are consonant with the earlier findings of Haagen-Smit and Fox (*60, 61*), showing production of excess ozone.

Finally, cases have been reported (*68,* cf. *81*) in which the rate of production of nitrogen dioxide is over five times the calculated rate of photodissociation of nitrogen dioxide and therefore five times the rate of generation of oxygen atoms. In such a case, if the chains can be started only by encounters of oxygen atoms with hydrocarbon molecules, since most of the oxygen atoms generated by photolysis must react with molecular oxygen to form ozone, the minimum chain length for oxidation of nitric oxide can be estimated as about 160. If only a fraction of the necessary encounters were effective in chain initiation, the chain yield would have to be even higher. The data of Cvetanović (*82*) suggest that the appropriate factor may be between 3 and 10 for 2-butene; this gives an estimated chain yield of 500 to 1600 in the instance at hand.

Regarding synergism or interaction between various compounds, chamber irradiation experiments show that, when hydrocarbons are mixed, their net individual rates of reaction in the photooxidation system are modified. From the point of view of chemical kinetics, this is to be expected because the various compounds compete in the production and consumption of the reactive intermediate species, such as oxygen atoms, ozone, and free radicals. In careful studies with mixtures of the olefins, ethylene, propylene, and *cis*-2-butene, it has been shown (*78*) that the rate of consumption of the less reactive constituent is increased and that of the more reactive constituent is decreased, relative to the rates found at equivalent reactant concentrations when each is studied alone.

Figure 16 shows the fractional rate of consumption of ethylene as a function of time. When the very reactive butene is present, the consumption of ethylene proceeds much more rapidly than when ethylene is alone, although in each case the maximum fractional consumption rate is about the same. In each case, also, the maximum fractional consumption rate occurs at about the time when the most reactive hydrocarbon is substantially depleted, and this is much later than the time of maximum nitrogen dioxide, as can be seen in Figure 17. From the kinetic point of view, these observations indicate that the ethylene reacts most readily with neither oxygen atoms nor ozone, but with reactive intermediates—probably free radicals—which are most copiously produced by initial reactions involving the more reactive hydrocarbon.

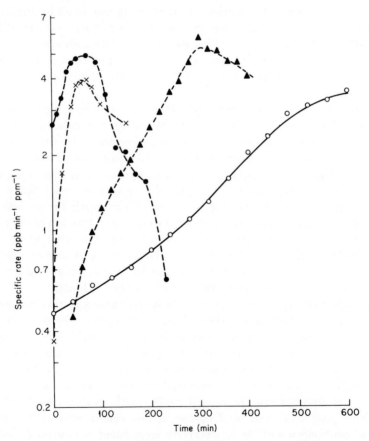

Figure 16. Specific rate of ethylene consumption (78). ○: C_2H_4 only; ▲: with C_3H_6; ×: with C_4H_8; ●: with $C_3H_6 + C_4H_8$. (Copyright by the American Chemical Society.)

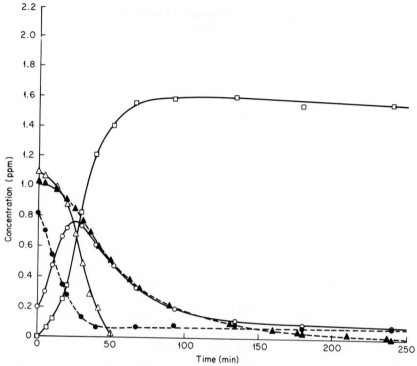

Figure 17. Concentration changes on irradiation of a mixture of NO + NO₂ + C₃H₆ + C₄H₈ (78). □: RCHO; ▲: C₃H₆; △: C₄H₈; ●: NO; ○: NO₂. (Copyright by the American Chemical Society.)

Various chain mechanisms have been proposed which will in some measure account for photooxidation phenomena. Among the first was that shown in Table IV, given by Saltzman (42), in which hydrocarbon radicals generated by hydrogen abstraction react with oxygen molecules to yield ozone, and the product radicals continue the chains by abstracting hydrogen from fresh hydrocarbon molecules. A mechanism owing to Leighton (83), Table V, invokes the formation of organic radicals by reaction of olefins with oxygen atoms, which breaks the carbon skeleton of the molecule into two fragments. A mechanism involving chain-branching steps where free radicals react with olefin molecules in a process which multiplies the number of chains is presented in Table VI. (See also Table IV, Chapter 10.)

Further development of the chemical kinetics approach has been stimulated by the availability of computer programs for the numerical integration of the simultaneous differential equations which represent the chemical kinetics of complicated reaction systems. Briefly, this approach

Table IV   A Suggested Photooxidation Mechanism Involving Hydrogen Abstraction by Oxygen Atoms (42) (Copyright by the American Chemical Society.)

Inorganic
1. $NO_2 + h\nu = NO + O$
2. $O + O_2 = O_3$
3. $2NO + O_2 = 2NO_2$
4. $O_3 + NO = O_2 + NO_2$
5. $O_3 + 2NO_2 = N_2O_5 + O_2$
6. $O + H_2O(g) = 2HO\cdot$

Generation of organic free radicals
7. $O_3 + olefins = products$
8. $O + RH = R\cdot + OH\cdot$
9. $OH\cdot + RH = R\cdot + H_2O$

Organic chain reactions
10. $R\cdot + O_2 = RO_2\cdot$
11. $RO_2\cdot + O_2 = RO\cdot + O_3$
12. $RO\cdot + RH = ROH + R\cdot$
13. $RO_2\cdot + RH = ROOH + R\cdot$
14. $2RO_2\cdot = 2RO\cdot + O_2$

Consumption of organic free radicals
15. $2RO\cdot = aldehyde + alcohol$
16. $RO\cdot + NO = RONO$
17. $RO_2\cdot + NO = RO_2NO$
$\qquad\qquad\quad = RO\cdot + NO_2$

Table V   A Suggested Photooxidation Mechanism Involving Reaction of Oxygen Atoms with Olefins (83)

1. $NO_2 + h\nu = NO + O$
2. $O_2 + O + M = O_3 + M$
3. $O_3 + NO = O_2 + NO_2$
4. $O + C_4H_8 = CH_3\cdot + C_3H_5O\cdot$
5. $CH_3\cdot + O_2 = CH_3O_2\cdot$
6. $CH_3OO\cdot + O_2 = CH_3O\cdot + O_3$
7. $CH_3O\cdot + NO = CH_3ONO$
8. $CH_3ONO + h\nu = CH_3O\cdot^* + NO$
9. $CH_3O\cdot^* + O_2 = H_2CO + HOO\cdot$
10. $HOO\cdot + C_4H_8 = H_2CO + (CH_3)_2CO + H\cdot$
11. $H\cdot + O_2 = HO_2\cdot$
12. $HOO\cdot + NO = OH\cdot + NO_2$
13. $OH\cdot + C_4H_8 = (CH_3)_2CO + CH_3\cdot$
14. $2HOO\cdot = H_2O_2 + O_2$
15. $2OH\cdot = H_2 + O_2$

Table VI   A Suggested Photooxidation Mechanism Involving Branching Chains (81)

1. $NO_2 + h\nu = NO + O$
2. $O + O_2 + M = O_3 + M$
3. $O_3 + NO = NO_2 + O_2$
4. $Ol^a + O = OlO*$
5. $OlO* + O_2 = OlO_3*$
6. $Ol + O_3 = OlO_3*$
7. $OlO_3* = $ aldehyde $+ RO\cdot + RCO\cdot$
8. $RO\cdot + O_2 + NO = RO_2\cdot + NO_2$
9. $RCO\cdot + O_2 = RCO_3\cdot$
10. $RCO_3\cdot + NO = RCO_2\cdot + NO_2$
11. $RO_2\cdot + NO = RO\cdot + NO_2$
12. $RCO_2\cdot + NO = RCO\cdot + NO_2$
13. $Ol + RO_2\cdot = OlO* + RO\cdot$
14. $Ol + RCO_2\cdot = OlO* + RCO\cdot$
15. Aldehyde $+ RO\cdot = RCO\cdot + ROH$ (alcohol)
16. $RO\cdot + NO_2 = RONO_2$ (alkyl nitrate)
17. $RCO_3\cdot + NO_2 = RCO_3NO_2$ (peroxyacyl nitrate)
18. $RO\cdot + RCO = $ ketone $+$ alcohol

^{a} Ol represents an olefin molecule.

amounts to a rapid reiteration of a multitude of simple calculations. A mechanism (i.e., a set of postulated elementary reactions) is formulated, and reaction rate constants are estimated (one for each reaction) in accord with the best available evidence. Given a selection of initial concentrations for each of the reacting components of a mixture, the program computes the rates of each of the elementary reactions and, from these, the net rates of change of concentration of the components. It next calculates the net changes in component concentrations for a small time interval and applies these changes to produce an estimate of the altered concentrations at the end of the selected time interval. This set of newly estimated concentrations is then used for computing new values of the rates of the elementary reactions, and the process is repeated as often as is necessary to simulate the passage of an experimentally significant reaction time.

Such simulation provides an excellent means of testing to what extent a postulated mechanism is consistent with kinetic behavior observed in laboratory experiments. Furthermore, since the rate constants associated with elementary reactions are invariant with respect to the chemical composition of the gas phase in which the reactions occur, it follows that a mechanism, to be considered satisfactory, must simulate experiments over a wide range of contaminant concentrations, using an invariant set of estimated rate constants. In principle, therefore, numerical experimentation of this sort, in conjunction with irradiation chamber studies in the laboratory and theoretical exploration of the characteristics of relevant elementary reactions, can provide a very powerful tool for the elucidation of the chemical dynamics of photochemical smog.

An excellent illustration of the use of this technique has been furnished by Niki, Daby, and Weinstock (84). Using a set of 60 postulated elementary reactions, they simulated experimental data for a photooxidation system with initial concentrations near 2 ppm of propylene, 1 ppm nitric oxide, and 0.05 pphm nitrogen dioxide. Figure 18 shows the agreement between experimental data for this system and the computer-generated concentration profiles for the selected experiment.

Of particular interest, in the case illustrated, is the fact that the mechanism chosen contained only three reactions involving propylene, viz, reactions with oxygen atoms, ozone, and hydroxyl radicals. The rather good agreement exhibited in Figure 18, therefore, constitutes support for the authors' hypothesis that reactions of propylene with radicals other than OH, or with other reactive intermediate species present in the system, may be less important than these three reactions.

A further step in the exploration of this mechanism is shown in Figure 19, where the profiles shown in Figure 18 are compared with corresponding profiles obtained from a simulation in which the reaction between propylene and oxygen atoms is forbidden (i.e., its rate constant is set equal to zero). The rather small difference in detail of the two sets of

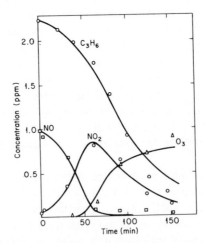

Figure 18. Concentration changes on irradiation of a mixture of NO + NO₂ + C₃H₆; comparison of model predictions with experiment (84, 84a). Solid line: ref. calculated; points: experimental data. (Copyright by the American Chemical Society.)

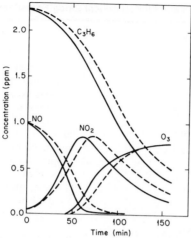

Figure 19. Effect of chain initiation by oxygen atoms, as tested by comparison of model predictions *(84)*. Solid line: ref. calculated; dashed line: No reaction between O and C_3H_6 assumed in calculating curve. (Copyright by the American Chemical Society.)

curves clearly demonstrates that, in the chosen mechanism, oxygen atom attack is not a critical factor controlling the progress of the reaction.

The simulation approach, although it can test a proposed mechanism for consistency with experimental data, is not capable of testing it for uniqueness. It is not surprising, therefore, that several mechanistic schemes, differing somewhat in details of the assumed reaction chains, have been tested with a somewhat similar degree of success—a mechanism published by Hecht and Seinfeld *(85)*, for instance, comprises eighty-one reactions for simulating propylene photooxidation. These are divided by the authors into the following major categories:

(a) 19 reactions in the $NO-NO_2-O_2-CO-H_2O$ system
(b) 19 reactions for oxidation of propylene by O and O_3
(c) 16 reactions for attack on propylene by OH
(d) 13 reactions for formation and destruction of pernitrates
(e) 8 reactions for decomposition of aldehydes
(f) 6 additional reactions involving free radicals

Table VII, by Hecht *(86)*, summarizes the status of knowledge about the rates of 42 reactions or types of reactions thought to be both elementary and important in the photooxidation process. A further very detailed review of the chemistry of this system has been given by Demerjian, Kerr, and Calvert *(87)*.

The complexity of atmospheric photochemical reaction systems stems partly from the multiplicity of organic pollutant species, partly from the

Table VII   Reaction Rates in Photochemical Smog Formation (86)

Reaction	*Comments*
1. $NO_2 + h\nu \rightarrow NO + O$	$k_a\emptyset$ uncertain by $\sim 20\%$; difficulties encountered in trying to relate measured k_d to $k_a\emptyset$
2. $O + O_2 + M \rightarrow O_3 + M$	Well known; $k = 2.0 \times 10^{-5}$ ppm^{-2} min^{-1}
3. $O_3 + NO \rightarrow O_2 + NO_2$	$20 \leq k \leq 40$ ppm^{-1} min^{-1}
4. $O_3 + NO_2 \rightarrow NO_3 + O_2$	Two measured values, 0.11 and 0.048 ppm^{-1} min^{-1}, both of which may be too high
5. $NO_3 + NO_2 \rightarrow N_2O_5$	k_5 and k_6 uncertain by $\sim 20\%$; equilibrium constant is very temperature dependent
6. $N_2O_5 \rightarrow NO_3 + NO_2$	
7. $N_2O_5 + H_2O \rightarrow 2HNO_3$	Reaction in gas phase is negligible; reaction occurs only on surfaces
8. $NO_3 + NO \rightarrow 2NO_2$	k factor of 10^2 uncertainty; most important NO_3 removal process
9. $NO_3 + h\nu \rightarrow NO_2 + O$	Importance not determined
10. $NO + NO_2 + H_2O \rightarrow 2HONO$	Rate may be higher than first order with respect to H_2O; heterogeneous contributions to this rate not determined
11. $2HONO \rightarrow NO + NO_2 + H_2O$	
12. $HONO + h\nu \rightarrow OH + NO$	$k_{12} \sim 0.1\ k_a\emptyset$; photolysis of HONO in early morning may initiate smog formation
13. $OH + NO_2 + [M] \rightarrow HNO_3 + [M]$	k uncertain; order of reaction depends on pressure of M; probably second order in atmosphere
14. $OH + NO + [M] \rightarrow HONO + [M]$	k uncertain; pressure dependence not determined
15. $OH + CO \rightarrow CO_2 + H$	$230 \leq k \leq 280$ ppm^{-1} min^{-1}; reaction is too slow to compete with olefin–OH reactions
16. $H + O_2 + M \rightarrow HO_2 + M$	Only atmospheric reaction of importance for H atoms
17. $HO_2 + NO \rightarrow HO + NO_2$	$10^2 \leq k \leq 10^3$ ppm^{-1} min^{-1}; most important HO_2 removal process
18. $2HO_2 \rightarrow H_2O_2 + O_2$	k uncertain by $\sim 50\%$
19. $H_2O_2 + h\nu \rightarrow 2OH$	Photolysis rate in sunlight not well known
20. Paraffins $+ O \rightarrow R + OH$	k uncertain by $\sim$ factor of 10
21. Paraffins $+ OH \rightarrow R + H_2O$	k uncertain by $\sim$ factor of 10
22. Olefins $+ O \rightarrow$?	k factor of 2 uncertainty; products of reaction under atmospheric conditions unknown

Table VII (*Continued*)

Reaction	Comments
23. Olefins + OH → ?	Most important olefin reaction; products of reaction unknown
24. Olefins + O_3 → aldehyde + zwitterion	k factor of 2 uncertainty; reactions of zwitterion unknown
25. Olefins + H_2O → ?	Probably unimportant
26. Aromatics + O → R + OH	k's for most reactions unknown; $k_{26} \sim k_{20}$
27. Aromatics + OH → R + H_2O	k's for most reactions unknown
28. RCHO + $h\nu$ → R + HCO	Quantum yields as function of $h\nu$ not well known; k_{28} may be an important chain initiation step in atmosphere
29. RCHO + $h\nu$ → RH + CO	
30. RCHO + OH → RCO + H_2O	$k \sim 10^4$ ppm^{-1} min^{-1}
31. RCHO + O → RCO + OH	Slow compared to R_x 30
32. HCO + O_2 → HO_2 + CO	k high; only important HCO reaction
33. RCO + O_2 → RC(O)OO	Only RCO reaction of importance
34. RC(O)OO + NO → R + CO_2 + NO_2	k unknown
35. RC(O)OO + NO_2 → RC(O)OONO$_2$	k unknown
36. R + O_2 → RO_2	Very fast
37. RO_2 + NO → RO + NO_2	k unknown
38. RO_2 + NO_2 → RO_2NO_2	k unknown
39. RO_2 + HO_2 → ROOH + O_2	k unknown
40. RO + O_2 → RCHO + HO_2	k unknown
41. RO + NO → RONO	k unknown
42. RO + NO_2 → $RONO_2$	k unknown

large numbers of possible elementary reactions, and partly from the difficulties of observation and analysis of components, products, and intermediate compounds at the very low concentrations encountered.

Regardless of complexity, the behavior of such systems may have important implications for policies, regulations, and methods in air pollution control. For example, in setting standards for the quality of ambient air, the United States Environmental Protection Agency has recommended certain limitations on the concentrations of hydrocarbons and oxides of nitrogen, based (at least in part) on the potential interaction of these contaminants to yield photochemical oxidants. The interaction of oxides of nitrogen with hydrocarbons from motor vehicle exhausts has been studied in irradiation chamber experiments by various investigators, especially by Dimitriades (*87a*), who systematically explored the maximum oxidant levels achieved in 6-hour irradiations as a function of the initial concentrations of the precursors. This work was utilized by the Committee on Motor Vehicle Emissions of the United States National

Academy of Sciences (87b) in further efforts to estimate what levels of emissions from automobiles would be consistent with achieving ambient air quality standards in metropolitan areas in the United States.

Further technical discussion relating to air quality standards is to be found in Chapter 11, Vol. V, while motor vehicle emission standards are discussed in Chapter 13, Vol. V. Extended discussion of the photochemical aspects is included in the Air Quality Criteria documents published by the United States Environmental Protection Agency (87c, 87d) and in reports of the State of California Department of Public Health (88).

With the use of modern techniques, these systems should increasingly prove susceptible to both experimental and theoretical approaches. Gas chromatography and infrared spectroscopy have already contributed mightily to the solution of experimental difficulties, and high-speed computers have abolished most obstacles to the simulation of rate processes in consecutive and simultaneous reactions. It is to be expected that these and other modern techniques will soon provide answers to many of the problems outlined above.

III. Atmospheric Scavenging Processes

In principle, the available routes for scavenging of foreign material from the atmosphere may be separated into (a) deposition and (b) conversion to normal atmospheric constituents. Any pollutants not eliminated by one or the other of these mechanisms must accumulate in the atmosphere. Chemical reactions facilitate both processes; the former principally in the lower atmosphere by oxidation and combination of pollutants to give solid or liquid particles, or adsorbed phases on such particles; the latter in the upper atmosphere, through disruption of more complex gaseous or vapor molecules by high energy, short wavelength radiation from the sun. The residence time of various pollutants in the atmosphere have been estimated and range from a few hours as in the case of ozone, to years for carbon monoxide and carbon dioxide (89a).

A. Sulfur Oxides

Most of the sulfur that contaminates urban atmospheres is emitted in the form of gaseous sulfur dioxide; the total world production of this material has been estimated at 100 million tons per year. If sulfur dioxide were to remain in the air and be distributed evenly over the globe, the increase per year would be about 0.006 ppm. However, sulfur dioxide is subject to reactions that remove it from the atmosphere; the turnover time has been estimated to be in the order of a few days (29a, 90).

A large part of the sulfur dioxide in the air oxidizes to sulfur trioxide, which reacts with water vapor to form sulfuric acid mist. The sulfuric acid in the aerosol reacts with other materials in the air and forms sulfates, especially ammonium and calcium sulfates. Available evidence (91) suggests that another substantial portion of the atmospheric sulfur dioxide is directly neutralized by ammonia, calcite dust, or other airborne alkalis and is then rapidly oxidized by the air to the corresponding sulfates. Precipitation finally removes these salts from the air.

B. Fluorides, Sulfides, and Ammonia

Hydrogen fluoride and the hydrolyzed silicofluorides readily attack a wide variety of materials—carbonates, silicates, and organic compounds. Neither ammonia nor its unpleasant-smelling organic homologues, the amines, absorb light in the range of 2900–8000 Å, and they are removed from the atmosphere by reaction with acids or acid-forming oxides. Consequently, ammonium sulfate is a common pollutant in urban areas.

Hydrogen sulfide is emitted from natural sources at a rate of about 300 million tons per year (29a, 92). There is no evidence of accumulation of hydrogen sulfide in the global atmosphere; its sources are so diffuse, and its lifetime in the atmosphere so short, that perceptible concentrations are seldom encountered. Hydrogen sulfide, transparent to the spectral range available at ground level, readily participates in chemical reactions, as, for example, in the combination of hydrogen sulfide with white lead in paint. The sulfides and their organic homologues, the mercaptans, as well as the amines, are subject to the oxidizing action of the atmosphere. These reactions are of importance in daylight when photochemically active components degrade the amines and mercaptans. In laboratory experiments their unpleasant odors disappear when these compounds are exposed to irradiation in air in the presence of a few tenths of a part per million of oxides of nitrogen. Hydrogen sulfide, itself, is probably oxidized to sulfates, although small amounts of elemental sulfur found in polluted atmospheres indicate that relatively stable intermediates may be formed.

C. Carbon Dioxide and Monoxide

Both biological and geochemical processes provide a natural system for disposal, as well as replenishment, of carbon dioxide. Green plants use carbon dioxide in photosynthesis, but most of the plant material formed is reoxidized within a few years, returning carbon dioxide to the atmosphere. Carbon dioxide is also absorbed, slowly, in the weathering of rocks, and it is emitted by volcanoes and mineral springs. The potential of these processes to stabilize the global atmospheric concentration of carbon dioxide has not been accurately assessed, but it may be considerable.

Another major influence on the carbon dioxide concentration of the world's atmosphere is probably the exchange of carbon dioxide with the oceans, which contain, by estimate (93), some 60 times as much carbon dioxide as the atmosphere. Studies of the concentration of radioactive carbon in surface seawater have indicated an average residence time of 10 years for carbon dioxide in the atmosphere before transfer to ocean waters; however, the approach to equilibrium between the atmosphere and the ocean may be much slower than this. Keeling (94) estimates that from 30 to 50% of the carbon dioxide so far released by the burning of fossil fuels may be now remaining in the atmosphere. (See also Chapter 1, Vol. II.)

Incomplete combustion of gasoline is responsible for most of the 270 million tons of carbon monoxide released into the atmosphere each year. This amount is more than half of the 530 million tons of carbon monoxide normally present in the troposphere. In the absence of scavenging processes, the emissions of carbon monoxide (94a) would be enough to raise the atmospheric concentration of this substance by 0.04 ppm per year (94, 95). In the absence of measurements sufficiently accurate to detect such concentrations, it is conceivable that such accumulation is, in fact, presently occurring. It is, however, not unlikely that oxidation processes in the upper atmosphere, which receives very short-wave ultraviolet radiation from the sun, convert carbon monoxide to the dioxide. An important role in this conversion is played by hydroxyl radicals (95a, 95b). These reactions would reduce the rate of accumulation of carbon monoxide and possibly result in an upper concentration limit, providing global emissions were to remain steady for a long period.

D. Oxides of Nitrogen and Photochemical Products

It appears likely that the oxides of nitrogen released to the atmosphere by industrial processes and other urban activities are so short-lived that no detectable accumulation in the general atmosphere can be expected. As indicated above, practically all of the nitric oxide and nitrogen dioxide involved in photochemical episodes is converted to nitrogen-containing organic compounds within hours of their exposure to sunlight. Even if these compounds are included, the total of nitrogen compounds emitted from such sources is doubtless far less than 100,000 tons per day, which would increase their concentration in the general atmosphere (when uniformly distributed) by less than 10^{-5} ppm per year.

Various mechanisms seem likely to consume an appreciable proportion of the oxides of nitrogen (excluding nitrous oxides) which might reach the general atmosphere. Nitric oxide reacts slowly with oxygen, even in the dark, giving nitrogen dioxide; nitrogen dioxide appears to have an autocatalytic effect on this reaction (*96*).

At higher levels of the atmosphere, ozone is photochemically generated, and this ozone can react with nitrogen dioxide to form nitrogen pentoxide and trioxide. In this process ozone is consumed and concern has been expressed about the injection of the oxides of nitrogen into the stratosphere by high flying aircraft, which may induce possible changes in spectral distribution of sunlight at ground level (*96a*). At lower altitude, the anhydrides of nitrous and nitric acid (like sulfur dioxide) combine with airborne alkalis, forming nitrates which are washed out by precipitation. It is possible, also, that some of the nitrogen dioxide may react directly with ammonia, releasing molecular nitrogen and water vapor. Apparently, molecular nitrogen is also generated in the photooxidation of ethylene which occurs in urban atmospheres (*97*). Again, like carbon dioxide, nitrogen dioxide is somewhat water soluble and can be transferred from the lower atmosphere to the ocean and other surface waters as inorganic nitrates and nitrites.

The organic molecules photochemically produced in urban atmospheres, as well as organic primary pollutants which may reach the general atmosphere without participating in photochemical reactions, are still subject to the degradative oxidation processes described above (Section II,B,1). These processes eventually degrade such substances to carbon dioxide, carbon monoxide, and water. However, the carbon chains of some of the organic compounds will be relatively slowly attacked, and the lower paraffins may be expected to accumulate in the atmosphere. Methane in particular is a well-recognized component of the general atmosphere, although, like hydrogen sulfide, it is principally from natural sources. Its oxidation to carbon monoxide and carbon dioxide takes place through the action of hydroxyl radicals. The average residence time for carbon monoxide is variously estimated to be 0.1 to 0.3 years and that of methane, 1.5 years (*94a, 95a, 95b*).

E. Particulate Matter

The principal mode of removal of particles from tropospheric air (see Chapter 3) is gravitational settling, which prevents the largest particles, like those of fly ash and wind-driven soil, from traveling very far from their sources. Somewhat smaller particles, which settle more slowly, may be removed from the lowest atmospheric layer by impaction on the surfaces of obstacles such as buildings and trees. However, particles as large as 10 μm in diameter are kept airborne by turbulent diffusion for extended periods of time (*98, 99*).

However, clouds and rain cleanse the higher layers of the troposphere quite efficiently, as liquid droplets form on the small particles and, in growing, gather others. Further, the falling drops of rain accumulate particles from the lower levels. This "washout" mechanism is ineffective for removing particles smaller than about 2 μm in diameter. Nevertheless, soluble particles such as sea salt and nitrates grow substantially in size under conditions of high relative humidity, due to their own accumulation of water from the vapor phase. They may thus be brought to the ground by rain much more effectively than by gravitational settling.

The importance of rainfall as a scavenging agent has been discussed by Smith (*100*), whose calculations are illustrated by Figure 20. The figure shows hypothetical curves for concentration as a function of time for three different effluents, one a gas unaffected by rainfall, one a somewhat soluble gas, and one a particulate material of uniform particle size. Assumed conditions are uniform wind velocity, constant effluent rate, constant dispersal volume with complete mixing, and constant rainfall after a given time. While these conditions are highly idealized, the estimates may indicate the order of magnitude of the washout effect.

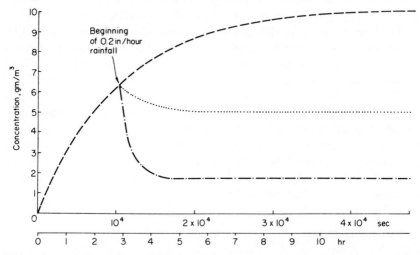

Figure 20. Pollutant concentrations as a function of time, for a hypothetical model of an urban atmosphere. Dashed line: Pollutant unaffected by rain; dotted line: SO₂ vapor, dotted-dashed line: Particles with 0.15-cm/second settling speed.

While particles may remain airborne for only a few days on the average, under some circumstances they may be transported over great distances. If a particular source is very large or intense—for example, an erupting volcano, a forest fire, or a large metropolitan area, international and even intercontinental transport of particulate matter is to be expected.

REFERENCES

1. "Second Technical and Administrative Report on Air Pollution in Los Angeles County," Annu. Rep. 1950–1951, p. 41. Air Pollution Control District, County of Los Angeles, California, 1952.
2. E. R. Weaver, E. E. Hughes, S. M. Gunther, S. Schuhmann, N. T. Redfearn, and R. Gordon, Jr., *J. Res. Nat. Bur. Stand.* **59**, 383 (1957).
3. F. E. Blacet, *Ind. Eng. Chem.* **44**, 1339 (1952).
4. L. E. Miller, *J. Air Pollut. Contr. Ass.* **8**, 138 (1958).
4a. E. K. Peterson, *Environ. Sci. Technol.* **3**, 1162–1169 (1969).
5. F. Daniels, "Outlines of Physical Chemistry," 1st ed. Wiley, New York, New York, 1951.
6. P. A. Leighton and W. A. Perkins, *Air Pollut. Found., Rep.* No. 14 (1956).
7. P. A. Leighton and W. A. Perkins, *Air Pollut. Found., Rep.* No. 24 (1958).
8. P. A. Leighton, "Photochemistry of Air Pollution." Academic Press, New York, New York, 1961.
9. N. A. Renzetti, *Air Pollut. Found., Rep.* **9**, 200 (1955).
10. R. Stair, *Proc. Nat. Air Pollut. Symp., 3rd, 1955* p. 48 (1955).
11. A. M. Adams and E. J. Schneider, *Air Pollut. Smoke Prev. Ass. Amer., Proc.* **45**, 61–63 (1952).
12. H. S. Johnston, *Ind. Eng. Chem.* **48**, 1488 (1956).
13. J. G. Calvert, *Air Pollut. Found., Rep.* **15**, 89–112 (1956).
14. M. C. Markham and K. J. Laidler, *J. Phys. Chem.* **57**, 363 (1953).
15. G. V. Elmore and H. A. Tanner, *J. Phys. Chem.* **60**, 1328 (1956).
16. F. N. Frenkiel, *Smithson. Inst., Annu. Rep.* pp. 296–299 (1956).
17. L. G. Wayne, A. Kokin, and M. I. Weisburd, Final Report, Contract No. 68-02-0345. Pacific Environmental Services, Inc., Santa Monica, California, 1973; United States Environmental Protection Agency Document No. R4-73-013a, Research Triangle Park, North Carolina, 1973.
18. P. Roth, M. K. Liu, S. D. Reynolds, T. A. Hecht, and P. J. W. Roberts, Final Report, Contract CPA 68-02-0339. Systems Applications, Inc., Beverly Hills, California, 1973, for the United States Environmental Protection Agency, Research Triangle Park, North Carolina, 1973.
19. A. Q. Eschenroeder, J. R. Martinez, and R. I. Nordsieck, Final Report, Contract No. 68-02-0336. General Research Corp., Santa Barbara, California, 1972; United States Environmental Protection Agency Document No. R4-73-0120Vol-a, Research Triangle Park, North Carolina, 1973.
20. E. R. Gerhard and H. F. Johnstone, *Ind. Eng. Chem.* **47**, 972 (1955).
21. N. A. Renzetti and G. J. Doyle, *Int. J. Air Pollut.* **2**, 327 (1960).

22. H. W. Sidebottom, C. C. Badcock, G. E. Jackson, J. G. Calvert, G. W. Reinhardt, and E. K. Damon, *Environ. Sci. Technol.* **6**, 72 (1970).
23. M. J. Prager, E. R. Stephens, and W. E. Scott, *Ind. Eng. Chem.* **52**, 521 (1960).
24. H. F. Johnstone and D. R. Coughanowr, *Ind. Eng. Chem.* **50**, 1169 (1958).
25. G. J. Doyle and N. A. Renzetti, *J. Air Pollut. Contr. Ass.* **8**, 23 (1958).
26. E. A. Schuck, H. W. Ford, and E. R. Stephens, *Air Pollut. Found., Rep.* **26** (1958).
27. Sixth Technical Report, *Air Pollut. Found., Rep.* **30** (1960).
28. R. A. Cox and S. A. Penkett, *Nature (London)* **230**, 321 (1971); **229**, 486 (1971).
29. C. Hellner and R. A. Keller, *J. Air Pollut. Contr. Ass.* **22**, 959 (1972).
29a. W. W. Kellogg, R. D. Cadle, E. R. Allen, A. L. Lazrus, and E. A. Martell, *Science* **175**, 587–596 (1972).
30. A. Adel, *Science* **113**, 624 (1951).
31. H. S. Johnston and D. M. Yost, *J. Chem. Phys.* **17**, 386 (1949).
32. H. S. Johnston and H. J. Crosby, *J. Chem. Phys.* **19**, 799 (1951).
33. J. K. Dixon, *J. Chem. Phys.* **8**, 157 (1940).
34. T. C. Hall, Jr. and F. E. Blacet, *J. Chem. Phys.* **20**, 1745 (1952).
35. R. D. Cadle and H. S. Johnston, *Proc. Air Pollut. Symp., 2nd, 1952* pp. 28–34 (1952).
36. L. H. Rogers, *J. Air Pollut. Contr. Ass.* **8**, 124 (1958).
37. L. M. Richards, *J. Air Pollut. Contr. Ass.* **5**, 216 (1956).
38. A. J. Haagen-Smit, *Ind. Eng. Chem.* **44**, 1342 (1952).
39. E. R. Stephens, P. L. Hanst, R. C. Doerr, and W. E. Scott, *Ind. Eng. Chem.* **48**, 1498 (1956).
40. C. Brown, K. Franson, and A. Miller, *Air Pollut. Found., Rep.* **17**, 97–99 (1957). (Report on project with Armour Research Foundation on "Effect of Trace Materials in Air on Photochemical Formation of Oxidant.")
41. R. D. Cadle, *Air Pollut. Found., Rep.* **15**, 29–59 (1956).
42. B. E. Saltzman, *Ind. Eng. Chem.* **50**, 677 (1958).
43. J. T. Middleton and A. J. Haagen-Smit, *J. Air Pollut. Contr. Ass.* **11**, 129 (1961).
44. A. J. Haagen-Smit, E. F. Darley, M. Zaitlin, H. Hull, and W. M. Noble, *Plant Physiol.* **27**, 18 (1952).
45. E. F. Darley, E. R. Stephens, J. T. Middleton, and P. L. Hanst, *Proc. Amer. Petrol. Inst., Sect.* **38**, 313 (1958).
46. N. N. Arnold, *Int. J. Air Pollut.* **2**, 167 (1959).
47. E. F. Darley, E. R. Stephens, J. T. Middleton, and P. L. Hanst, *Int. J. Air Pollut.* **1**, 155 (1959).
48. E. R. Stephens, E. F. Darley, O. C. Taylor, and W. E. Scott, *Proc. Amer. Petrol. Inst., Sect. 3* **40**, 111 (1961); also *Int. J. Air Pollut.* **4**, 79 (1961).
49. E. R. Stephens, *in* "Chemical Reactions in the Lower and Upper Atmosphere" (R. D. Cadle, ed.), p. 51. Wiley (Interscience), New York, New York, 1961.
50. W. L. Faith, N. A. Renzetti, and L. H. Rogers, *Air Pollution Found., Rep.* **22**, (1958).
51. G. J. Doyle and N. A. Renzetti, *J. Air Pollut. Contr. Ass.* **8**, 23 (1958).
52. P. P. Mader, R. D. MacPhee, R. T. Lofberg, and G. P. Larson, *Ind. Eng. Chem.* **44**, 1352 (1952).
53. C. Ellis, "The Chemistry of Petroleum Derivatives," Vol. 2. Van Nostrand-Reinhold, Princeton, New Jersey, 1937.
54. I. R. Cohen, T. C. Purcell, and A. P. Altshuller, *Environ. Sci. Technol.* **1**, 247 (1967).

55. J. J. Bufalini and K. L. Brubaker, *in* "Chemical Reactions in Urban Atmospheres" (C. S. Tuesday, ed.), p. 225. Amer. Elsevier, New York, New York, 1971.

56. O. C. Taylor, *in* "Impact of Air Pollution on Vegetation" (S. M. Linzon, ed.), p. 1. Air Pollution Control Association and Air Management Branch, Ontario Department of Energy and Resources Management, Toronto, Ontario, 1970.

57. J. M. Heuss and W. A. Glasson, *Environ. Sci. Technol.* **2**, 1109 (1968).

58. E. A. Schuck, G. J. Doyle, and N. Endow, *Air Pollution Found., Rep.* **31** (1960).

59. A. J. Haagen-Smit, *Ind. Eng. Chem.* **48**, 65A (1956).

60. A. J. Haagen-Smit and M. M. Fox, *J. Air Pollut. Contr. Ass.* **4**, 105 (1954).

61. A. J. Haagen-Smit and M. M. Fox, *Ind. Eng. Chem.* **48**, 1484 (1956).

62. M. W. Korth, A. H. Rose, Jr., and R. C. Stahman, *J. Air Pollut. Contr. Ass.* **14**, 168 (1964).

63. W. A. Glasson and C. S. Tuesday, *J. Air Pollut. Contr. Ass.* **20**, 239 (1970).

64. S. L. Kopczynski, A. P. Altshuller, and F. D. Sutterfield (in press).

65. A. P. Altshuller and J. J. Bufalini, *Environ. Sci. Technol.* **5**, 39 (1971).

66. C. K. K. Yeung and C. R. Phillips, *Atmos. Environ.* **7**, 551 (1973).

67. E. E. Harton, Jr. and C. C. Bolze, *Air Pollution Found., Rep.* **23**, (1958).

68. C. S. Tuesday, *in* "Chemical Reactions in the Lower and Upper Atmosphere" (R. D. Cadle, ed.), p. 15. Wiley (Interscience), New York, New York, 1961.

69. W. A. Glasson and C. S. Tuesday, *J. Amer. Chem. Soc.* **38**, 2901 (1963).

70. C. S. Tuesday, *Arch. Environ. Health* **7**, 188 (1963).

71. A. P. Altshuller, I. R. Cohen, S. F. Sleva, and S. L. Kopczynski, *Science* **138**, 442 (1962).

72. A. P. Altshuller and I. R. Cohen, *Int. J. Air Water Pollut.* **7**, 787 (1963).

73. J. J. Bufalini and A. P. Altshuller, *Int. J. Air Water Pollut.* **7**, 769 (1963).

74. A. P. Altshuller and I. R. Cohen, *Int. J. Air Water Pollut.* **8**, 611 (1964).

75. A. P. Altshuller and P. W. Leach, *Int. J. Air Water Pollut.* **8**, 37 (1964).

76. S. L. Kopczynski, *Int. J. Air Water Pollut.* **8**, 107 (1964).

77. A. P. Altshuller and J. J. Bufalini, *Photochem. Photobiol.* **4**, 97 (1965).

78. C. K. K. Yeung and C. R. Phillips, *Environ. Sci. Technol.* **9**, 732 (1975).

79. P. A. Leighton, "Photochemistry of Air Pollution," pp. 263 et seq. Academic Press, New York, New York, 1961.

80. J. Heicklen, K. Westberg, and N. Cohen, *in* "Chemical Reactions in Urban Atmospheres" (C. S. Tuesday, ed.), p. 55. Amer. Elsevier, New York, New York, 1971.

81. L. G. Wayne, *Arch. Environ. Health* **7**, 229 (1963).

82. R. J. Cvetanović, *Can. J. Chem.* **36**, 623 (1958).

83. P. A. Leighton, "Photochemistry of Air Pollution," p. 269. Academic Press, New York, New York, 1961.

84. H. Niki, E. E. Daby, and B. Weinstock, *Advan. Chem. Ser.* **13**, 16–57 (1972).

84a. A. P. Altshuller, personal communication (1969) to Niki *et al.* (*84*).

85. T. A. Hecht and J. H. Seinfeld, *Environ. Sci. Technol.* **6**, 47 (1972).

86. T. A. Hecht, as reported in M. C. Dodge, "Workshop on Mathematical Modeling of Photochemical Smog: Summary of the Proceedings," United States Environmental Protection Agency Doc. No. R4-73-010. Nat. Environ. Res. Cent., Research Triangle Park, North Carolina, 1973.

87. K. L. Demerjian, J. A. Kerr, and J. G. Calvert, *in* "Advances in Environmental Science," Vol. 3, p. 1. Wiley, New York, New York, 1973.

87a. B. Dimitriades, *Environ. Sci. Technol.* **6**, 253 (1972).

87b. "Joint Report of Panel on Emission Standards and Panel on Atmospheric Chemistry," Committee on Motor Vehicle Emissions of the National Academy of Sciences, Washington, D.C., 1973.

87c. "Air Quality Criteria for Photochemical Oxidants," Publ. No. AP-63. U.S. Department of Health, Education and Welfare, Public Health Service, Environmental Health Service, National Air Pollution Control Administration, Washington, D.C., 1970.

87d. "Air Quality Criteria for Hydrocarbons," Publ. No. AP-64. U.S. Department of Health, Education and Welfare, Public Health Service, Environmental Health Service, National Air Pollution Control Administration, Washington, D.C., 1970.

88. State of California, Department of Public Health, "The Oxides of Nitrogen in Air Pollution." Berkeley, California, 1966.

89. A. P. Altshuller, D. L. Klosterman, P. W. Leach, I. J. Hindawi, and J. E. Sigsby, Jr., *Int. J. Air Water Pollut.* **10**, 81 (1966).

89a. E. Robinson and R. C. Robbins, "Sources, Abundance and Fate of Gaseous Atmospheric Pollutants," Rep. Stanford Res. Inst., Stanford, California, 1968.

90. C. E. Junge and R. T. Werby, *J. Meteorol.* **15**, 417 (1958).

91. A. P. Van Den Heuvel and B. J. Mason, *Quart. J. Roy. Meteorol. Soc.* **89**, 271 (1963).

92. E. Erikksson, *J. Geophys. Res.* **68**, 4001 (1963).

93. H. E. Suess, *Bull. At. Sci.* **17**, 374 (1961).

94. C. D. Keeling, *Tellus* **12**, 200 (1960).

94a. T. H. Maugh, II, *Science* **177**, 338–339 (1972).

95. "Restoring the Quality of Our Environment. Report of the Environmental Pollution Panel, President's Science Advisory Committee," p. 111. U.S. Govt. Printing Office, Washington, D.C., 1965.

95a. B. Weinstock, *Science* **176**, 290 (1972).

95b. J. C. McConnell, M. B. McElroy, and S. C. Wofsy, *Nature (London)* **233**, 187 (1971).

96. E. M. Morrison, R. G. Rinker, and W. H. Corcoran, *Ind. Eng. Chem., Fundam.* **5**, 175 (1966).

96a. H. S. Johnston, *Science* **173**, 396 (1971).

97. J. J. Bufalini and J. C. Purcell, *Science* **150**, 1161 (1965).

98. S. Twomey, *J. Meteorol.* **12**, 81 (1955).

99. H. R. Byers, J. R. Sievers, and B. J. Tufts, *in* "Artificial Stimulation of Rain" (W. Weickmann and W. E. Smith, eds.), pp. 47–70. Pergamon, Oxford, England, 1957.

100. M. E. Smith, *in* "Chemical Reactions in the Lower and Upper Atmosphere" (R. D. Cadle, ed.), p. 155. Wiley (Interscience), New York, New York, 1961.

7

Global Sources, Sinks, and Transport of Air Pollution

D. M. Whelpdale and R. E. Munn

I. Introduction

Air pollution in the first instance is a local problem. However, the trace substances gradually mix into larger and larger volumes of air until they are distributed around the world. This dilution of pollutants is accelerated by a number of so-called *"sink mechanisms"*—deposition of particles by gravity forces, scavenging by precipitation, absorption at the surface of the earth, and chemical transformations in the atmosphere itself. In the last case, the concentrations of a particular compound (such as sulfur dioxide) may decrease at the expense of an increase in the concentrations of a related substance (such as sulfate particles).

On the continental and global scales, the assimilative capacity of the atmosphere is immense but not unlimited. However, the recent evidence for a background increase in carbon dioxide concentrations suggests the need for careful examination of the global budgets of pollution. The problem here is that the strengths of the natural sources are so large, the biogeochemical variability of the biosphere is so great, and the concentrations of trace gases are so low (often near the threshold of chemical detection) that there is considerable difficulty in determining the existence of trends and in relating them to changes in human activities. In this connection, the point should be made that not all man-made sources of pollution are associated with industry, transportation, and space heating. Agriculture and forestry make a substantial contribution on the global scale, and the emissions resulting from these activities are often as difficult to estimate as are those from natural sources such as sea spray.

In order to determine a mass budget of air pollution for the world, several kinds of information are required:

(a) Inventories of emission rates, both natural and man-made
(b) Estimates of atmospheric chemical reaction rates
(c) Estimates of sink strengths, i.e., of the rates at which pollutants are removed from the atmosphere
(d) Estimates of atmospheric concentrations suitably averaged in space and time
(e) Estimates of transport and diffusion rates

The last type of information is necessary because source strengths are not distributed uniformly around the world, and because the atmosphere is not a well-mixed reservoir; for example, the exchange rates between the northern and southern hemispheres are rather slow. Thus it is not immediately obvious in many cases how to obtain the most relevant estimates of atmospheric concentrations in (d) above, for verifying the predicted trends given by the differences between source and sink strengths.

To simplify the problem, global budgets of elemental substances such as sulfur are often considered rather than those of compounds such as sulfur dioxide. This may seem to remove the difficulty of obtaining information of type (b) above. However, because precipitation scavenging of particulates is more efficient than that of gases, the atmospheric chemical reaction rates must be known in order to estimate the wet deposition rates.

The study of global air pollution is important for four reasons:

(a) Substances such as carbon dioxide and suspended particulates affect the climate.

(b) The atmosphere is an efficient distributor of potentially toxic substances to all parts of the world. Although concentrations of atmospheric trace gases and aerosols are too low at remote locations to have any direct effects on life, there is a possibility of gradual accumulation in other media (e.g., mercury in lakes).

(c) In order to develop alternate environmental strategies for the world, simulation models are required, which can only be validated with data from networks of stations that include remote as well as impact locations.

(d) The global patterns of air pollution sometimes are useful indicators of large-scale atmospheric motions and of turbulent mixing, assisting in the development of predictive models, particularly in the case of interhemispheric exchanges of gases.

In this chapter, the sources of air pollution will be discussed in Section II, followed by consideration of the major sinks in Section III. A brief description of the global transport and diffusion processes in Section IV will lead to the topics of the observed concentration patterns in Section V, of synoptic modeling in Section VI, and of climatological modeling in Section VII. Finally, the ecological significance of global air pollution will be discussed in Section VIII.

II. Global Sources of Pollution

A. Urban and Industrial Emissions

Methods for estimating urban and industrial emission strengths have been described in Chapter 16, Vol. III. The techniques are reasonably well established and can be used to estimate total global emissions from these sources within an order of magnitude or so. Alternatively, indirect indicators such as regional or national consumptions of fuel oil or gas can be used to infer emission estimates for particular substances.

B. Agricultural and Forest Emissions Resulting from Human Activities

The principal man-induced processes contributing to air pollution in the countryside are

(a) *Soil erosion:* Soil erosion causes wind-blown dust during dry weather. For example, Bryson and Baerreis (*1*) have suggested that the Rajasthan desert of northwestern India and southeastern West Pakistan is the result of agricultural mismanagement more than 3000 years ago.

(b) *Slash burning:* In many parts of the world, slash burning is a recommended procedure to prevent development of disease organisms, and to reduce fire hazards.

(c) *Fertilizers:* Large quantities of fertilizers, particularly nitrates and phosphates, enter the atmosphere every year.

(d) *Pesticides:* Pesticides are used in many agricultural and forested parts of the world. Aircraft spraying (e.g., for spruce budworms in New Brunswick) is common.

(e) *Decaying farm wastes:* Animal and vegetable wastes release a wide variety of substances into the atmosphere. The list includes ammonia, hydrogen sulfide, methane, sulfides, mercaptans, carbon monoxide, and carbon dioxide.

Although each of these processes has been investigated on the local scale, the problem of extrapolation to obtain global estimates is often difficult.

C. Natural Emissions

A number of natural sources of pollution are of importance on the world scale.

(a) *Blowing dust:* For example, deep layers of dust originating in the Sahara desert have been detected in the Bahamas (*2*), carried there by the Trade Winds.

(b) *Forest, bush, and grass fires:* Some of these fires are caused by human carelessness but others are caused by lightning; in any event, they release large quantities of smoke and trace gases into the atmosphere.

(c) *Volcanoes:* Volcanic releases of sulfur dioxide and particulates are substantial. Because some of the plumes penetrate into the stratosphere, the residence times can be long, the atmospheric radiation balance can be affected significantly, and the world climate may be influenced. Lamb (*3*), for example, has found a weak correlation between strong volcanic activity and below-normal surface annual temperatures averaged around the world since the year 1500.

(d) *Forest terpenes:* Forests release large amounts of volatile hydrocarbons (terpenes) which may participate in photochemical reactions during sunny weather.

(e) *Sea spray:* Sea spray is a major source of particulate matter in the atmosphere.

(f) *Evaporation:* Trace gases are sometimes transferred from the oceans to the atmosphere by evaporation, although at other times the movement is in the opposite direction (through absorption). For example, the ocean is generally a sink for carbon dioxide but during periods of intense phytoplankton blooming, the ocean becomes a source.

D. Techniques for Obtaining Estimates of Global Emissions

In general, the methods used for estimating global emissions of air pollution are very crude. Examination of the literature reveals many cross references, which tend to lend an air of authority to the quoted emission values, although the underlying methodologies may consist of only some elementary arithmetic and/or some unconfirmed extrapolations. The scientific community has a considerable need for better estimates, particularly of the natural sources. Global models of air pollution are, in fact, often difficult to verify simply because of inaccurate source inventories.

In some cases, the total mass of a pollutant in the atmosphere is so great, and the secular trends are so small, that the sources and sinks may be assumed to balance as a first approximation. The order of magnitude of the natural sources can then be inferred, or at least adjusted in such a way that the sink strengths also seem reasonable. In addition, if the mass M (in grams) of pollution in the atmosphere is divided by the removal rate R (in grams per year), then the resulting ratio M/R has the dimensions of "years," and is called the *turnover time*. For a well-mixed reservoir, the turnover time is the same as the *residence time* (the average length of time spent by a substance in the atmosphere), and this assumption is often employed to obtain a rough indication of whether the estimates of the M/R ratio are reasonable. In some cases, there are independent methods of estimating the residence time (through the use of radioactive species, for example), which may be used to confirm the order of magnitude of M/R. The cross-checks described in this paragraph apply only to a volume as large as the northern hemisphere troposphere or stratosphere, and cannot be employed readily on the continental scale.

E. Estimates for Some of the Major Trace Substances

1. Suspended Particulates

Recent estimates of global tropospheric aerosols have been presented by Hidy and Brock (4). In summary, they believe that more than half of the aerosols come from secondary sources such as chemical reactions in the gas phase (e.g., conversion of sulfur dioxide to sulfate particles). Their detailed estimates are given in Table I which indicates that man-

Table I   Estimates of Tropospheric Aerosol Production Rates (4)

Source	Production rate (metric tons per day)	% by weight of total
A. Natural sources		
1. Primary		
Wind-blown dust	2×10^4 to 10^6	9.3
Sea spray	3×10^6	28
Volcanoes	10^4	0.09
Forest fires	4×10^5	3.8
2. Secondary		
Vegetation	5×10^5 to 3×10^6	28
Sulfur cycle	10^5 to 10^6	9.3
Nitrogen cycle	2×10^6	14.8
Volcanoes (gases)	10^3	0.009
Subtotals	10×10^6	94
B. Man-made sources		
1. Primary		
Combustion and industrial	1×10^5 to 3×10^5	2.8
Dust from cultivation	10^2 to 10^3	0.009
2. Secondary		
Hydrocarbon vapors	7×10^3	0.065
Sulfates	3×10^5	2.8
Nitrates	6×10^4	0.56
Ammonia	3×10^3	0.028
Subtotals	6.7×10^5	6
Total	10.7×10^6	100

made processes contribute only about 6% by weight of the total tropospheric burden.

Joseph et al. (5) have estimated that the injection of dust into the atmosphere from deserts is at least 128 ± 64 million metric tons per year $(0.4 \times 10^6$ tons per day), which is within the range given in Table I by Hidy and Brock. Joseph et al. obtained their estimate in the following way. First, they measured the atmospheric turbidity during dust storms (Khamsins) in the Middle East. Next, they used an empirical relation between turbidity and the mass of suspended particulates to estimate the mass in selected size ranges. From estimates that about 10 Khamsin episodes per year occur over an area of about 10^6 km^2 in the Middle East, an annual source strength for that part of the world could be computed. Finally, an extrapolation to obtain the global source strength was made from a knowledge of the total area of arid zones around the world.

2. Carbon Dioxide, Carbon Monoxide, and Hydrocarbons

Probably the most reliable estimates of man-made global carbon dioxide emissions from fossil fuels and kilning of limestone are by Keeling (6). Keeling's paper is recommended reading on the methodologies available for obtaining such estimates. One of the primary data sources is an annual (since 1951) United Nations publication on world energy supplies.

Keeling's estimates for carbon dioxide are given for selected years in Table II, from which it is seen that production of carbon dioxide has

Table II   World Production of CO$_2$ Released into the Atmosphere by the Burning of Fossil Fuels and Manufacturing of Limestone (6)[a]

	Year				
	1929	1937	1949	1961	1969
Coal	964	959	1008	1381	1591
Petroleum[b]	163	225	371	879	1621
Natural gas	30	41	89	264	514
Cement	—	—	16	46	74
	1157	1225	1483	2570	3801

[a] Values are given in 10^6 tons of carbon per year.
[b] In a note added in proof, Keeling suggests that these petroleum emissions should be increased by 5.7% to account for the flaring of waste gases during the processing of petroleum.

more than tripled since 1929. Keeling suggests that his error limits of ±13% can only be reduced by a coordinated intergovernmental effort to collect more reliable data on the consumption of fossil fuels.

Keeling has also made estimates of the 1968 production of carbon monoxide and gaseous hydrocarbons from fossil fuels. He computes values of 256×10^6 tons of carbon monoxide per year and 79×10^6 tons of hydrocarbon per year.

The only primary natural sources for carbon dioxide are volcanoes and forest fires. These emissions are relatively small in comparison with those from man-made sources, probably not more than 5%. However, there are secondary sources, mainly the oxidation of methane and formaldehyde to carbon monoxide and the subsequent conversion to carbon dioxide, which take place relatively rapidly [turnover time for carbon monoxide of 2 years or less (6)]. The magnitude of these secondary sources of carbon dioxide is not well established.

The major natural sources of carbon monoxide (in 10^6 tons per year) are as in the following tabulation (7):

Oxidation of methane and formaldehyde	3000
Decay and synthesis of chlorophyll	90
Photochemical oxidation of terpene	54
Oceans	220
	3364

For hydrocarbons, the major natural sources are terpene-type releases from vegetation (170×10^6 tons per year) and methane from swamps (310×10^6 tons per year).

3. Sulfur Compounds

Estimates of the world emissions of sulfur (in 10^6 tons sulfur per year) have been given by Robinson and Robbins as in the following tabulation (8):

Industry, space heating and transportation (mainly in the form of sulfur dioxide and hydrogen sulfide)	70
Fertilizer applications to soil (sulfates)	11
Rock weathering (sulfates)	14
Biological decay (hydrogen sulfide):	
Continents	68
Oceans	30
Sea spray (sulfates)	44
Volcanoes (sulfur dioxide, hydrogen sulfide, sulfates)	small
	237

Robinson and Robbins note that man-made emissions of sulfur compounds have been increasing during this century, with a fivefold strengthening of sulfur dioxide sources between 1900 and 1965.

4. Nitrogen Compounds

Here again the Robinson and Robbins estimates (for the year 1967) are quoted (in 10^6 tons per year of nitrogen dioxide equivalent) (8).

Nitrous oxide (bacterial action)		592
Nitrogen dioxide (forest fires)		0.8
Subtotal		592.8
Nitrogen dioxide (industrial)		
Burning of coal	26.9	
Burning of petroleum	22.3	
Burning of natural gas	2.1	
Incineration	0.5	
Burning of wood	0.3	
Subtotal	52.1	52.1
Total		645

For ammonia, the man-made sources (10^6 tons per year) are:

Coal	3
Oil and gas	1
Other sources	0.2
	4.2

Natural sources of ammonia are very large but the global strengths are difficult to estimate. A production rate of about 10^9 tons per year has been suggested.

F. The Stratosphere

The stratosphere is still almost in its natural state. There is a slow exchange of trace gases and aerosols with the troposphere (particularly at latitudes 30 to 40 deg, where there are breaks in the tropopause), and there are occasional violent intrusions from volcanoes and thunderstorms. For example, Castleman et al. (9) have found evidence for increased sulfate concentrations in the stratosphere (northern hemisphere) within a year following the eruption of Mt. Agung (southern hemisphere). Nevertheless, experimental data all confirm that there is a sharp drop in concentrations of trace gases between the troposphere and the stratosphere

(*10, 11*), indicating that the tropopause is a rather effective lid on ground-based emissions.

There has been concern about the possible future contamination of the stratosphere by aircraft (particularly the emissions of oxides of nitrogen). For this reason, the need is emphasized for immediate organized monitoring programs to determine the concentrations of trace constituents in the stratospheric natural state. Jocelyn et al. (*12*) have made estimates of aircraft traffic in the year 1990. They suggest that emissions from civil supersonic aircraft will be about equal to those from subsonic aircraft flying in the stratosphere.

III. Sink Mechanisms

A. Removal Processes

For the removal of *particles* from the troposphere, the important processes are

(a) Wet removal by precipitation
(b) Dry removal by sedimentation
(c) Dry removal by impaction on vegetation

For the removal of *gases* from the troposphere, the important ones are

(a) Wet removal by precipitation
(b) Absorption or reaction at the earth's surface
(c) Conversion into other gases or particulates by chemical reaction within the atmosphere
(d) Transport into the stratosphere

Wet removal by precipitation, or precipitation scavenging, is one of the most effective atmospheric cleansing mechanisms for both particles and gases. Contaminants may be incorporated into precipitation elements within clouds, in which case the processes are referred to as *rainout* or *snowout*, and below clouds, in which case the processes are called *washout*. (See also Chapter 8, this volume.)

The removal of particles by gravitational sedimentation is an effective process only for particles of radius larger than approximately 10 μm; the fall velocity of smaller particles becomes insignificant in comparison with atmospheric vertical motions. Impaction on vegetation can be an effective removal process near the ground. The scavenging efficiency of vegetation strongly depends on such factors as particle size, and the shape, size, and wetness of the collecting surfaces. (See also Chapter 3, this volume.)

The chemical reactions of trace gases within the atmosphere to form either new gaseous compounds or particles is of great importance as a sink for specific compounds, for example, sulfur dioxide. The efficiency of the subsequent removal by the other sink processes is then affected. (See also Chapter 6, this volume.)

The removal of gaseous pollutants by absorption or reaction at the surface of the earth is a sink mechanism about which little is known for many substances. Uptake at the earth's surface depends on many factors: meteorological parameters, the physical, chemical, and physiological nature of the surface, and the properties of the gas itself. (See also Section III,B.)

Finally, it should be mentioned that the stratosphere may be a sink for some tropospheric constituents. Measurements by Seiler and Junge (13) in the tropopause region of the northern hemisphere show rapid decreases of carbon monoxide concentrations above the tropopause indicating that the stratosphere is a sink. Reaction with the hydroxyl radical seems to be the mechanism for removing carbon monoxide there.

B. Removal Processes at the Surface of the Earth

Gaseous pollutants and particles small enough not to be appreciably affected by sedimentation can be transported to the earth's surface by atmospheric motions and removed there by absorption or chemical reaction. Two factors which affect the efficiency of uptake by this process are

(a) The rate at which the substance is supplied to the surface
(b) The rate at which the substance is bound or transformed by the surface

The transfer of pollutants through the surface boundary layer by turbulent diffusion is analogous to the turbulent transfer of heat, water vapor, and momentum as described, for example, by Businger (14). Meteorological parameters such as wind speed, turbulence intensity, or the stability of the lower atmosphere are governing factors. Transfer across the viscous sublayer (within a few millimeters of the surface) is by molecular diffusion. In experiments over snow with tritiated water vapor and iodine-131, for example, Barry and Munn (15) have ascribed differences in resistance to vertical transfer to differing molecular diffusivities of the two gases.

The nature of the sink surface itself is also critical to the removal process. Certain surfaces may be perfect sinks for specific gases, irreversibly absorbing the gas. For example, snow is a perfect sink for the radioactive gases mentioned above and seawater is an extremely efficient, if not perfect, sink for sulfur dioxide (*16, 17*). Physical properties of the surface such as roughness (*18*) and wetness (*19*) have been shown to affect uptake. Chemical properties of the surface such as pH of soil also affect the uptake of sulfur dioxide (*20*). The physiological state of a vegetative surface is important: Spedding (*21*) found that sulfur dioxide uptake by barley leaves increased by a factor of 6 when the stomata were open compared with stomata closed. Excessively high concentrations of pollutants often result in physiological changes in the plants which decrease uptake rates (*22*).

The flux of a pollutant to a surface may be determined from measurements of vertical concentration gradient and eddy diffusivity in the first several meters above a surface during steady-state conditions over a uniform surface. The vertical flux F is

$$F = K \, \Delta\chi/\Delta z \tag{1}$$

where K is the eddy diffusivity of the gaseous pollutant, and $\Delta\chi$ is the vertical concentration difference over the thickness Δz. Using this approach, it is possible to estimate vertical fluxes without having detailed information on the surface capture processes. The main source of uncertainty is the value of K, which must be inferred from vertical profiles of wind or temperature.

Alternately, it is possible to determine uptake by certain surfaces from direct measurements of surface loading. Chamberlain and Chadwick (*23*) used this approach with radioactive iodine-131 over grass as did Barry and Munn (*15*) over snow. However, with most pollutants this is not usually a satisfactory approach because the flux is small compared to the background level of contaminant or because the compound is chemically altered on uptake.

A second, very useful approach to surface uptake of pollutants is in terms of a resistance to transfer r or its inverse, a transfer velocity v. These are defined by

$$r = 1/v = \Delta\chi/F \tag{2}$$

where $\Delta\chi = \chi_1 - \chi_0$, the vertical concentration difference between a reference level (often 1 m) and the surface. For a perfect sink, $\chi_0 = 0$, and the problem is simplified considerably. In other cases, the advantage to a formulation in terms of Equation (2) is that the resistances of the different parts of the boundary layer are assumed to be additive:

$$r = r_a + r_b + r_s \tag{3}$$

where r_a, r_b, and r_s are the resistances of the surface boundary layer, the viscous sublayer, and the surface itself, respectively. In any particular situation, any term on the right-hand side of Equation (3) may be dominant. For example, r_a will be limiting for sulfur dioxide uptake by a basic soil surface under stable atmospheric conditions; whereas, r_s would be limiting for the uptake of carbon monoxide by a sterile soil surface.

The transfer velocity v or deposition velocity, as it is often called, is frequently used as a means of estimating pollutant fluxes to various types of surfaces on a regional or global scale. Assuming as an approximation that the surface is a perfect sink, Equation (2) may be expressed in the form

$$F = v\chi_1 \qquad (4)$$

Thus, fluxes may be estimated from a knowledge of published values of v and concentration values which are representative of a standard height and for the area in question. Transfer velocities determined from experiments with iodine-131 by Chamberlain and Chadwick (23) are near 1 cm second⁻¹, as are those found by Garland et al. (24) in experiments with sulfur dioxide. Such values have been used in many global budget calculations, particularly for sulfur (25, 26).

C. Known Sinks for Common Air Pollutants

Table III lists many of the known sinks for some common gaseous pollutants (8, 13, 16, 17, 19–22, 24, 27–45). The listing is not meant to be complete; however, the references include some works of a review nature. [See also a recent review by Rasmussen et al. (26a).] Study of the references indicates that of the common air pollutants, sulfur dioxide has had its sinks most fully investigated. On the other hand, information on the stratospheric and microbiological sinks of such an abundant pollutant as carbon monoxide is far from complete.

D. Sink Strength Estimates for Some Major Pollutants

As has already been pointed out in Section II,D, there are many uncertainties inherent in estimates of global emissions; for sinks, the available information is also inadequate. Estimates of annual quantities removed from the atmosphere are frequently obtained by choosing a value which will yield a balanced budget for a specific pollutant. Often removal rates, particularly in the case of surface absorption, are extrapolations from limited laboratory investigations where gas-phase pollutant concentrations are unrealistically high or where gas-phase transfer conditions do not simulate those in the real atmosphere. In summary, less confidence

Table III   Summary of Sinks of Selected Gaseous Air Pollutants

Substance	Sinks	References
Sulfur dioxide	Precipitation scavenging: washout, rainout	(27)
	Oxidation in gas and liquid phase to sulfate	(28–30)
	Soil: microbial degradation, physical and chemical reaction, absorption	(20, 31)
	Vegetation: sorption on surfaces, stomatal intake	(21, 22, 24)
	Oceans, lakes: absorption	(16, 17, 19)
Hydrogen sulfide	Oxidation to sulfur dioxide	(28, 32)
Ozone	Chemical reaction on vegetation, soil, snow and ocean surfaces	(22, 33–35)
Nitrous oxide	Soil: microbiological destruction	(36, 37)
	Stratosphere: photodissociation	(37, 38)
	Ocean: absorption	(36)
Nitric oxide/ nitrogen dioxide	Soil: chemical reaction	(31)
	Vegetation: sorption, stomatal uptake	(22)
	Chemical reaction in gas and liquid phase	(36)
Ammonia	Chemical reaction to ammonium in gas and liquid phase	(28, 39)
	Precipitation scavenging: washout, rainout	(36)
	Surface uptake: physical and chemical reaction, absorption	(28, 39, 40)
Carbon monoxide	Stratosphere: reaction with hydroxyl radical	(13, 37)
	Soil: microbiological activity	(41)
Carbon dioxide	Vegetation: photosynthesis, absorption	(8, 42)
	Oceans: absorption	(43)
Methane	Soil: microbiological activity	(37)
	Vegetation: chemical reaction, bacterial action	(8)
	Troposphere and stratosphere: chemical reaction	(44)
Hydrocarbons	Chemical reaction to particulates	(8)
	Soil: microbiological activity	(31)
	Vegetation: Absorption, stomatal intake	(45)

can be placed in global annual sink-strength estimates than in emission inventories.

Some of the difficulties associated with making sink estimates on a global scale are

(a) Quoted "representative global pollutant concentrations" are frequently averages of a few, nonrepresentative measurements. Most available data are "surface data"; data are lacking over the oceans and in the upper troposphere.

(b) The efficiency of removal by different surfaces is often not considered. For example, little is known about scavenging by forests.

(c) Although the chemistry and microphysics of precipitation scavenging processes are understood, quantitative estimates of washout and rainout coefficients are not available for many substances.

(d) Removal of particles by sedimentation is impossible to assess because of the difficulty in measuring dry particulate deposition.

(e) Sink processes themselves are in many cases poorly understood. Nevertheless, numerous estimates are available for some of the major pollutants. Figure 1 is a schematic representation of the sulfur cycle, and Table IV lists the sinks of sulfur compounds as proposed by Kellogg *et al.* (*26*). Sulfate removal by precipitation is based on observed concentrations in precipitation obtained from networks of stations, whereas sulfur dioxide removal is determined from a theoretically derived flux and assumed average hemispheric concentrations. The mass of sulfate deposited by impaction on surfaces is assumed to equal the mass of sulfur dioxide absorbed.

The Robinson and Robbins estimates of sink strengths in the nitrogen cycle (*36*) are summarized in Table V. Although neither nitrous oxide nor nitrogen are considered to be pollutants, they are included to provide a more complete picture of the cycling of nitrogen compounds through the atmosphere. Sink strength estimates for both nitrogen and sulfur compounds have been made within the context of a total cycle for the element.

Fabian and Junge (*46*) and Junge and Czeplak (*47*) have made calculations of the ozone and the carbon dioxide continental sink strengths, respectively, based on measured uptake rates and representative gas concentrations of ozone, the yearly production of plant material for carbon dioxide, and the distributions of various surfaces by latitude belts. On this basis the global ozone sink is estimated to be between 4.0 and

Table IV   Sink Strengths of Sulfur Compounds (*26*)[a] [Reprinted with permission from W. W. Kellogg, R. D. Cadle, E. R. Allen, A. L. Lazrus, and E. A. Martell, *Science* **175**, 587 (1972). Copyright 1972 by the American Association for the Advancement of Science.]

Sink	Land	Ocean	Total
A. Northern hemisphere			
Sulfate in precipitation	64	43	107
Sulfur dioxide absorption at surface	12	0	12
Sulfate impaction at surface	8	0	8
B. Southern hemisphere			
Sulfate in precipitation	22	30	52
Sulfur dioxide absorption at surface	3	0	3
Sulfate impaction at surface	2	0	2
	111	73	184

[a] Original values are in units of 10^6 tons calculated as sulfate per year. Values given are in 10^6 tons per year as sulfur.

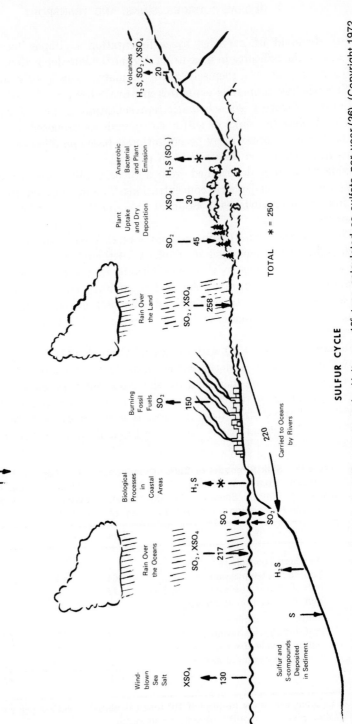

SULFUR CYCLE

Figure 1. Sources and sinks of atmospheric sulfur compounds. Units are 10⁶ tons calculated as sulfate per year (26). (Copyright 1972 by the American Association for the Advancement of Science.)

Table V   Sink Strengths of Nitrogen Compounds (36)[a]

Substance	Sinks	Sink strength
Nitrous oxide	Stratosphere	24
	Surface: biological	352
Nitrogen	Soil: biological	130
Ammonia	Surface absorption	749
	Reaction to ammonium with	
	Subsequent precipitation scavenging	166
	Dry deposition	44
Nitric oxide/	Surface absorption of nitrogen dioxide	145
nitrogen dioxide	Oxidation to nitrate aerosol with	
	Subsequent precipitation scavenging	83
	Dry deposition	22

[a] Original values are in units of tons of nitrogen components per year. Values given are in 10^6 tons per year as nitrogen.

7.1×10^8 tons year^{-1}, while for carbon dioxide the total rate of vegetation uptake is 141×10^{15} gm year^{-1}.

Finally, two last examples—methane and carbon monoxide—are of interest. It has been estimated that the total destruction rate of methane in the atmosphere by chemical reaction is approximately 3×10^{15} gm year^{-1}, with 10% occurring in the stratosphere (44). In the case of carbon monoxide, based on an extrapolation from laboratory experiments (41), it is estimated that the absorptive capacity of the total soil surface of the continental United States is 6.5 times the annual estimated production of carbon monoxide in that country and three times the annual estimated worldwide production. Although there are no data on long-term trends in carbon monoxide concentrations at remote stations, it is unlikely that there is a worldwide increase as in the case of carbon dioxide.

IV. Global Transport and Diffusion Processes

When pollution is released from a point or volume source into a turbulent atmosphere, the pollution is carried forward by the wind and expands in the lateral and vertical directions. The cross-wind spread is not bounded, but in the vertical direction there are two strata of limited permeability that constitute significant barriers to upward mixing:

(a) The top of the surface mixed layer (near the surface during a nighttime radiation inversion; 500 m to several kilometers during daytime convection)

(b) The tropopause (about 10 km)

Thus for sufficient downwind distance, the vertical dimension of a plume is limited and can often be estimated from synoptic weather charts: a uniform vertical distribution of concentrations is then often assumed in the lowest layers.

If a turbulent flow is horizontally homogeneous and not changing with time, the crosswind spread approaches Fickian diffusion for sufficiently long travel times, i.e.,

$$\sigma_y{}^2 = 2Kt \tag{5}$$

where σ_y is the standard deviation of the crosswind spread, K is the atmospheric diffusivity, and t is travel time. This condition is approached when the width of the plume expands to about the size of the largest turbulent eddies in the flow.

There are three important scales of turbulence in the atmosphere:

(a) *Microscale:* These are the turbulent fluctuations that cause most of the mixing in the first hour or so of travel of pollution away from a source.

(b) *Mesoscale:* These eddies have dimensions of a few kilometers. They may be random (as in a thunderstorm) or organized (as in a sea breeze or valley flow).

(c) *Macroscale:* These eddies have dimensions of 500 km or more. They include the large-scale systems that dominate the weather in the temperate zones. They also include disturbances embedded in the Northeast and Southeast Trades, and in the ITC (Intertropical Convergence) Zone.

Attempts to model large-scale diffusion by the Fickian assumption have not been entirely successful, even when long climatological averaging times are used, because the macroscale eddies are not completely random. Although the weather is noted for its variability, there are many well-defined and persistent centers of activity in the atmospheric general circulation. For example, the subtropical anticyclones are coherent and show only a slow seasonal latitudinal shift. The continental–oceanic monsoons are well developed in many parts of the world, while mountain chains cause significant anomalies in the general circulation, which means that the global wind patterns cannot be assumed to approximate a random field of turbulence. Consequently, even if the sources and sinks of pollution were distributed uniformly around the world, the global concentration patterns would be distorted. The background values quoted in Section V are to be interpreted in this light.

Of particular interest to atmospheric chemists is the study of exchange rates between the northern and the southern hemispheres. The transfer of matter across the meteorological equator is in fact so slow that the two hemispheres can be assumed quite appropriately to behave as separate reservoirs for modeling purposes. Characteristic background concentrations of trace substances in the two hemispheres are likely to be different for three reasons:

(a) The inputs of man-made pollutants are higher in the northern hemisphere than in the southern.

(b) The ratio of ocean/land areas is different, which has a significant effect on the strengths of the natural sources and sinks.

(c) The seasons are reversed, which means that the annual cycles in the strengths of many types of sources (e.g., emissions due to space heating) and of sinks (e.g., photosynthesis) are also reversed in the two hemispheres.

These examples illustrate the difficulties involved in the interpretation of air chemistry data from background stations, and in the development of predictive models. One further problem should be mentioned here—the interpretation of annual trends in regional air quality. For an area the size of the United States or Western Europe, trends observed over a few years may be due to two factors:

(a) Trends in source and/or sink strengths

(b) Long-term fluctuations in the atmospheric general circulation, resulting in anomalies in the frequencies of stagnation or other weather patterns that create high pollution potential

An improvement in air quality over a period of a few years therefore indicates but does not prove that control programs have been effective. The possibility must always be investigated that there has been a significant change in the frequencies of various types of weather patterns.

V. Observed Global Concentration Patterns

A. Synoptic Events

Sometimes an anomalous large-scale wind field (e.g., easterly winds over the North Atlantic for a few days) or an unusual pollution release (e.g., a volcanic eruption or a nuclear detonation) provides a special opportunity for studying the transport and diffusion of pollution. Munn (*48*) and Munn and Bolin (*49*) have given a number of examples, including the forest fires in Alberta, Canada in September 1950 that caused smoke palls over Western Europe; the Windscale reactor accident in the United Kingdom in October, 1957, that was detectable on the continent and in Scandinavia; and a number of unusual migrations of insects and birds associated with periods of anomalous wind flows. These curiosities illustrate the ability of the atmosphere to transport substances very long distances indeed, but they often occur so unexpectedly that adequate supplementary monitoring programs cannot be initiated.

B. Spatial Patterns of Concentrations of Trace Substances

Fragmentary data are available on world background concentrations of trace substances. There are a few fixed stations such as Mauna Loa (on a mountain top, above the surface mixed layer in Hawaii) where sufficient observations are made to permit the calculation of annual mean values and trends. In addition, there are a large number of individual observations (made from cruising ships or aircraft, for example), which when taken together reveal reproducible patterns such as the exponential decrease of concentrations eastward from Japan and also from North America.

Spatial patterns of particle concentrations provide an informative example of the global spread of a substance with a relatively short atmospheric residence time. Away from sources where large particles will already have been removed by sedimentation, particle residence times are of the order of a few weeks. Thus, they remain in the atmosphere long enough to be transported thousands of kilometers from sources but not long enough to become uniformly mixed throughout the global troposphere. The average exchange time between the northern and southern hemispheres is about a year and the two hemispheres are therefore independent as far as particle loading is concerned. The horizontal distribution is controlled by the location of source and sink areas and by the large-scale wind systems. Atmospheric electrical conductivity measurements (50), for example, indicate that although particulate levels are unaffected by man's activities at Mauna Loa and over most of the oceanic regions of the world, paths of particulate pollution extend eastward from the United States across the Atlantic almost to Europe, eastward from Japan in the north Pacific, and southward from Asia in the northern Indian Ocean. These regions cover less than about 10% of the earth's ocean area.

Because sources of particles are predominantly over land, there is a generally pronounced decrease in concentration with increasing altitude over continental areas and adjacent ocean areas. Above altitudes of approximately 5 km, the influence of surface sources becomes negligible and there is a uniform concentration of about 300 particles cm⁻³ (*28*).

Even for such a thoroughly investigated substance as sulfur dioxide, our knowledge of worldwide concentration patterns is limited. Although its sources are dominantly anthropogenic, sulfur dioxide has been found throughout continents, coastal areas, and over large portions of the ocean surface. In general, concentrations range from a few μg m⁻³ in clean background tropospheric air up to several hundred μg m⁻³ in polluted air. Vertical profiles taken over central Europe (*29*) indicate concentrations which are highly variable, typically in the neighborhood of 20 μg m⁻³ in the surface layer below a haze layer, dropping to a uniform value near 5 μg m⁻³ above 3 km. Other clean air data from the continental United States and Panama indicate levels < 1–5 μg m⁻³ (*26, 51*). There are few data available for the upper troposphere or stratosphere.

Figure 2 (*29*) indicates that over the mid-Atlantic (30°W), sulfur dioxide concentrations are relatively low. Presumably because of the relatively short sulfur dioxide atmospheric residence time (about 4 days) and because of the large volume of air over the oceans available for mixing, the influence of continental sources is not great. Surface-layer background concentrations over the oceans are estimated to be 0.5 to 1 μg m⁻³ in the tropics; these are perhaps typical of large areas of the earth's surface. North of 30°N, concentrations were found to vary from about 2 to 4 μg m⁻³, indicating that some sulfur dioxide had come from remote combustion sources in North America.

A final remark about spatial distribution of pollutants concerns atmo-

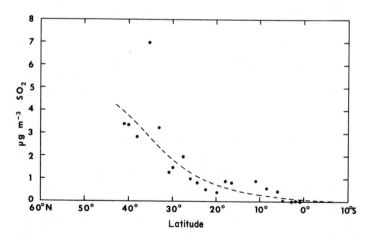

Figure 2. Meridional profile of sulfur dioxide concentrations over the mid-Atlantic (30°W) (*29*).

spheric residence times. The strength and location of sinks for specific pollutants control their atmospheric residence time. It is apparent that a short residence time, hours or a few days, will result in the limited spread of a substance by winds. Therefore observed concentrations will be high near the sources and low or zero elsewhere. On the other hand, a substance with a long residence time, carbon dioxide for example, will have the opportunity to become widely distributed over the globe with a resultant more uniform level of concentration. Junge (*37*) suggests that an approximately inverse relationship holds between the tropospheric residence time of a trace gas and its variability, expressed in terms of the global tropospheric standard deviation of its mixing ratio.

C. Secular Time Trends in Background Concentrations

The most widely discussed time trends are those for carbon dioxide, amounting to between 0.7 and 1.0 ppm per year since 1958. Figure 3 (*52*) shows atmospheric carbon dioxide concentrations in parts per million by volume measured at Mauna Loa, Hawaii between 1958 and 1973. The circles are monthly mean values. Also shown on the abscissa are the annual changes in parts per million. The annual cycle in concentration is quite evident: concentrations are highest in May and lowest in October as a result of uptake by vegetation in the summer months, combined with year-long processes of respiration and decay. However, more important is the continuing increase in concentrations over the past 15 years. Between 1958 and 1968, the average annual increase was 0.7 ppm, and between 1969 and 1973 it has exceeded 1.0 ppm. Although the absolute calibrations of the nondispersive infrared sensors are still being questioned, there is little doubt as to the validity of these trends. Measurements made at Point Barrow, Alaska, over Scandinavia, at Mauna Loa, and at the South Pole all show similar increasing trends (*52*).

There is also a body of evidence indicating that increases in atmospheric particulate loading have been occurring in some areas of the world. Table VI shows long-term trends in atmospheric turbidity and direct solar radiation at a number of stations. The Soviet, European, Japanese, and North American stations show upward trends in particulate loadings (i.e., increasing turbidity or decreasing solar radiation) whereas, the Mauna Loa, Aspendale, and Antarctic stations reveal no long-term changes. Such a picture points to increasing particulate loading in the northern hemisphere, but not in the southern, with Mauna Loa being free of the influence of northern hemispheric anthropogenic emissions. The study by Munn (*53*) of haziness in the Canadian Atlantic provinces over the last 20 years, suggests that the additional particulate

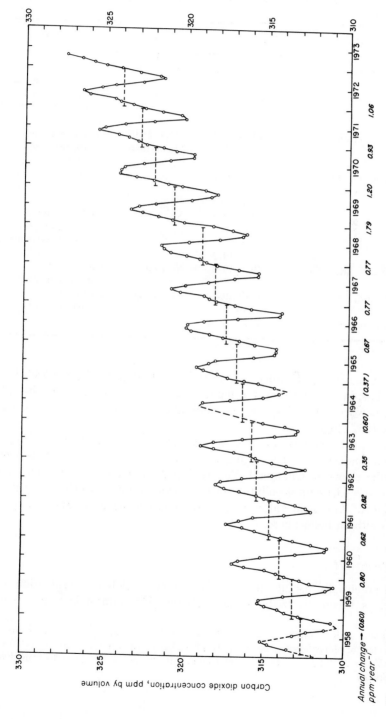

Figure 3. Mean monthly carbon dioxide concentrations at Mauna Loa. Horizontal dashed lines indicate average yearly value. Annual changes in parenthesis are based on incomplete record (52).

Table VI Long-Term Trends in Atmospheric Turbidity and Direct Solar Radiation (52)

Station	Period of record	Period of trend	Remarks
USSR, 8 stations	~1900–1967	~1940–1967	~10% reduction of normal radiation
Davos, Switzerland	1914–1926 vs 1957–1959	—	~20% increase in turbidity per decade
Japan, 24 stations	1936–1955	1936–1945 1948–1955	~20% increase, ~30% increase in turbidity
Washington, D.C., United States	1903–1907 vs 1962–1966	—	~10% increase in turbidity per decade
Mauna Loa, Hawaii	1958–1971 1958–1962 vs 1970–1971	No trend	1963–1969 influence of Mt. Agung; direct solar measurements
Aspendale, Australia	1955–1970	No trend	Direct solar measurements
Maudheim, Antarctica	1949–1952 vs	No change	Turbidity measurements
McMurdo, Antarctica	1966–1967		

matter may be of photochemical origin. In that part of the world, haziness has, in fact, been decreasing during the winter months; in the summer, however, there has been a doubling of the number of hours of haze since 1952 (Fig. 4).

Electrical conductivity measurements of Cobb and Wells (54) also provide an indication of trends in global levels of particles in the atmosphere. Results shown in Chapter 1, Vol. II, Figure 18 support the contention of increased particulate pollution in the north Atlantic, which is influenced by North American anthropogenic sources, and no noticeable change in the southern hemisphere.

D. Freon 11*

A potentially useful substance for observing global concentration patterns is Freon 11, trichlorofluoromethane. This almost inert gas is one

* Freon 11 is a commonly used trade name of the DuPont Company. It is but one of a class of chlorofluorocarbons, the two best known of which are Freon 11 ($CFCl_3$) and Freon 12 (CF_2Cl_2). These are referred to interchangeably as chlorofluoromethanes or Freons.

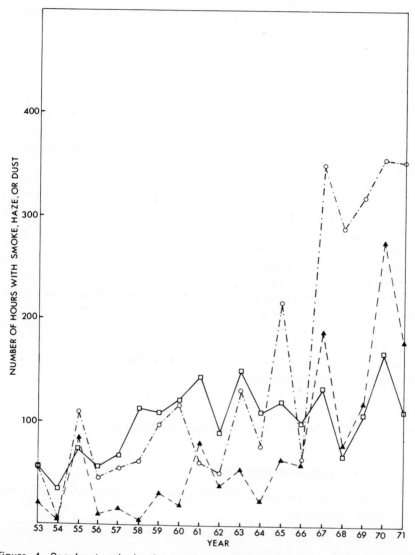

Figure 4. Secular trends in the number of hours of smoke, haze, or dust at Fredericton, Gander, and Mont Joli Airports for the months May to October for the period 1953 to 1971 *(53)*. ○: Fredericton Airport; ▲: Mont-Joli Airport; □: Gander International Airport.

of the two most frequently used propellant solvents for aerosol dispensers such as for spray paints, etc. Freon 11 is a useful tracer because

(a) It has no natural sources and thus its presence in the atmosphere is indicative of transport from areas of anthropogenic emissions

(b) It is physically and chemically inert under conditions found in the troposphere and lower stratosphere

(c) There are no known strong sinks for the gas except in the upper stratosphere

Recent measurements (55) of atmospheric Freon 11 concentrations between the United Kingdom and Antarctica during the voyage of the RRS Shackleton yielded a mean aerial concentration of approximately 50×10^{-12} by volume. Concentrations were found to increase in the northern latitudes reflecting the presence of sources in the industrialized north temperate zone.

In a study of the variations of atmospheric turbidity with season and with wind direction at two sites in the British Isles, Lovelock (56) has made use of the fact that Freon 11 concentration and summertime turbidity vary with wind direction in an almost identical way to suggest possible sources of the turbid air. The knowledge that Freon 11 is of anthropogenic origin leads to the conclusion that the turbid air also is from industrialized areas, in this case continental Europe. On the global scale, Machta (57) has proposed a simple two-dimensional model (vertical and north–south) of diffusion of an inert tracer and has compared the predicted concentrations with measured Freon 11 concentrations over the Atlantic Ocean. While discrepancies exist, the agreement in latitudinal trend in concentration provides encouragement for the use of such substances in model verification.

VI. Synoptic Models of Global Transport and Diffusion

A. Relevance of Point-Source Models

The classic Gaussian models for diffusion from a point source are based on several assumptions:

(a) A homogeneous, turbulent wind flow, i.e., no changes in the meteorological fields in the horizontal

(b) Steady-state conditions, i.e., no changes in the meteorological fields with time

(c) Passive pollutants, i.e., no loss of material by the various sink mechanisms

These assumptions are often justified for diffusion over a uniform surface for periods of a few hours during settled weather conditions. Occasionally, in fact, the classic approach is useful for downward distances of 200 km or so. On this scale, there have been attempts to assume that a city is a point source, and to model the urban plume (*58, 59*) ; however, extrapolation to the continental and global scales has not yet been attempted.

An alternative approach is to solve numerically the Navier–Stokes equations or their finite-difference approximations, i.e., to keep track of the inputs, storage, and outputs of pollution entering and leaving a series of connected boxes.

B. Trajectory Analysis

The direction of motion of a plume or cloud of pollution can often be estimated from a knowledge of the wind fields on successive surface and upper-air weather charts. The technique, which has been described by Munn (*48*), is most frequently used to explain the occurrence of a rare event such as an unusual invasion of pollen, spores, insects, or birds, or a notably high concentration of some pollutant at a remote monitoring station. In this way, Robinson and Robbins (*60*) inferred that a threefold increase in carbon monoxide concentrations on July 29, 1967 at a sampling site on the Greenland ice cap was associated with an air mass that had moved northward from the eastern United States.

Trajectory analysis is most successful when there is a well-defined wind field that is changing only slowly with time. However, there are two types of errors that often make the analysis very uncertain after a day or so of tracking on successive weather charts:

(a) Insufficient information about the initial height of the center of gravity of the cloud of material

(b) Insufficient information about the three-dimensional wind fields, including their changes with time (rawinsonde stations are at least several hundred kilometers apart)

In a survey paper, Danielson (*61*) has examined this latter type of error and has given examples of rather substantial deviations, which are caused by several factors:

(a) The network of upper-air stations is too coarse in space to permit detection of subsynoptic-scale perturbations in the wind field

(b) The 12-hour intervals between rawinsonde releases are too great to permit interpolations on occasions when the wind field is changing significantly

(c) The vertical fields of motions are often ignored, with the assumption that the center of gravity of the pollution cloud remains at some standard level such as 850 mbar

(d) The rotational and divergent ageostrophic components of the wind are often neglected. Particularly in the surface mixed layer during periods of unstable stratification, there is insufficient information to evaluate the wind divergences.

Trajectory analyses are therefore only indicative and must at the very least include a statistical treatment of a number of cases, as has been demonstrated by Rodhe (62). In addition, it is evident that the information content of successive weather maps is not sufficient to give any indication of the rate of lateral dispersion of pollution plumes. Thus a trajectory analysis predicts only the position of the plume and not the concentration field.

C. Diffusive Transport Models

The transport and diffusion of pollutants over moderate and large travel distances can be successfully predicted from numerical models during slowly changing meteorological conditions, given adequate information on source and sink strengths. In an early attempt, Reiquam (63) examined the possibility of the transport of sulfur compounds on a continental scale over western Europe using a simple atmospheric transport and accumulation model, with mean wind data and an estimated source distribution, and concluded that observed concentrations in northern Europe could be the result of emissions in Great Britain and central Europe. However, his model contained no sink term and therefore was indicative only.

One of the shortcomings of this type of modeling has been a lack of data for verification. An OECD (Organisation for Economic Cooperation and Development) program has been undertaken by ten western European countries to study the long-range transport of air pollutants (64). An important element of this program is the development of atmospheric dispersion models for both prediction of sulfur dioxide and particulate sulfate levels during expected high-concentration episodes, and for a comprehensive statistical evaluation of the long-range transport of pollutants when sufficient data are available. An extensive ground sampling network exists to provide verification of the model predictions.

A model based on Eulerian integration of the continuity equation is being extensively used (64, 65); it is based on sectorial displacements in the eight main directions of a regular grid system. Input to the model consists of wind information from the world meteorological network (based on a grid size of 127×127 km) and emission data obtained by inventory (grid size of $\frac{1}{2}°$ longitude $\times$ $1°$ latitude or approximately

50×60 km). The pollutants are assumed to be uniformly distributed in the vertical beneath the mixing height and losses by chemical transformation and deposition are accounted for. The continuity equation for sulfur dioxide is modeled as follows:

$$\frac{\partial C}{\partial t} = -k_0 C - k_1 C^2 + Q - \nabla \cdot (CV) + \nabla \cdot (K \nabla C) \tag{6}$$

where C is the concentration of sulfur dioxide, t is time, Q is the source strength, V is velocity, and K is the diffusion constant. The sink expression, $k_0 C + k_1 C^2$ includes the usual term which is proportional to concentration and a second quadratic term which is supposed to account for the presence of reactive components other than sulfur dioxide by assuming their concentrations to be proportional to C. Values of k_0 and k_1 are chosen to be 10^{-6} second^{-1} and 0.25×10^{-6} m^3 second^{-1} μg^{-1} which lead to decay times that are in reasonable agreement with existing values in the literature.

Testing of such a model is best accomplished when there are no large sources between a region of strong emissions and a distant observation area. Close to a strong source, the assumption of instantaneous mixing through the surface layer will be in error. Scandinavia is therefore ideally situated as a measurement site for detecting pollutants transported across the North Sea from areas of heavy industry in Great Britain. For this type of situation, Nordø (65) compared model predictions with concentrations of sulfur dioxide from aircraft measurements over the Skagerak during an episode in March 1972 and, in general, found fairly good quantitative agreement. In addition, model concentration isopleths tended to form plumes which coincided with computed 850-mbar wind trajectories from source areas to Scandinavia. This agreement provides confidence in the use of diffusive models and should encourage further development.

However, there are certain aspects of the models which require further attention:

(a) The assumption of a uniformly mixed layer is often unrealistic; aircraft observations suggest that pollutants are frequently transported in layers for considerable distances. The development of two-layer models will improve the description of mixing-height variability and layering.

(b) More realistic modeling of chemical transformation and removal processes is required. In addition, it is necessary to take account of removal by precipitation during the course of transport over large distances.

(c) Improved temporal and vertical resolution of emissions is necessary as input to episode modeling. Frequently source data are available only quarterly or even annually with no separation of high- and low-level emissions.

D. A Numerical Simulation of Global Transport

A final interesting example of synoptic modeling is the application of a general circulation model to the problem of global transport of carbon monoxide from automotive sources (66). The model, based on a grid size of 5° longitude and 4° latitude, is permitted to run for 24 simulated days to establish a realistic circulation pattern before carbon monoxide sources are turned on. Advection by the simulated wind is the means of transport of the gas; subgrid-scale diffusion is not modeled. Vertical transport results from large-scale advection and cumulus convection. The global carbon monoxide distribution is then allowed to develop over a 4-week period (Fig. 5). This is one of the first studies to apply a general circulation model in this way, and although neither turbulent diffusion nor removal processes are modeled, nevertheless it is capable of providing an insight into the development of global patterns of carbon monoxide concentrations.

VII. Climatological Models of Global Air Pollution

For the interpretation and prediction of averaged patterns and long-term trends in global-scale air pollution, and particularly of interhemispheric differences, climatological models are employed. The simplest type is the box model in which certain natural reservoirs, for example, the oceans and the atmosphere, or selected portions of them, are represented by boxes in which the tracer substances of interest are assumed to be uniformly distributed. By correctly choosing boundaries of the boxes and determining exchange rates between adjacent boxes, a study of the transfer of a tracer such as radiocarbon (67, 68) can provide an insight into pollutant cycles in nature and large scale mixing. However, box models do not provide the resolution required for the investigation of pollutant distributions in the global troposphere.

A more satisfactory approach is through the use of semiempirical models, in which all turbulent motions are parameterized on the basis of empirically determined eddy diffusion coefficients. Czeplak and Junge (69) have used a one-dimensional, global diffusion model of this type to investigate the exchange time between the northern and southern hemispheres. This parameter is of particular interest to meteorologists and atmospheric chemists because when used along with measured interhemi-

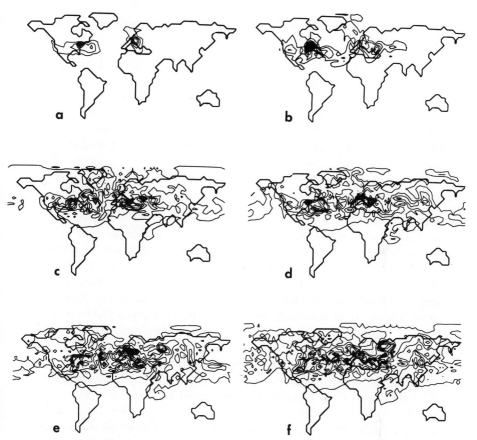

Figure 5. Simulated CO concentration contours following the inclusion of sources in the model: (a) after 2 days, (b) after 6 days, (c) after 16 days, (d) after 20 days, (e) after 24 days, and (f) after 28 days. The contour interval is 10 ppb CO (66).

spheric pollutant concentration differences it provides information about the importance of sources and sinks on the global scale. The model is based on large-scale Fickian diffusion in a vertically and zonally well-mixed troposphere which is symmetric about the equator. Tropopause height and eddy diffusion coefficient variations with latitude are accounted for; the latter is assumed to be proportional to the variance of the meridional wind component which is weighted with respect to air density in the vertical to obtain representative mass transport. It is thus possible to examine the development of global distributions as a function of time, eddy diffusion coefficient field, and chemical lifetime of the substance as a result of continuous or one-time tracer injections in the northern hemisphere. The following predictions may be made from model calculations:

(a) Uniform meridional mixing in the northern hemisphere is approached after 2 to 3 months

(b) The interhemispheric exchange time, approximately 1 year, is controlled by the value of the eddy diffusion coefficient at the equator

(c) The low value of the diffusion coefficient at the equator, $\sim 0.3 \times 10^{10}$ cm^2 second^{-1}, results in a pronounced separation of the two reasonably well-mixed hemispheres

(d) As a result of (c), the troposphere may be approximated by a two-box model for the purpose of interhemispheric exchange

The two-dimensional model of Machta (57) referred to briefly in Section V,D is also of the semiempirical type; it is able to approximate reasonably well the coarse features of tracer distributions on a global scale. However, as Machta emphasizes, a more complete body of tracer data is required before these models can be satisfactorily verified. Because the locations and magnitudes of sinks for most pollutants are poorly known, it is necessary to use as tracers a few substances such as radon, argon-37, krypton-85, Freon 11, and carbon dioxide whose sinks are either negligible or exactly known. Unfortunately, the available information on global distributions of these substances is quite inadequate. Freon 11, discussed earlier, appears to be one of the most promising.

The third and most complex type of model in use is the large-scale, general circulation model [e.g., Hunt and Manabe (70)]. Such models are not strictly climatological; however, tracer distributions, averaged over periods of several days, are obtained after they have reached a quasi-steady-state condition. Global-scale dispersion is predicted from instantaneous, three-dimensional wind fields based on a grid whose horizontal dimension is typically a few hundred kilometers. However, at subgrid scales turbulent diffusion must still be parameterized. This type of model is potentially most useful for modeling transport and dispersion on a global scale, subject of course to the cost limitations imposed by the use of large computers. The difficulty of model verification still exists, however, and it is likely that the lack of knowledge of sinks and the inadequate data on global distributions of tracers will remain major limitations on the usefulness of global modeling for some time.

VIII. The Ecological Significance of Global Air Pollution

The importance of trace substances (mainly carbon dioxide and particulates) in modulating world climate is discussed in Chapter 8, this volume. A few words should be added here concerning the ecological significance of global air pollution. Near strong urban and industrial sources, there is, of course, evidence for biological effects. At remote locations, however, atmospheric concentrations of pollution are usually so low as to be near the limit of chemical detection, and no direct effects on life have been observed. For this reason, the statement is sometimes made that there is no need to monitor air quality away from cities and industrial sources. However, this point of view overlooks the fact that because the troposphere participates in biogeochemical cycling, the concentrations of trace substances in other media cannot be interpreted or predicted without a knowledge of atmospheric processes.

The world environment may be considered as a series of connected reservoirs—the stratosphere, the troposphere, the oceanic surface mixed layer, the deep ocean, snow, ice, soil, vegetation, lakes, rivers, and groundwater. Trace substances are continually moving through these reservoirs, sometimes diluting (as in the case of the troposphere) and sometimes concentrating (as in the case of certain food chains). Until the arrival of man, the system was in near equilibrium. Particularly since the last century, however, anthropogenic outputs of waste substances have sometimes exceeded the self-cleansing capabilities of one or more of the reservoirs. It should be noted here that the troposphere is particularly effective on the global scale in assimilating and disposing of pollution, because

(a) The volume of air available for mixing is very large

(b) The turbulent winds are effective in transporting and diffusing pollution through the hemispheric volume of air

(c) Strong sink mechanisms are operating continually, transferring trace substances to other reservoirs in the biogeochemical chain

If all anthropogenic emissions were suddenly turned off, the atmosphere would return rather quickly to its natural state. In contrast, polluted groundwater may take centuries to recover.

In the above context, a knowledge of tropospheric transport and diffusion processes and of the pathways of trace substances to other reservoirs is of fundamental importance in developing simulation models of the behavior of the world biosphere. Some preliminary models have already been developed, e.g., for DDT (71). However, one of the obstacles to further progress is the interdisciplinary nature of the subject, and the general lack of communication among researchers who are modeling particular parts of the world ecosystem. The atmospheric global modelers are, of course, interested in the interfaces between the troposphere and other reservoirs; however, inputs and outputs are usually considered as boundary conditions without much attention being given to the ultimate fate of the contaminants. A total ecosystem approach is required, as can be illustrated with two examples:

(a) Some terrestrial and marine organisms (e.g., lichens) are not only accumulators of such substances as heavy metals, pesticides and, polychlorinated biphenyls, but they are also in human food chains. The trace substances are carried to remote areas, and although the atmospheric concentrations may often be below the threshold of detection, accumulation in living organisms over decades may be highly significant. On the other hand, essential nutrients may also be carried to remote areas by the wind, contributing usefully to the local materials balance.

(b) As described in the Swedish Case Study for the Conference on the Human Environment (*72*), precipitation scavenging of industrial sulfates may cause acidification or soils (particularly podsols) at very great distances from the emission sources, resulting in leaching and reduced agricultural and forest productivity.

For reasons such as these, tropospheric transport and diffusion processes should be viewed in the very broad context of modeling of the world biosphere.

REFERENCES

1. R. A. Bryson and D. A. Baerreis, *Bull. Amer. Meteorol. Soc.* **48**, 136 (1967).
2. T. N. Carlson and J. M. Prospero, *J. Appl. Meteorol.* **11**, 283 (1972).
3. H. H. Lamb, *Phil. Trans. Roy. Soc. London, Ser. A* **266**, 425 (1970).
4. G. M. Hidy and J. R. Brock, *Proc. Int. Clean Air Congr., 2nd, 1970* pp. 1088–1097 (1971).
5. J. H. Joseph, A. Manes, and D. Ashbel, *J. Appl. Meteorol.* **12**, 792 (1973).
6. C. D. Keeling, *Tellus* **25**, 174 (1973).
7. L. S. Jaffe, *J. Geophys. Res.* **78**, 5293 (1973).
8. E. Robinson and R. C. Robbins, *in* "Air Pollution Control" (W. Strauss, ed.), Vol. II, p. 1. Wiley (Interscience), New York, New York, 1972.
9. A. W. Castleman, H. R. Munkelwitz, and Manowitz, *Nature (London)* **244**, 345 (1973).
10. H. W. Georgii and D. Jost, *Nature (London)* **221**, 1040 (1969).
11. W. Bischof, *Tellus* **23**, 558 (1971).
12. B. E. Jocelyn, J. F. Leach, and P. Wardman, *Water, Air, Soil Pollut.* **2**, 141 (1973).
13. W. Seiler and C. Junge, *J. Geophys. Res.* **75**, 2217 (1970).
14. J. A. Businger, *in* "Workshop on Micrometeorology" (D. A. Haugen, ed.), p. 67. Amer. Meteorol. Soc., Boston, Massachusetts, 1973.
15. P. J. Barry and R. E. Munn, *Phys. Fluids, Suppl.* S263 (1967).
16. S. Beilke and D. Lamb, *Tellus* **26**, 268 (1974).

17. D. J. Spedding, *Atmos. Environ.* **6**, 583 (1972).
18. A. C. Chamberlain, *Proc. Roy. Soc., Ser. A* **290**, 236 (1966).
19. D. M. Whelpdale and R. W. Shaw, *Tellus* **26**, 196 (1974).
20. Massachusetts Institute of Technology, "Inadvertent Climate Modification," p. 197. MIT Press, Cambridge, Massachusetts, 1971.
21. D. J. Spedding, *Nature (London)* **224**, 1229 (1969).
22. A. C. Hill, *J. Air Pollut. Contr. Ass.* **21**, 341 (1971).
23. A. C. Chamberlain and R. C. Chadwick, *Tellus* **18**, 226 (1966).
24. J. A. Garland, W. S. Clough, and D. Fowler, *Nature (London)* **242**, 256 (1973).
25. E. Eriksson, *Tellus* **11**, Part I, 375 (1959); **12**, Part II, 63 (1960).
26. W. W. Kellogg, R. D. Cadle, E. R. Allen, A. L. Lazrus, and E. A. Martell, *Science* **175**, 587 (1972).
26a. K. H. Rasmussen, M. Taheri, and R. L. Kabel, *Water, Air, Soil Pollut.* **4**, 33 (1975).
27. S. Beilke and H. W. Georgii, *Tellus* **20**, 435 (1968).
28. C. E. Junge, "Air Chemistry and Radioactivity." Academic Press, New York, New York, 1963.
29. H. W. Georgii, *J. Geophys. Res.* **75**, 2365 (1970).
30. M. Bufalini, *Environ. Sci. Technol.* **5**, 685 (1971).
31. F. B. Abeles, L. E. Craker, L. E. Forrence, and G. R. Leather, *Science* **173**, 914 (1971).
32. R. D. Cadle and E. R. Allen, *Science* **167**, 243 (1970).
33. I. E. Galbally, *Quart. J. Roy. Meteorol. Soc.* **97**, 18 (1971).
34. V. H. Regner and L. Aldaz, *J. Geophys. Res.* **74**, 6935 (1969).
35. L. Aldaz, *J. Geophys. Res.* **74**, 6943 (1969).
36. E. Robinson and R. C. Robbins, *J. Air Pollut. Contr. Ass.* **20**, 303 (1970).
37. C. Junge, *Quart. J. Roy. Meteorol. Soc.* **98**, 711 (1972).
38. K. Schütz, C. Junge, R. Beck, and B. Albrecht, *J. Geophys. Res.* **75**, 2230 (1970).
39. T. V. Healy, H. A. C. McKay, A. Pilbeam, and D. Scargill, *J. Geophys. Res.* **75**, 2317 (1970).
40. G. L. Hutchinson and F. G. Viets, *Science* **166**, 514 (1969).
41. R. E. Inman, R. B. Ingersoll, and E. A. Levy, *Science* **172**, 1229 (1971).
42. H. Leith, *J. Geophys. Res.* **68**, 3887 (1963).
43. B. Bolin and W. Bischof, *Tellus* **22**, 431 (1970).
44. D. H. Ehhalt, *Tellus* **26**, 58 1974.
45. R. A. Rasmussen and R. S. Hutton, *Chemosphere* **1**, 47 (1972).
46. P. Fabian and C. E. Junge, *Arch. Meteorol., Geophys. Bioklimatol., Ser. A* **19**, 161 (1970).
47. C. E. Junge and G. Czeplak, *Tellus* **20**, 422 (1968).
48. R. E. Munn, "Biometeorological Methods." Academic Press, New York, New York, 1970.
49. R. E. Munn and B. Bolin, *Atmos. Environ.* **5**, 362 (1971).
50. W. E. Cobb, *J. Atmos. Sci.* **30**, 101 (1973).
51. R. J. Breeding, J. P. Lodge, Jr., J. B. Pate, D. C. Sheesley, H. B. Klonis, B. Fogle, J. A. Anderson, T. R. Englert, P. L. Haagenson, R. B. McBeth, A. L. Morris, R. Pogue, and A. F. Wartburg, *J. Geophys. Res.* **78**, 7057 (1973).
52. L. Machta, "Man's Influence on the Climate—a Status Report." Scientific Lecture CAS-VI/INF. 4 (Item 20) (World Meteorological Organization Commission for Atmospheric Sciences, Sixth Session, Versailles, 1973).
53. R. E. Munn, *Atmosphere* **11**, 156 (1973).

54. W. E. Cobb and H. J. Wells, *J. Atmos. Sci.* **27**, 815 (1970).

55. J. E. Lovelock, R. J. Maggs, and R. J. Wade, *Nature (London)* **241**, 194 (1973).

56. J. E. Lovelock, *Atmos. Environ.* **6**, 917 (1972).

57. L. Machta, *Advan. Geophys.* **18B**, 33 (1974).

58. D. Trout and H. A. Panofsky, *Advan. Geophys.* **18B**, 151 (1974).

59. R. E. Munn, *Advan. Geophys.* **18B**, 111 (1974).

60. E. Robinson and R. C. Robbins, *J. Geophys. Res.* **74**, 1968 (1969).

61. E. F. Danielsen, *Advan. Geophys.* **18B**, 73 (1974).

62. H. Rodhe, *Advan. Geophys.* **18B**, 95 (1974).

63. H. Reiquam, *Science* **170**, 318 (1971).

64. B. Ottar, *Proc. Int. Clean Air Congr., 3rd, 1973* p. B102 (1973).

65. F. J. Nordø, *Proc. Int. Clean Air Congr., 3rd, 1973* p. B105 (1973).

66. H. C. W. Kwok, W. E. Langlois, and R. A. Ellefsen, *IBM J. Res. Develop.* **15**, 3 (1971).

67. R. Nydal, *J. Geophys. Res.* **73**, 3617 (1968).

68. D. Lal and Rama, *J. Geophys. Res.* **71**, 2865 (1966).

69. G. Czeplak and C. Junge, *Advan. Geophys.* **18B**, 57 (1974).

70. G. B. Hunt and S. Manabe, *Mon. Weather Rev.* **96**, 503 (1968).

71. J. Cramer, *Atmos. Environ.* **7**, 241 (1973).

72. "Sweden's Case Study for the United Nations Conference on the Human Environment." Royal Ministry for Foreign Affairs, Stockholm, Sweden, 1971.

Part C

THE
TRANSPORT
OF
AIR
POLLUTANTS

8

The Meteorological Setting for Dispersal of Air Pollutants

Raymond C. Wanta and William P. Lowry

I. Introduction

The spreading, intermingling, and dilution of air pollutants by stirring motions and turbulent diffusion, their photochemical transformation by solar radiation, and their removal by rainout, washout, agglomeration, fallout, and surface chemical and physical action are the most significant meteorological processes in the dispersal of air pollution.

Diffusion is the exchange of fluid parcels, including their conservative contents and properties, between neighboring portions of the atmosphere, in apparent or presumed random motions, on a scale too small, or considered too small, to be treated by the equations of motion (1). A useful assumption is that the net flux of a property in a given direction is proportional to the gradient in the same direction. Diffusion tends to make uniform the distribution of the property diffused, in the process reducing its space-averaged gradient. Atmospheric motions which diffuse such interior properties as momentum, heat, water vapor, or pollutants have been studied extensively. Their significant motions are much larger than molecular in scale.

Air parcels exchange in irregular patterned motions called eddies. The molecular diffusion equation is often extended by analogy to such turbulent diffusion. Each application to turbulent diffusion requires observational verification. Nonrandom motions, called stirring, may lead to increase of the average gradient in some region, followed by its subsequent reduction by mixing or diffusion (2, 3).

Atmospheric stirring motions and turbulent diffusion take place in a highly variable setting, such that steady-state conditions of more than several hours' duration occur only infrequently or at special locations. Situations vary from the long polar night when the primary solar energy source is continually absent, to the diurnal rhythms which the sun imposes on atmospheric processes at lower latitudes. Solar energy—impinging on a rotating, gravitationally attracting, atmosphere-bearing, irregularly covered earth that revolves about the sun with its rotational axis inclined—is the prime mover for practically all atmospheric processes. This energy is accepted, transferred, and dissipated by the earth–atmosphere system. Resulting temperature and wind fields varying in space and time are intimately connected with stirring motions and turbulent diffusion.

II. Atmospheric Radiant Energy Transfer and Exchange

Air pollutants embedded in the atmosphere play a role in the radiative processes which are the primary determinants of atmospheric behavior. The atmosphere, like any fluid, may be expected to move or to become stagnant according to the times, places, and degrees to which it is differentially heated. Whether the atmosphere is or is not in motion depends on the impulses on it to equalize densities of neighboring volumes of air that differ in density mainly because of differences in heat content. Radiant heat transfer plays a major role in producing these differences in heat content, and a minor role in reducing them. Thermal conduction plays a much smaller role.

A. Blackbody Radiation

A blackbody is an imaginary material, describable in theory, which absorbs all radiant energy falling upon it. A blackbody at a given temperature emits the greatest amount of energy possible at that temperature. No known natural material qualifies as a perfect blackbody, though some materials approach it over large parts of the electromagnetic spectrum. A blackbody exhibits "standard" radiative behavior and is the basis by which the behavior of natural materials is compared.

The three laws describing the radiative behavior of a blackbody (4) are those of Planck, Wien, and Stefan and Boltzmann. Planck's law describes the blackbody's distribution of radiant intensity across the wavelength or frequency spectrum as a function of its Kelvin temperature T. This intensity distribution for a given temperature is unimodal. Wien's law states that the wavelength of the maximum is inversely related to the Kelvin temperature. The product of the wavelength of maximum intensity and the temperature is close to 2900 $\mu m°K$. Thus the hotter a blackbody, the shorter the wavelength of maximum radiant intensity.

The Stefan–Boltzmann law says that the integral of the radiation over all wavelengths, i.e., the total radiation, is proportional to the fourth power of the Kelvin temperature. Thus, doubling the temperature multiplies the total radiant intensity by 16; increasing it by 1% increases total intensity by about 4% and increasing it by 8% increases total intensity by about 36%. Typical midlatitude changes in Kelvin temperature are, for land surface, 4–7% between day and night, and 8% between summer and winter; for open water, less than 1% between day and night, and perhaps 3% between summer and winter. Typical land surface temperature differences beneath scattered clouds are 3–5% depending on the season.

B. Solar and Terrestrial Radiation

The sun behaves approximately as a blackbody at temperature 6000°K. The flux density of its radiation is reduced according to the inverse square principle as it travels through space toward the earth–atmosphere system some 1.5×10^8 km distant. The total radiant flux of the solar beam, incident on a normal unit surface at the "top" of the earth's atmosphere, is called the solar constant. Its value is approximately 2.0 cal cm^{-2} min^{-1} (1.4×10^3 W m^{-2}). While the total flux density decreases during passage across interplanetary space to earth, the relative flux densities do not—i.e., the quality of the radiation remains the same—with the maximum intensity per unit wavelength at a wavelength of about 0.5 μm, which corresponds to a frequency of about 6×10^{14} Hz. The maximum intensity per unit frequency occurs at a wavelength of about 0.85 μm which corresponds to a frequency of about 3.5×10^{14} Hz.

The equivalent blackbody temperature at which the earth–atmosphere system dissipates the solar energy it has absorbed is near 252°K. This is called the earth's effective temperature. Wien's law puts the maximum intensity of this terrestrial radiation near 11.5 μm. Since the solar beam at any one time is incident upon a "disk" whose area is one-fourth that of the earth's total surface, the mean flux density of the solar beam over the disk at any moment must be $\frac{1}{4} \times 2.0$ or 0.50 cal cm^{-2} min^{-1}. Recent measurements of the mean flux density of radiation from the earth–atmosphere system, obtained from satellites (5), place it near 0.33 cal cm^{-2} min^{-1}. Striking a radiant energy balance for the system then requires that $(0.33/0.50) = 0.66$ be the fraction of solar radiation which, on the average, is absorbed by the system of the earth's surface, atmosphere, and clouds. The effective reflectivity of the system, $(1 - 0.66)$, or 0.34, is called the global albedo for the earth–atmosphere. Figure 1 shows the approximate spectra of solar radiation arriving at the earth's atmosphere

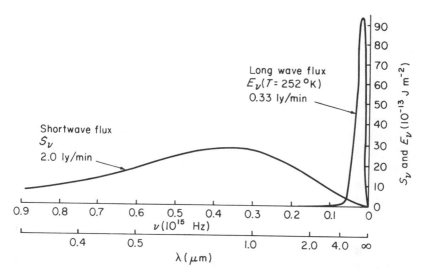

Figure 1. Distribution of solar flux and global flux on a linear scale of frequency (1 langley/minute = 1 cal cm^{-2} min^{-1} = 6.97×10^2 W/m^2).

and of the back radiation to space. The nearly complete separability of the two beams at about 4 μm is the basis for the customary treatment of the two beams as if they were two different kinds of radiation. Solar radiation is usually called "shortwave" radiation, and terrestrial radiation, "long wave." More loosely, they are sometimes called "light," and "thermal" radiation, respectively.

C. Absorptivity, Emissivity, and the "Greenhouse" Effect

If an incident beam of radiation may be distributed only among the three pathways of absorption, transmission, and reflection, then the relative amounts of energy in these pathways may be described by the absorptivity α, the transmissivity τ, and the reflectivity r. Then $\alpha + \tau + r = 1$. Each of these three numbers, $0 \leq \alpha, \tau, r \leq 1$, may be evaluated for individual wavelengths, for groups or bands of wavelengths, or for entire spectra of wavelengths. For example, the global albedo just described is the effective reflectivity of the earth–atmosphere with respect to the solar energy beam. Using subscripts "S" for shortwave and "L" for long wave, the passive behavior of a substance in response to incident radiation may be symbolized, for example, in terms such as the absorptivity with respect to shortwave radiation α_S, transmissivity with respect to long wave radiation τ_L, and the like.

The facts that a blackbody is the most efficient emitter and that it absorbs all incident radiant energy suggest the content of Kirchhoff's law: for any combination of temperature and wavelength a material which is an efficient emitter relative to a blackbody is also an efficient absorber relative to a blackbody at the same temperature. Thus peaks in an absorption spectrum for a material also define the peaks in its emission spectrum.

The emissivity ϵ of a material with respect to some stated part of the spectrum is defined as the radiant flux intensity relative to that of a blackbody at the same temperature. The emissivity of a blackbody is by definition unity; and if the total emissivity of a material is known, its total flux density may be expressed in the "gray body" form of the Stefan–Boltzmann law as

$$F = \epsilon \sigma T^4 \tag{1}$$

where σ is the Stefan–Boltzmann constant. We may then express Kirchhoff's law as either

$$[\alpha_\lambda = \epsilon_\lambda]_T \tag{2}$$

or

$$[\alpha_{\nu_1-\nu_2} = \epsilon_{\nu_1-\nu_2}]_T \tag{3}$$

where λ is a given wavelength and $\nu_1 - \nu_2$ is a given frequency band.

Most natural materials behave differently in the presence of incident shortwave radiation than they do with long wave radiation. In particular, the atmospheric gases taken together, including water vapor, have a much larger shortwave transmissivity τ_S than long wave transmissivity τ_L. Ozone, a strong shortwave absorber, affects mainly the ultraviolet wavelengths smaller than about 0.3 μm well to the side of the radiant intensity maximum. The result is that the mean temperature of the inner atmosphere, near the earth's surface, is more nearly equal to 290°K than to the value of 252°K mentioned earlier as the effective mean radiant temperature for the earth–atmosphere system inferred from outside the atmosphere. This warming of the inner atmosphere relative to the outer, caused by the differential transmissivity of the atmosphere itself, has long been called the "greenhouse" effect. A case has been made (6, 7) for calling it the "atmosphere" effect. By whatever name, its nature may be deduced from a simple radiation balance of the earth–atmosphere system, according to Table I.

In the table F_S' is the remainder of the mean solar energy flux density which actually enters the earth–atmosphere system from space without reflection. Absorptivities, and thus emissivities, are expressed as complements of transmissivities, making all components nonreflective for sim-

Table I Rudimentary Radiation Balance of the Earth–Atmosphere System

Radiation to	Radiation from		
	Space	Atmosphere	Earth
Space	—	$(1 - \tau_L)F_a$	$\tau_L F_e$
Atmosphere	$(1 - \tau_S)F_S'$	—	$(1 - \tau_L)F_e$
Earth	$\tau_S F_S'$	$(1 - \tau_L)F_a$	—

plicity of the model. Subscripts, aside from the "S" and "L" already given for shortwave and long wave, are "a" for the atmosphere and "e" for the earth. Flux density F_a due to the atmosphere is

$$F_a = \sigma T_a^4 \tag{4}$$

and flux density F_e due to the earth is

$$F_e = \sigma T_e^4 \tag{5}$$

where T is Kelvin temperature.

From the balance equations for atmosphere and earth

Atmosphere: $(1 - \tau_S)F_S' + (1 - \tau_L)F_e = 2(1 - \tau_L)F_a \tag{6}$

Earth: $\tau_S F_S' + (1 - \tau_L)F_a = F_e \tag{7}$

and recalling that τ is transmissivity of the atmosphere, one may obtain an expression for the "greenhouse" or "atmosphere" effect in the form

$$F_e/F_S' = (1 + \tau_S)/(1 + \tau_L) \tag{8}$$

For the case of the earth's atmosphere, the fact that $\tau_S > \tau_L$ yields in this simple model $F_e > F_S'$; i.e., the radiant flux density outward from the earth's surface exceeds that inward from space to the atmosphere, the more so as τ_S grows larger relative to τ_L.

D. Transmissivity, Depletion, and Beer's Law

The manner in which a radiant beam is depleted as it passes through a partially absorbing medium, such as the atmosphere, is of considerable importance in air pollution meteorology. The amount of depletion is proportional to the amount of incident energy, the path length, and the concentration of depleting material in the medium. More exactly, it has been found empirically that for both individual wavelengths and the solar beam itself, the change in intensity dS_0 of the incident beam S_0, as it passes through a layer of thickness dz at an angle Z from the normal to the layer, is

$$dS_0 = (-k_S \, dz \sec Z) S_0 \qquad (9)$$

where the absorption coefficient k_S for the atmosphere with respect to the solar beam (units of length^{-1}) describes the physical nature of the atmosphere for this process.

The path length through the layer is $dz \sec Z$. Here the zenith angle Z is measured from the vertical, and integration through an entire atmosphere of thickness z gives the fractional depletion by Beer's law:

$$S/S_0 = \exp(-k_S z \sec Z) \qquad (10)$$

If the effective transmissivity of the atmosphere with respect to the shortwave solar beam, for the zenith path $(Z = 0)$, is taken to be $\tau_S = \exp(-k_S z)$, then

$$\ln(S/S_0) = \sec Z \ln \tau_S \qquad (11)$$

Beer's law also describes the depletion of sunlight by passage through clouds and through polluted air. Values of k_S for clouds are typically in the range of 0.005 to 0.02 m^{-1}, depending on the kind of cloud and its structure. The physical meaning is that each meter-thick layer of the cloud depletes the beam entering it by between 0.5 and 2%. Polluted air in which the visual range is only 1 km would have the same depleting ability for sunlight as a typical cloud.

E. The Global Heat and Radiant Energy Balance

Pollutants and water clouds act qualitatively in roughly the same way with respect to radiant energy transfer in the atmosphere. Both reduce the incoming shortwave radiation which reaches the earth's surface, and both reduce the outgoing long wave radiation from the lower atmosphere. This, on balance, is a thermostatic effect for the earth's surface and the air layers near it. Thus, both the mean temperatures found in the lower atmosphere and the amount of variability in these temperatures are affected by the aerosol content of the atmosphere, acting through the turbidity, expressed by an effective value of the absorption coefficient. Time and space differences in turbidity are related in principle to pressure differences and thus to the atmospheric motion which results. Therefore, pollutants affect not only visual range in the atmosphere, but also the motion of the atmosphere. The significance of this effect is an object of current research. Furthermore, as will be seen presently, pollutants affect the processes governing atmospheric precipitation in its various forms.

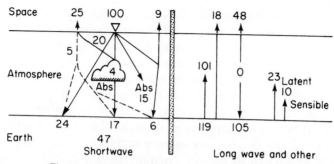

Figure 2. Global radiation and heat balance (4).

Figure 2 shows the global heat and radiant energy balance in terms of the relative amounts of energy in vertical pathways, both radiative and convective (4). The numbers in this balance are estimates of the annual mean values averaged over the entire earth's surface. While these values will certainly be different at particular places and times, the complexity of the energy system will be at least as great. Since these values are for net vertical fluxes, the horizontal fluxes which are referred to as "advection" do not appear (8, 9).

The connections between the gross radiative and convective balances on the global scale may be seen from a rudimentary model (Fig. 3) (10). A hemisphere is divided into two parts at some selected latitude, e.g., 50°. The model says simply that the difference in radiative balance between the two parts is made up by the advective transfer of heat between the two parts, and that the intensity of this transfer is proportional to the temperature difference between the two parts. Thus, specifying solar intensity and albedos for the two parts, and the transfer coefficient between them, also specifies a pair of equilibrium temperatures for the two parts. In this simple context, pollutants have their effect by altering effec-

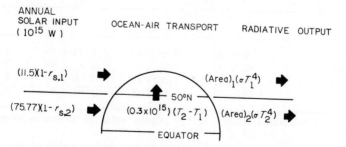

Figure 3. Schematic model of radiative and convective heat balance between two units of a hemisphere. Albedos of the units are $r_{s,1}$ and $r_{s,2}$ (10). (Reprinted with permission from The Ecological Society of America.)

tive values of the two albedos, thus altering the thermal balance, and perhaps in the process, the vigor of advective intrahemispheric heat transfer by means of large-scale storm eddies in the turbulent atmosphere.

A sense of the potential effects of aerosols acting through the two albedos is found in Table II. In the table, pairs of relative numbers are given for each of 16 pairs of albedos: the vigor of intrahemispheric advection (a kind of storminess index) and the mean system temperature, both relative to present values. For example, this simple model predicts that changes in the two albedos over the tabulated ranges would produce changes of 59% (1.25 to 0.66) in "middle latitude storminess" and only 6% (1.03 to 0.97) in hemispheric mean temperature. The purpose for inclusion of Table II is a reminder that changes in the vigor of large-scale motion may be a more important result of changes of aerosol content than are changes in hemispheric mean temperature. In this sense, the table is intended to be diagnostic and tentative rather than definitive.

Additional complications in the question of effects of man-made pollutants on the global energy balance and atmospheric motion may be seen in Table III. Changes in three characteristics of the earth–atmosphere system are described qualitatively as outcomes of four processes involving changes in atmospheric gases and aerosols.

Processes I and II involve increases in atmospheric carbon dioxide and water vapor unaccompanied by any change in mean cloud cover. Feedback between these two processes is positive, so that the increases in the two gases produce increases in surface temperatures and in vertical motion leading to further increases in these gases. In process III, however,

Table II Relative Changes in Intrahemisphere Heat Transfer[a] (as an index of storminess in middle latitudes) and in Mean Earth–Atmosphere Temperature[b] (°K) Calculated by the Balance Model in Figure 3

Albedo north of 50°N[c]	Albedo south of 50°N latitude[d]			
	0.20	0.25	0.30	0.35
0.30	a/b = 0.92/1.03	0.84/1.01	0.75/1.00	0.66/0.98
0.40	1.03/1.02	0.94/1.01	0.84/0.99	0.78/0.98
0.50	1.14/1.02	1.05/1.00	0.96/0.99	0.87/0.97
0.60	1.25/1.01	1.16/1.00	1.07/0.98	0.98/0.97

[a] Transfer is calculated relative to the present value of advection estimated as 4.0×10^{22} cal/year (or) 5.3×10^{15} W (10).

[b] Mean temperature is calculated relative to the present value of mean system temperature estimated as 251.8°K.

[c] Present value of northern albedo estimated as 0.50 (10).

[d] Present value of southern albedo estimated as 0.27 (10).

Table III Probable Changes in Three Characteristics of the Earth–Atmosphere System as the Result of Four Processes Involving Atmospheric Gases, Clouds, and Aerosols (10)

	Change in		
Process	*Mean surface temperature*	*Vigor of mean vertical motion*	*Mean system temperature*
I. Increase in CO_2/H_2O; no change in cloud	Increase	Increase	None
II. Increase in CO_2/H_2O due to increase in surface temperature; no change in cloud	Increase	Increase	None
III. Increase in cloud due to increases in H_2O and vertical motion from I and II	Probable decrease[a]	Probable decrease[a]	Decrease
IV. Increase in particulate aerosol load	Probable decrease[b]	Probable decrease[b]	Possible decrease[b]

[a] Depends on relative changes in shortwave and long wave fluxes.
[b] Depends upon relative changes in scattering and absorption.

clouds form as the result of processes I and II. This additional cloudiness tends to counteract I and II in negative feedback, but also tends to reduce the mean temperature of the earth–atmosphere system. Process IV involves increases in aerosol loadings of the atmosphere. The greatest likelihood is that this increase will act, as noted previously, in a manner like that of increases in cloud cover. However, the validity of this speculation is under scrutiny (11–15) in the face of information about the different behavior of scattering relative to absorption for different aerosols. Other research is attempting to explore various aspects of global climate by modeling (16–18). The contents of Table III, as with Table II, should be taken as diagnostic and tentative rather than definitive, the purpose being to remind the reader of the interactive complexity of the system.

III. Vertical Structure of Temperature and Stability

Familiarity with the concept of atmospheric stability is a most useful aid in penetrating the complexity of environmental air quality. Variations in atmospheric stability serve to explain qualitatively much of the variation in the power of the atmosphere to dilute pollutants, and stability is an excellent parameter for stratifying a variety of meteorological and other statistics relating to air pollution. Stability and wind interact with each other (*19*).

The observational network for measuring temperature and wind aloft is much less dense in space and time than that for measuring these parameters at the earth's surface. This partly explains why intuitions based upon surface wind or temperature behavior are sometimes misleading with respect to winds or temperatures even a mere 100 m aloft.

A. Temperature Lapse Rates, Stability, and Potential Temperature

The rate of decrease of temperature with increase in height is called the temperature lapse rate. If the rate of decrease refers to the air environment, it is called the environmental lapse rate; if to a parcel of air moving within the air environment, it is called the process lapse rate (*1*). Since, on the average, the air temperature below the stratosphere decreases with height, the average environmental lapse rate in the troposphere is positive (Fig. 4).

Two significant lapse rates are

1. The adiabatic lapse rate for dry air—this is the process lapse rate of a parcel of dry air as it moves upward in a hydrostatically stable environment and expands slowly to lower environmental pressure without exchanging heat with its environment; it is also the rate of increase in temperature for a descending parcel. If the environmental lapse rate hap-

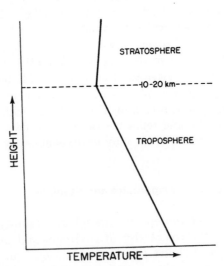

Figure 4. Temperature change with height, illustrating positive lapse in the troposphere.

pens to be adiabatic, then such parcels at any height in that environment are in neutral equilibrium. As long as the air is unsaturated, the approximation of the adiabatic lapse rate by the dry-adiabatic lapse rate is satisfactory for most purposes. It can be shown that the dry-adiabatic lapse rate is given by the ratio of the gravitational attraction to the specific heat at constant pressure, i.e., 9.8×10^{-3} °C/m. The approximate values 10°C/km or 5.5°F/1000 ft for the dry-adiabatic lapse rate are in common use.

2. The adiabatic lapse rate for saturated air (known as the pseudo-adiabatic or saturation lapse rate)—this process lapse rate is smaller than the dry-adiabatic lapse rate because of the release of latent heat as the air parcel ascends and cools. The pseudoadiabatic lapse rate varies with temperature and height (20, 21).

When the environmental temperature decreases faster with height than the adiabatic rate, the environmental lapse rate is superadiabatic. A rising air parcel, cooling at the adiabatic rate, becomes warmer and less dense than its environment and therefore buoyancy tends to accelerate it upward (Fig. 5). In such an environment the parcel is in unstable equilibrium. Vertical motions upward or downward are reinforced. When the environmental lapse rate is less than adiabatic ("subadiabatic") or negative, a rising air parcel becomes cooler and more dense than its environment and tends to return to its starting point. The parcel is in stable

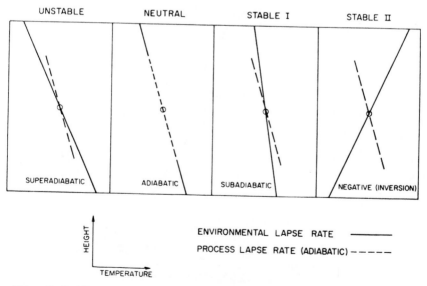

Figure 5. Stability of an air parcel, determined by the environmental lapse rate.

equilibrium. Vertical motions are resisted. Instability and stability are therefore defined with reference to a neutral equilibrium, present when the environmental lapse rate is dry-adiabatic for dry or unsaturated air, and pseudoadiabatic for saturated air.

Environmental lapse rates in stable layers may have small positive (subadiabatic), zero, or negative values. The lapse rate in an isothermal layer is characterized by the heights of its base and its top, together negative, and the condition is termed inversion (Fig. 5). An inversion layer is characterized by the heights of its base and its top, together with the magnitude of the negative lapse rate (Fig. 6).

Vertical temperature gradients are sometimes grouped into two classes, lapse and inversion. A third intermediate group may be added. Lapse is usually defined to include superadiabatic and adiabatic lapse rates and some or all subadiabatic lapse rates. Inversion is defined as above, although when an intermediate transitional class between lapse and inversion is added, this class usually includes both weak inversions and weak subadiabatic conditions. It will be obvious therefore that lapse and inversion are not strictly synonymous with instability and stability. Classifications of lapse rates are not widely standardized; standardization of diffusion categories which include stability as one of the determining parameters is making progress (see Chapter 9, this volume), but the difficulties presented by variations of stability with time, location, and especially height remain to be surmounted.

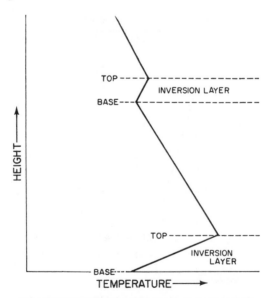

Figure 6. Surface inversion and inversion aloft.

By glancing again at Figure 5 the reader can verify that a parcel of air having a process lapse rate that is adiabatic will stay on the same adiabat (i.e., the same dashed line) as it rises or descends. Its temperature at one height defines its temperature at all heights reached, barring the occurrence of nonadiabatic processes such as condensation, evaporation, and thermal radiation. The temperature of this air parcel when at the height where the environmental pressure is 100 kPa (1000 mbars) is defined as its potential temperature. Whatever the purely adiabatic processes to which the parcel is subjected, e.g., lifting to lower pressure or horizontal advection into a region of high pressure, its potential temperature will remain constant.

The potential temperature θ (°K) of a dry-air parcel is a function of its actual temperature T (°K) and its pressure p:

$$\theta = T(p_0/p)^{2/7} \tag{12}$$

where p_0 is the standard pressure defined above (100 kPa or 1000 mbar), and p and p_0 are measured in the same units. An approximation for the potential temperature at height z above sea level is

$$\theta = T + \gamma z \tag{13}$$

where z and the dry-adiabatic lapse rate γ are in consistent units. This approximation depends on the fact that the average sea level pressure exceeds the standard pressure by only about 1%.

The vertical gradient of potential temperature $d\theta/dz$ is given by $(dT/dz) + \gamma$. When the environmental lapse rate is adiabatic, both dT/dz and γ have the same magnitude but opposite sign and the vertical gradient of potential temperature is then zero.

Note that when $d\theta/dz$ in a layer is zero (i.e., θ is constant) we have the neutral condition illustrated in the second panel of Figure 5. If the change of potential temperature with height is negative (first panel), the layer is unstable; if positive (third and fourth panels), the layer is stable.

Meteorologists make use of potential temperatures to keep track of air parcels moving not only vertically but also horizontally. The student of this technique should compare time cross sections in which the actual and potential temperatures are both shown as functions of height and time (22, 23).

B. Influences Contributing to the Vertical Structure of Temperature

If the earth's atmosphere were in purely radiative equilibrium, its average surface temperature would be tens of degrees Celsius higher than the observed temperature (24). Compared with the theoretical lapse rate which would exist in the presence of purely radiative equilibrium, the observed rate is steeper in the lower troposphere and less steep in the upper. What is actually observed in the troposphere is a roughly linear decrease of temperature with height (Fig. 4) at a rate about two-thirds of the dry-adiabatic rate. These departures in average surface temperature and tropospheric lapse rates are produced primarily by the large-scale planetary circulation and by condensation of water vapor (see also Chapter 12, this volume).

The temperature in the upper stratosphere rises to a maximum at about 50 km height because of absorption there of solar radiation in the near ultraviolet (wavelengths 0.2–0.3 μm) by ozone. After falling off above that level, it rises again to values exceeding even those at the earth's surface because of absorption of solar radiation in the far ultraviolet (wavelengths less than 0.1 μm) by a greatly modified atmosphere. Enough stirring and mixing takes place in the troposphere and stratosphere so that relative concentrations of the great majority of the gases in the atmosphere remain constant on the average (25), two exceptions being water vapor, which is not conserved in vertical mixing because of precipitation, and which therefore decreases with height; and ozone, which increases with height to a maximum at roughly 25 km.

The four principal processes which significantly alter the lapse rate from the average value near two-thirds of the dry-adiabatic rate or, from another point of view, jointly act to maintain it at that value, are radiation, advection, vertical motion, and the state changes of water. Of the three classic modes of heat transfer, conduction is absent. In the atmosphere it is the least significant, although it is involved in heat transfer at the very bottom of the atmosphere. Convection is for convenience divided into horizontal and vertical components. Heat transfer through state changes of water, especially from vapor to liquid, and the reverse, also significantly affects the vertical temperature structure. Liquid–solid changes of state are, of course, important in some precipitation processes, and the amount of polar ice seems to play a major role in climatic change (26). Such changes have relatively local effects on the dispersion of air pollution, except as an ice- or snow-covered surface requires extra energy to melt and thus contributes to the prolongation and even steepening of inversions during advection (inflow) of warm air. These four processes can promote either instability or stability depending on the circumstances. Other things being equal, the turbulent diffusion of air pollutants is faster in an unstable layer than in a stable layer.

Incoming solar radiation (i.e., irradiation) absorbed at the earth's surface tends to produce instability which is greatest near the surface, where the lapse rate can be superadiabatic, and can affect a deep layer, 2 km or more, establishing a neutral lapse rate in its upper portion. The increase of the temperature of the earth's surface in response to irradiation depends upon the thermal properties of the surface. It can be large for a dry, poorly conducting soil; and small for a reflecting surface such as ice, or when the acquired thermal energy can be distributed fairly rapidly, as in the ocean surface layer. Absorption of solar radiation directly by the air at low altitudes is quite small so that the initial transfer of heat at the surface is by conduction to a thin, sometimes stagnant layer or blob which grows and detaches itself in convection—which, in stronger winds, may be partially or almost wholly forced convection rather than free convection. This action transports heat upward and makes it possible for air to be freshly brought to the earth's surface and warmed.

Long wave radiation outgoing from the earth's surface creates a stabilizing tendency which is greatest near the radiating surface. Curiously, the minimum temperature is observed to be just above rather than at the surface (27). Surface-based inversion layers are commonly observed over land at night with clear skies and weak winds, due to the cooling of the land surface by long wave radiation and exchange of heat between the ground and the lower air layers. The decrease in temperature of the earth's surface is large when the cooling effect is not distributed through a deep layer of atmosphere, earth, or water. Thus intense surface inversions develop when the wind is light or calm over poorly conducting soil. An additional requirement for an intense surface inversion is a minimum of back radiation from clouds or water vapor in the atmosphere. On a clear morning at sunrise, with these conditions present, an inversion may extend upward from the surface to 150 m or more. Under somewhat special circumstances, nocturnal inversions in Arizona, for example, reach to well over 300 m (28) and in the polar night regions to well over 1000 m. In the winter, over the northern interior portions of northern hemisphere continents, the time-averaged temperature lapse rate in the lowest kilometer or so may be negative. The radiating (cooling) surface can also be a cloud, a fog bank, or a smoke haze layer, with the production of a thin inversion layer on top.

Long wave back radiation to the earth's surface from clouds or water vapor, and to a lesser degree from carbon dioxide and particulate pollutants, promotes a tendency away from stability extremes, in that it limits the temperature drop at the surface due to outgoing radiation. This process, however, contributes to the production of deep polar and arctic air masses during periods of stagnation or slow movement in their source regions (29, 30). Back radiation is responsible for most of the average flux of radiation outward from the earth's surface, being about $2\frac{1}{2}$ times the shortwave flux directly absorbed at the surface. Long wave back radiation from a polluted urban atmosphere can roughly compensate for the depletion of solar irradiation (*31*).

Advection can promote instability or stability depending upon whether the temperature of the incoming air mass is cooler or warmer than the surface over which it is passing. If the thermal lag of the surface is large, as in the case of ice or water, pronounced alteration of the lapse rate is observed. During invasions of cold air masses, when, on a large scale, cold air is advected over a relatively warm surface, the effect may be intense, especially in passage over warm bodies of water or a populous industrial city. Instability during the night is commonly due to such cold advection. Typically a fairly deep layer is affected, so that the air does not rapidly warm up as a result. Cool air from suburban or rural areas, if driven to move over the relatively warm urban surface (heat island) at night, produces the same tendency toward instability on a smaller scale. A variation of the cold advection effect on stability occurs when cold air aloft is differentially advected over warm air below, a situation not uncommon as an upper pressure trough passes some time after a cold front.

Warm advection tends to produce stability. This stabilization effect typically extends through only a shallow depth of atmosphere. The thermal lag of relatively cold water currents, such as are observed on the eastern side of continents at moderate and high latitudes, can result in great stability even in the presence of a strong flow of tropical air poleward; regions of oceanic or lake upwelling similarly present a relatively cold surface. Fog is then often present (e.g., the foggy Grand Banks and coastal California) which is also the case when there is warm advection over snow and ice surfaces. Warm air, meeting cold air at fronts and elsewhere, characteristically overrides it with shallow slopes of the order of 1:150. As a result we observe elevated inversions on large (synoptic) scales and also on local scales, e.g., when air from a warm lake passes over colder land air, or when air from a cold lake undercuts warmer land air. Pronounced stability is also observed when air passes over relatively colder air pooled in valleys or along the lower slopes of broken terrain, as in the Great Basin region of the western United States in winter. Where warmer urban air is driven in local patterns of circulation to move over relatively colder suburban and rural areas, a tendency toward stability should be expected.

Vertical motions affect stability by way of the process lapse rate. For saturated air this is the pseudoadiabatic lapse rate, which is close in numerical value to the average observed in the troposphere, thus indicating the important role of this process rate, through condensation and precipitation, in maintaining the observed rate.

Lifting of a body of unsaturated air in which initially the lapse rate is subadiabatic (cf. the average observed in the troposphere), produces a less stable lapse rate. This occurs, for example, when the air is being forced up a mountainside. A zone in which stability is enhanced may then be found at the upper boundary of the lifted air. If cooling to condensation does not occur, then on descent the tendency in the layer involved is back toward the initial lapse rate, i.e., toward stability.

Descent of an air mass on a large scale aloft in the free atmosphere is an ordinary meteorological feature called subsidence, and is found wherever a cold air mass spreads out horizontally, typically away from a low pressure center following a cold front passage and lasting until after passage of a high pressure center or ridge. It may be viewed as a compensatory motion, downward slowly in the extensive high pressure areas to balance faster upward motion in less extensive low pressure areas. Conservation of vorticity as an air mass moves toward the equator may contribute (32). In the vertical, because of the closer spacing of pressure surfaces nearer the earth's surface, the overall lapse rate in a descending layer will tend toward stability because parcels of air at all levels in the layer are warming at the process lapse rate. In this manner subsidence inversions are formed. A single temperature sounding may show both a surface-based nocturnal inversion and an elevated subsidence inversion (Fig. 6). The trade inversions, maintained by subsidence on the eastern sides of the semipermanent low-latitude anticyclones, have bases which decrease in height eastward from above 1000 m to below 500 m (32). Southern California, Cape Verde, north-central Chile, and southwestern Australia are examples of regions situated during at least part of the year under such semipermanent trade inversions.

Air motion over a rough surface generates mechanical turbulence including vertical fluctuations which propagate upward, increasing with roughness of the surface and wind speed. The stable environmental lapse rate then tends toward the dry-adiabatic process rate, i.e., toward instability. One of the best times to observe a neutral environmental lapse rate of some hours duration is during a period of strong winds, especially with thick clouds overhead or at night. Fast moving currents above relatively slower moving or calm air produce wind shear that can also generate mechanical turbulence.

A surface layer of vigorous mixing, such as occurs during strong cold advection, may be capped by a turbulence inversion with base height ranging from roughly 100 to 1000 m or more. This inversion may be thought of as a consequence of overshooting by the rising air parcels (which are cooling at the adiabatic rate) into a more stable environment.

State changes of water have a pronounced effect on tropospheric stability. Once a rising moist air parcel in a mixed layer reaches saturation and water condenses, its process lapse rate becomes the smaller saturated rate, and relative to the environmental (dry) adiabatic rate becomes unstable in a manner analogous to the first panel in Figure 5. This is a well-known route to the formation of clouds, showers, and thunderstorms. The clouds produced are generally topped by stable layers, which can be quite intense in the case of thunderstorms (cf. spreading anvil tops). The layer-type clouds distributed haphazardly in a thundery sky are usually associated with locally stable laminas of air. An inversion tends to form at the cloud base when the generating moist upward current is prolonged (*25*). For an introduction to the thermodynamic diagrams which meteorologists use to analyze convective processes, a number of textbooks are available (*4, 30, 33*).

Formation of fog and ground fog is accompanied by a tendency away from extremes of either stability or instability, i.e., by a tendency toward the saturated adiabatic lapse rate (*6*).

Evaporation from a wet surface, as after a shower, can produce an inversion in the layer of air near the ground, but this tendency toward stability is only of local importance in the present context. See Sellers (*8*), Deacon (*27*), and Geiger (*34*) for details of processes which occur in the lowest few meters of air.

C. Variation of Environmental Lapse Rate with Height, Time, and Location

Figure 6 shows a vertical temperature structure having a mixture of stability types that is not unusual. Mixed structures of this and other kinds are more often present than not in the lowest 300 to 500 m, in many locations of the globe and for one or more seasons, contrary to the impression one may gain from simple atmospheric models. The prototype unstable temperature profile—associated with midday or early afternoon, little or no cloud cover, the absence of strong winds or local mesoscale circulations, and an absorbing surface that does not transport heat rapidly into its interior—consists of a superadiabatic (unstable) portion at the surface in which the lapse rate decreases smoothly from perhaps 100 or more times the adiabatic rate below 1 m to near the adiabatic (neutral) rate by around 100 m, and then gradually becomes slightly subadiabatic above that. The midday (12-hour) and afternoon (16-hour) curves of Figure 7 are typical of the prototype unstable profile and consist of an average of profiles measured over a 2-month period at Oak Ridge, Tennessee (36°N) during weather that was not especially cloudy, windy, or rainy (*23*). The prototype unstable temperature profile

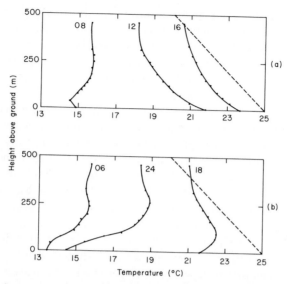

Figure 7. Vertical temperature profiles based on average lapse rates, 28 August–3 November 1950, Oak Ridge, Tennessee. Dashed line represents the adiabatic process lapse rate. Times are E.S.T. After Holland and Myers (23).

is commonly represented by from one to three superposed layers, each of constant stability class.

The prototype stable temperature profile—associated with the last part of the night and sunrise, little or no cloud cover and fairly dry air, light winds or calm, the absence of local mesoscale circulations, and a long wave radiating surface that does not quickly replenish its heat loss from within—consists of a stable portion with negative lapse rate at the surface, i.e., a surface inversion, with the rate changing sign at about 100 m or above, and in the subadiabatic range (stable) or adiabatic (neutral) in the upper portion. It is easy to see that it presents a mixed profile. Note the midnight and early morning (06-hour) curves of Figure 7; the weak stability observed near the surface at 06 hours was associated with the occurrence of fog, and the tendency toward stability in the upper portion of the profile is associated with the location of the soundings in a rough-surfaced broad valley and the occurrence of subsidence aloft. The prototype stable temperature profile is commonly reduced to only one stability class, reflecting the stability of the bottom of the profile, but such a characterization is often inappropriate when dealing with an elevated source of pollutants.

The profile for 08 hours in Figure 7 illustrates a significant vertical temperature structure which is a transient feature of the postsunrise change from a stable to an unstable condition, but also appears on a larger scale as a semipermanent feature in some seasons under the trade inversion. In the postsunrise case, the surface layer is unstable (or neutral), and a part of the preexisting surface inversion remains aloft, somewhat intensified at its underside, capping the growing surface convective layer up to 1–4 hours following sunrise (*19*).

An hour or two before sunset the air near the ground begins to cool rapidly, and by 18 hours a mixed profile appears with stable below unstable air, opposite to that at 08 hours. The inversion grows in intensity and depth during the night, on the average rapidly at first and then more slowly, reaching maximum thickness near sunrise. After sunrise the cycle may repeat. Note that the diurnal cycle indicated in Figure 7 may be interrupted or altered at any time by such processes as stirring or mixing due to strong winds aloft or by the appearance of thick clouds or fog. The characteristic cycle of temperature profiles is also modified especially near the surface within and sometimes immediately downwind of urban areas, mainly because of more mixing over the rough, warmed, and heat-generating surface in the daytime and because of such mixing and a reduced cooling rate at night.

It is evident from Figure 7 that a large daily (i.e., 24-hour) range of near-surface temperature should be highly correlated with the occurrence of unstable daytime and very stable nighttime conditions, and that

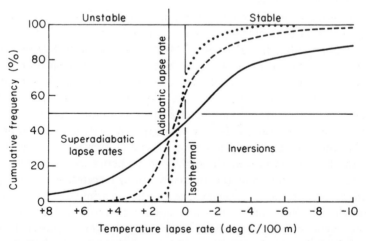

Figure 8. Frequency distribution of temperature lapse rates in three superimposed layers, 1945–1948, at Rye, England. Dotted line, 106.7–47.2 m; dashed line, 47.2–15.2 m; solid line, 15.2–1.1 m. From Best *et al.*, in R. Geiger, "The Climate Near the Ground" (translation of the 4th German ed., 1961). Harvard Univ. Press, Cambridge, Massachusetts, 1965. (Copyright 1965 by the President and Fellows of Harvard College.)

this daily range decreases with increasing height. Figure 8 shows the cumulative distributions of lapse rates over a 3-year period at Rye, England (51°N) for three superposed layers based at 1.1 m above grass. A less maritime location would tend to yield smaller central slopes and even greater extremes in such distributions.

For similar wind conditions, instability is more frequent and of greater intensity in the daytime than at night, on summer days than on winter days, on clear days than on cloudy days, and on cloudy nights than on clear nights. Greater daily range of stability occurs in summer than in winter. A maximum in the autumn occurs where the frequent presence of dry air masses enhances the daytime solar irradiation and likewise the net nocturnal outgoing long wave radiation from the land.

With nocturnal cooling and the formation of a surface inversion, fog forms in moist air cooled below its dew point. Fog within an inversion layer tends to make the layer of fog less stable, as indicated above. A deep fog layer, however, will restrict the solar radiation reaching the ground, because of absorption in the fog layer and reflection and scattering from the fog droplets. Therefore the usual daytime increase in instability, and hence mixing, can be thereby delayed or reduced. Under these conditions convection does not assume its normal intensity for a given time of day. During conditions of air stagnation, fog and pollutants together similarly affect the receipt of solar radiation at the surface, and may cause the lowest air layer to remain stable for longer periods than otherwise—a positive feedback (35). Subsidence aloft may compound the stabilizing process.

The selection of temperature data presented above is intentionally biased toward land and open country. It can be inferred from the discussion that conditions over water, and thus at times over the portions of land or urban areas near major bodies of water, will be notably different. Basically because of the change of radiant and other thermal properties at the surface, the stability of the lower troposphere over deep, open water is greater in the daytime than at night, in direct opposition to the situation with respect to land and open country (36). The relevant properties of different soils and ground cover should be considered when estimating or attempting to explain observed stability conditions (27, 34). The literature is rich with examples of measurements made in various parts of the globe and over differing terrains and surfaces (23, 25, 34, 37).

D. Mixing Height and Other Stability Parameters and Indices

A parameter useful in the study of air pollution is the thickness of the convective layer or the "mixing height" (38, 39). Its magnitude for a particular time of day is usually calculated with the aid of a nocturnal or early morning temperature sounding. Figure 9 illustrates the method (*40*). On a chart of height or pressure versus temperature showing the latest sounding, the surface temperature at the time of interest is used to construct a dry-adiabatic process curve up to the height at which the environmental temperature sounding is intersected. The point of intersection defines the thickness of the convective layer. It may be used to compare different times of day, one day with another, or one place with another relative to dilution capacity of a pollutant emitted near the surface, for a given wind speed (*41*). An early morning urban mixing height has been defined similarly, but in terms of the minimum surface temperature at an outlying airport, increased by 5°C to allow arbitrarily for the urban "heat island." (See also Chapter 12, this volume.)

The nature of mixing height as an index is evident from its definition and applications. The reference sounding is held artificially constant; the actual temperature profile changes, generally in such a manner as to give the actual mixing height à larger value than estimated from the earlier reference sounding. Moreover, some portion of the pollutant load reaching the stable top of the convective layer remains aloft, so that a variable small loss of the layer's load occurs at the upper surface of the layer. Finally, pollutants which are introduced above the convective layer can only "leak" to the surface through a stable layer, in correspondingly small amounts on a time average, even though brief periods of intense vertical flow occur.

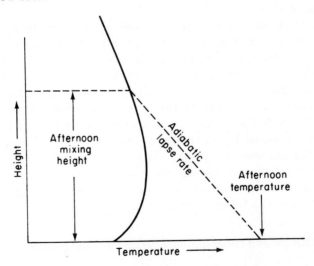

Figure 9. Determination of afternoon mixing height from morning upper-air sounding and afternoon surface temperature (*40*).

As implied in the preceding subsection, the daily temperature range can be a useful index of the stability regime. Since a prototypical "day" consists of two quite disparate parts as far as stability is concerned, this index applies only in the sense that a night of great stability, and to a lesser degree a day of great instability, is often implied by a large daily range. Thunderstorms, land–sea breezes, mountain–valley winds, precipitation, and strong warm or cold advection or frontal passages can alter the circumstances. But this index is attractive in spite of its deficiencies because of the much greater availability of surface temperature data than of upper air data. Some of the uncertainty can be eliminated by considering other weather elements and the synoptic weather maps for the day.

Another index of stability is the degree of unsteadiness of the weather elements, including temperature, wind direction, wind speed, humidity, and others. For example, the air temperature record shows fluctuations within limits of a few degrees if measured with small-lag sensors. This fluctuation compared between time periods can be used as an index of stability since the main cause for relatively large swings is the presence of vertical mixing.

One can also use the expected solar irradiation in comparing places at different seasons and latitudes. Ground fumigations, which occur downwind of power plant chimneys in the southeastern United States when the surface convective layer reaches the body of the plume of pollutant aloft, typically occur some $2\frac{1}{2}$ to 4 hours after sunrise. The cumulative solar irradiation at Oak Ridge, Tennessee (36°N) corresponding to this time lag is roughly 2×10^6 J/m² (50 langleys). In Figure 10 we see how day length, i.e., the interval from sunrise to sunset, varies with latitude and month over the earth. Since sunrise and sunset are closely symmetrical about local noon, the times of sunrise and sunset can be roughly estimated from this diagram for any date and latitude. More precise values are tabulated elsewhere (20). The annual variation of insolation at the top of the atmosphere and the irradiation of the surface assuming a transmission coefficient of 0.3 may be found in Figure 11 for several latitudes.

A rough estimate of the time interval from the sunrise maximum of stability to the early afternoon maximum of instability on a prototypical "day," and also of the subinterval during which elevated stable layers left over from the morning surface-based inversion are eliminated, can be gained from Figures 10 and 11. It is convenient to add here that thermographs of surface temperature exhibit a noticeable reduction in the rate of temperature increase when such elevated stable layers are eliminated and mixing through a deep layer starts. The shape of the

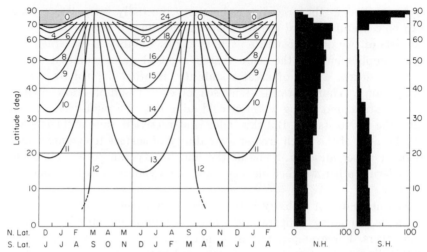

Figure 10. Interval in hours between sunrise and sunset as a function of latitude and month of the year. The ordinate is scaled to the sine of the latitude and this is proportional to the global area affected. The percentage of land area (shaded) by 5-degree zones in the two hemispheres after Kossinna is illustrated in the panels at the right. Based on data in List (20).

thermograph daytime maximum differs if a capping or low subsidence inversion persists.

IV. Atmospheric Motions

A thin skin of atmosphere covers the earth: the tropical troposphere at its highest contains nearly 95% of the total mass but measures only $\frac{1}{300}$th of the earth's radius; the polar troposphere may be less than half as high. Solar energy (230 W/m²) received by the rotating and revolving atmosphere and earth is distributed unevenly with latitude, especially in the winter hemisphere, with both polar regions representing major heat sinks (Figs. 10 and 11). The earth's atmosphere as a whole, in contrast to the lowest layers above land areas, does not respond markedly to the daily alternation of day and night, so that the effect of the poleward-directed thermal gradient dominates (24). The ensemble of resulting atmospheric motions thus redistributes the thermal energy—some $2\frac{1}{2}$ times as much falling on the atmosphere at the equator as at the poles—before its eventual return to space via long wave radiation. Reservoirs of internal energy, latent heat, and mechanical energy are built up in the atmo-

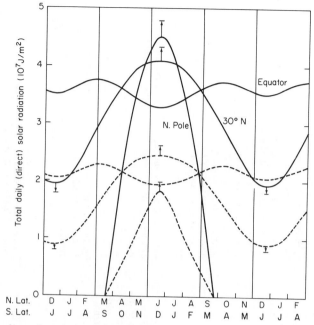

Figure 11. Annual variation of total daily (direct) solar radiation on a horizontal surface arriving at the atmosphere (solid), and at the earth's surface assuming a transmission coefficient of 0.7 (dashed). Arrows indicate southern hemisphere maxima and minima. Based on data in List (20).

sphere, estimated to total about 1.9×10^9 J/m² in the approximate ratios $60^2:12^2:1$, respectively (42, 43). With the incoming energy turned off while rates of energy consumption remained the same, the internal energy, latent heat, and mechanical energy reservoirs was estimated to last roughly 100, 12, and 3 days, respectively. The motions of redistribution in the troposphere look turbulent on a large scale, whether one views weather charts or satellite cloud photographs. Rossby early pointed out that the general configuration and quasi-stability of the atmospheric field of motion are both due to the earth's rotation, a view supported later by rotating-disk experiments (44).

When the motions are time averaged over the earth we observe the so-called general circulation, with wind belts and "semipermanent" high and low pressure areas. In addition, there are secondary circulations, e.g., the migratory high and low pressure areas seen on daily weather charts, and small-scale circulations, e.g., sea–land breezes, mountain–valley winds, and thunderstorms. Descriptive features of atmospheric motions at various scales are presented in Table IV, abridged from Lamb (45).

Table IV Descriptive Features of Atmospheric Motions at Various Scales, Abridged from Lamb (45)[a]

System	Cross measurement	Vertical extent	Winds	
			Horizontal component	Vertical component
Jet stream	500–5000 km long	1–5 km (centered at heights about 10–15 km above sea level)	Commonly 30–70 m/second. Extremes >100 m/second in many parts of the world outside the tropics, up to 150 m/second over Japan and southern Indian Ocean near 35–40°S, possibly also at same latitude near Atlantic seaboard of North America	
	50–500 km broad			
Anticyclone (a) In the subtropical belt (or extension thereof)	3000–4000 km long	Throughout troposphere (12–20 km); leans towards equator (or towards the warmest side) with increasing height	Mostly <10 m/second at sea level, but stronger near the periphery	Subsidence rate of order of 1 cm/second
	750–1500 km broad			
(b) Blocking anticyclone (warm cutoff anticyclone) in higher latitudes	1500–3000 km (occas. 4000 km) major axis (in various orientations)	As for (a)	Mostly 0–10 m/second but stronger near periphery	As for (a)
	750–1500 km minor axis			
(c) Cold, polar anticyclones and ridges	1000–3000 km (roughly circular). Occas. linked with neighboring warm blocking anticyclones into a very large system	Commonly 1–5 km	Mostly 0–10 m/second (occas., locally and at periphery, up to 20 m/second)	As for (a)

Wave on the polar front (incipient frontal cyclone)	200–500 km long	Commonly 3–6 km	Commonly 5–10 m/second at sea level	Rates of order of 5 cm/second in up-gliding near frontal surface, but much greater in cumulonimbus clouds q.v.
Polar front cyclone	50–200 km broad 1000–2500 km, occas. 3000 km over oceans in seasons of maximum vigor of circulation	From mean sea level up to 5–15 km as a complete whirl. Commonly throughout the troposphere, up to 15 km at occluded (cold center) stage, as a stationary cyclone in high latitudes	10–25 m/second at surface in strongest wind-streams, occas. up to about 50 m/second in gusts	
Cold cutoff cyclone	500–1500 km	From mean sea level commonly up to 10–15 km 5–15 km	7–20 m/second at sea level	Strongest in cumulonimbus clouds q.v.
Tropical cyclone	300–500 km	Of order of 5 km	25–50 m/second common at sea level	As in cumulonimbus clouds elsewhere q.v. Extremes possibly >50 m/second
Tornado (accompanied by waterspout if over water or dust whirl over land)	Up to 200 m in max. velocity ring; diameter varies rapidly, 1–10 m commonest	Commonly 1–10 km. Extremes circa 15 km in Europe, 20 km in tropics.	10–100 m/second at surface in max. velocity ring, but over a breadth of the order of ($\leq$) 1 m	
Cumulus convection cells, cumulonimbus clouds	5–50 km (occas. neighboring cells arranged in line along a front or line squall, giving congested cells merging together along a line up to 500–1000 km long but only 20–50 km broad) Up to 400 km across in congestions of multiple-cell cumulonimbus in afternoons over some islands in the tropics		5–50 m/second in gusts and squalls	Mostly 1–5 m/second but 10–50 m/second in active cumulonimbus clouds with squalls and thunderstorms

Table IV *(Continued)*

System	Cross measurement	Vertical extent	Winds	
			Horizontal component	Vertical component
Land and sea breezes	Commonly 10–20 km on either side of the coast. Occas. linked with other regional circulations (e.g., mountain winds) into one wind system 100–200 km across	0.5–1.5 km	Mostly 1–5 m/second in the temperate zone, up to 10 m/second in warmer climates or where linked into a bigger circulation	
Mountain and valley winds (anabatic, up-slope and katabatic, down-slope breezes)	Seldom more than 10–15 km from the mts. and valleys concerned, often only within a few meters of the slope	Very restricted if air thermally stable; if unstable up to max. cumulonimbus heights (but strong winds only among the topography, especially in narrow valley channels, at mountain-tops and along the edge of the mts.; also in very localized lee-wave concentrations down-wind).	Very various. Anabatic breezes up open mountain slopes commonly 1–2 m/second Katabatic winds 'funnelled' in Greenland fjords up to 50–60 m/second (also very strong where the mistral blows through the Rhone gap). Strongest winds just over the mountain tops and along mountain sides often occurring just when thermally stable stratification checks wider deflection of the wind	Commonly ≤1 m/second. Cumulonimbus rates apply to extreme cases of ascent where such clouds form over the ridges. Extreme rates of descent vary widely

a max.: Maximum; mts.: mountains; occas.: occasionally.

A. The General Circulation

The principal features of the general circulation as observed at the earth's surface are huge anticyclonic whirls (clockwise in the northern hemisphere, counterclockwise in the southern) centered at about one-third of the distance from equator to pole over the eastern Pacific (north and south), eastern Atlantic (north and south), and south Indian oceans. Their eastward and equatorward sides are characterized by massive, steady, converging streams which meet as trades at the intertropical convergence zone (or front). This zone, strung like a necklace about the earth, lies exactly on the geographical equator only by way of exception, its position being influenced by the appearance over midlatitude continental masses of anticyclonic whirls in the winter half of the year and cyclonic whirls in the summer. This winter–summer alternation is most marked over the extensive Asian land mass. The result is that, on the average, the intertropical convergence zone is found everywhere to the north of the equator in the northern summer, even reaching middle latitudes in eastern Asia; while it averages roughly on the equator in the southern summer. East of the Pacific and Atlantic Oceans, it bulges southward in Africa and South America, but, at maximum penetration, only half as far as the Asian summer bulge, because of the much smaller area of the heated land mass. At high latitudes we find in the northern winter a constancy of flow directions, approaching that of the trades, associated with the anticyclonic whirl that lies centered on Greenland and the Baffin Bay portion of the cyclonic whirl to the west. Flow is mainly from the west in a zone poleward of the southern continents at all longitudes in both halves of the year. Generally winds tend to be easterly (from the east) at the very high latitudes, especially in the northern summer, opposing westerly currents equatorward in zones that are well-marked southeast of Greenland, and in the far north Pacific in the northern winter and north of Alaska in the northern summer.

The global wind field outlined above is conceived to be produced by two directly driven large-scale convective cells in each hemisphere. The tropical cells show motion upward at the atmospheric equator and downward at roughly 30 deg of latitude, giving a component toward the equator to the surface winds, and a component toward the pole to the winds aloft. The polar cells show descending motion over the heat sink at the poles, and an equatorward component to the surface wind, extending down to roughly 60 deg of latitude. An intermediate cell with components of surface wind from both the south and the north is located between the directly driven cells. On the average all these meridional wind components are relatively small increments added to the surface wind streams from the east in the tropics and polar regions, and from the west in the intermediate regions. The direction of flow aloft is mainly from the west except for easterly flow in an equatorial belt. The interested reader should consult Lamb (45), Rossby (46), and Reiter (47).

B. Wind and Pressure in Equilibrium; Some Conservation Laws

The earth's rotation at an angular velocity ω of 7.28×10^{-5} rad/second increasingly affects the apparent motion of a parcel of air the farther it is from the equator, as if acted upon by a force proportional to its speed and directed to the right of the wind vector in the northern hemisphere and to the left of the wind vector in the southern hemisphere. This is the Coriolis force (C.F.), the magnitude of which per unit mass is the product of the wind speed $|\mathbf{V}|$ and the Coriolis parameter f:

$$\text{C.F.} = f|\mathbf{V}| \tag{14}$$

The Coriolis parameter f is equal to twice the product of the earth's angular velocity ω and the sine of the latitude ϕ:

$$f = 2\omega \sin \phi \tag{15}$$

The effect of such tendency to turn is to pile up air on the side toward which the parcel turns. When such parcels are in equilibrium we have the balance of forces shown in Figures 12a and 12b. The force per unit mass vectors (P.G.F.) due to the pressure gradient (i.e., the pressure gradient dp/ds divided by the air density ρ) balances that of the Coriolis force (C.F.) [Equation (14)] appropriate to the magnitude of the wind vector $\mathbf{V}$. $V = |\mathbf{V}|$, i.e., the magnitude of the wind vector, is the geostrophic wind speed.

If the isobars p and $p + \Delta p$ are curved, a centrifugal force directly proportional to the square of the wind speed and inversely proportional to the radius of curvature enters the balance of forces. Larger pressure gradients are required for the same wind speed in low pressure areas, other conditions remaining the same. This is the gradient wind speed, for which the geostrophic wind approximation can often be used. A notable exception is the case of strong winds, particularly at lower latitudes (e.g., in tropical storms) where the centrifugal force can dominate the Coriolis force.

In Figures 12c and 12d we see that the equilibrium effect of the frictional drag (F.D.F.) near the surface is to turn the wind vector toward low pressure. When instead of isobars we have lines of equal height, as on the constant pressure charts usually drawn for the atmosphere aloft,

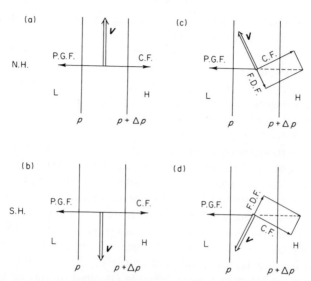

Figure 12. Balance of pressure gradient, Coriolis, and frictional drag forces (per unit mass of air) for a flow with and without friction in the northern and southern hemispheres. P.G.F. is pressure gradient force, C.F. is Coriolis force, F.D.F. is frictional drag force, and **V** is the geostrophic wind vector. Vector forces in balance are represented by single-width arrows; the flow vector, by double-width arrows. H, high; L, low.

the pressure gradient force term transforms to $g \, dZ/ds$, where g is the gravitational acceleration and dZ/ds is the topographic gradient of the constant pressure surface.

With such background information, one may look at a weather chart or upper air chart and be able to estimate the vector wind field at any point. Conversely, from the observed global time-averaged wind field described above one may anticipate that the average pressure field will show belts of minimum pressure at the atmospheric equator, 10–20°N in July and 0–10°S in January; maximum pressure at about 25–40 deg of latitude; minimum pressure again at about 60–70 deg of latitude; and maximum pressure in the polar regions.

It should be noted that the oceans exhibit giant gyres apparently driven by the semipermanent anticyclones. The Coriolis effect acting on the surface water dragged by the wind tends to pile water to the right of the wind in the northern hemisphere and to the left in the southern hemisphere, and an equilibrium condition similar to that described for the atmosphere is often approached. The cold water surfaces characteristic of regions of upwelling, as off the California coast and elsewhere, which assist in the maintenance of local inversions, occur with winds paralleling the coast in a manner such that the piling up of water takes place in a direction away from the coastline.

Air parcels moving poleward from lower latitudes show excess angular momentum arising from the reduction of the radius of rotation about the earth's axis with increasing latitude. The narrow belt of strong west winds, the subtropical jet, observed over the subtropical high pressure belt is thought to result from the conservation of momentum of air moving poleward aloft (*45*). Conversely, air parcels moving equatorward are deflected into motions from the east in the absence of other influences.

Another conservation law applies to large-scale atmospheric motion, and in particular to large-scale divergence. The semipermeable barriers to vertical flux of pollutant load which are presented by the stable subsiding layers of air associated with divergence can significantly limit pollutant dispersal. When the large-scale divergence is negative, i.e., with large-scale convergence, these limitations on vertical flux are removed. The applicable law is conservation of vorticity, which combines the effects of angular momentum and shear, whether acquired by the airstream from latitude change or from rotation relative to the air beneath. For example, a current of air moving equatorward, i.e., to a belt where the vorticity due to the earth's rotation is less, tends to curve eastward or develop cyclonic shear (or both); moving poleward, it tends to curve eastward or develop anticyclonic shear. Changes in vertical thickness are implied. A large column of air of a specified mass moving equatorward, if it is to conserve its absolute vorticity, must exhibit an increase in vorticity relative to the earth's surface or a decrease in height (divergence and subsidence) or both. The ratio of absolute vorticity to depth of column remains constant (see *24, 30, 33, 45, 46*).

C. Secondary Circulations

The redistribution of thermal energy on the rotating earth evidently requires those large-scale eddies called migratory cyclones (lows) and anticyclones (highs) seen on the daily weather charts and visualized through satellite photography mainly in the belt of westerlies. These eddies appear within a zone of large average meridional (poleward) temperature gradient, in which air masses of tropical and polar origins meet. Associated with this zone is a narrow belt of strong winds aloft, the polar jet stream. The warmer tropical air tends to overrun the cold polar air, often along relatively narrow frontal zones (ideally, surfaces). Migratory cyclones form when this zone is perturbed under suitable conditions. Latent heat contributes energy to their formation, and the ensuing convergence at low levels and divergence at high levels is compensated by the converse motions of adjacent anticyclones. Each place in the middle latitudes is subjected to the passage of such lows and highs at intervals of a few days, particularly in the cold half of the year when thermal gradients are large.

Tropical cyclones, the genesis of which differs somewhat, represent another agency of exchange necessitated by the ineluctable latitudinal redistribution of thermal energy.

The central regions of anticyclones are characterized by very light winds or calms and by subsiding air aloft. If a migratory anticyclone stalls, the surface beneath may experience air stagnation in a deep layer for several days with an intensification of air pollutant concentrations and adverse effects (see Chapters 2 and 3, this volume). These same effects of stagnation occur also, with high frequency, beneath the large semipermanent anticyclones. It is the persistent residence of a large body of air in such places which leads to characteristic vertical profiles of temperature, water vapor, and pollutants over extensive horizontal areas. As such air masses move away from their source regions and participate in mixing, e.g., within migratory cyclones, certain thermodynamic functions of pressure, temperature, and humidity remain quasi-constant. Meteorologists track such air masses in this way (30, 33).

The alternate passage of migratory high and low pressure areas affects spectra of measured winds, vertical temperature lapse rates, and pollutant concentrations (48). Peaks in such spectra at about 3 days in late fall and early winter have been reported over eastern Canada (48), and at nearly 4 days for the period August to February, over Long Island, New York (49).

Circumpolar planetary waves, a prominent feature of the wind field above the first 1–2 km, have essential associations with the secondary circulation features described above (30, 33, 45). Large-scale cyclonic curvature of height lines (concave toward low pressure) on a constant pressure surface is broadly associated with lower level convergence and rising motion of air; anticyclonic curvature, with lower level divergence and subsidence. For example, the curvature just above 2 km is a rough indicator of regional-scale stability.

D. Smaller-Scale Circulations

The smaller-scale circulations include mountain–valley winds, land–sea breezes, thunderstorms, and urban–exurban flows. Each of these systems adds vectorially to the concurrent regional motion. Environmental conditions may mask or prevent development of these smaller circulations, as for example in the cases of strong winds nullifying the typical mountain–valley flow or, in some cases, strong increase of wind speed with height inhibiting or modifying the development of a thunderstorm.

Mountain–valley winds are characterized by fully developed up-slope and up-valley flows during and immediately following the hours of peak solar irradiation, and down-slope and down-valley flows at dawn. The slope flows often precede the corresponding valley flows so that transitional periods are present (50).

Sea breezes generally penetrate inland some 10–20 km, though they may lose identity as they merge into a large-scale monsoonal flow from ocean to continent as, e.g., in the northeastern United States. They seldom exceed 1 km in height and exhibit a diffuse but deeper return flow aloft. Land breezes are typically shallower than sea breezes and reach less than 10 km out to sea. Lakes, ponds, and islands commonly experience qualitatively similar circulations (34). The reinforcing or cancelling combinations of mountain–valley and sea–land breeze circulations can often produce unexpected results, of some importance to air pollution control. One of the authors (R.C.W.) has observed inland flow at midday several kilometers up a broad valley in direct opposition to a prevailing trade of some 7 m/second (Jamaica, 1973).

Thunderstorms, and tornadoes occasionally embedded in or near severe thunderstorms, are examples of intense convection cooperating with instability produced by release of latent heat of condensation in the rising current. The environmental lapse rate of temperature will be greater than pseudoadiabatic. Local surface wind patterns associated with thunderstorms include inflow into the regions of rising current, and downflow and divergence in small cold domes associated with heavy precipitation (30).

V. Vertical Structure of Wind and Shear

Air in motion very near the earth's surface is retarded by friction which varies directly with surface roughness. The effect of surface friction is transmitted upward, but diminishes with height. The air layer so influenced (called the planetary boundary layer) is on the order of 1 km in thickness, variations being dependent on the stability of the layer near the ground and the change of stability with height. In the presence of stable layers the degree of coupling is reduced between motions above and below so that marked differences of wind speed are observed across stable layers (47). The planetary boundary layer is thicker in unstable conditions, when the loss of momentum at the surface is more readily propagated upward, then in stable conditions. Viewed alternatively, surface winds are the result of a driving wind in the free atmosphere just above the planetary boundary layer, modulated by a "leaky-membrane" property of the layer between there and the surface, and by surface friction.

The average wind profile in the troposphere shows an increase with height, the rate of which is marked near the ground and in the lower portion of the planetary boundary layer. Above the boundary layer in the middle latitudes, wind speed increases to a maximum beneath the tropopause near the top of the troposphere, consistent with an average temperature gradient directed poleward. The rate of increase near the ground is great enough to have produced agreement among national weather services as to standard height(s) for wind measurement (see also Chapter 11, this volume). Rapidly moving cores of air aloft (jets) are usually associated with large transverse temperature gradients there. The rate of change of wind in the vertical is generally greater than in the horizontal, with the possible exception of transitory conditions associated with squall lines, thunderstorms, fronts, and sea or valley breezes.

A. Profiles of Wind Speed near the Surface

Expressions of the steady-state variation of mean wind speed with height must allow for the effects of surface roughness, stability, and wind speed at the top of the layer of frictional influence, in addition to the Coriolis effect already discussed. Additional complications enter when, as is often the case, these parameters are changing in space or time. The power law for wind speed change with height has been found useful up to 100 m and more, over a wide range of temperature lapse rates. Other speed profile expressions are discussed in Chapter 9, this volume.

For a (horizontal) wind speed u at height z and wind speed u_1 at a reference height z_1, the power law is expressed by

$$u/u_1 = (z/z_1)^p \qquad (16)$$

where the positive exponent p has a value between 0 and 1, and most often less than 0.5. Note that the value 0 applies to a wind constant with height. If the environmental lapse rate is dry-adiabatic and the terrain is fairly level with low surface cover, p is approximately $\frac{1}{7}$. The exponent p is observed to increase, i.e., the wind profile is steeper, with increasing stability or away from the immediate surface, with increasing surface roughness.

B. Effects of Horizontal Temperature Gradient and Friction

The geostrophic wind, for which the pressure gradient force and Coriolis force balance each other, will change with height in the presence of a horizontal temperature gradient. The vector change upward in a layer is called the thermal wind, so that the wind vector at the top of the layer is the sum of the wind vector at the bottom and the thermal wind vector. This thermal wind, or shear vector, is oriented perpendicular to the temperature gradient and directed so that in the northern hemisphere the warm air lies to the right and in the southern hemisphere to the left. The magnitude of the thermal wind (T.W.) is approximately

$$\text{T.W.} = (g/fT)(dT/ds) \tag{17}$$

where T is the absolute temperature. In general the observed wind crosses the isotherms so that, from the observed shear in a layer, one may infer the rate of warm or cold advection. In the northern hemisphere, the wind veers, and in the southern hemisphere, backs with height with warm advection and does the opposite with cold. The vector wind profile (wind hodograph) sometimes shows layers of opposite vector turning with height, one above the other. From the differential thermal advection implied, one can infer areal changes of stability.

Figures 12c and 12d show how friction at the surface unbalances the geostrophic equilibrium by slowing up the wind. The resultant wind has a component across the isobars toward low pressure. With increase of height above the surface the influence of surface friction diminishes, the wind attaining its geostrophic or gradient orientation and speed at a height which is a function of the intensity of vertical mixing. For a constant mixing profile, the locus of the endpoints of the wind vector describes an Ekman spiral in the hodograph. The planetary boundary layer is also known as the Ekman layer. The pressure gradient force, the Coriolis force, the frictional force from below, and the drag force from above are in balance at any given level in the Ekman layer (51). The thickness of the boundary φayer decreases with increasing stability, decreasing surface roughness, and increasing latitude.

The change of direction with height in the lowest 100 m is at most a few degrees in strong winds in the absence of local topographical influences. However in light to moderate winds, under about 6 m/second, appreciable direction changes at lower levels may be produced by surface friction, distortion of constant-pressure surfaces by local temperature influences, etc. The frictional influence on wind direction is supplemented in warm air advection and opposed in cold air advection. Typical values of the angle between surface and geostrophic wind because of friction alone are 5–15 deg over the ocean and 25–45 deg over land (43). Over rough terrain the speed may be as little as one-third of the geostrophic or gradient speed whereas over water it would be nearer two-thirds. At

Oklahoma City, Oklahoma, 80% of the vector differences between 7 and 444 m in the course of a year were to the right of the flow (veering), and the median change varied with wind direction between 15 and 37 deg (*52*).

C. Variation of Wind and Wind Profile with Height, Time, and Location

The diurnal variation of wind speed near the surface (the lowest 70–100 m) exhibits maximum speed in the early to middle afternoon, when instability and the vertical eddy diffusion of momentum are a maximum, and a minimum in the evening and early morning hours (Fig. 13). At greater heights a minimum appears near midday and a maximum during the night, when the coupling with the surface is a minimum because of greater stability and consequent reduced vertical diffusion of momentum. With stability present, energy lost by the surface air to friction is replaced slowly if at

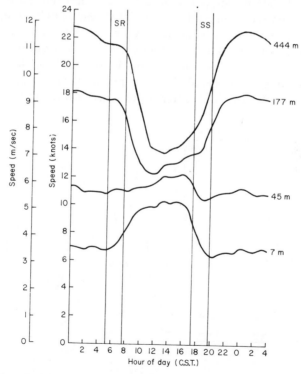

Figure 13. Diurnal variation of wind speed at four levels up to 444 m, June 1966–May 1967, Oklahoma City, Oklahoma. After Crawford and Hudson (*52*). SR: Sunrise; SS: sunset.

all, and a calm surface layer may develop even with moderate wind speeds aloft. In the daytime, when vertical coupling improves because of convection, surface wind speeds are higher, turbulence is greater, and convective motions enhance the vertical exchange of properties. The typical cycle of wind speed described above and the associated changes in vertical temperature structure described earlier are further illustrated by Figures 14 and 15, which show the average height and time cross sections of the lowest 1500 m at Oak Ridge, Tennessee during a 2-month period (23). Beneath and within the dashed line appearing first near the ground before sunset, the temperature lapse rate is negative, i.e., an inversion is present. Note that the associated decrease in vertical coupling between layers produces an increase in the vertical gradient of speed, while the speed at the surface falls to low values. A number of the features of the wind and temperature profiles discussed here and in Section III may be identified in these figures. Since they represent an average of many measurements, some of the features described are even more prominent in individual cases.

Because of the diversity of natural terrain, local flow patterns exhibit deviations from the regional flow, especially near shorelines, over uneven ground, or in the regions influenced by urban heat emission and roughness, e.g., heat islands. When the regional wind is not too strong remarkable changes of wind direction and speed with height or horizontal distance may often be observed. The air pollution investigator may find detailed study of local flow patterns quite necessary. Topography modi-

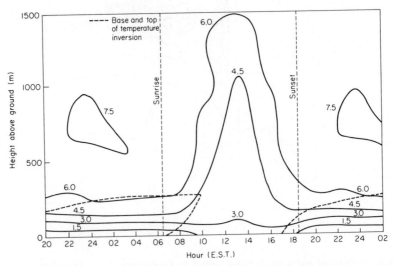

Figure 14. Time cross section of average wind speed (m second⁻¹) up to 1500 m, September–October, 1950, Oak Ridge, Tennessee. After Holland and Myers (23).

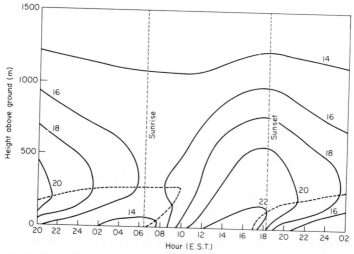

Figure 15. Time cross section of average temperature (°C) up to 1500 m, September–October, 1950, Oak Ridge, Tennessee. After Holland and Myers (23). Dashed line: Base and top of temperature inversion; cf. Fig. 14.

fies both temperature and wind, and their vertical profiles, because of the combined effects of surface friction, radiation, and cold air drainage. As has been mentioned, radiative and thermal properties of surface features influence the range of heating and cooling of the surface. Thermal effects due to both incoming and (especially) outgoing radiation are enhanced at high surface elevation where the absorbing mass of air overhead is reduced. Temperature lapse rates and wind profiles over built-up areas may differ markedly from those in open country. For example, instability may be observed to a height of three times the mean roof height in what would otherwise be stable air. Even the heat exchange accompanying freezing, melting, condensation, and evaporation at the surface will affect the temperature profile and hence the wind profile in the lowest layers. These changes of water state are often not distributed uniformly over an area. For example, evaporation after a heavy rainfall will have a cooling effect that may significantly change the stability of the lowest layers, though quite locally.

D. Profiles of Gustiness and Turbulence

Gustiness is the manifestation of eddies and turbulence displayed in the fluctuations of a continuous record of wind or temperature measured with sensors of relatively short response time. The most important mixing process in the atmosphere is called eddy diffusion; it involves scales of length considerably larger than that of molecular diffusion. Because of the vital role of eddies in atmospheric mixing processes, the spectral distribution of the energies of the many sizes of eddies occurring concurrently are always under investigation. The presence of eddies in the atmosphere is indicated not only by short-period fluctuations, but on a larger scale by changes of, say, the wind from day to day.

Eddy or turbulent diffusion is most efficient when the length scale of the eddy is similar to that of the body of air being diluted. This is readily visualized by considering that much smaller eddies should be effective only on the margins of a quasi-homogeneous mass of air, whereas much larger eddies would move the mass of air as a whole. Dilution of a property by the atmosphere is a combination of turbulent diffusive mixing proportional to the gradient of the property being diluted, and of stirring which stretches and distorts the diluting body, thereby increasing its surface area and making diffusion by small eddies more effective.

The presence of eddies implies turbulence and gustiness. Qualitatively the intensity of turbulent mixing may be estimated from an examination of fluctuations evident in continuous meteorological records made with fast-response sensors. Eddy or gustiness types have been classified on the basis of their effect on wind direction fluctuations at a given height above ground (53).

A classification of profiles which also takes some account of the change in gustiness with height has been applied to balloon-borne measurements up to 500 m at Voyeykovo, U.S.S.R. (60°N) (Table V) (37). There are four categories, each with two divisions based on the wind speed at 100 m. The first two categories have the common property of a constant or, more often, steadily diminishing gustiness with height associated with a positive lapse throughout the profile; lapse is greater for class 1 than for class 2. The last two are mixed profiles—not uncommonly observed—class 3 being an inversion below, temperature lapse above, and class 4 being the reverse. Gustiness in class 3 profiles is small throughout, whereas in class 4 it is moderate or large near the surface, but decreases, usually very abruptly, when the elevated inversion is reached. Table V also gives the times of day at which the various classes were most frequently observed in the warm and cold halves of the year; power law exponents [see Eq. (16)] based on average wind speeds at 25, 100, and 500 m are also shown.

A ratio of functions of the vertical temperature gradient and the wind speed gradient called the Richardson number is used to characterize profiles. The ratio varies with height. Its various forms, and other related ones including stability length, are described in Chapter 9, this volume.

Table V Classification of Wind and Temperature Profiles up to 500 m in the Warm and Cold Halves of the Year, Voyeykovo, USSR [a,b]

	Class stability	Temperature lapse rate (°C/100 m)	Wind speed at 100 m (m/second)	Time of day (hour) most frequently observed (%); percent of total — April–October	November–March	Power law exponent, p A–O 25–100 m	N–M 25–100 m	A–O 25–500 m	N–M 25–500 m
1	Superadiabatic in layer 2–100 m	>1	≤6	11(41); 07(39) 19	11(33); 07(33) 5	0.07	0.19	0.08	0.31
1a			>6	07(49); 11(35) 8	07(42); 11(29); 15(29) 4	0.13	0.18	0.12	0.22
2	Subadiabatic	0 – 1	≤6	03(26); 07(25) 20	11(26); 15(26) 15	0.13	0.17	0.15	0.26
2a			>6	07(35); 11(21) 8	07(40); 03(20); 11(20); 15(20) 3	0.18	0.18	0.16	0.21
3	Inversion in surface layer to 100–200 m, lapse above[c]	>0 above; <0 below	≤6	03(32); 23(24); 19(23) 28	07(27); 11(27) 38	0.22	0.21		
3a			>6	19(40); 23(37) 14	07(27); 11(21); 19(21) 15	0.39	0.31		
4	Lapse in surface layer to 100–200 m, inversion above	<0 above; >0 below	≤6	03(41); 11(34) 2	07(23); 11(23) 12	0.18	0.29		
4a			>6	03(50) 1	11(26); 15(26) 8	0.20	0.43		
	No. of ascents			100 541	100 184				

[a] Power exponents were calculated from tabulated averages.
[b] Note: ascents were made mainly in anticyclonic weather.
[c] A few cases of inversion to 500 m were included.

VI. Self-Cleansing of the Atmosphere

Through eons of time, the earth's surface and its atmosphere have accommodated each other, reaching equilibria of energy and mass exchanges appropriate to the changes taking place on the geological time scale with divisions of the order of millions of years. Very recently, from this perspective, man's activities have emerged as a potentially significant part of the system of energy and mass exchanges between earth and atmosphere. Whether his "emissions" represent a major perturbation, or whether they are a minor "signal" added to the natural "noise" is very much an open question (54). There is little doubt that, locally and for relatively short periods, the man-made components of mass exchange are detectable as signal, but even under these acute conditions, there is a pathway of exchange from the atmosphere to the earth which constitutes part of the continual self-cleansing of the atmosphere. Intuition tells us, moreover, that the ambient rate of self-cleansing, and perhaps even the maximum potential rate, is itself a function of the rate at which pollutants enter the atmosphere. Information on the self-cleansing system is relatively sparse. While the broad outlines of the subject can be sketched only in theory (55, 56), details on some aspects are well understood and are found here and elsewhere in these volumes.

The pathways along which pollutant materials leave the atmosphere are various, admitting various kinds of classification. For example, pollutants may depart as primary materials, in the same form that they entered the atmosphere; or they may depart as secondary or tertiary materials, having been altered while airborne. For another example, materials may depart along a biochemical pathway—plant, animal, or human—acting as sink; or, along a pathway not involving biochemical reaction. Yet a fourth classification is that between dry removal through fallout or impaction, or wet removal through natural processes of precipitation formation. In this section, the major aspects of dry and wet removal will be considered.

A. Dry Removal

Eddy diffusion and stirring do not remove pollutants; they only dilute them. In the process some fraction of the pollutant load may be brought near enough to a surface for capture by impaction or chemical reaction, or two or more initially separated pollutants capable of reacting with each other may be brought together. Discussion of chemical interactions among pollutants in the air, between pollutants and other constituents of the air, and adsorptive and contact removal at the surface may be found in Chapters 3 and 6, this volume.

Particulate pollutants are effectively removed from the atmosphere by gravitational settling if the particle size is sufficiently large (Fig. 16) (*20, 57*). Such pollutants include dust raised by high winds passing over dry surfaces. The velocity of settling is a function of shape, density, and size of particle. For equivalent sphere diameters less than 20–30 μm, the settling speed is negligible compared with atmospheric vertical velocities. These small particles are removed by impaction with surface cover and structures, and by aggregation to form larger particles which can be effectively removed by settling.

The growth of particles by aggregation, electrostatic forces, or by attachment of water vapor molecules is beyond the present scope [for

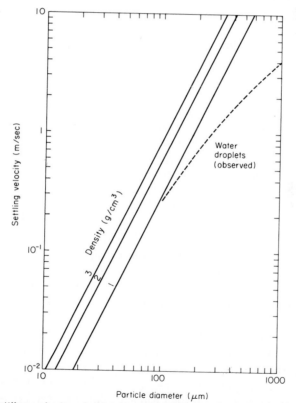

Figure 16. Settling velocity of spherical particles in quiet air as a function of size and density. Based on Stokes' law and Gunn and Kinzer data (*20*). (20°C, 101.32 kPa.)

further discussion, see Green and Lane (*58*)]. For discussion of the history of volcanic dust, forest fire smoke, etc., in the atmosphere, see Lamb (*45*). For discussion of the distribution of fallout from atmospheric nuclear bursts, see Matveev (*25*).

B. Wet Removal

1. Natural Formation of Precipitation

Atmospheric precipitation is the arrival at the earth's surface of water substance from above in any number of forms, such as rain, drizzle, snow, and sleet. Precipitation requires clouds, but not all clouds produce precipitation. Common to all precipitation are (a) a supply of water (vapor or liquid), (b) microscopic sites for condensation called nuclei, and (c) updraft motion of proper magnitude and persistence. The nuclei provide locations for conversion of water vapor into liquid or solid, the water supply provides the reservoir sustaining the growth of droplets and crystals, and the updraft provides the time and often the turbulence for collisions to take place between existing droplets and drops, or between droplets and crystals and flakes, causing additional growth. When drops or flakes have grown to sizes beyond the ability of the updraft to support them aloft, they fall relative to earth and become precipitation (*59–61*).

Only a few percent of all nuclei present ordinarily become condensation sites for the vapor at temperatures above freezing. The greater the solubility of the nucleus material and the greater the size of the particle, as a general rule, the more rapidly the condensation and growth proceeds, and the larger the resulting droplet may become by condensation alone. A population of aerosol particles may be characterized by spectra of both solubility and of size. Thus, a cloud whose droplets have formed primarily by condensation has a recognizable spectrum of droplet sizes. As a general rule, spectra of oceanic clouds are broader and have larger mean sizes than continental clouds, but with smaller numbers of droplets in a unit volume. Even so, only a small fraction of the aerosol particles present ever becomes activated and forms droplets.

In temperatures above freezing, updrafts support droplets of different sizes while their differential fall rates result in the collisions which enhance growth by producing coalescence, a process many times more effective than simple vapor-to-droplet condensation. Coalescence is most effective with certain combinations of droplet sizes and with certain updraft speeds. Finally, subcloud erosion of drops occurs, which depends on the height of cloud bases and the humidity of the environmental air. Thus, the production of drops in clouds is not, of itself, sufficient to result in significant ground-level precipitation.

In temperatures below freezing, formation of ice crystals is the analog of the formation of cloud droplets in warm clouds. The crystals form on ice nuclei, which are many times less numerous in ordinary air samples than are condensation nuclei. Thereafter, the crystals grow by both condensation of vapor and agglomeration analogous to coalescence. In the temperature range $-12°$ to $-17°C$, the crystals which form on ice nuclei have a dendritic, spiderlike shape which provides a large surface per unit mass for further growth. At both colder and warmer temperatures, crystals are in the forms of needles or plates with small areas per unit of mass. Further, whatever their shapes, ice crystals grow more rapidly into snow crystals and then into snow flakes when the growth takes place in temperatures between $-5°$ and $0°C$. This is so because of the enhanced bonding of colliding masses when liquid films on their surfaces freeze on impact.

2. Intentional Modification of Natural Processes

So-called "cloud-seeding" or "weather modification" is an intentional intervention by man in the natural processes of precipitation formation. It has the goal of changing the rate at which droplet or crystal formation takes place, and also the rate of subsequent growth. Moderate increases in numbers of ice nuclei by artificial additions will enhance the production of drops, under the right conditions, during the time available from the updrafts present. Larger additions of nuclei may so reduce the amount of vapor per nucleus as to forestall growth to larger sizes, thus retarding growth by coalescence. In both cases, moderate and larger additions, the behavior of the supporting updrafts is often affected by the intervention, adding another dimension to the outcome. Within this general framework, then, man intervenes for a variety of reasons: to increase or decrease the "yield" from a cloud system, or to change the form of the precipitation, most notably from damaging hail into valuable rain. The ability to match the goals of intervention with the proper outcome—the skill of the "cloud seeder"—is a subject of continuing debate *(62, 63)*.

3. Pollutants and Precipitation

Materials added intentionally to modify natural processes of precipitation formation qualify as air pollutants, though they are seldom perceived as such. On the other hand, pollutants may act to alter these same processes when they are added unintentionally.

Removal of pollutants by precipitation results from both particulates acting as nuclei and being carried to earth, and from particulates being collected by drops and crystals during collisions. In either case, the natural processes of precipitation, which would take place even in the absence of pollutants, result in the scavenging of particulate pollutants and thus in self-cleansing. Scavenging of gaseous pollutants is less well understood.

4. Types of Precipitation-Producing Storms

The precipitation rate observed at the earth's surface depends much more on the sizes of the drops than on their numbers; and in turn the sizes are a reflection primarily of updraft speeds when air is moist below a cloud base. Thus, updraft speeds during precipitation formation are highly correlated with observed rainfall rates (64).

Gentle, widespread uplift of an entire air layer moving upslope over terrain or over a denser air mass will support large droplets of the maximum size resulting from condensation alone. For example, motion of 1 m/second (2 mph) up a 1% slope will produce an updraft of about 10^{-2} m/second which in turn will support droplets of several tens of micrometers in diameter.

Updrafts of several meters per second are common when airstreams converge in rough and differentially heated terrain, or when cold air moves over surfaces several tens of degrees warmer producing vigorous cellular motion. Experiments show that raindrops several hundred micrometers in diameter can be supported in such updraft speeds (Fig. 16). Thus, these products of the early stages of coalescence may commonly be found under conditions of "scattered light showers."

The vigorous updrafts within thunderstorms, measured commonly at several tens of meters per second (20–40 mph) are quite compatible with the centimeter-sized hail stones from them. Drops which grow to such sizes rarely escape breaking up, thus rarely reach earth as large as 1 cm in diameter. Typical combinations of updraft speed, median drop diameter, and measured rainfall rate are shown in Table VI (60).

5. Scavenging of Particulate Pollutants

Particles are removed from the atmosphere by precipitation processes either after they have become wetted nuclei or before; if after, then the particles will have been among the small fraction of all particles in a typical air sample to become nuclei. They most probably will have been collected as part of a droplet by a larger drop, which itself will include particles which were the nuclei of all the droplets previously collected.

Table VI   Typical Combinations of Updraft Speed, Median Drop Diameter, and Measured Surface Rainfall Rate (60)

Rainfall type	Updraft (m/second)	Median drop size (μm)	Rainfall rate	
			mm/hour	inch/hour
Drizzle	1.0	300	0.2	0.008
Light shower	3.5	850	5.0	0.20
Heavy shower	5.0	1,200	20.0	0.80

Thus, these particle nuclei congregate by coalescence until there are many thousands in a raindrop by the time it reaches earth. Though they are a small fraction of the particles aloft, their probabilities of being scavenged increased severalfold the moment they were wetted, because of the relative ease with which a rain drop (typically 500 μm) can capture a cloud droplet (typically 30 μm) compared with that for an aerosol particle (typically 0.5 μm) which easily avoids the drop aerodynamically. Thus, particles which act as nuclei are fewer in number but have a much greater chance of being scavenged than the much more numerous dry particles which do not act as nuclei. Pollutants are removed by scavenging in a typically logarithmic fashion—less and less effectively through a rain period—based on a proportionality of the removal rate dC/dt to the concentration C of the pollutant:

$$dC/dt = -WC \qquad (18)$$

which is equivalent to

$$C/C_0 = \exp(-Wt) \quad \text{or} \quad \ln(C/C_0) = -Wt \qquad (19)$$

where C_0 is the aerosol concentration in air before scavenging begins, C the concentration after time t, and W the "washout coefficient" (60, 64). The coefficient has the units of (time)$^{-1}$. If W has the value 2×10^{-5} sec^{-1} for 1-μm particles, half will be washed out in about 11 hours. Figure 17 shows greater detail on the relationships among particle size, precipitation rate, and washout effectiveness as expressed by the coefficient for particles, W_p.

Capture of particles by raindrops (and snowflakes) is usually treated as an essentially irreversible process which is a function of the rate of encounter of particles by drops (or flakes). The washout coefficient for particles W_p is related to the raindrop space or number density $N(D)$, and the collision efficiency $E(D,d)$, for the combination of drop and

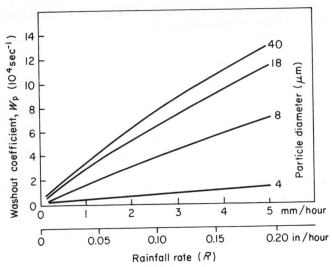

Figure 17. Typical values of washout coefficient as a function of rainfall rate and diameter of the scavenged particle. After Chamberlain (*64*).

particle diameters (*60*). W_p is also a function of the vertical (terminal) velocity of the drop relative to that of the particle. All these dependencies reflect the essential nature of the capture as a collision, or encounter, of particle and drop (flake). Once the particle is captured after encounter with probability E, it is effectively removed from the pollutant field and treated as irreversibly scavenged, except in special cases. These special reinjections of particles may occur following evaporation of a drop either below precipitating clouds or above vigorous cumulus columns.

The general relationships in Figure 17 may be formulated as follows. The rainfall rate R may be expressed as

$$R \sim \int_0^\infty (\tfrac{1}{6}\pi D^3) V_D N(D) dD \approx (\tfrac{1}{6}\pi D^3) VN \qquad (20)$$

where D is the drop diameter, V_D the vertical velocity of the drop, and $N(D)$ the space density of the drop. In the rightmost expression the integral expression is simplified by considering D to be a constant appropriate "median drop" size (*65*).

The physical collisional removal mediated by the coefficient W_p may be expressed by

$$W_p \sim E(D,d)(\pi D^2) VN \qquad (21)$$

where terms are as defined before. The terminal velocity of a small drop is proportional to D^2 by Stokes' law, and to D for larger drops by experiment (Fig. 16). Thus, for a "median drop" we may say V is related to D^p where $1 \leq p \leq 2$. Rewriting Equations (20) and (21), R is proportional to D^{3+p} and W_p is proportional to D^{2+p}, so that $W_p \sim R^r$, where $r = (2 + p)/(3 + p)$. Within the bounds for p, the bounds for r become $0.75 \leq r \leq 0.80$. In Figure 17, the value of r is approximately 0.80.

6. Scavening of Gaseous Pollutants

The scavenging of gaseous pollutants is an order more complex than that for particles, primarily because the "capture" of a gas molecule by a raindrop is to some extent reversible. It is best viewed as a diffusional, rather than a collisional, phenomenon. Here the capture is a function not of the rate of encounter so much as of the concentration gradient of the gas across the liquid–air interface. Thus the gas analog to W_p is related to the space density of the drops and their diameter, to the concentration gradient, and to a measure of the diffusivity of the gas in air. Because the washout depends on the gradient, the capture may be reversed either by the pollutant concentration being increased within the drop because of wet chemical transformation to another species, or by the drop's falling into "clean" air (*66, 67*). Because of these several possibilities for reversing the gradient, the scavenging of gaseous pollutants is much more strongly dependent upon the spatial distribution of the pollutant itself, and of related species, than is the case of particles.

Formulation of a relationship between the gaseous washout coefficient W_g and the rainfall rate R is analogous to that for particles. The coefficient may be expressed as

$$W_g \sim k(\pi D^2)N(C - C_e) \tag{22}$$

where k is a "gas-phase mass-transfer coefficient" which expresses the diffusivity, and C_e is an equilibrium gas-phase concentration of the pollutant which makes $(C - C_e)$ equivalent to the liquid–air gradient mentioned previously (*66*). The relationship $V \sim D^p$, from Stokes' law and experiment, combined with the relationship $k \sim V^{1/2}$ from the expression of the Sherwood number (*66*), implies that $W_g \sim D^{[2+(p/2)]}$. It follows that $W_g \sim R^s$, where $s = [2 + (p/2)]/[3 + p]$. Within the bounds for p, the bounds for s become 0.60 and 0.625. In the compilation of data presented in Figure 18, the value of s is empirically very near 0.60.

If one accepts scavenging as meaning removal to ground level, then on the whole gaseous pollutants are less well scavenged than particulate pollutants. This holds if the captures implied in the equations above are considered alone, as is appropriate for particles and for relatively unreactive gases. However, in view of the wide range of possible values for

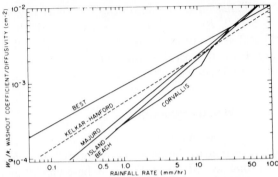

Figure 18. Washout coefficients divided by diffusivity for soluble gases as predicted from various rain spectra (64).

the efficiency E, one must say that some gases will sometimes be scavenged better than some particles. Furthermore, when liquid-phase transformations to other chemical compounds are contributing, as in the case of sulfur dioxide, particles may be removed less effectively than gases.

While in general gases are scavenged less well, they are more likely to be displaced vertically in the air mass, to levels and at rates dependent upon the initial spatial distribution of the gas itself. The complexity of the analytical problems of scavenging, and the recency of attacks on the theory and its experimental base, leave the subject poorly understood at this time, particularly with regard to its importance in the field of air quality management and in the assessment of environmental impacts.

VII. Some Meteorological Approaches in Air Pollution Analysis

Air pollution in a populated area, apart from the background pollution level of the incoming air, is a function of man's activities and of weather conditions, both of which are functions of time and location. The temporal and spatial variations of weather are no less important in the study of air pollution than the temporal and spatial variation of man's activities. Study of weather measurements made over a period of time, and distributed over an area, is therefore a desirable prerequisite for understanding most of the significant problems of air pollutant effects. Difficulties which might stand in the way of precise quantitative analysis should not be permitted to obscure the fact that qualitative understanding of the climatic features of a place leads to improved understanding of its air pollution and its effects and better judgments regarding efficient methodologies for its abatement.

A. Analogies in Eddy Diffusion of Momentum, Heat, and Pollution

The similarities of eddy diffusion of momentum, heat, and pollution have useful application. Any of the situations described in this chapter in which the vertical flux of momentum or heat varies may be assumed in first approximation to apply similarly to the vertical flux of pollution. The vertical eddy diffusion of any of these entities is so different between unstable and very stable conditions that averages including both should be treated and used with caution. The analogy between the vertical eddy diffusion of heat and of pollution permits one to infer from the hourly thermograph trace—if it exhibits a plateau or rise during nighttime hours when conditions are usually appropriate for falling tempera-ture—that vertical mixing took place (*40*). A similar inference holds true if the surface wind increases during the night when low speeds or calm are otherwise expected (*68*). Thus with air pollution aloft, in or above a stable surface layer and thus "insulated" from the surface, an increase of wind speed aloft, or other disturbance of the flow, may weaken the inversion and mix momentum, heat, and pollution downward (*19*). If the increase of speed aloft is only transient, then the antecedent degree of stability tends to be reestablished (Fig. 19), and the pollution brought down in the early phase will diffuse in the lower layer at the slow rate characteristic of stable air. More than one such local-scale shear destabi-lization may occur during a single night in a manner analogous to a flip-flop or multivibrator electronic circuit.

It is a popular misconception that the presence of a valley or bowl is a necessary adjunct to pollution buildup. Unfavorable topography is not a necessary condition for extreme pollutant concentrations provided the pollutant source strength is great enough compared with the eddy

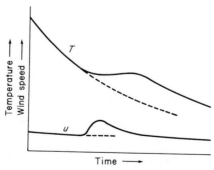

Figure 19. Typical nocturnal temperature (*T*) and wind speed (*u*) time profiles with shear destabilization. Dashed curves indicate probable trend without destabi-lization.

diffusion rate, or more generally, the total dispersal rate, at the time. Note that a pronounced urban heat island signals a relatively slow eddy diffusion of urban-generated heat, away from the island, and thus one might also expect slow dispersion of pollution on the urban scale, even though local convection does occur. A city has a characteristic surface roughness usually greater than that of its suburbs. The enhanced mixing in lower layers due to roughness is evident from the sharp horizontal temperature gradient on the edges of the heat island and the relatively flat gradient over the city (45).

It is not always appreciated how valuable pollutant measurements themselves—going beyond the point where the method of sensing, recording, and data handling is already accepted—e.g., their frequency distributions, and other statistics, can be in informing of meteorological processes that have acted, again by way of the three analogies. For example, the logarithmic normal form of the cumulative distribution of serial air pollutant concentrations first noticed by Wanta (69) and developed later by Larsen et al. (70, 71), Stern (72), and others has proved of considerable diagnostic help: if a particular measured distribution departs from that expected, the analyst can focus on what meteorological, instrumental, emission source, or other peculiarities, perhaps previously not considered or rejected, are affecting the record.

Figure 20 illustrates how strong mixing between two levels z and $z + \Delta z$ affects the profiles of temperature, potential temperature, and water vapor or pollutant concentration. The well-mixed portion of the temperature profile shows a lapse of temperature at the dry-adiabatic rate (the pseudoadiabatic rate if the air is saturated) ; the corresponding potential temperature profile shows constancy with height as does the pollutant profile. Under the conditions assumed the well-mixed layer would be bounded top and bottom by very stable zones.

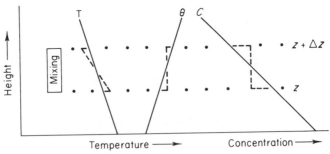

Figure 20. Effect of strong mixing confined to a previously stable layer on the distributions of temperature (T), potential temperature (θ), and water vapor or a pollutant (C).

A pollutant or trace substance, such as radon gas emitted at the earth's surface or carbon dioxide emitted from respiring vegetation, will under stable conditions show relatively high concentrations at the surface with a marked decrease with height, and under unstable conditions smaller surface concentrations and a lesser rate of decrease with height.

When the coupling between air layers is reduced relatively rapidly, the wind above may accelerate, and speeds approaching twice that expected from geostrophic balance are observed (73). Beneath this low-level jet, the wind shear may become great enough locally, especially near surface obstacles such as low hills, to produce destabilization and intense mechanical turbulence accompanied by thorough mixing through all the surface layers (19, 74).

B. Comparison of Sequences of Weather Categories and Actual Weather

Some degree of practical success is realizable with many air pollution problems by considering the continuously varying weather in a place as a succession of discrete stages, e.g., nocturnal inversion, morning inversion breakup, daytime instability, nocturnal inversion, and so on. The sequence is interrupted by frontal passages, wind changes, precipitation, etc. The consequences of neglecting important changes in wind, stability, or turbulence over relatively small vertical distances has already been noted. There is, also, a tendency among untrained analysts to overemphasize those weather and diffusion conditions one most frequently sees, e.g., daytime rather than nighttime patterns, and persistent conditions of quasi-equilibrium, rather than the atmospheric dynamism which so often governs. Careful analysis of several independent adequate sets of pollutant measurements made in all kinds of weather is a useful antidote to bias. Part of the dynamism of the atmosphere yields periods of stagnation, a not crisply definable weather condition. The usual idea of "downwind" may fail under these circumstances, especially in uneven topography. Under stagnation conditions, the same point source may cause simultaneous fumigations in all directions to several kilometers from the source (75). Attempts to accomplish meteorological control of emissions based partly on current measurements of pollutant concentrations often run squarely into the question of atmospheric dynamism and can make the task of the forecaster anything but routine. An example of atmospheric dynamism differing from set categories, such as those in the discrete sequence suggested above, is the stable layer which may represent a lid on convection. The lid is sometimes "hard," as in the case of the turbulence inversion topping a layer of strong cold advection. It is sometimes "soft," as in the case of tropopause and some other inversion surfaces, which are from time to time stirred, distorted, and ruptured, and which then no longer suppress exchange of air masses and their contents. A logical result is the observed homogeneity of air composition from the surface up to well into the stratosphere.

One of the open questions in air pollution is how precisely the profiles and fluxes must be known in each individual case of flow and pollutant dispersion on urban and larger scales. Such middle- and large-scale conditions are by definition nonhomogeneous with respect, at least, to surface roughness and thermal properties. Empirical expressions for profiles and transfer can be of practical value and may be, in particular instances, superior to exact laws for which input data are unavailable. The power-law increase of wind with height is considerably less than the most sophisticated state of the art of profile and exchange theory (76), but may prove to be a step forward compared with a constant wind.

C. Cross-Sectional Analysis

In studies where the transport of a property through the atmosphere is of concern, it is often useful to express the varying state of the atmosphere by measures which permit direct recognition of the gross features of movement of that property through the medium. Reasonable account should be taken of (a) the possibilities for temporal changes in the "tracer," and (b) the essential spatial differences in the tracer irrespective of time.

In the first case, the "aging" of a tracer material being transported would confound the study if only the initial form of the material were monitored and used as the measure of its presence. In the second case, spatial interactions among atmospheric properties could confound the study unless appropriate measures of the tracer were employed.

An atmospheric property which does not change under the influence of some process is said to be conservative with respect to that process (77). For example, the potential temperature θ of an air parcel is conservative with respect to dry-adiabatic expansion and compression as was discussed above. Thus, use of the potential temperature as a tracer of heat through the atmosphere would be useful if vertical motions were of primary concern.

Since relative humidity is a function of both moisture content and heat content, it would not serve well as a tracer of moisture content where vertical motions are a principal transport mechanism. However, the mixing ratio (mass of water per mass of dry air) is conservative with respect to dry-adiabatic motion, thus making it a more suitable tracer of moisture.

With respect to isobaric warming and cooling, potential temperature is not conservative, but the mixing ratio is, as long as no condensation takes place. Where only limited vertical motions are involved, dewpoint temperature is nearly conservative with respect to both dry-adiabatic changes and isobaric changes without condensation.

As already noted, observations of atmospheric properties are much more numerous in both time and space at the earth's surface than aloft. However, even those sparse once- or twice-daily radiosonde flights sampling the profiles of wind, temperature, and moisture on a 250-km horizontal grid and an approximately 0.5-km vertical grid (*78, 79*) produce data which are more dense, both temporally and spatially, than those from any existing air pollution monitoring network. Without a substantial network for monitoring of pollutants, it is nearly impossible to infer and estimate pollutant trajectories, on either a mesoscale or a macroscale. Fortunately, the travel of identifiable air masses, in which the pollutants are embedded, may be estimated by means of mapping and cross-section analyses of conservative properties such as potential temperature and mixing ratio.

Air parcels tend to move, at least over time periods of tens of hours, along isentropic surfaces, along which appropriate functions of heat content and moisture content taken together are everywhere the same. Early studies of the possibilities for weather forecasting by "isentropic analysis" (*80, 81*) were followed by a period in which these ideas were set aside. In the late 1960's and early 1970's a revival of such concepts appeared to be taking place, first with attention to motion on the mesoscale (*22, 82*) and then on the macroscale (*83*). Figure 21 shows an example of macroscale cross-sectional analysis (*84*) and Figure 22 an example of mesoscale analysis (*82*). In these two figures and in Figure 23 (*85*), vertical isentropes mark areas of thorough vertical mixing where lapse rates are dry-adiabatic. Areas in which θ decreases with altitude exhibit superadiabatic lapse rates, usually associated with strong heating of the land surface.

Some suggestion of the utility of cross sections in the study of pollutant trajectories may be seen in a consideration of Figure 23. If pollutants are injected into the atmosphere at point A (above shoreline) during the day, and move along an isentrope to point B before nighttime, they will probably move along the same isentrope to point C during the following night. However, if the pollutants are injected at point A' during the night, and move to point B before daybreak, their expected movement during the following day will be to the surface at C'.

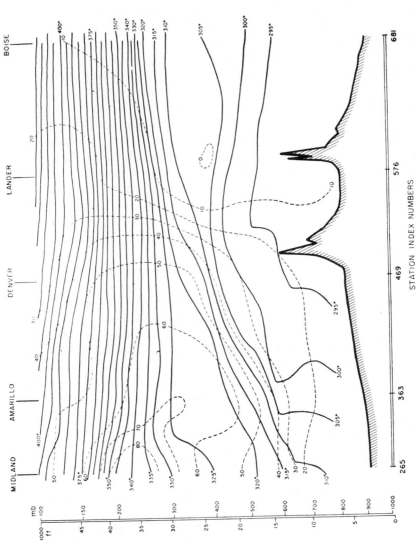

Figure 21. Large-scale cross-sectional analysis, 0000 GMT, 19 April 1963. Isopleths are potential temperature (solid) in °K and westerly component of wind (dashed) in knots. After Duquet, Danielsen, and Phares (84). (Reprinted with permission from J. Appl. Meteorol., Amer. Meteorol. Soc.)

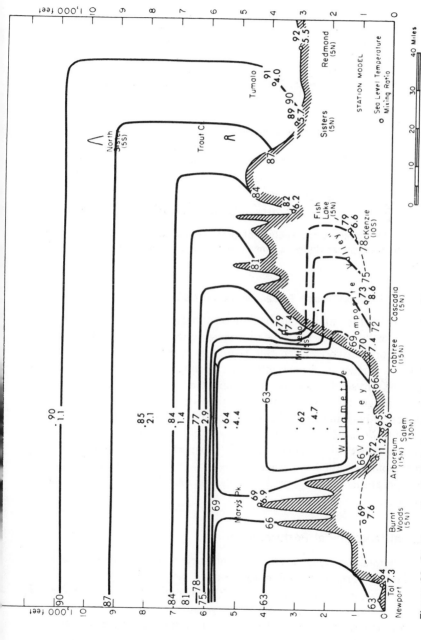

Figure 22. Cross-sectional analysis, 1600 PST, 1 July 1960. Distances from center line are indicated in parentheses. Isopleths are potential temperature in °F. After Cramer and Lynott (82). (Reprinted with permission from *J. Appl. Meteorol.*, Amer. Meteorol. Soc.)

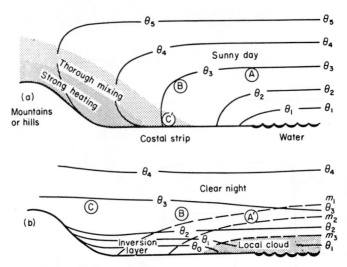

Figure 23. Mesoscale cross section of potential temperature across a shoreline and mountain range (a) during a sunny day and (b) the following night; m is mixing ratio (85).

D. Airflow through Rough Terrain and Built-Up Areas

Because of the fundamentally indeterminate nature of fluid flow, only "typical" or "average" patterns of atmospheric behavior can be explained and used predictively. Even this ability is compromised in studies near the earth's surface, and particularly so in rough terrain. Although certain recognized physical principles play important roles in determining flow through rough terrain, the infinite range of possibilities for placement of topographic elements, and for their shapes and sizes, makes each locality essentially unique. It is no historical accident that studies of atmospheric behavior with respect to local transport of pollutants in hilly and mountainous terrain have been pursued by means of intensive observational programs *in situ* or by physical modeling in "wind tunnels" (*86–89*).

1. Local Mechanical Perturbations

Requirements of mass continuity and such resulting phenomena as the behavior of vortices combine to produce several basic forms of flow around obstacles in the windstream (*90*). Where surfaces depart abruptly from being "streamlined," continuity produces "separation" along surfaces, and counterflows result because there is no component of flow across the surfaces (*91*). If there are prominent features on the terrain which are abnormally abrupt, such as cliffs or sharp ridgelines, margins of quasi-horizontal separation surfaces tend to be anchored at these "salients." Under such circumstances, counterflows will be reasonably persistent when separation is present. In the absence of salients, however, the intersections of separation surfaces and terrain will tend to wander upslope and down, in response to subtle changes in cross-ridge wind speed, thermal stability of the windstream, and local variations of surface roughness. Downwind of such isolated elements as hills or knobs, quasi-vertical separation surfaces may enclose cavities in which circulation is vigorous but scarcely connected with the general flow. Again, the presence of salients tends to anchor the surfaces, thus producing a less variable local wind field.

2. Local Thermal Perturbations

Differences in heat content between adjacent columns of air produce horizontal differences in pressure, which are translated into airflow in the atmosphere's attempts to balance pressure, and thus mass, at all levels. This fundamental causal chain operates at all scales. The local, or mesoscale, is no exception.

Sunny slopes usually experience up-slope winds in the absence of vigorous regional flow, which if present would facilitate balance and suppress the differential heating which drives the local circulation. Shaded slopes often experience down-slope winds when they are paired with an opposing sunlit slope across a valley. Such a down-slope wind is likely to be caused by continuity of flow within the valley rather than by differential heating as such. Down-slope flows at night, however, appear reliably on slopes of intermediate steepness (92), when clear skies and regional calms permit rapid cooling of the slope surfaces. These circulations are driven by gravity in the presence of differential cooling. These "mountain–valley" wind systems are relatively well understood, both theoretically and observationally, for simple terrain types (50). As the terrain becomes more complex, however, the theory becomes less useful and the observational requirements become nearly prohibitive. Nonetheless, the same principles are at work.

3. Combinations of Perturbations

Table VII and Figure 24 (93) suggest the general features of some of the various combinations of mechanical and thermal perturbations and their resulting flow patterns. These combinations of circumstances are classified according to ridgeline orientation, regional wind direction rela-

Table VII   Generalized Mesoscale Wind Flow Patterns Associated with Different Combinations of Wind Direction and Ridgeline Orientation (93)

Wind direction relative to ridgeline	Time of day	Ridgeline orientation	
		East–West	North–South
Parallel	Day	South-facing slope is heated—single helix	Up-slope flow on both heated slopes—double helix
	Night	Down-slope flow on both slopes—double helix	Down-slope flow on both slopes—double helix
Perpendicular	Day	South-facing slope is heated. North wind—stationary eddy fills valley. South wind—eddy suppressed, flow without separation	Up-slope flow on both heated slopes—stationary eddy one-half of the valley
	Night	Indefinite flow—extreme stagnation in valley bottom	Indefinite flow—extreme stagnation in valley bottom

tive to the ridgelines, and time of day. These, of course, suggest the simplest kinds of topography, but the variety of results possible even under such uncomplicated circumstances reinforces the notion that flow is essentially indeterminate in naturally occurring terrain of even moderate complexity.

The four basic forms of flow depicted in Figure 24 may be crudely classified as: (a) and (b)—helical; (c) and (d)—isolated valley eddy; (e)—quasi-laminar; and (f)—indefinite. Helical flows may be single or double, and isolated eddies may fill all or only part of the valley.

Applicability of these concepts to air pollution depends upon the nature of the problem. For example, preferred wind speeds and directions on the regional scale (prevailing winds), when associated with essentially linear ridgelines, will produce higher probabilities of some types of flow relative to others, thereby implying that some topographic sites are distinctly better than others with respect to the likelihood of effective local pollutant dispersion from sources at those sites. These preferred combinations of wind direction and terrain orientation ought to suggest useful guidelines for source siting while installations are still in the planning stage, or when land use plans are being formulated.

Another application might be rule making about the times for permissible open burning, proscribing those combinations of time and wind which have a high probability of poor removal of effluents from the local area.

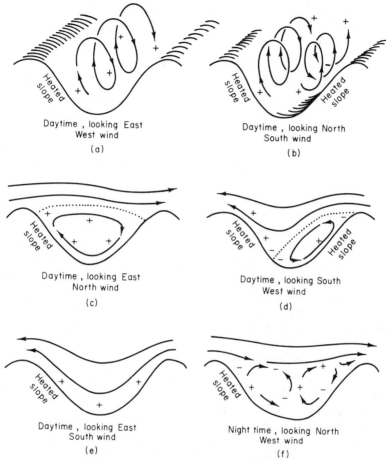

Figure 24. Mesoscale flow patterns described in Table VII. Reinforcement of thermal and mechanical circulations is marked $+$; opposition, $-$ (*93*).

Research on wind flow in rough terrain has not been more than exploratory, with the result that generalization with confidence is not possible. Increased interest in intensive land use planning would seem to indicate the need for appropriately designed research efforts.

The combinations of mechanical and thermal perturbations associated with shorelines have been better studied than those in rough terrain. Windflow across a shoreline will experience abrupt changes of surface roughness as a rule, the land usually being aerodynamically rougher than the open water. However, the land may produce thermal perturbations under some circumstances; the water under others. Observational studies and qualitative theory on these matters are available (94, 95) as guidance in various applications.

E. Effects of Urbanization on Local and Regional Atmospheric Behavior

It has been recognized for many decades that the presence of an urbanized surface produces a variety of changes in the climate of the city (96). These departures from the preurban climate of the place constitute the "urban effect on climate," which is usually estimated by obtaining measures of differences in atmospheric variables between the urban area and some typical, contemporary, nearby rural area. Landsberg reports a program to measure the urban effect itself, by records of meteorological variables in the planned "new town" of Columbia, Maryland (97). This program may well be unique. The urban–rural differences generally reported as constituting the urban effect are only approximations, because they include effects of other factors such as specific local topography, major watercourses, and vegetative mosaics, which predated the urbanization and are still present on the landscape.

Whatever the outcome of the study of disparities between these two versions of urban effects on climate—the changes from preurban to contemporary local climate, and the contemporary urban–rural differences— the disparities will probably be small relative to the true effects of urbanization, which in turn are small when expressed as percentage changes in the averages of climatic elements (54). While averages may be changed only slightly by urbanization, significant changes have been detected for some kinds of weather. Furthermore, changes in some variables "downwind" of the urban area are possibly a good deal greater than changes in those same variables in the city itself. These "regional" effects of urbanization were recognized later than the local effects (96).

In broadest terms, urbanization produces local mechanical and thermal perturbations, much as topographic elements do. Extending this analogy, a linear megalopolis of cities merging along some major urbanized corridor, such as those shown in Table VIII (98), acts thermodynamically as a low mountain range with respect to atmospheric motion. Thus, as with true mountains, there are effects which are local and others which are regional, some effects being traceable largely to mechanical causes and others largely to thermal causes.

1. Urban Effects on Temperature and Airflow

Nearly without exception, cities are warmer than their surroundings, and presumably warmer than the same locations were before the cities

Table VIII   Projected Megalopolises in the United States in A.D. 2000 (98)

Location	Name	Population in A.D. 2000	
		Millions	Percent of national total
Boston to Washington	BOSNYWASH	80	25
Chicago to Pittsburgh	CHIPITTS	40	12
San Francisco to San Diego	SANSAN	20	6

were built and used. Figure 25 suggests the form and cause of this result in a schematic relationship between a typical diurnal temperature record in a rural site and in a nearby urbanized site. There are two thermal effects applied to the rural record to produce the urban record: (a) a thermostatic reduction of the amplitude of the diurnal wave without a change of mean (curve 1 to curve 2), and (b) a general warming which produces the same temperature increment at each hour (curve 2 to curve 3). This rudimentary model produces a relationship between the two curves which is commonly accepted as typical: a smaller diurnal temperature range, and a higher mean temperature in the city, resulting in a temperature difference which is distinctly larger at night than during the day. Typically for large cities, the difference in annual mean temperatures is the order of 1°C, while on calm, clear nights the urban–rural difference between sites with locally extreme records is 10°C.

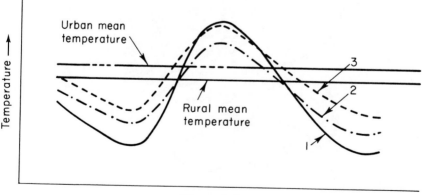

Figure 25. Urban effects on temperature. Comparative diurnal temperature variation in the city (curve 3) and environs (curve 1), as the results of effects due to clouds, pollutants, and artificial heat sources in the city. Curve 2 is a smaller amplitude version of curve 1; see text for further description.

Figures 26 (*99*) and 27 (*100*) depict several aspects of the "urban heat island," a characteristic pattern of warmer temperatures having a margin closely related to the boundaries of the urbanized parts of the landscape. Figure 26 shows two areas of higher temperature associated with higher elevation in addition to the warmer area associated with urbanization. Outside the small city, the drainage of cold air into lower lying areas produces an inversion with hilltops showing relative warmth. The separation of urban effect and topographic effect is not always so clearcut as in this case. Figures 26 and 27 together exhibit the well-documented increase in local temperature difference (the "height" of the island) with increase in city population noted in midlatitude industrialized cities of occidental culture (*101*). Differences have been detected in "cities" no larger than shopping malls (*102*).

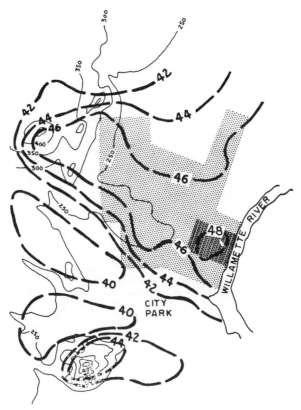

Figure 26. Topographic and isothermal maps of the Corvallis, Oregon area for 2200 PST on the night of 18 April 1966. Elevations are in feet above mean sea level; temperatures are in °F. After Hutcheon, Johnson, Lowry, Black, and Hadley (*99*). (Reprinted with permission from *Bull. Amer. Meteorol. Soc.,* Amer. Meteorol. Soc.)

Figure 27. Distribution of minimum temperature in London, England, 14 May 1959, in °C with °F in brackets (*100*).

In the vertical, the urban heat-island circulation may be toroidal under conditions of light regional wind: inflow at the surface towards the city center, and outflow aloft at a height about 300–500 m. With a gentle regional wind, the heat island suggests a "plume" (*103*) shaped much like that of bonfire smoke, first rising and then flowing relatively un-diffused downwind. In strong regional flows, the temperature differential virtually disappears, and the "plume" is all but eradicated.

The explanation of these general patterns of thermal differentiation is usually assigned to five kinds of causes: (a) increased roughness of the urban surfaces compared with rural; (b) greater conductive flux of heat into and out of the denser surface materials of the city; (c) the reduced evaporative cooling in cities which exposes relatively little open water and vegetation to the atmosphere; (d) so-called "artificial" sources of heat related to activities of the populace (heating, cooling, manufacturing, transportation, etc.); and, (e) the greater turbidity of the urban air, due primarily to air pollutants.

It may be seen that the thermostatic and the warming components suggested in Figure 25 are each net effects of some causes which tend to warm, some which tend to cool, some which tend to be thermostatic, and some which do not. Rudimentary heat balance models bear out these explanations (104, 105). Other rudimentary models of motion also bear out the broad speculations about boundary layer behavior over cities (106, 107). In general, it is agreed that the effects of urbanization and attendant industrialization on local temperature fields and flow fields are reasonably well understood, with the exception of cases in which substantial cloud and precipitation activity is present. The general dearth of observational relative to theoretical information about temperatures and winds above the urban surface, however, may hide a complete understanding of these effects (108–110).

2. Urban Effects on Precipitation

As noted, urban effects on cloud- and precipitation-related processes are not as well understood as effects on temperature and windflow. The idea that urbanization produces effects downwind, on a regional scale, first drew widespread attention when Changnon's study of precipitation records at LaPorte, Indiana and vicinity led to several published responses (111–116).* Whether or nor the reality of anomalies in the precipitation field was established by these papers, the mechanisms for such changes were not. Robinson provides an effective review of the "LaPorte controversy" in Chapter 1 of Vol. II.

Speculation about probable mechanisms for urban-induced changes in the amount and timing of precipitation has centered on (a) thermal enhancement of updrafts, (b) mechanical enhancement of updrafts by roughness of urban surfaces, (c) modifications of particulate aerosol loadings, and (d) modifications of the patterns of moisture injection into the boundary layer above urbanized and industrialized areas.

Two major research efforts to establish both the real and the probable causes of a variety of urban effects, including those on hail and thunderstorms, have followed the LaPorte exchanges. The first is a study of eight American cities, through climatological records from their regions (117, 118), by Huff and Changnon. The second is a multiagency program

---

* The six references each include both an initiating paper and the subsequent comments and replies, in the interest of a more concise reference list

called METROMEX (*96*), or *Metro*politan *M*eteorological *E*xperiment. METROMEX involves seven research agencies, four sources of funds, and three research modes, the most visible of which is the field program in Greater St. Louis, Missouri and Illinois. The field program involves ground-based networks, ground-based remote sensing of the boundary layer (up to about 1500 m), and aircraft operations.

The tentative conclusions being drawn from the programs of METRO-MEX up to the time of this writing, while not definitive, are supportive of the concept of urban effects on regional precipitation. For example, injection of tracer chemicals into cloud systems at St. Louis followed by capture of these tracers in networks of recording rain gauges (Fig. 28) has shown that the time–space components of the processing of the chemicals by cloud systems are consistent with the notion that urban effluents can cause alteration of precipitation patterns at distances 25–50 km downwind of cities (*119*).

For another example, it appears (as suggested by Robinson in Chapter 1 of Vol. II) that giant nuclei in urban effluents must more than make up for the smaller nuclei, which by themselves would suppress rather than enhance the production of precipitation (*120*).

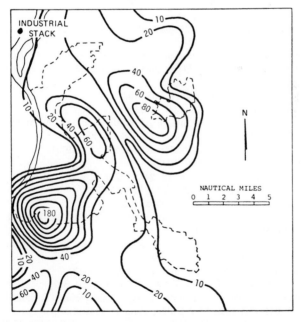

Figure 28. Deposition pattern in rain for lithium tracer released from industrial stack on July 1971. Units are pg Li/cm^2. After Gatz (*119*). (Reprinted with permission from *Bull. Amer. Meteorol. Soc.*, Amer. Meteorol. Soc.)

Whatever the ultimate picture of how urban effluents affect the precipitation climate in and near cities, indications are rapidly accumulating that they do produce such changes. More than that, cities may have a major influence on the times, places, and rates at which pollutants in the boundary layer are injected by deep convection into the upper troposphere, where if relatively inert and resistant to scavenging they can circle the hemisphere and become part of the global atmosphere for relatively long residence times.

REFERENCES

1. R. E. Huschke, ed., "Glossary of Meteorology." Amer. Meteorol. Soc., Boston, Massachusetts, 1959.
2. C. Eckhart, *J. Mar. Res.* **7**, 265 (1948).
3. J. G. Edinger, *Bull. Amer. Meteorol. Soc.* **36**, 211 (1955).
4. A. C. Stern, H. C. Wohlers, R. W. Boubel, and W. P. Lowry, "Fundamentals of Air Pollution," Chapter 17. Academic Press, New York, New York, 1973.
5. T. Vonder Haar and V. Suomi, *J. Atmos. Sci.* **28**, 305 (1971).
6. R. Fleagle and J. Businger, "Introduction to Physical Meteorology," p. 153. Academic Press, New York, New York, 1963.
7. R. Lee, *J. Appl. Meteorol.* **12**, 556 (1973).
8. W. D. Sellers, "Physical Climatology." Univ. of Chicago Press, Chicago, Illinois, 1965.
9. M. I. Budyko, "The Heat Balance of the Earth's Surface." U.S. Department of Commerce, Washington, D.C., 1958. (Translated by N. Stepanova.)
10. W. P. Lowry, *Ecology* **53**, 908 (1972).
11. M. Atwater, *J. Appl. Meteorol.* **10**, 205 (1971).
12. M. Atwater, *J. Atmos. Sci.* **28**, 1367 (1971).
13. R. W. Bergstrom and R. Viskanta, *J. Appl. Meteorol.* **12**, 901 and 913 (1973).
14. S. Rasool and S. Schneider, *Science* **173**, 138 (1971).
15. G. W. Paltridge and C. Platt, *J. Atmos. Sci.* **30**, 734 (1973).
16. P. Stone, *J. Atmos. Sci.* **30**, 521 (1973).
17. W. D. Sellers, *J. Appl. Meteorol.* **12**, 241 (1973).
18. W. Washington, *J. Appl. Meteorol.* **11**, 768, (1972).
19. R. C. Wanta, *J. Geophys. Res.* **74**, 5536 (1969).
20. R. J. List, ed., "Smithsonian Meteorological Tables," 6th rev. ed. Smithsonian Inst., Washington, D.C., 1949.
21. D. H. McIntosh, "Meteorological Glossary," 5th ed. (1st American ed.). Chem. Publ. Co., New York, New York, 1972.
22. J. Z. Holland, *Bull. Amer. Meteorol. Soc.* **33**, 1 (1952).
23. J. Z. Holland and R. F. Myers, "A Meteorological Survey of the Oak Ridge Area," Rep. ORO-99. U.S. Atomic Energy Commission, Technical Information Service, Weather Bureau, Oak Ridge, Tennessee, 1953.
24. R. M. Goody and J. C. G. Walker, "Atmospheres." Prentice-Hall, Englewood Cliffs, New Jersey, 1972.

25. L. T. Matveev, "Osnovy obschey meteorologii Fizika atmosfery" (Fundamentals of General Meteorology Physics of the Atmosphere). Gidrometeoizdat, Leningrad, USSR, 1965. (Translation by I. Schechtman, TT67-51380, available from Clearinghouse for Federal and Scientific Information, Springfield, Virginia, 1967.)

26. M. I. Budyko, "Atmosfernaya uglekislota i klimat" (Atmospheric Carbon Dioxide and Climate). Gidrometeoizdat, Leningrad, USSR, 1973.

27. E. L. Deacon, in "World Survey of Climatology" (H. Flohn, ed.), Vol. 2. Elsevier, Amsterdam, Netherlands, 1969.

28. S. B. Idso and P. C. Kangieser, J. Geophys. Res. 75, 2179 (1970).

29. H. Wexler, Mon. Weather Rev. 64, 122 (1936).

30. H. R. Byers, "General Meteorology," 3rd ed. McGraw-Hill, New York, New York, 1959.

31. W. R. Rouse, D. Noad, and J. McCutcheon, J. Appl. Meteorol. 12, 798 (1973).

32. H. Riehl, "Tropical Meteorology." McGraw-Hill, New York, New York, 1954.

33. D. H. McIntosh and A. S. Thom, "Essentials of Meteorology." Wykeham Publications, London, England, 1969.

34. R. Geiger, "The Climate Near the Ground." Harvard Univ. Press, Cambridge, Massachusetts, 1965. (Translation of the 34th German ed., 1961.)

35. H. H. Schrenk, H. Heimann, G. D. Clayton, W. M. Gafafer, and H. Wexler, U.S., Pub. Health Serv., Pub. Health Bull. No. 306 (1949).

36. A. A. Miller, "Climatology," 9th ed. Methuen, London, England, 1961.

37. P. A. Vorontsov, "Turbulentnost' i vertikal'nyye toki v pogranichom sloye atmosfery" (Turbulence and Vertical Currents in the Boundary Layer of the Atmosphere). Gidrometoizdat, Leningrad, USSR, 1966. (Translation FTD-HT-23-1332-68, available from Clearinghouse for Federal and Scientific Information, Springfield, Virginia, 1967.)

38. R. C. Wanta, W. B. Moreland, and H. E. Heggestad, Mon. Weather Rev. 89, 289 (1961).

39. P. W. Summers, J. Air Pollut. Contr. Ass. 16, 432 (1966).

40. R. C. Wanta, in "Air Pollution Manual, Part 1—Evaluation," 2nd ed., p. 218. Amer. Ind. Hyg. Ass., Akron, Ohio, 1972.

41. G. C. Holzworth, Mon. Weather Rev. 92, 235 (1964).

42. H. Lettau, Arch. Meteorol., Geophys. Bioklimatol., Ser. A 7, 133 (1954).

43. S. L. Valley, ed., "Handbook of Geophysics and Space Environments." McGraw-Hill, New York, New York, 1965.

44. D. Fultz, Advan. Geophys. 7, 1 (1961).

45. H. H. Lamb, "Climate: Present, Past and Future," Vol. 1. (Table IV abridged from App. 1.12, pp. 487–490.) Methuen, London, England, 1972.

46. C. G. Rossby, in "Climate and Man" (Yearbook of Agriculture), p. 599. U.S. Dept. of Agriculture, Washington, D.C., 1941.

47. E. Reiter, in "World Survey of Climatology" (D. F. Rex, ed.), Vol. 4. Elsevier, Amsterdam, Netherlands, 1969.

48. M. A. Tilley and G. A. McBean, Atmos. Environ. 7, 793 (1973).

49. I. Van der Hoven, J. Meteorol. 14, 160 (1957).

50. F. Defant, in "Compendium of Meteorology" (T. F. Malone, ed.), p. 655. Amer. Meteorol. Soc., Boston, Massachusetts, 1951.

51. R. S. Scorer, "Natural Aerodynamics." Pergamon, Oxford, England, 1958.

52. K. C. Crawford and H. R. Hudson, "Behavior of Winds in the Lowest 1500

Feet in Central Oklahoma June 1966–May 1967," ESSA Tech. Mem. ERLTM-NSSL 48, p. 20. U.S. Dept. of Commerce, National Severe Storms Laboratory, Norman, Oklahoma, 1970.

53. I. A. Singer and M. E. Smith, *Meteorol. Monogr.* 1, 50 (1951).
54. H. E. Landsberg, *Science* 170, 1265–1274 (1970).
55. S. S. Butcher and R. J. Charlson, "Introduction to Air Chemistry," Chapter 9. Academic Press, New York, New York, 1972.
56. G. M. Hidy, *in* "Chemistry of the Lower Atmosphere" (S. Rasool, ed.). Plenum, New York, New York, 1973.
57. R. Gunn and B. B. Phillips, *J. Meteorol.* 14, 272 (1957).
58. H. L. Green and W. R. Lane, "Particulate Clouds, Smokes and Mists." Spon, London, England, 1964.
59. R. R. Braham, *Bull. Amer. Meteorol. Soc.* 49, 343–353 (1969).
60. A. C. Stern, H. C. Wohlers, R. W. Boubel, and W. P. Lowry, "Fundamentals of Air Pollution," Chapter 19. Academic Press, New York, New York, 1973.
61. H. Byers, "Elements of Cloud Physics." Univ. of Chicago Press, Chicago, Illinois, 1965.
62. M. Neiburger, "Artificial Modification of Clouds and Precipitation," Tech. Note No. 105. World Meteorol. Organ., Geneva, Switzerland, 1970.
63. R. G. Fleagle, ed., "Weather Modification: Science and Public Policy." Univ. of Washington Press, Seattle, Washington (n.d.).
64. R. Englemann, *in* "Meteorology and Atomic Energy" (D. Slade, ed.), pp. 208–221. U.S. At. Energy Comm., Washington, D.C., 1968.
65. J. Hales, M. Wolf, and M. Dana, *Amer. Inst. Chem. Eng., J.* 19, 292–297 (1973).
66. J. M. Hales, *Atmos. Environ.* 6, 635–659 (1972).
67. W. Slinn and J. Hales, *J. Atmos. Sci.* 28, 1465 (1971).
68. R. C. Wanta, *Atmos. Environ.* 8, 687 (1974).
69. R. C. Wanta, *Bull. Amer. Meteorol. Soc.* 37, 186 (1956).
70. R. I. Larsen, C. E. Zimmer, D. A. Lynn, and K. G. Blemel, *J. Air Pollut. Contr. Ass.* 17, 85 (1967).
71. R. I. Larsen, *J. Air Pollut. Contr. Ass.* 19, 24 (1969).
72. A. C. Stern, "Multiple-Source Urban Diffusion Models" (A. C. Stern, ed.), Air Pollut. Contr. Office Publ. No. AP-086. United States Environmental Protection Agency, Research Triangle Park, North Carolina, 1970.
73. A. K. Blackadar, *Bull. Amer. Meteorol. Soc.* 38, 283 (1957).
74. C. S. Durst, *Quart. J. Roy. Meteorol. Soc.* 59, 131 (1933).
75. T. W. Kleinsasser and R. C. Wanta, *Proc. 49th Annu. Meet. Air Pollut. Contr. Ass.* p. 7–1 (1956); also *AMA Arch. Ind. Health* 14, 307 (1956).
76. D. A. Haugen ed., "Workshop on Micrometeorology." Amer. Meteorol. Soc., Boston, Massachusetts, 1973.
77. A. C. Stern, H. C. Wohlers, R. W. Boubel, and W. P. Lowry, "Fundamentals of Air Pollution," pp. 244–250. Academic Press, New York, New York, 1973.
78. A. C. Stern, H. C. Wohlers, R. W. Boubel, and W. P. Lowry, "Fundamentals of Air Pollution," p. 233. Academic Press, New York, New York, 1973.
79. R. A. McCormick and G. C. Holzworth, Chapter 12, this volume.
80. H. Byers, "General Meteorology," pp. 389–405. McGraw-Hill, New York, New York, 1944.
81. V. Oliver and M. Oliver, *in* "Compendium of Meteorology" (T. F. Malone, ed.), p. 725. Amer. Meteorol. Soc., Boston, Massachusetts, 1951.
82. O. Cramer and R. Lynott, *J. Appl. Meteorol.* 9, 740–759 (1970).

83. R. Bleck, *J. Appl. Meteorol.* **12**, 737–752 (1973).
84. R. Duquet, E. Danielsen, and N. Phares, *J. Appl. Meteorol.* **5**, 233 (1966).
85. A. C. Stern, H. C. Wohlers, R. W. Boubel, and W. P. Lowry, "Fundamentals of Air Pollution," p. 249. Academic Press, New York, New York, 1973.
86. B. Davidson, *J. Appl. Meteorol.* **2**, 463 (1963).
87. F. Pooler, *J. Appl. Meteorol.* **2**, 446 (1963).
88. J. Marwitz, *J. Appl. Meteorol.* **13**, 450 (1974).
89. M. Orgill, J. Cermak, and L. Grant, Technical Rep. CER70-71MMO-JEC-LOG40. Fluid Dynamics and Diffusion Laboratory, Colorado State University, Ft. Collins, Colorado, 1971.
90. J. Halitsky, *in* "Meteorology and Atomic Energy" (D. Slade, ed.), pp. 221–255. U.S. At. Energy Comm., Washington, D.C., 1968.
91. R. Scorer, "Air Pollution," pp. 107–123. Pergamon, Oxford, England, 1968.
92. R. Fleagle, *J. Meteorol.* **7**, 227–232 (1950).
93. A. C. Stern, H. C. Wohlers, R. W. Boubel, and W. P. Lowry, "Fundamentals of Air Pollution," pp. 290–291. Academic Press, New York, New York, 1973.
94. E. Hewson and L. Olsson, *J. Air Pollut. Contr. Ass.* **17**, 757–761 (1967).
95. W. Lyons and H. Cole, *J. Appl. Meteorol.* **12**, 494 (1973).
96. W. Lowry, *Bull. Amer. Meteorol. Soc.* **55**, 87–88 (1974).
97. H. E. Landsberg, *in* "Weather and Climate Modification" (W. H. Hess, ed.), pp. 726–763. Wiley, New York, New York, 1974.
98. H. Kahn and A. J. Wiener, *Daedalus (Boston)* **96**, 705 (1967).
99. R. Hutcheon, R. H. Johnson, W. P. Lowry, C. H. Black, and D. Hadley, *Bull. Amer. Meteorol. Soc.* **48**, 7 (1967).
100. T. Chandler, "The Climate of London." Hutchinson, London, England, 1965.
101. T. R. Oke, *in* "Preprints of Conference on Urban Environment," pp. 144–146. Amer. Meteorol. Soc., Boston, Massachusetts, 1972.
102. J. R. Norwine, *Bull. Amer. Meteorol. Soc.* **54**, 637–641 (1973).
103. A. C. Stern, H. C. Wohlers, R. W. Boubel, and W. P. Lowry, "Fundamentals of Air Pollution," p. 294. Academic Press, New York, New York, 1973.
104. D. O. Myrup, *J. Appl. Meteorol. Soc.* **8**, 908–918 (1969).
105. E. Miller and W. Lowry, *in* "Preprints of Conference on Urban Environment," pp. 77–82. Amer. Meteorol. Soc., Boston, Massachusetts, 1972.
106. D. Olfe and H. Lee, *J Atmos. Sci.* **28**, 1374–1388 (1971).
107. R. Bornstein, *in* "Preprints of Conference on Urban Environment," pp. 89–94. Amer. Meteorol. Soc., Boston, Massachusetts, 1972.
108. J. Angell, D. Pack, O. Dickson, and W. Hoecker, *J. Appl. Meteorol.* **10**, 194–204 (1971).
109. B. Ackerman, *J. Air Pollut. Contr. Ass.* **24**, 232–236 (1974).
110. R. Bornstein, A. Lorenzen, and D. Johnson, *in* "Preprints of Conference on Urban Environment," pp. 28–33. Amer. Meteorol. Soc., Boston, Massachusetts, 1972.
111. S. Changnon, *Bull. Amer. Meteorol. Soc.* **40**, 4–11 (1968). Comment: B. Holtzman and H. Thom, *ibid.* **51**, 335–337 (1970). Reply: S. Changnon, *ibid.* p. 337–342.
112. T. L. Ogden, *J. Appl. Meteorol.* **8**, 585–591 (1969); Comment: S. Changnon, *ibid.* **10**, 165–168 (1971); Reply: T. Ogden, *ibid.* p. 168.
113. B. Holtzman, *Science* **171**, 847 (1971); Reply: S. Changnon, *ibid,* **172**, 987–988 (1972); also J. Hidore, *ibid.* p. 988.
114. W. Seidel, *Bull. Amer. Meteorol. Soc.* **52**, 105 (1971); Reply: S. Changnon, *ibid.* p. 105.

115. J. Hidore. *Bull. Amer. Meteorol. Soc.* **52**, 99–103 (1971); Comment: B. Holtz-mann, *ibid.* pp. 572–573 (1971); Reply: J. Hidore, *ibid.* pp. 573–574.
116. W. Ashby and H. Fritts, *Bull. Amer. Meteorol. Soc.* **53**, 246–251 (1972); Comment: F. Charton and J. Harman, *ibid.* **54**, 26 (1973); Reply: H. Fritts and W. Ashby, *ibid.* pp. 26–27.
117. F. Huff and S. Changnon, *J. Appl. Meteorol.* **11**, 823–842 (1972).
118. F. Huff and S. Changnon, *Bull. Amer. Meteorol. Soc.* **54**, 1220–1232 (1973).
119. D. Gatz, *Bull. Amer. Meteorol. Soc.* **55**, 92–93 (1974).
120. R. Braham, *Bull. Amer. Meteorol. Soc.* **55**, 100–106 (1974).

9

Transport and Diffusion of Stack Effluents

Gordon H. Strom

Nomenclature

The following symbols are used in various equations of this chapter.
Those omitted from this list are defined where they are employed.

A	Reference area for building wake diffusion [Eq. (104)]
A_s	Stack outlet area
A_w	Building wake cross-sectional area [Eq. (105)]
c	Effluent wake diffusion parameter [Eq. (106)]
c_p	Specific heat of air at constant pressure
C_y, C_z	Sutton diffusion parameters [Eqs. (81) and (82)]
C_1	Plume rise equation constant [Eq. (22)]
C_2	Plume rise equation constant [Eq. (47)]
d_s	Internal diameter, stack outlet
F_b, F_{ba}	Buoyancy flux parameters [Eqs. (19), (20), and (21)]
F_m	Momentum flux parameter [Eq. (62)]
Fr	Modified Froude number [Eq. (125)]
Fr'	Froude number [Eq. (124)]
g	Acceleration due to gravity
h_e	Effective plume height, center line above ground level [Eq. (14), Figs. 6 and 7]
h_s	Height of stack outlet above ground level (Figs. 6 and 7)
h_i	Height of base of inversion or stable atmospheric layer
Δh	Plume rise, height of plume center line above stack outlet (Figs. 6 and 7)
k	von Kármán constant [Eq. (2)]
K_w	Effluent concentration coefficient for building wakes [Eq. (104)]
l	Reference or characteristic length
l_b	Buoyancy scaling length [Eq. (36)]
L	Monin–Obukhov stability length [Eq. (6)]
n, n_y, n_z	Sutton turbulence indices [Eqs. (80), (81), and (82)]
p	Velocity profile power law exponent [Eq. (13)]
p	Concentration ratio for plume edge definition [Eqs. (75) and (76)]
Q	Stack effluent emission rate for selected property [property $\times$ time^{-1}] [Eq. (66)]
Q_h	Stack effluent heat emission rate relative to ambient air [heat $\times$ time^{-1}] [Eq. (16)]
Q_m	Stack effluent mass emission rate [mass $\times$ time^{-1}] [Eq. (17)]
r	Power law exponent, concentration ratio vs sampling time ratio [Eq. (114)]
r_s	Internal radius, stack outlet
Re	Reynolds number [Eq. (130)]
Rf	Flux Richardson number [Eq. (8)]
Ri	Gradient Richardson number [Eq. (7)]
Ri'	Bulk Richardson number [Eq. (11)]
Ri'', Ri'''	Variations of gradient and bulk Richardson numbers [Eqs. (85) and (131)]
R_v	Stack effluent velocity ratio [Eq. (122)]
s	Atmospheric stability parameter [Eq. (48)]
t	Time
T	Absolute temperature of ambient atmosphere at given elevation
T_s	Absolute temperature of stack effluent at stack outlet
u	Time mean velocity at given elevation or time and spatial mean as defined for an application
u_w	Velocity for a building wake [Eq. (105)]
u_*	Friction velocity [Eq. (3)]
x, y, z	Rectilinear coordinates centered at stack base (Fig. 7)

y_e, z_e	Coordinates of plume outline [Eqs. (77) and (78)]
z_0	Surface roughness length [Eq. (2)]
Γ	Dry adiabatic atmospheric temperature gradient with elevation (lapse rate)
θ	Absolute potential temperature of ambient atmosphere at given elevation
θ	Wind direction angle
$\partial\theta/\partial z$	Potential temperature gradient [Eq. (1)]
μ	Fluid dynamic viscosity
ρ	Mass density of ambient atmosphere [mass $\times$ vol^{-1}]
ρ_s	Mass density of effluent at stack outlet [mass $\times$ vol^{-1}]
σ_y, σ_z	Standard deviations of plume concentration distribution in y and z directions [length] [Eq. (66)]
σ_θ	Standard deviation of wind direction angle fluctuations
χ	Concentration of plume effluent at given location [property $\times$ vol^{-1}] [Eq. (66)] (Same property as used for Q)
ω	Deposition of particulate effluent on horizontal surface [property $\times$ area^{-1} $\times$ time^{-1}] [Eq. (118)]

I. Introduction

A. Single Stack and Multiple Source Pollution

Air pollution caused by effluents from a single or small group of stacks is a local problem. Effluent concentrations of concern generally occur at distances ranging from the immediate vicinity of the stack to those on the order of several kilometers. Maximum ground-level concentrations tend to occur within this range. At greater downwind distances plumes formed by stack effluents become so diluted by diffusion in the ambient atmosphere that concentrations may become negligibly small.

Combined effect of a large number of stack effluents spread over a large area (usually urban) produces a different type of air pollution problem which may have its principal area of concern at much greater distances. While the local pollution from any one stack may not be a problem, the combined effect of large numbers produces the serious urban pollution. The latter is the multiple source urban diffusion problem treated in Chapter 10, this volume.

Variables which affect the single and miltiple-source problems have differing effects on the resulting pollutant concentrations. In some cases they may be diametrically opposite. Increase in wind speed tends to increase ground-level concentration from a single stack in the low-speed range, while any increase in speed tends to reduce concentration from multiple sources. Initial stack effluent characteristics other than emission rate have little effect on multiple source problems but may be of major importance for single stacks.

B. Temperature Profiles

Transport and diffusion of stack effluents occur within the atmospheric boundary layer where most of the local effects of earth–air interface occur. The complex energy transfer processes determine the vertical distribution of temperature, velocity, and turbulence. Temperature may increase or decrease with elevation depending on the direction of surface heat transfer and advection of air masses of differing thermal characteristics. Many of the striking features of plume diffusion are closely related to vertical temperature gradients.

Temperature change with elevation tends to be linear in a substantial portion of the lower atmosphere. Departures from linearity do occur, particularly near the surface and in regions influenced by marked variations in surface configuration as with buildings and topographical features. Linear variation is assumed over limited ranges of elevation significant to plume diffusion. Sometimes the nonlinear aspect is ignored and temperature difference is given for a specified difference in elevation.

Production of convective (thermally induced) turbulence is related to vertical temperature gradient. This is demonstrated with an elementary consideration of fluid forces which leads to introduction of potential temperature θ as a variable. A small volume or parcel of air undergoes changes in elevation due to dynamics of motion. For short distances it tends to act as if thermally insulated. It will, therefore, undergo adiabatic temperature change. If the ambient atmosphere has a nonadiabatic gradient, the parcel will experience a force due to buoyancy (positive or negative) which tends to increase or reduce vertical motion. If the ambient gradient is less negative than adiabatic or is positive (inversion), motion is suppressed, and the parcel tends to return to its initial equilibrium position. This condition is termed stable. It results in reduction of mechanical turbulence which would otherwise occur in an adiabatic atmosphere. The unstable condition occurs when the ambient gradient is more negative than adiabatic. Initial motion of a parcel is amplified and it diverges from equilibrium. This adds turbulence of the convective type. Adiabatic gradients are also termed neutral. For purpose of classification a small range on either side of adiabatic is included in the neutral category. Expressed as the derivative with elevation, $\partial T/\partial z$, the adiabatic gradient in dry air has the numerical value of $-0.986°C/100$ m, often approximated at $-1°C/100$ m. The adiabatic gradient or lapse rate is generally designated by Γ and given as a positive value.

Owing to the importance of nonneutral gradients in relation to adiabatic, a fictitious temperature is used, namely, potential temperature θ. It is the temperature a parcel of air would have if brought adiabatically to a standard sea level pressure of 1000 mbar. Potential temperature gradient $\partial\theta/\partial z$ is related to actual gradient $\partial T/\partial z$ by the following equation:

$$\frac{\partial\theta}{\partial z} = \frac{\partial T}{\partial z} + \Gamma \tag{1}$$

It is evident that the neutral gradient is given by $\partial\theta/\partial z = 0$, stable with $\partial\theta/\partial z$ positive, and unstable with $\partial\theta/\partial z$ negative.

C. Plume Types

The wide range of shapes and time variations of plumes formed by visible stack effluents is evident to the most casual observer. Analytical techniques seek to predict concentrations of effluents in the variable plume, visible or not, in the entire downwind field of action. Degree of accuracy varies widely from one of relatively high degree for simple uniform conditions to crude estimates for highly nonuniform conditions. What may appear to the casual observer as an irregular plume may still be susceptible to a useful analytical treatment.

The widely varying plume characteristics are due in large part to turbulence of the ambient atmosphere. Qualitative aspects categorized by plume types are often described by geometric features of visible plume outline shown in Figure 1. These occur in the absence of marked local effects of surface configuration such as buildings and terrain. Vertical variation of atmospheric temperature plays an important role in identifying the type of plume that is likely to occur. Representative temperature gradients are included in Figure 1 for the corresponding plume types.

In the absence of thermal effects, the temperature gradient tends to be adiabatic (neutral), and the coning-type plume is formed. Its cross section is elliptical with the horizontal axis generally larger than the vertical. Gradients less than adiabatic (more negative) are thermally unstable. They are indicative of greater turbulence and plume diffusion. The turbulence elements or eddies may be larger than the plume cross section and, as a result, form a looping plume. Temperature gradients greater than adiabatic (more positive) indicate a suppression of turbulence, particularly in the vertical direction. The result is a fanning plume whose horizontal axis may be much larger than the vertical.

The three remaining plume types are associated with nonuniform temperature gradients having discontinuities which result from layers of differing stability. The fumigating and trapping plumes may be the most critical of all types for ground-level pollutant concentrations. A thermally stable layer above suppresses upward diffusion while it is enhanced down-

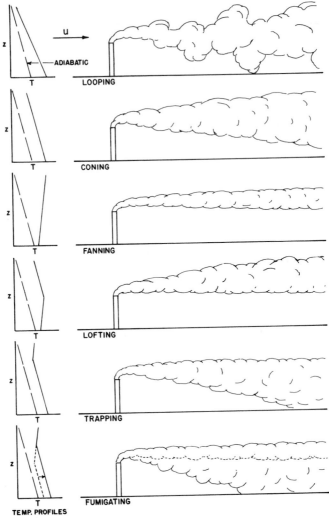

Figure 1. Plume types for various thermal stabilities. The top three occur with uniformly varying temperature with elevation: unstable, adiabatic (neutral), and stable. The bottom three are caused by discontinuities in stability of atmospheric layers.

ward by the neutral or unstable layer at and below the plume. This restricted layer is sometimes termed a limited mixing layer. The fumigating plume is the result of transition of a surface-based stable layer to a developing neutral or unstable layer below the plume. As the nonstable layer envelopes the plume, high ground-level concentrations may be produced for a fraction of an hour or more. It is caused by surface solar heating during morning hours following nighttime stable conditions. It may also be caused by advection of air from cool to warm surfaces as from water to land or rural to urban regions. The trapping plume results from the longer period occurrence of a stable layer at elevations above the plume. The neutral or unstable layer enveloping the plume down to ground level may produce higher ground-level concentrations for hours. The lofting plume is the most favorable from an air pollution viewpoint. The stable layer below plume elevations suppresses downward diffusion.

II. Atmospheric Characteristics Significant to Transport and Diffusion of Stack Effluents

A. Atmospheric Boundary Layer over Flat and Level Regions

The various motions of the atmospheric boundary layer determine plume transport and diffusion. Temperature profiles discussed above are the result of the action of various energy processes involving these motions. Air motion at a given location is described in terms of the time mean value and the turbulent component. The terms wind velocity and wind speed usually refer to the time mean value. Wind velocity generally increases with elevation. Turbulence, the principal diffusing mechanism for problems considered in this chapter, is usually more intense at lower elevations; but the scale of turbulence tends to increase with elevation. Energy for maintaining boundary layer motion is supplied principally by the overlying atmosphere. Large-scale atmospheric motions are discussed in Chapter 8, this volume.

The height of boundary layer is variable with various surface and atmospheric variables including wind speed. It is difficult to obtain precise determination owing to widely varying fluid forces and to the asymptotic nature of the mergence of boundary layer variables into those of the overlying atmosphere. The entire layer, often termed the planetary boundary layer, has heights varying up to 1000 m (1, 2). Davenport (3) gives values ranging from 274 to 518 m for various surface configurations under neutral conditions.

Complex interaction of fluid forces in the boundary layer makes it impossible to accurately represent velocity (time mean) profiles with simple analytical expressions. This subject is covered in considerable detail in the work of Plate (2) and in many texts (1, 4–6). For purposes of stack plume diffusion studies the simple expressions given below are generally adequate. Turbulence is even more difficult for obtaining simple formulation and also for experimental determination. Most techniques for diffusion calculations rely on indirect specification of turbulence through velocity and temperature profiles.

Of the various analytical expressions for velocity profiles the most frequently used are the logarithmic, log-linear, and power law. The first two have foundations in physical phenomena while the last is empirical. For neutral stability the log-linear equation reduces to the logarithmic form.

1. Logarithmic Velocity Profile

Derivation of the logarithmic profile equation is based on the assumption of constant shear stress with elevation. This is substantially true for only the surface layer which has a height of the order of tens of meters. The equation is, however, applied to a considerable portion of the atmospheric boundary layer. Neutral stability is also required, but non-neutral conditions are handled with the log-linear form given below.

One form of the equation is as follows:

$$u = (u_*/k) \ln(z/z_0) \tag{2}$$

k is the von Kármán constant having an experimentally determined value of 0.4, z_0 is roughness length discussed below, u_* is friction velocity which is related to surface shear stress τ_0 by the following equation:

$$u_* = \sqrt{\tau_0/\rho} \tag{3}$$

Direct experimental determination of u_* is difficult. It is usually found from velocity profile data.

Roughness length z_0 is related to surface roughness. It depends on size, distribution, and geometry of roughness elements. It is obtained by extrapolation of velocity profile data for neutral conditions by Equation (2). A roughness element can be anything from minute irregularities in naturally smooth surfaces, such as ice, to buildings in an urban complex. Table I summarizes z_0 values for various surface configurations (1, 3, 4, 7–10). Values for grasses decrease with increasing wind speed because they bend over in the wind.

Based on extensive experimental studies, Lettau (11) developed the following equation for estimating z_0 values in terms of dimensions and distribution of roughness elements.

$$z_0 = 0.5h^*s/S \tag{4}$$

h^* is average height of elements; s is cross-sectional (silhouette) area on the vertical plane normal to wind direction for an average element;

Table I   **Parameters for Logarithmic Velocity Profile Equation**

Surface configuration	z_0 (cm)	u_*/u	u (m/sec)	z (m)	Ref.
Very smooth–mud flats and ice	0.0|01	0.08	5	2	(1)
	0.001	0.03		2	(4)
Even snow cover of great depth	0.05				(7)
Snow surface–natural prairie	0.10	0.053		2	(4)
Desert (Pakistan)	0.03	0.045		2	(4)
Semidesert	0.3				(7)
Bare solid ground	1.0				(7)
Ploughed land	2.0				(7)
Flat open country	3.0	0.071		10	(3)
Mown grass, h = 1.4–4.5 cm	0.2–2.4	0.06–0.09	2–8	2	(4)
Various long grasses—thin and thick, h = 50–70 cm	3.7–9.0	0.1–0.32	1.5–6.2	2	(1, 4)
Woodland forest	30	0.12		10	(3)
Town	100				(7)
City	200				(7)
Urban area	300	0.22		10	(3)
City (Liverpool)	123				(8)
City (Philadelphia—varying terrain)	22–310				(9)
City (central Copenhagen)	750				(10)

S is the specific area, i.e., horizontal area per element (total horizontal area/number of elements). His tabulated estimates of z_0 range from 0.025 cm for sand grains to 12.5 and 125 cm for city houses in loose and dense array. Corresponding values of the ratio h^*/z_0 are 4, 40, and 4, respectively. House height h^* for the latter two is 5 m. In his study of urban diffusion, Lettau (12) gives typical calculated values of z_0 ranging from 5 to 1000 cm for buildings in urban settings ranging from low- ($h^* = 4$ m) to high-rise ($h^* = 100$ m). Corresponding values of h^*/z_0 are 80 and 10, respectively. For houses in urban areas, Jensen and Franck (10) give h^*/z_0 as 2 and 3 for surface area densities (fraction of area occupied by houses) of 0.25 and 0.14, respectively. Lettau's equation is not in terms of surface area densities, but from the other dimensions it appears that the area densities of Lettau's examples are somewhat lower than those of Jensen and Franck.

Other forms of the profile equation [Eq. (2)] add a constant to the numerator of z/z_0 to form $(z + z_0)/(z + d)/z_0$. These change the reference elevation at which the velocity becomes zero. In Equation (2), values of z smaller than z_0 give negative velocities which are meaningless. The constant d is treated as zero plane displacement to account for difference in elevation between the surface on which the roughness elements are resting and that at which the velocity is effectively zero as represented by the logarithmic profile. This is discussed by Lettau (11). The profile equation should not be applied at small values of z comparable with z_0 since local irregular flow around elements will not be accounted for. In applications to plume diffusion, z values are not likely to reach small values where above considerations become a problem.

2. Log-Linear Velocity Profile

The simplest modification of the logarithmic equation for treating nonneutral conditions is the addition of a term linear with elevation as follows:

$$u = \frac{u_*}{k}\left(\ln\frac{z}{z_0} + \alpha\frac{z}{L}\right) \tag{5}$$

Accuracy of the above equation decreases with increasing departure from neutral conditions. The Monin–Obukhov stability length L is defined by the following equation:

$$L = -u_*{}^3 T\rho c_{\mathrm{p}}/kgq_{\mathrm{h}} \tag{6}$$

q_{h} is heat transfer at the surface, positive upward. Unfortunately, heat transfer data are difficult to obtain and may not be readily available for routine plume diffusion analyses. The effort required for evaluation of parameters in Equation (5) makes it less attractive than the power law equation for many applications. The needed data may be available in an urban pollution study.

The numerical value of α in Equation (5) has been the subject of much research and controversy. A value of 6.0 is suggested in Slade (13).

Richardson numbers, Ri and Rf, are used as stability parameters in analyses of boundary layer and diffusion data. The following gradient form depends on velocity and temperature gradients:

$$\mathrm{Ri} = \frac{(g/T)(\partial\theta/\partial z)}{(\partial u/\partial z)^2} \tag{7}$$

Flux form of Richardson number, Rf, is given as follows:

$$\mathrm{Rf} = -\frac{gq_{\mathrm{h}}}{c_{\mathrm{p}} T\tau_0(\partial u/\partial z)} \tag{8}$$

It is related to the gradient form as follows:

$$\mathrm{Rf} = (K_{\mathrm{h}}/K_{\mathrm{m}})\,\mathrm{Ri} \tag{9}$$

K_h and K_m are diffusivities of heat and momentum, respectively. For small departures from neutral conditions it is approximated with the following equation [see Munn (6)]:

$$Rf = z/L \tag{10}$$

In a simplified version of the gradient Richardson number Ri' termed bulk Richardson number, the velocity gradient $\partial u/\partial z$ is replaced with $u(z_b)/z_b$. The temperature gradient $\partial\theta/\partial z$ is replaced with $\Delta\theta/z_b$ where $\Delta\theta$ is a potential temperature difference between surface and elevation z_b:

$$Ri' = gz_b\,\Delta\theta/Tu_b^2 \tag{11}$$

Stability ratio SR is another simplified variation of Ri. The following form is given by Munn (6), except for omission below of a constant (10^5) included in Munn's original equation.

$$SR = [T(z_b) - T(z_a)]/u^2 \tag{12}$$

Temperatures are at elevations z_a and z_b ($z_b > z_a$). Velocity u is an average taken as that value obtained at an elevation midway between z_a and z_b on logarithmic scale.

3. Power Law Velocity Profile

The power law velocity profile equation given below is used in many plume diffusion studies. While it is empirical in formulation, numerous experimental studies show it to be effective in covering a wide range of conditions when the parameters are properly evaluated.

$$u = u_1\,(z/z_1)^p \tag{13}$$

where u_1 and z_1 are reference velocity and elevation, respectively, selected for the particular application or from available data.

Determination of values for exponent p has been the subject of much research. It is dependent on surface configuration and stability. A given site will show a range of values due to variation in these two variables. Change in wind direction may influence its value because of varying upwind surface configuration. The value of $\frac{1}{7}$ is often quoted for neutral stability and flat surfaces of low roughness. This value is obtained in laboratory boundary layer and pipe flow experiments. Exponent p generally increases with stability and surface roughness.

The exponent has also been found to show dependence on elevation range over which it is evaluated. This and above factors were studied extensively by De Marrais (14) with field studies at the Brookhaven National Laboratory and analysis of data from other sites. He also found that averaging time had little effect, which is in contrast with its effect on plume concentrations (see Section III,D).

Exponent values from several sources are given in Table II for a range of conditions. Sutton's values (15) obtained at an earlier time are for relatively smooth surface configurations. Davenport's values are from his analysis of data from various sources for high winds for which neutral conditions usually prevail. The table includes his heights of the planetary boundary layer up to the elevation of gradient wind. Thus his exponent values apply to the entire planetary boundary layer. Only ranges of values are quoted from De Marrais (14). Details are too extensive to include herein. The values of Jones et al. (9) obtained from measurements up to a 330-m elevation are a little low for urban areas. They suggest this may be due to incomplete development of the boundary layer and is a problem for any site that does not have a long upwind homogeneous surface configuration.

Some departures from the consistent trend of decreasing p with instability were found by De Marrais (14) and Jones et al. (9). They have some results which show p variation with instability to reverse at high instabilities.

Profile Equation (13) is useful for interpolating and extrapolating available site data to needed elevations not included in the field measurements. For a given set of wind conditions, the three parameters u_1, z_1,

Table II   Exponent p for Power Law Velocity Profile Equation (13)

Surface configuration	Stability	p	Ref.
Smooth open country	Unstable	0.11	(15)
	Neutral	0.14	
	Moderate stability	0.20	
	Large stability	0.33	
Nonurban—varying roughness and terrain	Daytime—unstable and neutral	0.1–0.3	(14)
	Nighttime—stable incl. inversion	0.2–0.8	
Urban (Liverpool)	Unstable, $\Delta\theta^a < 0$	0.20	(8)
	Neutral, $\Delta\theta = 0$	0.21	
	Stable, $0 \leq \Delta\theta < 0.75$	$0.21 + 0.33\,\Delta\theta$	
Flat open country, $z_G{}^b = 274$ m	Neutral	0.16	(3)
Woodland forest, $z_G = 396$ m	Neutral	0.28	
Urban area, $z_G = 518$ m	Neutral	0.40	

^{a} $\Delta\theta$ = potential temperature difference between 162- and 9-m elevations.
^{b} z_G = height of planetary boundary layer to gradient wind.

and p can be evaluated with a minimum of three elevations. There is the usual danger of inaccuracy inherent in extrapolation of an empirical equation with limited data.

4. Wind Direction Variation with Elevation

Change of wind direction with elevation produces crosswind horizontal shear which effectively increases horizontal dispersion of a plume. An initially circular plume becomes elongated with its long axis in a diagonal direction, which becomes more nearly horizontal with increasing shear.

The entire planetary boundary layer experiences change in wind direction with elevation due to large-scale dynamics of motion including Coriolis acceleration caused by earth's rotation. This is shown graphically by the Ekman spiral in Slade (*13*) and many texts on dynamic meteorology. Except for large power plants, plume dimensions are not normally so large as to make this an important factor in plume diffusion. It may be automatically accounted for in experimental determination of plume diffusion parameters.

At low winds and stable conditions, horizontal variations in atmospheric thermal characteristics may cause large and variable crosswind shear. Its effect on plume diffusion becomes increasingly important at large downwind distances, 10 km and more.

B. Airflow over Buildings and Topography

1. Airflow around a Building

Patterns of air motion around isolated buildings have some common qualitative features. Quantitative properties vary widely with building geometry and wind direction. Figure 2 shows qualitative features de-

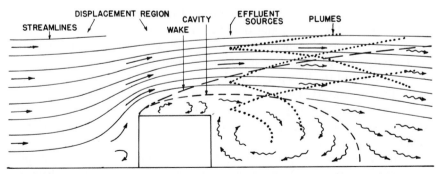

Figure 2. Flow pattern, wake, and cavity regions around a building. Diffusion of simple nonrising stack effluent sources is shown for various elevations.

scribed by various authors including Halitsky (*16*), Halitsky, in Slade (*13*), and Task Group (*17*). (Halitsky includes three-dimensional aspects.) The building in Figure 2 is approximately cubical in shape with the upstream wall normal to the airstream. The first effect of a building on the approaching stream is deflection of streamlines (local directions of time mean velocity) as the air moves into the region of displacement flow. This envelopes the building and early part of the downwind region. Local velocity varies in magnitude and direction but shows no major change in level of turbulence preexisting in the approaching airstream. Local velocity increases above the building as indicated by closer spacing of streamlines.

Within the displacement region is the wake, a region of higher turbulence level. It originates at upstream edges and surfaces of the building. Lateral wake dimensions (horizontal and vertical) increase with downwind distance, but intensity of turbulence decreases. Eventually, the wake becomes indistinguishable from the surrounding region, a downwind distance of the order of ten or more building heights. Diffusion of a plume emitted in the wake region is much greater than in upwind or surrounding regions, but it has continuous downwind motion unless diffused into the cavity.

The cavity is within the wake as shown in Figure 2. It has higher levels of turbulence and shows marked variations in local flow direction including reversal. It usually originates with the upstream edge of the wake. A plume emitted within the cavity tends to become diffused throughout the region.

Figure 2 shows only the flow pattern on a vertical plane through the center of the building. Similar features are found on horizontal planes through the building but with symmetrical flow about the longitudinal axis. One-half of such a pattern would appear similar to that in Figure 2 with the ground line of the figure as the axis of symmetry.

Boundaries between the above-described regions are not sharp as may be implied by the figure, and they are time dependent. Dimensions vary widely with three-dimensional geometry and wind direction. Large wake and cavity size is formed with a two-dimensional (infinitely long) vertical wall normal to the wind. Length of cavity is of the order of ten times wall height. Cavity height (above ground level) is about twice wall height. For the more nearly cubical shape of Figure 2, cavity length downwind of the building is two to three times building height. Cavity height is about $1\frac{1}{2}$ times building height.

Wind direction along the diagonal axis of a cubical building will produce smaller cavities. Vortices with axes along the downwind direction are formed at upwind diagonal edges. For a rectangular building at a diagonal wind direction the longitudinal vortex formed at the long up-stream edge may be large enough to cause a pronounced rotational flow in the cavity and wake regions. Wake and cavity dimensions may be larger than for other directions. A rectangular building with its long dimension parallel to the wind may have smaller cavity than a cube of same width and height. Evans (18) gives schematic diagrams of cavity sizes and shapes for a large number of building shapes and orientations. These are based on wind tunnel experiments.

Approximate values for wake size and velocity are given by Halitsky (16) for the usual blunt-shaped building. Cross-sectional area of wake is twice the area of largest side of the building. This is taken shortly downwind where the wake is well developed. Wake velocity is one-half of wind velocity.

2. Airflow around Urban Structures

The complicated flow patterns of individual buildings are compounded in an urban setting where configurations have endless variations. The specification of urban configuration by roughness length in velocity pro-file equations of Section II,A,1 gives no information on local flow among buildings. Experimental data from a given urban location lack generality of application to others. In addition to the variations of configuration are the variable thermal characteristics. Building and street surfaces of varying heat transfer properties and man-made heat sources cause genera-tion and suppression of air motions.

Local air velocities may be greater or lesser than those of the free stream above urban structures. Higher velocities may result from a chan-nel effect between adjacent structures. Lateral locations at the sides of an exposed building have higher velocities similar to those in the displace-ment region above the building of Figure 2.

In a uniform array of closely spaced similar buildings, the regions be-tween buildings have lower velocities and reduced ventilation. The most critical occur over streets crosswise to the wind direction. Georgii et al. (19) found with their studies in Frankfurt/Main a horizontal vortex-type motion shown in Figure 3 for winds above 2 m second⁻¹. This extended down to street level where local wind direction is opposite to that of the overlying free stream. Below 2 m second⁻¹ this pattern did not exist, being replaced with more complex patterns and poorer ventilation. Some re-versed flow at street level still existed.

Wind tunnel scale-model experiments of Hoydysh and Chiu (20) showed intermittent corner vortices with vertical axes as in Figure 3. They tended to occur alternately at opposite crosswind corners of the

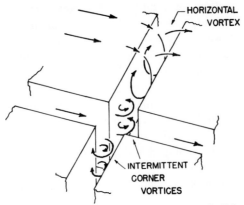

Figure 3. Some details of air flow among buildings.

building but did not have definite periodicity. Vortex velocities were higher than those of other motions in the crosswind street. Velocities in the streamwise streets were relatively high but lower than free stream.

3. Effect of Topography

Topographical variations in surface configuration affect both the centerline trajectory and the diffusion of stack plumes. The plume centerline may be highly distorted and have directions completely different from that of the wind aloft above topographical influences. This is often the case for valley regions. Data on wind speed and direction at one location in a mountainous region may be quite different from those at another nearby if not taken at sufficiently high elevations. The effect of topography on diffusion about the plume centerline is determined by more localized features of surface configuration. This is discussed in Section III,C,3,d.

While many analytical and field studies have been made on air movement in regions of marked topography, the amount of available data is small compared with the limitless variations of surface configuration. Data for a given region of concentrated study may also be limited because of enormous problems of instrument logistics and personnel movement. Valley regions have been the subject of much study since they have higher population concentration and potentially more critical air pollution problems. Effect on airflow patterns of simple configurations are discussed in various references (*13, 17, 21*).

Atmospheric and surface thermal characteristics play a major role in development of air motions in a given topographical situation particularly at low wind velocities. Local velocities may be greater or lesser than would otherwise occur in the absence of thermal influences. Diurnal variation of surface heating and cooling and advection of air masses of varying properties cause widely differing patterns of motion. The presence of an urban complex with the usual heat emission and surface roughness will produce air movement different from that in a rural setting. While the troposphere (elevations up to about 11 km) is normally stable, surface conditions have a strong influence on thermal stability in the surface layer.

Gently rolling surfaces of Figure 4a show typical variation in velocity profile for an undulating surface of low amplitude. Flow is continuous in the streamwise direction at all elevations. Velocities are a little higher and lower at the tops and bottoms, respectively, of the undulations. The abrupt configuration of Figure 4b (a long ridge normal to the wind) has some of the features of flow about a building shown in Figure 2. Flow separation occurs at the upstream surface discontinuity and forms a turbulent wake and stationary eddy the axis of which is parallel to the ridges. Figures 4a and 4b are based on velocity profiles and flow patterns found by Blenk and Trienes (22) with water-channel experiments.

Some pronounced effects of stable and neutral thermal conditions for flow over simple mountain configurations are illustrated by the flow patterns of Figure 4. These types of patterns are shown in Slade (13) and analyzed by Scorer (21). On the windward side of the mountain, flow is similar for neutral and stable conditions. Stack plumes emitted in this region will travel up the slope in either case. Flow separation occurs at the ridge. Under neutral conditions a large stationary eddy is formed as in Figure 4c. It fills much of the lee region. Plumes emitted near the surface will travel upslope, then downslope in the upper part of the eddy. With stable atmospheres, a small region of flow separation occurs but no large stationary eddy. Foehn-type surface winds are downslope on the lee side, as in Figure 4e. Yabuki and Suzuki (23) found lee winds higher than those on the windward side. Their extensive studies of mountain airflow cover various mountain ranges in Japan and other parts of the world. A phenomenon of stable flow is the formation of lee waves which may extend to considerable elevations above the mountain top. At high stability, wave amplitude may be high enough to form a rotor, a stationary eddy-type flow shown in Figure 4e. Wind direction at the surface is opposite to that of the downslope winds. The studies of Yabuki and Suzuki (23) include extensive water tank experiments on flow patterns over a scale-model mountain with lee wave formations for a wide range of fluid stability.

The marked influence of stability continues in the valley region be-

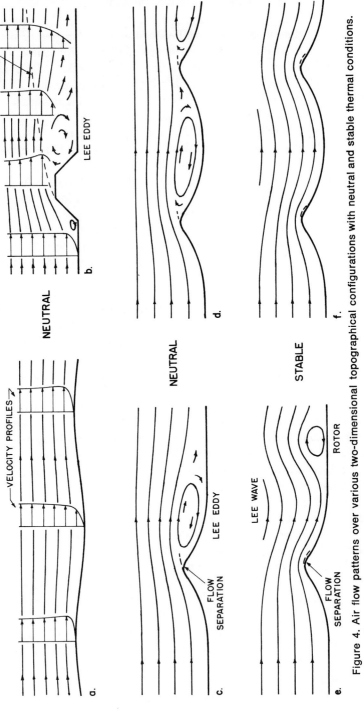

Figure 4. Air flow patterns over various two-dimensional topographical configurations with neutral and stable thermal conditions.

tween two ridges as illustrated in Figure 4f. With neutral conditions in Figure 4d the stationary lee eddy may fill most of the valley. On the windward side of the second ridge, rapid changes in local surface winds occur with relatively small changes in eddy size. Near the region of flow division in the diverging flow, wind direction may quickly change 180° as the division point shifts.

Climatological data on valley wind directions show the channeling effect of valley walls to produce dominant wind directions along the valley axis, upslope and down. This is illustrated in Slade (*13*) with wind roses at valley floor and elevated locations. At elevations, the channeling effect is reduced or nonexistent. Even though winds aloft may not be parallel to the valley axis, they will induce axial winds when off the direction normal to the axis.

With low winds aloft and stable conditions, valley circulations and axial winds are determined largely by valley conditions. These are most pronounced at nighttime. Strong inversion above a valley may completely suppress ventilation out of the valley both day and night. This has led to acute pollution episodes in industrialized valley regions. Heat emission from the urban–industrial complex causes local circulations of the form shown in Figure 5d. Valley sides are cooler than urban surfaces on the valley floor. This results in downslope winds and rising flow over the urban area. A great deal of mixing occurs throughout the region. Temperature gradients are neutral or unstable.

Superimposed on cross valley circulations, if they exist, are winds along the axis caused by valley temperatures and slope of valley floor. These are more pronounced at night with clear sky. Radiation cooling of valley surfaces produces a drainage flow which tends to move in the direction of water drainage as shown in Figures 5a and 5b. Smith (*24*) found that an industrial heat source blocked drainage flow in one branch of a valley system in which airflow, temperature, and diffusion studies were made. Figure 5a shows an interpretation of the flow pattern resulting from a large heat source. Part of the valley flow is lifted out as a heat plume.

Differential heating and cooling of valley sides may produce rotational circulation about a longitudinal axis as shown in Figure 5c. Solar heating of one side of the valley will induce this circulation. More nearly equal heating of valley sides will cause upslope winds on both sides which gives a circulation in reverse of that shown in Figure 5d. Continuation of surface heating or cooling will produce broad changes in valley air temperature which will induce axial winds upslope and drainage, respectively. Defant (*25*) includes a number of diagrams showing combinations of the diurnal cycle of cross valley and axial winds.

Velocity profiles for nighttime downslope winds are quite different for

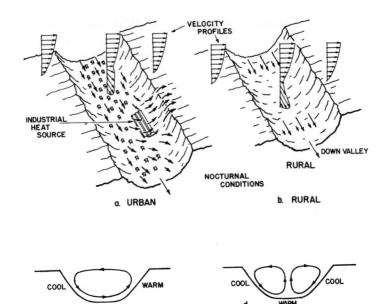

Figure 5. Air flow and velocity profiles in valley configurations.

urban and rural configurations as shown in Figures 5a and 5b. Smith found velocities substantially constant with elevation within the urban boundary layer, compared with winds at higher elevations. In their studies of valley–plain winds in rural areas, Davidson and Rao (*26*) obtained triangular-shaped velocity profiles with maximums near mid-elevation of the valley drainage layer.

III. Stack Plume Characteristics

A. Analysis Technique by Plume Rise and Diffusion

The rise of a plume into the atmosphere is accompanied by diffusion about its centerline. While interrelated, diffusion and plume rise are treated somewhat independently in plume analyses. Parameters which determine plume diffusion about its centerline do not explicitly include variables that affect plume rise, with the exception of wind velocity. Analysis techniques have evolved which separately analyze plume rise and diffusion. The complete plume is formed by simple superposition of the two characteristics.

Geometric features of a plume given in Figure 6 include physical height of stack h_s. Effective plume height h_e (elevation of centerline relative to

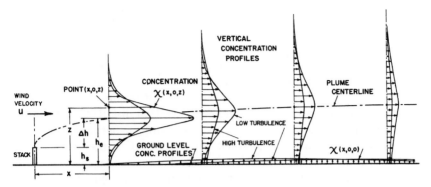

Figure 6. Plume concentration profiles and centerline location in the vertical xz plane.

ground level) is the sum of stack height h_s and plume rise Δh:

$$h_e = h_s + \Delta h \tag{14}$$

Plume rise calculations in Section III,B give the value of Δh. Calculation of concentrations in Section III,C requires effective height h_e. It appears in equation terms which account for effect of ground surface as a reflector in effluent diffusion. Plume rise values vary widely with initial stack effluent characteristics at stack outlet, and with atmospheric variables. Plume diffusion depends primarily on turbulence of the ambient atmosphere.

B. Plume Rise

1. Introduction

The characteristics of the stack effluent at stack outlet and those of the ambient atmosphere determine the formation and rise of a plume in the atmosphere. Effluent characteristics at the stack outlet are the result of the processes by which they are formed (such as burning of fuel) and transmitted to the stack outlet. Once the effluent characteristics at stack outlet are determined, process details are not important. Plume rise Δh is given by the elevation of the plume centerline above the stack outlet as a function of distance x downwind of the stack (Fig. 6). Prediction of the dependence of this relationship on various stack effluent and atmospheric variables is the purpose of plume rise equations and other techniques for plume rise determination.

Various definitions are given for the plume centerline location in the plume cross section. No difficulties arise for sections which have symmetrical distribution of mass or concentration about a horizontal axis. Symmetry is usually assumed for plumes at sufficient downwind distances well above ground effects. A logical location from a physical viewpoint is the center of mass, but this is difficult to determine for asymmetrical plumes. For experimental purposes the centerline location is usually taken midway between the observed top and bottom plume edges. When the concentration distribution is known, the point of maximum concentration is sometimes used. For symmetrical concentration distribution, the above three definitions give identical locations. Large departures from symmetry are found close to the stack where the initial interactions of the plume with the moving atmosphere occur. These are discussed in Section III,B,4. In a uniform atmosphere, departures from symmetry tend to smooth out with distance, and concentration distribution approaches the Gaussian type assumed in the theoretical concentration equations of Section III,C,1.

Most plume rise equations in general use were developed for uniform or smoothly varying atmospheric characteristics. They are also applied to cases of discontinuities in vertical temperature profile. Plume rise is dependent on plume growth which is caused by atmospheric turbulence whether preexisting in the ambient atmosphere or induced by interaction with the plume. The effect of turbulence is introduced in theoretical development of plume rise equations by some form of entrainment parameter. This does not allow for nonuniform variation of turbulence over plume cross section, but equations are successfully applied where departure from uniformity is not large. They must not be applied to regions of large local variation in turbulence as caused by buildings and other surface irregularities. A rule of thumb for stack height to avoid local effects is that the height be $2\frac{1}{2}$ times the height of associated buildings. This was developed early in British power plant practice to avoid adverse effects of buildings on plume diffusion. This rule does not account for variations in building configuration and may be overly conservative in specific instances with favorable wind directions. Plume diffusion in building wakes with vents and short stacks is discussed in Section III,C,3,c.

For purposes of analysis and for understanding of plume rise phenomena initial effluent variables (effluent temperature, emission speed, etc.) are treated in two broad categories, those which influence momentum and those which influence buoyancy. Upward momentum of the stack effluent at the stack outlet causes it to rise. Effluents having weight or mass density less than that of ambient air experience an upward force of buoyancy which also produces rise. The rate of rise induced by both momentum and buoyancy diminishes as entrained air increases total plume mass. Rise rate induced by momentum is initially high and rapidly decreases with distance. Forces due to buoyancy do not produce a high initial rise rate but continue to act. In most plumes encountered in industrial and domestic applications, rise is dominated by buoyancy, particularly for large rises. Because of this, the applicable plume rise equation may neglect the effect of momentum, or include it indirectly through the selection of experimental parameters. Such a plume will be termed a "buoyant plume." In other applications, there may be little or no buoyancy such as in venting of air close to the ambient temperature. For such plumes, the rise will be that due to their initial momentum. Such a plume will be termed a "jet." The initial rise of a buoyant plume near the stack will, if it has sufficient initial momentum, be essentially that of a jet. For this portion of its rise, a jet equation may be more accurate. Theoretical derivations of rise equations of the one-term type omit forces of momentum or buoyancy for buoyant plumes and jets, respectively.

The significance of both momentum and buoyancy to plume rise was recognized by Holland (27) in his early development of a plume rise equation [Eq. (40)]. His equation consists of two terms to separately account for momentum and buoyancy contributions to plume rise. Some later equations are based on modifications of Holland's equation.

Over the years many plume rise equations and techniques have appeared in the literature. This is in contrast with the relatively few equations given for prediction of plume dispersion about the plume centerline (Section III,C). Plume rise equations may be classified as empirical or theoretical, but the line of demarcation is not sharp. Empirical types rely heavily, if not exclusively, on experimental data for numerical parameters as well as functional form. While theoretical types depend on experimental data, their functional form is generally derived from application of the conservation laws of buoyancy and momentum or energy. Details and results vary with specific applications of these laws. Since the empirical types are designed to fit the data on which they are based, they are likely to be accurate for the specific conditions they represent, but may be quite inaccurate when extrapolated. The number of excellent plume rise studies appearing in the literature are too numerous to be included herein. For a specific application there may be available in the literature an equation or equation parameters for conditions close to those of the problem in hand. The reader is referred to papers cited below, each of which gives a number of references to previous studies.

Since large pollution sources such as those which may be produced by large fossil fuel power plants play an important role in air pollution studies on stack effluent diffusion, much research effort has been devoted to these areas. Accurate plume rise data are available for such sources. In most cases buoyancy plays the dominant role, particularly for the larger downwind distances where maximum ground-level concentrations occur. Jets and small buoyant plumes have been given much less attention. Analysis techniques are, therefore, less accurate. Plume rise equations included herein cover a wide range of effluent and atmospheric variables, but their accuracy varies.

Accuracy of plume rise equations is difficult to specify in broad terms. Field measurements of plume rise for a given stack will vary for identical values of variables selected for development of a plume rise equation. Variation of 25% about a mean value is not unusual. This is due to the fact that the other variables which affect atmospheric motions are too complex or numerous to be included in an equation. These include those peculiar to the particular site. A rise equation developed from data of a given site or group of sites may give different values from those based on other sites. There may be variations between equations of different investigators for the same set of data due to differences in interpretation and emphasis. Comparing equations is frustrating. Two equations of different functional form may give identical calculated rise for one set of input data and quite different values for another. Briggs (*28*) states that in the large number of equations he studied, calculated values varied by a factor of 10. Each equation is presumably accurate for the particular experimental data on which it is based. The group of equations Briggs studied includes some from earlier times when the amount of data was extremely limited and experimental techniques less accurate.

Regardless of potential inaccuracies, good results can be obtained with judicious application of plume rise equations. The above remarks are made to caution against their indiscriminate application. The various comments and discussions in this section are intended to emphasize possible areas of inaccuracy. With the rise equations given herein there is included information on limitations and ranges of applicability where available in simple form. It is suggested that in absence of a particular preference, Briggs' equations be applied first. This is in view of the wide range of field data on which they are based and the simple theoretical foundation which is in agreement with others. Application of additional equations may suggest modification of the first calculations and eliminate those obviously inappropriate for the given problem.

The reader desiring more information on the particular rise equations given herein, as well as others, is referred to the following: Briggs (*28*), as mentioned above; Carson and Moses (*29*), who analyzed 15 rise equations and compared them with various field data [these and additional equations are included in the Moses and Kraimer (*30*) study directed to a new technique of plume rise estimation]; and with particular regard to plumes from large fossil fuel power plants, the Tennessee Valley

Authority (TVA) studies reported by Carpenter *et al.* (*31*) and Thomas *et al.* (*32*).

Some equations give rise as a function of distance, but most give a constant value (for a given set of conditions) that the plume reaches at large downwind distances. Maximum ground-level concentrations are likely to occur for this value in many applications. This constant value is difficult to define accurately, particularly from an experimental viewpoint. It is described by various adjectives such as final (used in this chapter), maximum, asymptotic, leveled off, and ultimate. Briggs (*28*) defines final rise as "total plume rise after leveling off." Some definitions applicable to experimental data give a practical maximum, but the plume may eventually rise to a higher elevation. Carpenter *et al.* (*31*) and Thomas *et al.* (*32*) use the following definition: "Plume rise, Δh, was defined as elevation of the plume at the point in distance and space where rate of rise of the plume centerline as a function of distance reached a minimum value or became a constant." Other investigators include in their definition of where final rise is reached a specific value for the slope of the centerline.

Theoretically, a buoyant plume in a neutral or unstable atmosphere will rise indefinitely. In a real atmosphere it will eventually lose its identity owing to continuing diffusion. This potentially large rise is of limited practical value since the distance is likely to be well beyond the location of maximum ground-level concentration. In a stable atmosphere a buoyant plume will reach an elevation where its buoyancy vanishes, and it is in equilibrium with its surroundings. It may because of its upward momentum exceed this elevation and settle back. Thus a final rise equation may give values less than maximum attained in course of travel to this point. In stable atmospheres, jets will also tend to settle back to lower elevations after overshooting equilibrium elevation.

Theoretical studies of plume rise often treat rise and travel in two or three phases. They vary in phase definition. The need for such divisions arises from variation with distance in relative importance of momentum, buoyancy, and the effect of ambient atmospheric turbulence, generally in that order. The first or first two phases, during which most if not all rise occurs, appears to proceed largely independent of the turbulence preexisting in the airstream approaching the plume. Plume spread and the corresponding entrainment parameter is due primarily to turbulence caused by interaction of the plume with the ambient atmosphere. This was confirmed in the scale-model wind tunnel experiments of Hewett *et al.* (*33*) and by their correlations with field data. Thus the plume rise entrainment parameters are in many theoretical studies treated as constants except for the last phase.

Many plume rise equations are written in a form for which the numerical constants are nondimensional. Such equations may be applied for any set of consistent units of the dimensional variables. When not otherwise specified, the equations given herein are in that form.

A reference elevation must be specified for variables of the ambient atmosphere included in the equations. Unless otherwise indicated, stack outlet is the reference elevation. The reference elevation for wind velocity is most critical. The equations developed in TVA studies use an average wind velocity over the range of plume elevation. For other equations, the wind velocity at stack outlet is used. When plume rise studies are related to site measurements at other elevations, extrapolation must be made to reference elevation. In the absence of site velocity profile data, equations of the type given in Section II,A, are used.

2. Buoyant Plumes

Theoretical derivations of buoyant plume rise equations such as those of Briggs (*28*) do not include the rise due to upward momentum of the effluent at the stack outlet. Equation parameters are usually evaluated or confirmed with plume rise data from real stacks in which there is some momentum rise even though it may be small compared with the final rise due to buoyancy. To this extent buoyant plume rise equations include momentum rise, but it is not separately accounted for in the one-term equations. Rise data are usually obtained at distances beyond which the momentum rise contribution is substantially constant. Since the functional form of the momentum rise contribution (versus distance) is different from that for the buoyant part, buoyancy rise equations may be inaccurate close to the stack. For these locations the jet rise equations of Section III,B,3 will give more accurate results.

a. PLUME RISE $\Delta h(x)$ vs DISTANCE x. Most plume rise equations of recent years have a simple one-term power-law form with variables appearing with exponents at various values. The general form is

$$\Delta h(x) = K_1 Q_h{}^a x^b u^c \qquad (15)$$

where

x is the distance downwind of stack (Fig. 6)

u is the wind velocity

Q_h is the heat emission rate

K_1 is a constant dependent on other variables which affect Δh; it is dimensional except for certain combinations of exponents

a, b, c are exponents evaluated from theoretical or experimental considerations

Q_h is evaluated in various ways depending on available stack effluent data. For a pure source of heat (no effluent mass flow) it is actual heat emission. For a heated effluent, Q_h is equivalent heat emission relative to ambient air which would produce the same buoyancy. If the effluent has the same molecular weight and specific heat as air, Q_h is given by the following equation:

$$Q_h = Q_m c_p (T_s - T) \tag{16}$$

where

c_p is the specific heat at constant pressure
T is the absolute temperature of ambient atmosphere
T_s is the absolute temperature of effluent at stack outlet
Q_m is the effluent mass emission rate

Q_m may be expressed in terms of other characteristics as follows:

$$Q_m = \rho_s A_s V_s \tag{17}$$

where

ρ_s is the mass density of effluent at stack outlet
A_s is the stack outlet area $= \pi r_s^2$
r_s is the radius of stack outlet
V_s is the effluent emission velocity at stack outlet

Q_h may be expressed in terms of mass densities as follows:

$$Q_h = A_s V_s T c_p (\rho - \rho_s) \tag{18}$$

where ρ is mass density of ambient air.

Studies of Briggs (28), Bringfelt (34), Fay et al. (35), and others show on theoretical grounds and from much field data that good values for the exponents in Equation (15) are $a = \frac{1}{3}$, $b = \frac{2}{3}$, and $c = -1$. In this case K_1 is nondimensional. There is wide agreement on the value $c = -1$. Various values for a and b are found by various investigators as given in the following paragraphs.

i. *Briggs (28, 36).*   The following buoyancy flux parameters F_b and F_{ba} are used in Briggs' plume rise equations and others below.

$$F_b = g Q_h / \pi c_p \rho T \tag{19}$$

g is acceleration due to gravity. Other terms are as given above. If effluent has same molecular weight and specific heat as air, the following form is obtained in terms of temperatures:

$$F_{ba} = g V_s r_s^2 (T_s - T) / T_s \tag{20}$$

In terms of densities,

$$F_{ba} = gV_s r_s^2(\rho - \rho_s)/\rho \qquad (21)$$

F_{ba} is a good approximation for F_b in many applications involving combustion of fossil fuels.

Two plume rise equations are given to cover ranges of distance x separated by x^*. Beyond x^* ambient atmospheric turbulence plays a dominant role. For all atmospheric stabilities and $x < x^*$ Briggs found the following form with his theoretical development:

$$\Delta h(x) = C_1 F_b^{1/3} u^{-1} x^{2/3} \qquad (22)$$

Briggs (28) recommends $C_1 = 1.6$. Beyond x^* a more accurate form is the following:

$$\Delta h(x) = 1.6 F_b^{1/3} u^{-1} x^{*2/3} \left[\frac{2}{5} + \frac{16}{25} \frac{x}{x^*} + \frac{11}{5} \left(\frac{x}{x^*} \right)^2 \right] \left(1 + \frac{4}{5} \frac{x}{x^*} \right)^{-2} \qquad (23)$$

Distance x^* is a function of atmospheric turbulence characteristics and flux parameter. Briggs (28) gives the following relations as suitable approximations for use in Equation (23). h_s is stack height.

$$x^* = 0.52 F_b^{2/5} h_s^{3/5} \qquad (h_s < 1000 \text{ ft}) \qquad (24)$$

$$x^* = 33 F_b^{2/5} \qquad (h_s > 1000 \text{ ft}) \qquad (25)$$

Units are x^* [ft], F_b [ft^4 sec^{-3}], and h_s [ft]. Written in meter and second units, the above have the following numerical form:

$$x^* = 2.16 F_b^{2/5} h_s^{3/5} \qquad (h_s < 305 \text{ m}) \qquad (24a)$$

$$x^* = 67 F_b^{2/5} \qquad (h_s > 305 \text{ m}) \qquad (25a)$$

Units are x^* [m], F_b [m^4 sec^{-3}], and h_s [m].

Briggs (28) includes the following information on accuracy. Owing to normal variation in turbulence, x^* may vary by $\pm 20\%$. Because of lack of data at greater distances, Equation (22) should not be applied beyond $x = 5x^*$. Variation in turbulence may cause plume rise to vary from calculated values by $\pm 10\%$ for flat and uniform topography and $\pm 40\%$ for substantial topographical variations and the influence of large bodies of water.

Briggs (28) gives as a good approximation for the above rise equations for fossil fuel plants of 20 MW or more heat emission, the application of Equation (22) (with $C_1 = 1.6$) at distances up to 10 times stack height at which final rise is assumed to be reached.

$$\Delta h(x) = 1.6 F_b^{1/3} x^{2/3} \qquad (x < 10 h_s) \qquad (26)$$

While the above equations are applied for $\Delta h(x)$ at any stability, fluctuations about the mean will be larger for unstable conditions.

Briggs (*36*) later reviewed plume rise data in the form of Equation (22) from various papers. He found that values of C_1 ranging from 1.6 to 1.8 fit the bulk of data. He suggests 1.6 as a conservative value. Of the values he reviewed, those of earlier dates tended to be higher. This may be due to data sources having a larger proportion of shorter stacks.

ii. *Bringfelt (34, 37).* Bringfelt's study (*37*) covers various field data of the literature in addition to extensive plume rise measurements at industrial plants in Sweden. He summarizes his results with three equations for plume rise at three downwind distances as follows:

$$\Delta h = 103 Q_h^{0.39} u^{-1} \qquad (x = 250 \text{ m}) \tag{27}$$

$$\Delta h = 167 Q_h^{0.36} u^{-1} \qquad (x = 500 \text{ m}) \tag{28}$$

$$\Delta h = 224 Q_h^{0.34} u^{-1} \qquad (x = 1000 \text{ m}) \tag{29}$$

Units are Δh [m], Q_h [MW], and u [m sec^{-1}].

Bringfelt (*37*) includes some calculated values of x^b [Eq. (15)] from pairs of above equations as in the following tabulation. These show that the exponent decreases with distance.

	x range (meters)	
Q_h *(MW)*	*250–500*	*500–1000*
6.7	$x^{0.62}$	$x^{0.37}$
33.5	$x^{0.55}$	$x^{0.32}$

Bringfelt (*37*) includes comparisons with various plume rise equations of the past. He discusses in detail experimental techniques used for field measurements at Swedish industrial plants.

In a later paper Bringfelt (*34*) presents a theoretical development of plume rise equations which gives the same form as Equation (22). His analysis of industrial plant data obtained in Sweden gives C_1 values of 1.8 and 1.7 for slightly stable and very stable conditions, respectively.

iii. *Carpenter et al. (38) and Montgomery et al. (39)—Tennessee Valley Authority (TVA).* In their study of various plume dispersion models for TVA coal-burning power plants, Carpenter *et al.* (*38*) express plume rise in the form of Equation (22). Effect of atmospheric stability is included in the constant C_1 as shown below:

$$\Delta h(x) = C_1 F_{ba}^{1/3} u^{-1} x^{2/3} \tag{30}$$

Buoyancy flux parameter F_{ba} is defined as in Equation (21). C_1 is given in terms of potential temperature gradient as follows:

$$C_1 = 1.58 - 0.414 \frac{\partial \theta}{\partial z} \tag{31}$$

where $\partial \theta / \partial z$ is in units °C/100 m. Wind velocity u is the mean value between stack top and plume top.

Three stability categories are emphasized in the range covered by Equation (31). They are in decreasing order of stability and are shown in the following tabulation.

Stability class	$\partial \theta / \partial z$ (°C/100 m)	C_1
1	1.3	1.04
2	0.3	1.46
3	−0.06	1.60

In a later paper on TVA plume rise investigations, Montgomery et al. (39) gave separate rise equations for three stability conditions as follows:

Neutral stability ($-0.17 < \partial \theta / \partial z \leq 0.16$, av. $= 0.05$)

$$\Delta h(x) = 2.50 x^{0.56} F_{ba}^{1/3} u^{-1} \qquad (x < 3000 \text{ m}) \tag{32}$$

Moderate stability ($0.16 < \partial \theta / \partial z \leq 0.70$, av. $= 0.43$)

$$\Delta h(x) = 3.75 x^{0.49} F_{ba}^{1/3} u^{-1} \qquad (x < 2800 \text{ m}) \tag{33}$$

Very stable ($0.70 < \partial \theta / \partial z < 1.87$, av. $= 1.06$)

$$\Delta h(x) = 13.8 x^{0.26} F_{ba}^{1/3} u^{-1} \qquad (x < 1960 \text{ m}) \tag{34}$$

Units are $\Delta h(x)$ [m], x [m], F_{ba} [m⁴ sec⁻³], u [m sec⁻¹], and $\partial \theta / \partial z$ [°C/100 m].

Montgomery et al. (39) include a method for obtaining an equation for a selected distance to give plume rise for any stability. One is given for final rise in Section III,B,2,b,vii.

iv. *Fay et al. (35).* Fay et al. developed a plume rise equation which has the same functional form as Equation (22). They evaluated the equation constant on the basis of TVA and Bringfelt (37) data. The following is their recommended form for stable and neutral conditions expressed in terms of buoyancy scaling length l_b.

$$\Delta h(x) = 1.32 l_b^{1/3} x^{2/3} \tag{35}$$

where

$$l_b = g Q_h / \pi \rho_s c_p T u^3 \tag{36}$$

Rise is terminated in stable atmospheres at

$$x = 1.55 l_b S \tag{37}$$

where

$$S = \frac{u}{l_b \left(- \dfrac{g}{\rho_\theta} \dfrac{\partial \rho_\theta}{\partial z} \right)^{1/2}} \tag{38}$$

ρ_θ is atmospheric potential density related to potential temperature by the following equation:

$$- \frac{g}{\rho_\theta} \frac{\partial \rho_\theta}{\partial z} = \frac{g}{\theta} \frac{\partial \theta}{\partial z} \tag{39}$$

b. FINAL RISE Δh.   As pointed out earlier, final rise is difficult of exact definition. Theoretical studies such as those of Briggs (28) and others show that for stable atmospheres, the plume will overshoot the final rise elevation and settle back to equilibrium position. This has been observed but appears to be rare. Its occurrence may be masked by other variations in plume behavior. Overshoot is not accommodated in commonly used rise equations.

There is no theoretical limit to plume rise in nonstable atmospheres since buoyancy continues to act indefinitely. Final rise must be based on practical considerations derived from experimental data. These may be based on termination of a plume rise $\Delta h(x)$ equation at a selected downwind distance.

i. *Holland* (27).   Holland developed his equation at a time when little effort had been devoted to plume rise analyses. It has been included in many plume rise studies.

$$\Delta h = (1.5 V_s d_s + 4 \times 10^{-5} Q_h) u^{-1} \tag{40}$$

d_s is stack diameter at outlet. Units are Δh [m], V_s [m second^{-1}], d_s [m], and Q_h [cal second^{-1}].

The two terms separately account for momentum and buoyancy contributions to rise. The first term was based on laboratory experiments on jets. The second was derived from field data obtained at three TVA power plants. Various studies have shown the equation underestimates rise, but when multiplied by a factor ranging from 2 to 3 (from various studies) it gives results as good as others developed at later times. Some writers have suggested that Holland's buoyancy rise data were not obtained at sufficiently far downwind distances, thus the low predicted values.

Holland (27) suggests that stability be accounted for by increasing the calculated rise by adding 10–20% for unstable conditions and subtracting a like amount for stable conditions.

ii. *Moses and Carson (40)—Argonne National Laboratory (ANL).* Moses and Carson based their equations on field data from a wide range of stack sizes. Their analysis assumes the following general form of plume rise equation.

$$\Delta h = C_3 + C_4(V_s/u)^e d_s + C_5 Q_h{}^f/u \tag{41}$$

Plume rise data were divided into three stability groups for purpose of equation development. Among their results are the following regression equations for the stability groups as indicated:

Unstable $(\partial\theta/\partial z < -0.22)$

$$\Delta h = 3.47(V_s/u)d_s + 10.53 Q_h{}^{1/2}/u \tag{42}$$

Neutral $(-0.22 \leq \partial\theta/\partial z < 0.85)$

$$\Delta h = 0.35(V_s/u)d_s + 5.41 Q_h{}^{1/2}/u \tag{43}$$

Stable $(\partial\theta/\partial z \geq 0.85)$

$$\Delta h = -1.04(V_s/u)d_s + 4.58 Q_h{}^{1/2}/u \tag{44}$$

Units are $\partial\theta/\partial z$ [°C/100 m], Δh [m], V_s [m second^{-1}], u [m second^{-1}], and Q_h [kcal second^{-1}].

iii. *Briggs (28, 36).* Briggs' equations for final rise cover calm as well as the usual windy condition.

Neutral and unstable. These conditions are handled by applying Equation (22) at $x = 3x^*$ to give the following:

$$\Delta h = 1.6 F_b{}^{1/3} u^{-1}(3x^*)^{2/3} \tag{45}$$

For fossil fuel plants with heat emission rates of 20 MW or more, a good approximation is obtained with Equation (22) at $x = 10h_s$:

$$\Delta h = 1.6 F_b{}^{1/3} u^{-1}(10h_s)^{2/3} \tag{46}$$

Stable. Briggs (28) theoretical development for the windy condition leads to the following form of final rise equation similar to those of

Bringfelt (*34*) and Fay *et al.* (*35*):

$$\Delta h = C_2 (F_b / us)^{1/3} \tag{47}$$

Stability parameter s is given as follows:

$$s = \frac{g}{\theta} \frac{\partial \theta}{\partial z} \tag{48}$$

Briggs (*28*) recommends $C_2 = 2.9$. In his later review (*36*) of various final rise equations of the above form, he finds the bulk of data fits the C_2 range of 2.4 to 2.6. He suggests 2.4 as a conservative value.

For calm conditions, the following equation applies (*28*):

$$\Delta h = 5.0 F_b^{1/4} s^{-3/8} \tag{49}$$

Penetration of inversion. A plume will rise above a ground-based inversion of limited elevation if it has sufficient buoyancy. This will occur if the elevation the plume would reach, using Equations (47) and (49), is higher than elevation of the inversion layer.

A plume rising through a neutral layer may be stopped from further rise if it meets a sharp elevated inversion. Resistance of the sharp inversion to penetration is identified by the following parameter b_i:

$$b_i = g \, \Delta T_i / T \tag{50}$$

ΔT_i is temperature increase through the sharp inversion. The plume will penetrate the inversion and continue to rise in a neutral layer above if the height z_i of the inversion meets the following criteria for windy and calm conditions, respectively:

$$z_i \leq 2.0 (F_b / u b_i)^{1/2} \tag{51}$$

$$z_i \leq 4 F_b^{0.4} b_i^{-0.6} \tag{52}$$

iv. *Bringfelt (34).* Bringfelt's theoretical analysis leads to the final rise equation which has the form of Equation (47). α is an entrainment constant.

$$\Delta h = \left(\frac{3.61}{\alpha^2} \right)^{1/3} \left(\frac{F_b}{su} \right)^{1/3} \tag{53}$$

For slightly stable conditions, Bringfelt's arithmetic mean $\alpha = 0.53$ (Table I of the reference) gives the numerical factor in Equation (53) [equivalent to C_2 in Eq. (47)] a value of 2.34. The arithmetic mean $\alpha = 0.57$ for strongly stable conditions gives the factor a value of 2.23. Briggs (*36*) interprets a best fit for Bringfelt's strongly stable data to give a value of 2.4.

v. *Fay et al.* *(35)*. In their analysis covered above (Section III,B,2,a,iv) Fay *et al.* recommend the following equation for final rise. This is reached at the downwind distance of Equation (37).

$$\Delta h = 2.27 l_b S \tag{54}$$

The parameters are defined in Equations (36) and (38).

vi. *Task Group (17)—American Society of Mechanical Engineers (ASME) Recommended Guide.* The Task Group recommends the following plume rise equations for effluent temperatures 50°C or more above ambient and with volume flow rates 50 m³ second⁻¹ or more. The equations apply at the end of the phase where initial momentum and buoyancy are important. They may give plume rise values smaller than other equations. Momentum effect is included in buoyancy flux parameter F_{ba}.

The following equation is given for neutral and unstable conditions.

$$\Delta h = 7.4 h_s^{2/3} F_{ba}^{0.33} / u \tag{55}$$

Buoyancy flux parameter F_{ba} [Eq. (21)] is evaluated in terms of ambient air and stack effluent densities at stack top where wind speed u is also taken.

For stable conditions the Task Group *(17)* recommends the following rise equation.

$$\Delta h = 2.9 (F_{ba}/us)^{1/3} \tag{56}$$

s is given by Equation (48) where θ is evaluated at stack top and may be found in terms of ambient temperature T and atmospheric pressure P by the following equation. P_0 is standard pressure.

$$\theta = T(P/P_0)^{0.29} \tag{57}$$

The Task Group states that the above equations become unreliable when wind at stack top is less than 7 m second⁻¹. For light winds and neutral or unstable conditions at stack top, plume rise is probably dependent on the presence of stable layers at altitudes ranging from 300 to 1500 m.

vii. *Carpenter et al (38), Montgomery et al. (39, 41)—Tennessee Valley Authority (TVA).* Carpenter *et al.* *(38)* employ the following equations for final rise in their study of principal plume dispersion models at TVA power plants. This is obtained from rise $\Delta h(x)$ vs distance [Eq. (30)] terminated at about 1200 m where they found maximum rise is generally reached.

$$\Delta h = 114 C_1 F_{ba}^{1/3} u^{-1} \tag{58}$$

Units are Δh [m], F_{ba} [m^4 second^{-3}], and u [m second^{-1}]. F_{ba} is given by Equation (21) and C_1 by Equation (31).

Montgomery et al. (41) employ Equation (58) in their development of nomograms for estimating pollutant concentrations.

Montgomery et al. (39) apply their method for obtaining a stability equation at a selected distance. Their selected distance of 1824 m, applicable to TVA power plants, appears to give final rise. The resulting equation is as follows:

$$\Delta h = \frac{173 F_{ba}^{1/3}}{u \, \exp(0.64 \, \partial\theta/\partial z)} \tag{59}$$

Units are given following Equation (58).

viii. *Brummage (42)—The Oil Companies' International Study Group for Conservation of Clean Air and Water (Western Europe) (CONCAWE).* The various CONCAWE equations are of the empirical type based on a range of field data. Various one-term-type equations with empirical exponents for Q_h and u are given for adiabatic, slightly stable, and fairly stable categories. They recommend for near-neutral stability the following equation as a "simple relationship in best agreement with experimental data."

$$\Delta h = 0.175 Q_h^{1/2} u^{-3/4} \tag{60}$$

Units are Δh [m], Q_h [cal second^{-1}], and u [m second^{-1}]. They recommend the equation not be applied for heat emission rates greater than about 8×10^6 cal second^{-1}.

ix. *Moses and Kraimer (30), Tabulation Prediction Technique.* Moses and Kraimer developed a technique for presenting plume rise data in tabular form to include statistical properties not obtainable from the usual plume rise equations. They include mean, standard deviation, cumulative percentile frequency distribution, and interquartile range. Their tabular results are based on data from a wide range of stack sizes including those of TVA power plants.

Results are divided into stability categories, stable, neutral, unstable, and for all data. Independent variables are heat emission Q_h, wind velocity u, and product of emission speed and stack outlet diameter $V_s d_s$. Plume rise characteristics listed above are given for various ranges of the independent variables.

An example is taken from Moses and Kraimer (30), Table IX. Given a case there the input values fall in the following table ranges: Q_h from 100 to 999 kcal second^{-1}, $V_s d_s$ from 10 to 24 m^2 second^{-1}, u from 7 to 8 m second^{-1}, and neutral stability. The table gives mean plume rise as 27 m and standard deviation as 10 m in addition to other statistical characteristics.

Included in the study is an analysis of the comparative accuracy of a number of plume rise equations with various statistical criteria.

3. Jets

The following jet rise equations are those given by Briggs (*28*). They are based on the same form of theoretical development he used in buoyant plume analyses and on other equations and experimental data. Available information on jets is small compared with that for buoyant plumes. Briggs emphasizes that jet equations are less accurate. While jets have been the subject of much study in many technical areas, the available data are often outside the range of conditions applicable to stack effluents in the atmosphere.

Briggs developed the following equation for jet rise versus distance to include the effect of effluent density. It does not include bouyancy rise such as given by Equation (22).

$$\Delta h(x) = 2.3 F_m^{1/3} u^{-2/3} x^{1/3} \tag{61}$$

where momentum flux parameter F_m is given by the following equation:

$$F_m = V_s^2 r_s^2 \rho_s / \rho \tag{62}$$

Briggs recommended jet rise equation results from setting $\rho_s = \rho$ in Equation (61). The more general form is given above for possible application to jets having other than ambient air density, but its accuracy may be questioned.

Briggs suggests that the ambient form of Equation (61) ($\rho_s = \rho$) is accurate for rise values up to

$$\Delta h = 6(V_s/u)r_s \tag{63}$$

when $V_s/u \geq 4$. It appears that Equation (63) may be treated as final rise in neutral atmospheres. It gives higher values than some in the literature [twice the momentum term of Holland's Eq. (40)].

Briggs gives the following tentative equations for stable conditions, windy and calm, respectively:

$$\Delta h = 1.5(F_m u^{-1})^{1/3} s^{-1/6} \quad \text{(windy)} \tag{64}$$

$$\Delta h = 4(F_m s^{-1})^{1/4} \quad \text{(calm)} \tag{65}$$

Briggs recommends that in any application the most conservative value found from Equations (63)–(65) be used for final rise.

4. Nonuniform Plume Characteristics near the Stack Outlet

Near the stack outlet a plume in a crosswind may have highly asymmetrical concentration and velocity distributions. Except under certain conditions, such asymmetry tends to smooth out with distance and subsequent plume rise is largely independent of initial asymmetry.

Important to plume rise is the condition of downwash in which significant portions of the effluent are drawn into the low pressure wake immediately downstream of the stack. If the upward forces of momentum and buoyancy at the stack outlet are insufficient, downwash occurs in varying degrees of intensity. This effectively reduces plume rise in at least two aspects. Effluent drawn into stack wake leaves the stack region at elevations lower than stack outlet. This effectively lowers point of emission. Downwash also increases plume cross section which reduces plume rise. When downwash occurs, actual plume rise will be less than predicted by rise equations.

Some guidelines are available for estimating the conditions under which stack downwash will occur. Criteria for downwash were sought early in plume diffusion studies, often by wind tunnel experiments. The most commonly used parameter is the ratio of emission velocity to wind velocity V_s/u. This is a measure of the ratio of effluent momentum to that of the airstream. When $V_s/u > 1.5$ downwash is clearly avoided. For $V_s/u < 1.0$ it will definitely occur, probably to a serious degree. Intermediate values are given for criteria, but there is no sharp line of separation. Fay et al. (35) found that for $V_s/u < 1.2$, the effect of downwash on the entrainment parameter was significant. The above values are based on dominance of effluent momentum, but buoyancy aids in reducing downwash. Briggs (28) finds that if the effluent Froude number $Fr[= V_s{}^2 T/gd(T_s - T)]$ has a value of 1.0 or less, the above criteria may be relaxed; but available data do not indicate to what degree.

The characteristics of the plume cross section and the centerline shape near the stack outlet (distances within a few tens of outlet diameter) are not accurately predicted by the usual plume equations applicable at greater distances. This region has not been studied as extensively as those further downwind. Halitsky (43) in his studies of transverse jet plumes reviews various equations and data appearing in the literature.

One feature of plume cross section at near distances is the formation of two longitudinal vortices of opposite rotation, side by side. These give the plume a kidney-shaped cross section with a rounded top. The plume has an upward component of velocity in the central region and a downward component at its sides. Under some conditions, not well documented, the vortices may persist with distance and form a bifurcated plume.

Scorer (*21*) shows photographs of bifurcated plumes with the two parts completely separated. He finds that conditions favorable to bifurcation are a buoyant plume in nonstable low turbulence conditions as may occur in the evening when the stack top is still above the stable layer.

C. Effluent Concentration Distribution of Gaseous Plumes

A plume emitted into an atmosphere of relatively uniform turbulence diffuses in all directions and has properties of concentration distribution and geometry which are quite predictable by available analytical methods. Plume centerline location discussed in the preceding section is the reference position relative to which plume diffusion is expressed. Having centerline location and diffusion properties about centerline, plume concentration can be calculated to a useful degree of accuracy for any location in the three-dimensional field downwind of a stack. While turbulence is assumed uniform in the region occupied by the plume, its components may be different in the y and z directions. Diffusion about centerline in both y and z directions varies widely with atmospheric variables which affect turbulence, the basic mechanism of plume diffusion. Direct measurement of turbulence provides a fundamental characteristic by which diffusion may be predicted most accurately. While usually measured in research studies, it is more difficult to obtain than other relevant atmospheric variables such as velocity and temperature profiles. There are also problems of interpretation for turbulence data. Diffusion depends on motion of plume elements in space, a Lagrangian-type characteristic. Turbulence is usually measured by an instrument probe at a fixed location, an Eulerian-type characteristic. The difference must be resolved in analyzing turbulence data. Most techniques for routine prediction of plume diffusion are based on measurements other than turbulence.

As atmospheric motions (turbulence and time mean velocity) become more nonuniform in their spatial variation, prediction of concentrations becomes more difficult and inaccurate; but techniques have been developed for certain special cases important to air pollution. A common one is that of plume diffusion in a homogeneous neutral or unstable surface layer covered with a stable or inversion layer where diffusion is much less or negligibly small. Heat island effects for urban regions fall in this class.

Surface irregularities, mechanical and thermal, may produce nonuniformities in the surface layer which radically change plume behavior. Analytical techniques are of limited accuracy or are nonexistent. Plume centerline location and diffusion may both be affected but not necessarily to the same degree. Plume diffusion may have some regularity, but plume centerline location may be difficult to predict as may occur with topographical effects. In the extreme, there may be no well-defined plume in the usual sense. An example of the later is emission of an effluent near or into the turbulent wake downwind of a sharp and irregular surface configuration such as that caused by a building (Fig. 2) or discontinuity in topography. For near distances, plume geometry tends to be that of the wake.

While the diffusion equations given below are developed for gaseous effluents, they are applicable to small particulates whose fall velocity is small compared with other motions of the plume. Particle diameters less than 20 μm are included in this category.

1. Uniform Atmospheric Conditions

Plume diffusion equations resulting from assumption of uniformity of turbulence give a concentration distribution having Gaussian or "normal" profiles. While any one field experiment on plume diffusion may show significant differences, the Gaussian form has been found most effective in representing plume diffusion for a wide range of atmospheric conditions. The literature on the subject is voluminous (13). Sutton (15) was among the first to supply numerical values of diffusion parameters for several stability categories. Although Sutton's diffusion equation employs the Gaussian form of concentration distribution, it does not do so explicitly in terms of standard deviations of concentration distribution. Sutton's parameters have the advantage of being independent of downwind distance and permit the calculation of plume concentrations by equations only. This requires the standard deviations to be dependent on distance in a power law form. Removal of this restriction by explicit use of standard deviations allows more flexibility in expressing experimental data.

The dependence of standard deviations of pollutant concentration on standard deviations of wind-direction fluctuations in azimuth and elevation as measured with wind-direction instruments was brought out by Cramer (44) and Hay and Pasquill (45). Pasquill (46) used stability categories similar in principle to those of Sutton (15) in his graphical presentation of plume spread data in horizontal and vertical directions. This technique permitted the estimation of concentration distribution without direct measurement of turbulence but lost the accuracy inherent in direct inclusion of this fundamental diffusion characteristic. Gifford (47) reworked Pasquill's data into the form now in common use.

Various simplifying assumptions are made in derivation and application of diffusion equations in addition to that of turbulence uniformity.

Stack effluent is assumed to originate at a point, a virtual source. This is obviously unrealistic for regions close to the source, and the equations give erroneously high concentrations for real stacks at small distances. Limited correction for finite size of stack outlet has been made by locating the theoretical point source upstream of the real stack such that the theoretical plume will have at stack-outlet-location cross-section characteristics similar to those of the real plume. Techniques for this correction are not well established. It appears that the correction will vary with atmospheric and stack effluent variables. Halitsky (43) presents a method for estimating plume concentrations for regions close to the stack outlet. For most air pollution applications, downwind distances are sufficiently large that correction for virtual point source location is small compared with other dimensions of the problem. Furthermore, experimental data used for evaluation of parameters for the theoretical equations may have been obtained for real stacks where this correction is inherent in the data.

Another assumption of the theoretical plume equations is that the theoretical plume does not have initial momentum and buoyancy. Under this assumption the plume centerline will be horizontal, in the direction of local wind velocity, and there will be no plume rise. This is accomplished by locating the point source above the stack outlet a height equal to plume rise for the downwind distance x at which plume concentrations are to be determined (Fig. 7). Virtual source elevation, i.e., effective plume height h_e is variable for distances up to that at which final plume rise is reached.

a. GAUSSIAN FORM OF DIFFUSION EQUATIONS IN TERMS OF STANDARD DEVIATIONS σ_y AND σ_z. The following forms of the diffusion equation are presented in many references including Slade (13) and Turner (48). Plume concentration $\chi_{(x,y,z)}$ is given by Equation (66) below for a selected location (x, y, z) in the three-dimensional field downwind of the point source. The source has emission rate Q representing the pollutant of interest in the effluent of the actual stack. It is located above stack outlet at the elevation h_e (effective plume height). This is the sum of physical height of stack h_s and plume rise Δh given by Equation (14). The rectilinear coordinate system x, y, z is centered at the base of the stack. The various dimensions are shown in Figure 7 along with schematic representations of concentration profiles. The diffusion equation is as follows:

$$\chi_{(x,y,z)} = \frac{Q}{2\pi\sigma_y\sigma_z u} \exp\left[-\frac{1}{2}\left(\frac{y}{\sigma_y}\right)^2\right]$$
$$\times \left\{\exp\left[-\frac{1}{2}\left(\frac{z - h_e}{\sigma_z}\right)^2\right] + \exp\left[-\frac{1}{2}\left(\frac{z + h_e}{\sigma_z}\right)^2\right]\right\} \quad (66)$$

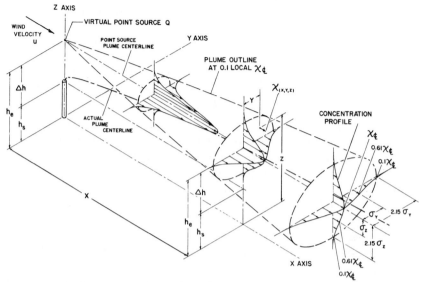

Figure 7. Three-dimensional concentration profiles of a nonrising point source and centerline of a rising plume for diffusion calculations at distance x.

The equation is valid in any consistent set of units for the various variables. Emission rate Q may be in any selected property per unit time. Concentration χ will then be in terms of the same property per unit volume. Commonly used forms of this property are mass and volume. Q applies only to the pollutant of interest in a given problem. If the entire stack effluent is of interest, then Q is equivalent to Q_m (property is mass) in Equations (16) and (17) in the plume rise discussions of Section III,B,2. Usually a component of the stack effluent is of interest, often a small fraction of total effluent. Q is then the emission rate of that component. Sometimes Equation (66) (or its variations) is written with a numerical multiplier when χ becomes very small. An example is the SO_2 component in power plant effluents which may be of the order of a few percent. Plume concentration χ at ground level may be of the order of millionths when the property is selected as volume. It is then convenient to include the multiplier 10^6 in Equation (66) in which case χ is in parts per million (ppm).

Since wind velocity varies with elevation, the single velocity u in Equation (66) must be selected as representative of the region of interest to a given problem. This is discussed in Section III,C,1,b. Other variables in Equation (66) have the dimension of length.

Development of Equation (66) assumes conservation of plume effluent in its passage through the atmosphere. The equation may be modified to account for loss of effluent at ground surface. The second term in the braces [$(z + h_e)$ term] is the ground reflection term which treats the ground boundary as a perfect reflector of the plume. It assumes that as the lower edge of the plume meets the ground surface it is reflected upward and adds to the downward diffusing plume as if a fictitious plume were located below ground boundary plane a distance h_e. If the lower plume edge is partially or completely absorbed at the ground surface, the reflection term should be reduced or omitted, respectively. This may occur with certain gaseous effluents and vegetation. Fine particulates treated as gases in Equation (66) may be partially removed by ground surfaces.

The ground reflection term may become negligibly small for certain combinations of z and h_e in relation to standard deviation σ_z. This will occur for significant plume elevations h_e and locations z well removed from ground surface. Inclusion of the reflection term destroys vertical symmetry about the plume centerline and shifts the maximum concentration below the centerline. The effect is negligbly small for conditions referred to above. Figure 7 is drawn as if there were no ground reflection contribution.

The following are applications of Equation (66) to special cases which often occur. For a given x, the maximum concentration in the plume cross section occurs near the centerline. If ground reflection is neglected, it occurs at the centerline and is given by the following equation:

$$\chi_{(x,0,h_e)} = \frac{Q}{2\pi\sigma_y\sigma_z u} \tag{67}$$

Neglect of the reflection term underestimates concentration. For very short stacks and small plume rise, the error may not be negligible; and Equation (66) should be used. The extreme is a ground-level source with no plume rise [Eq. (74)] for which centerline concentration is twice that given by Equation (67).

The following equation for ground-level concentration is obtained by setting $z = 0$ in Equation (66):

$$\chi_{(x,y,0)} = \frac{Q}{\pi\sigma_y\sigma_z u} \exp\left[-\frac{1}{2}\left(\frac{y}{\sigma_y}\right)^2\right] \exp\left[-\frac{1}{2}\left(\frac{h_e}{\sigma_z}\right)^2\right] \tag{68}$$

Perfect ground reflection is included. With surface absorption the equation should be reduced by a factor which has a value of $\frac{1}{2}$ for complete absorption.

Maximum ground-level concentration for a given x will occur directly under the plume centerline ($y = 0$, $z = 0$). The following equation is obtained assuming perfect ground reflection:

$$\chi_{(x,0,0)} = \frac{Q}{\pi\sigma_y\sigma_z u} \exp\left[-\frac{1}{2}\left(\frac{h_e}{\sigma_z}\right)^2\right] \tag{69}$$

In many applications, Equation (69) will reach a maximum value in the downwind region of interest. This is shown schematically in Figure 6. For the special, but not unusual, case in which σ_y and σ_z have a constant ratio as x varies, maximum ground-level concentration is given as follows:

$$\chi_{(x_{max},0,0)} = \frac{2Q}{\pi h_e^2 e u} \frac{\sigma_z}{\sigma_y} \tag{70}$$

x_{max} denotes the distance at which maximum concentration occurs. Its value is implicit in the following equation by which x_{max} can be determined:

$$h_e = 2^{1/2}\sigma_{z,x_{max}} \tag{71}$$

Having σ_z given as a function of x either graphically or analytically as in Section III,C,1,b, x_{max} can be found for the value of $\sigma_{z,x_{max}}$ obtained from Equation (71).

Another special case of Equation (66) is that of an effluent source located at ground level with no plume rise ($h_e = 0$) [Eq. (72)]. This is of interest to many field experiments on diffusion without the presence of various stack effects. It has also certain applications for short stacks and vents discussed in Section III,C,3,c. Perfect ground reflection is assumed, but surface absorption could be significant since the plume is in continuous contact with the ground throughout its travel. The plume centerline is at ground surface.

$$\chi_{(x,y,z,h_o=0)} = \frac{Q}{\pi\sigma_y\sigma_z u} \exp\left[-\frac{1}{2}\left(\frac{y}{\sigma_y}\right)^2\right] \exp\left[-\frac{1}{2}\left(\frac{z}{\sigma_z}\right)^2\right] \tag{72}$$

For ground-level concentration ($z = 0$), the following equation applies:

$$\chi_{(x,y,0,h_e=0)} = \frac{Q}{\pi\sigma_y\sigma_z u} \exp\left[-\frac{1}{2}\left(\frac{y}{\sigma_y}\right)^2\right] \tag{73}$$

Maximum concentration at a given x is found on the centerline and given by the following equation:

$$\chi_{(x,0,0,h_e=0)} = \frac{Q}{\pi\sigma_y\sigma_z u} \tag{74}$$

For a point source, σ_y and σ_z approach zero as x approaches zero. Maximum ground-level concentration will, therefore, approach an infinite value as x approaches zero.

The above equations give concentration at a selected location. Air pollution analyses may require extensive and repetitive applications for ranges of σ_y, σ_z, and h_e. Such are facilitated with the numerous graphical presentations of Turner (*48*). Isopleths of concentration are given for various stability categories. Numerical properties of Gaussian profiles are included.

An unrealistic feature of the diffusion equations is that they do not give a finite plume edge, one where concentration reaches a zero value for finite y and z. This does not pose a problem unless a finite boundary is needed in an analysis or application. In practice a finite edge is obtained by arbitrarily terminating the theoretical profile at a selected concentration. This is often taken as one-tenth of maximum concentration in the local plume cross section. This boundary is shown schematically in Figure 7. Location of the plume edge in the horizontal and vertical planes through the plume centerline is given by the following equations where p is plume edge concentration as a fraction of local (i.e., section) centerline concentration. The ground reflection term is neglected, but the equations apply to a ground-level source with no plume rise. y_e and z_e are the coordinates of plume edge:

$$y_e = \pm(-2\sigma_y{}^2 \ln p)^{1/2} \tag{75}$$

$$z_e - h_e = \pm(-2\sigma_z{}^2 \ln p)^{1/2} \tag{76}$$

y_e and $z_e - h_e$ are plume half-widths in the horizontal and vertical planes, respectively. The following equations are obtained for the often used $p = 0.1$:

$$y_e = \pm 2.15\sigma_y \tag{77}$$

$$z_e - h_e = \pm 2.15\sigma_z \tag{78}$$

The above equations for the plume edge may not be applicable to visible plumes, especially for large x. This will depend on visibility or light attenuation of plume effluent, background visibility, and lighting. Obscuration of light by plume effluent will be dependent on line integration of concentration χ along line of sight rather than its point concentration given by the diffusion equations. This is discussed in Slade (*13*). It is evident that a downwind distance may be reached where plume concentration falls below a visibility threshold and the plume will appear to terminate. This cannot be predicted by Equations (75) and (76).

Equation (66) and its variations have applications to problems of thermal pollution in which it is desired to find plume temperatures. There are, however, additional sources of inaccuracy; but estimates of plume temperatures may be made. In this application emission rate Q becomes equivalent to heat emission rate Q_h in the plume rise analyses of Section III,B. χ then becomes heat per unit volume in the plume. Temperature increment ΔT at the location (x,y,z) relative to that of the ambient atmosphere outside of the plume is given by the following equation:

$$\Delta T = \chi / c_p \rho_p \tag{79}$$

c_p and ρ_p are specific heat and density at the location in the diffused plume. These will be close to those of the ambient atmosphere except in regions close to the stack since the fraction of plume effluent in the effluent–air mixture is small. Consistent units must be used for the various quantities in Equation (79). Conservation of mass in the development of the diffusion equations now becomes conservation of heat. There may be radiative heat losses depending on effluent components. There will be heat transfer at the ground surface which will reduce the ground reflection term.

Sutton's diffusion parameters. Standard deviations σ_y and σ_z are directly related to Sutton's parameters. Sutton's diffusion equation is expressed in terms of three parameters: diffusion coefficients C_y and C_z and turbulence index n. In Sutton's treatment (*1, 15*) n depends on wind velocity profile in the power law form given by Equation (13). n is related to exponent p of Equation (13) as follows:

$$p = n/(2 - n) \tag{80}$$

Diffusion coefficients C_y and C_z depend on components of turbulence in the y and z directions, respectively. In application, C_y and C_z were determined by fitting the diffusion equation to concentration profile data. In later applications, the dependence of n on velocity profile data by Equation (80) was relaxed, and its value was also determined by fitting the diffusion equation to concentration data. Abandonment of any direct dependence of Sutton's parameters on velocity profile was completed by defining two values of n, namely, n_y and n_z associated with C_y and C_z, respectively. With this version the following equations relate σ_y and σ_z to Sutton's parameters:

$$\sigma_y = 2^{-1/2} C_y x^{(2-n_y)/2} \tag{81}$$

$$\sigma_z = 2^{-1/2} C_z x^{(2-n_z)/2} \tag{82}$$

Thus any diffusion data expressed in terms of Sutton's parameters can be converted to the standard deviation form. If expressed in terms of C_y, C_z, and n only, the conversion can be made by setting $n_y = n_z = n$ in the above equations. Furthermore, the various preceding diffusion equations for concentration χ can be converted to Sutton's original form by substitution of the above equations.

Discussion of diffusion equations. Inspection of the diffusion equations shows qualitative features of interest to various air pollution problems. For a given emission rate Q, concentration $\chi_{(x,y,z)}$ by Equation (66) is a function of σ_y, σ_z, u, and h_e. Diffusion of a plume about its centerline is determined by σ_y and σ_z. Results of field experiments in Section III,C,1,b below show σ_y and σ_z to increase with x and with turbulence and thermal instability. Surface roughness also increases σ_y and σ_z due to its effect on turbulence.

As shown in Figure 6, concentration at a given distance x decreases in the central region of the plume with increasing turbulence but increases in its outer regions. Plume widths by Equations (75) and (76) also increase with turbulence. Ground-level concentration increases with turbulence in the lower range of x shown in the figure. The trend reverses for large values of x, and the plume tends to have concentrations of a ground-level source given by Equations (72)–(74) (σ_y and σ_z become large compared with h_e).

Effect of distance x depends on relative magnitudes of the variables. Qualitative features are shown in Figures 6 and 7. Centerline concentration $\chi_{(x,0,h_e)}$ given by Equation (67) decreases with distance at any x (due to σ_y and σ_z). Effect on ground-level concentration by Equations (68) and (69) depends on x location. It increases with x until the maximum is reached beyond which it continues to decrease with any further increase in x. Location of maximum $x_{\max}$ is dependent on σ_z and h_e as given by Equation (71). Increased turbulence reduces $x_{\max}$ as illustrated in Figure 6. Smaller h_e also reduces $x_{\max}$. Magnitude of maximum ground-level concentration $\chi_{(x_{\max},0,0)}$ given by Equation (70) depends on the ratio σ_z/σ_y, as well as h_e and u. The ratio decreases with increasing stability as shown by experimental data in Section III,C,1,b. This reduces the ground-level maximum.

The importance of effective plume height h_e on maximum ground-level concentration is also brought out by Equation (70). h_e appears in the equation as the inverse square. Thus increase in h_e causes a large reduction in the ground-level maximum but does not hold at large values of x beyond $x_{\max}$. At such distances concentrations are, however, at much lower levels. Since h_e is the sum of physical height of stack h_s and plume rise Δh, increase in either contributes to reduction of ground-level concentration. Δh depends on the various stack effluent and atmospheric variables as covered in Section III,B. For buoyant plumes and jets the various plume rise equations show Δh to increase with total effluent emission rate. This contributes to reduction of concentration. Pollutant emission rate Q (generally proportional to total rate) appears in Equation (70) as a multiplying factor which increases concentration. The net result depends on the magnitudes of various quantities involved; but, with certain combinations, ground-level concentration may be reduced with increase in emission rate. This is likely to be true for only limited combinations of atmospheric and stack variables. It could be important in specific cases such as part load versus full load operation for some industrial operations. This effect diminishes with x beyond x_{max}. At large values of x (with σ_z also large) the exponential term in h_e of Equation (68) approaches a value of one, and the remainder of the equation is equivalent to ground source Equation (73). The plume rise effect is important in the matter of dividing total emission among two or more stacks as compared with one. This aspect is considered in the multiple stack discussion of Section III,C,2.

Consideration of a wind velocity effect on ground-level concentrations in view of plume rise shows an important difference between pollution caused by a single large stack and that from small sources representative of many in an urban region. σ_y and σ_z will be assumed unaffected by u. This is true only for neutral conditions, but the general conclusions are valid. Most plume rise equations show Δh to vary inversely with u. At low velocities, Δh is large and there will be little or no ground-level concentration. As u increases, Δh decreases rapidly and the h_e exponential term in Equation (68) may increase more than the decrease due to u^{-1} giving a net increase in concentration. A velocity will be reached where this effect reverses and further increase in u will reduce concentration approaching the u^{-1} variation. For small stacks, the h_e exponential term is, except for small x, more nearly constant approaching a value of one. The remainder of Equation (68) is that of a ground-level source (as in the preceding discussion) for which concentration varies as u^{-1}. Many urban type sources such as motor vehicles are essentially ground-level sources with little or no plume rise. Thus concentrations in urban regions tend to follow the u^{-1} variation except for large stacks which may give concentration increase with u in the low velocity range. The low velocity–low concentration characteristic of large stacks given above may not occur in many situations due to departure from uniformity in atmospheric characteristics. With a ground-based or elevated stable layer there is a limit to plume rise. With time, pollutants may accumulate at low winds or calm conditions. The effect of nonuniform atmospheric conditions is discussed in Section III,C,3,a. Multiple source urban diffusion is covered in Chapter 10, this volume.

b. DIFFUSION PARAMETERS FROM FIELD DATA. Results of several extensive field programs on plume diffusion in the form of standard deviations of concentration distribution σ_y and σ_z are included below for application of the diffusion equations of the preceding section. They cover a range of stabilities and surface configurations for which the assumption of uniformity in atmospheric conditions is met to an acceptable degree. These are most accurate for near-neutral conditions, higher wind velocities, and uniform surface configurations. Most data groups cover the range of stability for which reasonably uniform conditions prevail, but the extremes of stability and instability require special treatment. Non-uniform conditions in which there is a marked change in diffusive properties from one region of the atmosphere to another are also treated by other methods. These departures from uniformity and the stability extremes are covered in later sections.

Each group of data from a given data source represents one site or similar surface configuration and one type of effluent source configuration as listed in Table III. Within each group there is generally a range of stabilities. For each group the data are plotted in Figure 8 as a family of curves of varying stability. These are supplemented with additional data in Table III and the text.

σ_y and σ_z are plotted as functions of distance. The graphs cover distances up to 10^4 m. Some data sources cover larger distances which could be obtained by limited extrapolation of Figure 8, but the reference should be consulted.

Since diffusion shows dependence on various atmospheric variables which themselves may be interdependent, there is no unique manner by which diffusion must be characterized or identified. Selection of atmospheric variables or derived parameters by investigators for a given program depends on many factors such as objectives, degree of accuracy, or ease of application desired. Thus diffusion is characterized by various variables and parameters in the several data sources. The subject of variable selection is covered in Slade (13).

Since turbulence is fundamental to diffusion, it is a logical atmospheric variable and potentially most accurate as a predictor of diffusion. Although complex in its formulation, it is closest to a single type variable. It represents effects of surface configuration as well as atmospheric thermal structure. Regardless of its potential advantages, use of a turbulence parameter for diffusion estimation has its limitations. Since local site configuration as well as meteorological variables affect turbulence, related parameters should be obtained from site measurements rather than estimates if their potential value is to be realized. Their most effective use requires considerable competence in instrument selection and siting, program development, and data analysis as provided by an air pollution meteorologist.

For routine or preliminary diffusion estimation to which this section is primarily directed, other variables are more readily measured or estimated. These include profiles of wind velocity and temperature and vari-

Table III   Data for Figures 8a and 8b[a]

Fig. ref. symbol	Data source designation and Ref. no.	Class. desig. in figs.	Qualitative stability designation	Std. dev. wind dir. σ_θ (deg.)	Temp. grad. $\partial\theta/\partial z$ (°C/100 m)	Ri_B	Effluent source	Surface configuration
(a)	Pasquill–Gifford (13, 48)	A	Extremely unstable	25			Low level	Open country
		B	Moderately unstable	20				
		C	Slightly unstable	15				
		D	Neutral	10				
		E	Slightly stable	5				
		F	Moderately stable	2.5				
(b)	Brookhaven (50)	B$_2$	Unstable[b]	c			Mostly elevated (108 m)	Open country, partly wooded
		B$_1$	Unstable[b]	c				
		C	Neutral[b]	c				
		D	Stable[b]	c				
(c)	TVA (Tennessee Valley Authority) (38)	N[d]	Neutral		0		Large power plant stack	Open country
		S[d]	Stable		0.64			
		MI[d]	Moderate inversion		1.36			
(d)	St. Louis (52)	i[d]	Unstable[b]	30+		< −0.01	Low level	Urban, relatively flat topography
		ii[d]	Unstable[b]	24–29		< −0.01		
		iii[d]	Unstable[b]	18–22		< −0.01		
		iv[d]	Neutral[b]	15–20		±0.01		
		v[d]	Stable[b]	8–13		> 0.01		
(e)	Vandenberg (54)		Neutral (various stabilities included)				Low level	Mountainous topography

[a] See text for additional details.
[b] Approximate designation for reference only by chapter author.
[c] Subjective classification of wind-direction fluctuations based on wind-direction record.
[d] Symbols assigned by chapter author.

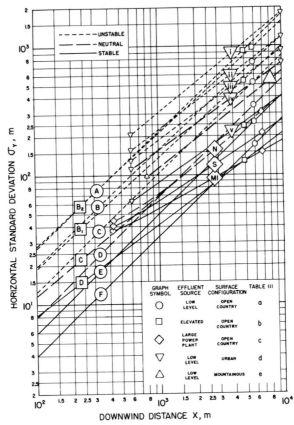

Figure 8a. Horizontal standard deviation σ_y of plume concentration distribution vs distance x for various surface configurations and effluent sources. Curves are not necessarily comparable due to differences in sampling time and other conditions.

ous parameters derived from or related to them. There seems to be no rational method for treating effects of surface configuration (other than turbulence measurement). Thus the data groups were selected to cover a range of configurations. The families of curves for σ_y and σ_z in Figure 8 are characterized by the various atmospheric variables used in the original data sources.

Various steps in the analysis of plume diffusion require selection of a wind velocity to represent the variable velocity (with elevation) in the real atmosphere. The velocity used in the plume rise equations of Section III,B is not necessarily appropriate for diffusion calculations since most or all plume rise may have occurred before reaching downwind distances of interest to concentration determination. Velocity is a vari-

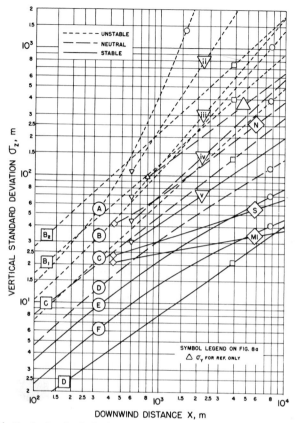

Figure 8b. Vertical standard deviation σ_z of plume concentration distribution vs distance x for cases in Figure 8a.

able in some of the methods for selection of diffusion parameters. Velocity enters directly in the diffusion equations as u^{-1}, representing the assumed uniform field of motion. It accounts for the diluting action of the moving volume of air into which the effluent is diffused. This single velocity must be a mean representative of the region of plume diffusion of interest to a particular application.

In the case where diffusion data are applied to an effluent source configuration and elevation similar to those of the experiments from which the data were obtained, it is evident that the original velocity reference elevation should be used. If the application differs only in velocity reference elevation such as due to availability of site velocity data at another elevation, correction for effect of elevation can be made with the velocity profile equations of Section II,A.

For applications to differing source configuration and elevation there is no general method for selection of reference velocity or elevation. In view of the fundamental difference between the assumed atmospheric uniformity in the diffusion equations and variable velocity of the real atmosphere, any method will give only an approximate value. Smith and Singer (49) developed a criterion for finding an elevation at which the wind velocity is a "mean equivalent wind" for low-level sources. This is based on integration of effect of variable velocity on concentrations. This elevation is approximately $0.62\sigma_z$. For elevated stack emissions the following are guides in the absence of specific information from the data sources. At limited downwind distances before substantial plume expansion has occurred, a mean velocity in the region between stack outlet and centerline elevation is representative of plume diffusion. For small plume rise, velocity at outlet elevation may be used. These velocities are also used for larger distances when plume central region is of interest. For ground-level concentrations a mean velocity of the region between plume centerline and the ground is more representative of velocity effect in the diffusion equations.

The σ_y and σ_z curves of Figure 8 apply to uniform atmospheric conditions as pointed out previously. They produce the coning-type plume of Figures 1 and 7. They also apply to some larger scale nonuniform conditions if the particular region occupied by the plume is uniform. An example is plume diffusion in the urban heat island surface layer capped with an inversion. The surface layer may be sufficiently uniform to be treated as such for plume diffusion analysis. This will hold as long as the plume is sufficiently small that it does not encounter boundaries of the uniform region. It will eventually become sufficiently large to be affected by region boundaries. Plumes in open country may become large enough to be affected by large-scale atmospheric stability variation with elevation. This occurrence for large power plant plumes is discussed by Carpenter et al. (38).

Important to plume concentrations in a given application is the sampling time or period over which concentrations are measured or calculated. Since atmospheric motions are time dependent and have a wide range of frequencies or wavelengths, these will influence concentrations obtained for a given period of measurement. Sampling period may be the characteristic of a concentration measuring instrument or the result of selection of a time interval over which concentration is obtained from concentration versus time data. Effluent emission is assumed to be continuous sufficiently long relative to period of concentration measurement that it is not affected by time of travel or longitudinal diffusion. These are not accounted for in the diffusion equations. In some cases effluent emission is made for a limited time period and concentration is measured over a longer period to cover longitudinal diffusion. The variable concentration is integrated to obtain total dosage [Turner (48)] and emission is likewise integrated to obtain total emission or release. The σ's are representative of the emission or release time. Effect of sampling time and methods for applying data for one time period to another are discussed in Section III,D. Such corrections are usually applied to concentration characteristics rather than σ_y and σ_z. The σ_y and σ_z data from the several field programs are for differing sampling or release times. They cover periods of approximately 1 hour and less. While the differing periods will have some effect on their values, they are viewed as covering similar periods for purposes of comparison to illustrate effects of other variables important to diffusion.

Sampling time is important and difficult to evaluate for the irregular looping plume (Fig. 1). It may produce high concentrations for extremely short periods, shorter than those for which σ_y and σ_z data are given. The looping plume occurs with highly unstable atmospheric conditions which, while uniform on a larger scale, produce the nonuniform smaller scale motions which cause looping. This is discussed in Section III,C,3,b.

The various data sources for the graphical presentations of σ_y and σ_z in Figure 8 and Table III are discussed briefly in the following paragraphs. While most of the data are derived from low-level effluent source experiments, they are treated as applicable to elevated stack effluent plumes. There are potential inaccuracies, but techniques for correction have not been developed. The letter symbols in parentheses for the following subsections refer to data groups in Table III and Figure 8.

(a) *Pasquill–Gifford data* (13, 48). The curves for the Pasquill–Gifford group in Figure 8 are based on those in Turner (48). Similar forms are given in Slade (13). They are derived from the earlier data of Pasquill (46) obtained for low-level sources in rural-type open country of relatively smooth surface. They apply to sampling periods of about 10 minutes. Stability categories are identified by letter designations A to F in increasing order of stability. Category D is neutral stability. Standard deviations of wind direction fluctuations σ_θ in Table III are from Slade (13).

Selection of stability categories for a given application is facilitated with Table IV taken from Slade (13) and Turner (48). Atmospheric variables are wind velocity and thermal conditions classified by day and night with subdivisions shown in Table IV. Night is taken as the period from 1 hour before sunset to 1 hour after sunrise. Strength of solar radiation depends on solar altitude and amount of cloudiness. Cloudiness is that fraction of the sky above the horizon covered with clouds. Neutral

Table IV   Guide for Selection of Stability Categories Pasquill–Gifford Data

| Wind velocity, u ($m\ sec^{-1}$) | Day, incoming solar radiation | | | Night, cloudiness | |
| | | | | Thin over-cast or $\geq 4/8$ cloudiness | $\leq 3/8$ cloudiness |
	Strong	Moderate	Slight		
<2	A	A–B	B		
2–3	A–B	B	C	E	F
3–5	B	B–C	C	D	D
5–6	C	C–D	D	D	D
>6	C	D	D	D	D

category D is used for overcast conditions at all velocities, day and night. Additional details are given in Turner (48).

Some estimates of accuracy are given by Turner (48). σ_y values are generally more accurate than those of σ_z. The more accurate values of σ_z are within a factor of 2 and occur at all stabilities out to a few hundred meters and for moderate instability out to a few kilometers. Accuracy decreases with distance and stability.

(b) *Brookhaven data (50)*. These data are based on numerous diffusion experiments performed at the Brookhaven National Laboratory, Upton, New York, over a period of many years. Most experiments were conducted with an oil–fog type effluent emitted at a height of 108 m from a meteorology tower. Sampling periods range from 30 to 90 minutes. Site configuration is partly wooded open country.

The results included herein are classified by gustiness categories designated B_2, B_1, C, and D. Category C appears closest to neutral and is so designated herein. Singer and Smith (50) do not employ stability designations. The categories are related to atmospheric turbulence by a subjective classification technique. It is based on wind-direction traces recorded by a Bendix–Friez Aerovane at 350-ft elevation. Details of the procedure are given in Singer and Smith (50) with samples of wind-direction traces to illustrate the technique.

Table V gives the equations for σ_y and σ_z from which the curves in Figure 8 are drawn. Included are wind velocities at two elevations associated with each gustiness category.

Singer and Smith (51) found in their analysis of the frequency of occurrence of the various categories that B_1 and D were dominant. Mean percentages were 42 and 40 for B_1 and D, respectively.

The generalized graphs and equations in Task Group (17) (ASME

Table V   Gustiness Categories—Brookhaven Data

Gustiness category	σ_y (m)	σ_z (m)	u (m sec^{-1})	
			$z = 9\ m$	$z = 108\ m$
B_2	$0.40x^{0.91}$	$0.41x^{0.91}$	2.5	3.8
B_1	$0.36x^{0.86}$	$0.33x^{0.86}$	3.4	7.0
C	$0.32x^{0.78}$	$0.22x^{0.78}$	4.7	10.4
D	$0.31x^{0.71}$	$0.06x^{0.71}$	1.9	6.4

Guide) are almost identical with the Brookhaven data. Smith and Frankenberg (*51a*) present a modification of the ASME horizontal standard deviation σ_y (apparently applicable to the Brookhaven data) by adding the term x tan ϕ to account for the effect of wind direction shear. ϕ is the angular wind direction change with height. This may become significant above 300 m for high stacks of large power plants. With large vertical plume thickness, the upper and lower portions may move in different directions which causes an increase in horizontal diffusion and reduction of concentrations. Smith and Frankenberg (*51a*) do not discuss the evaluation of ϕ. Their paper gives an extensive study of calculated and measured SO_2 concentrations for a large power plant before and after conversion from low to high stacks

(*c*) *Tennessee Valley Authority* (*TVA*) *data* (*38*). Plume dispersion for large coal-burning power plants has been investigated at TVA power plants in the Tennessee Valley region for many years. Carpenter *et al.* (*38*) report their analysis of various plume types They include graphical presentations of σ_y and σ_z for the coning plume from which the curves of Figure 8 are taken. The wind velocity used in their studies is the mean between stack top and plume top. The plants are located in open country. Sampling periods are very short.

Standard deviations σ_y and σ_z are given in Carpenter *et al.* (*38*) for six stability categories ranging from neutral to strong inversion. They are identified by potential temperature gradient $\partial\theta/\partial z$. Three of the categories were selected for presentation in Figure 8.

Montgomery *et al.* (*41*) employ equations for σ_y and σ_z at two stability conditions in their development of nomograms for a simplified technique of power plant plume analysis. For the coning plume they give the following equations which are identified for neutral conditions:

$$\sigma_y = 0.42x^{0.75} \qquad \sigma_z = 0.39x^{0.75} \tag{83}$$

σ_y, σ_z, and x are in meters. These equations are quite similar to the neutral condition curves of Figure 8. For stable conditions prior to formation of inversion breakup, they employ the following equations:

$$\sigma_y = 1.32x^{0.55} \qquad \sigma_z = 6.71x^{0.21} \tag{84}$$

Curves representing these equations would fall between the stable and moderate inversion curves of Figure 8.

Montgomery et al. (41) find that neutral conditions usually occur with moderate to strong winds greater than about 4 m second^{-1} and with cloudy skies in the day or with clear skies at night. At night to early morning with weak winds less than 1–2 m second^{-1} and with clear to partly cloudy skies, stable and inversion conditions usually occur. Unstable conditions usually occur with strong daytime solar radiation and weak winds less than 1–2 m second^{-1}.

(d) *St. Louis, Missouri data (52)*. McElroy (52) reports results of a number of urban diffusion experiments conducted over metropolitan St. Louis with low-level tracers. Pooler (53) discusses program plans, equipment, and experimental techniques along with early results. Topography is gently rolling. Winds and temperatures were measured at various elevations with an instrumented TV tower. Various other wind measurements were also made including pilot balloon observations. Tracer material was disseminated for periods usually 1 hour long.

Included in Figure 8 are σ_y and σ_z curves from McElroy (52) for a series of stabilities labeled herein with lower case Roman numerals as shown in Table III. They are in increasing order of stability. Meteorological variables used to identify the diffusion data are standard deviations of wind-direction fluctuations σ_θ measured at low levels and bulk Richardson number Ri'' [a variation of gradient Richardson number, Equation (7)] defined as follows:

$$\text{Ri}'' = g[\Delta T/\Delta z + \Gamma]z^2/Tu^2 \tag{85}$$

Ri'' is based on tower measurements at 39- and 140-m elevations. ΔT and Δz are determined from differences between the top and bottom of the 39- to 140-m layer. T is mean temperature through the layer and u is evaluated at upper elevation z. Since specific values of Ri'' are not given outside of the ± 0.01 range, they do not indicate the degree of stability or instability.

(e) *Vandenberg data (54)*. Effect on diffusion of the mountainous terrain of Vandenberg Air Force Base in southern California is reported by Hinds (54). While the regularity in diffusion data found in level regions are not expected for mountainous areas, limited data on σ_y and σ_z are included from Hinds (54) for comparison with other surface configurations.

The South Vandenberg area is a complex of ridge and canyon formations with ridges typically 300–500 m high. Tracer material was emitted at low levels for periods of 15–30 minutes duration. Effects of topography tended to reduce the significance to diffusion of meteorological variables found important in more uniform surface configurations. Hinds finds that for the mass of data obtained over many years for various conditions, the following equation for σ_y gives a reasonable fit with a scatter of about a factor of 2 above and below the mean:

$$\sigma_y = 0.65x^{0.75} \tag{86}$$

This equation is plotted in Figure 8a. The curve is designated neutral since it includes all conditions: stable, unstable, and neutral. Hinds does not give a corresponding equation for σ_z but shows a graphical presentation of σ_z data points compared with σ_y. For this purpose σ_y is shown on the σ_z graph of Figure 8b. Daytime values of σ_z scatter about the σ_y curve. Nighttime values are less than σ_y up to about 5000 m beyond which they tend to increase with some values of σ_z greater than σ_y. This may be due to specific topographical features. Hinds (54) also finds that concentrations in the valleys are much less than on ridges by a factor of about $\frac{1}{2}$.

Discussion of diffusion data. In comparing the σ_y and σ_z curves for various data sources in Figure 8, it must be remembered that there may be many factors contributing to variation in diffusion besides the variables emphasized in this presentation. Experimental techniques, data analysis, and sampling time as well as peculiar local conditions may have important influences. Effect of variations in σ_y and σ_z on concentrations at ground level and throughout the plume must be visualized in terms of the various diffusion equations discussed in Section III,C,1,a.

The more extensive and accurate data are those obtained for the uncomplicated open country with simple effluent sources as given by Pasquill–Gifford (a) and Brookhaven (b). These are sometimes used as a base relative to which others are compared. The Pasquill–Gifford data appear to be obtained with surface configuration a little smoother than that at Brookhaven although the difference is not emphasized in comparison studies. There is also the difference in source elevation. The Pasquill–Gifford data supplemented with Table IV for category selection have wide use. Strict application of the Brookhaven data requires wind direction records obtained with the Bendix–Frieze Aerovane. Singer *et al.* (55) have, however, developed a simplified technique for finding σ_y and σ_z for application to ground-level concentrations based on their extensive Brookhaven experience. In its general form the method requires measurement of various atmospheric variables including standard deviations of wind direction and azimuth angle fluctuations. Variations of method are included for more limited atmospheric data of various combinations. The simplified method given in the Task Group (*17*) (ASME Guide) appears to be the same.

There are certain features common to the various data groups in Figure 8. Variation with stability is larger for σ_z than for σ_y. This affects the ratio σ_z/σ_y in the maximum ground-level concentration Equation (70). For neutral stability, σ_z tends to be less than σ_y thus making the plume depth less than its width. With increased stability, σ_z/σ_y becomes smaller and approaches the ribbonlike fanning plume of Figure 1. The Pasquill–Gifford curves for σ_z show a greater spread with distance than those from Brookhaven. Smith and Singer (*49*) discuss this aspect and point out that differences in the unstable categories may arise from treatment of velocity variation with elevation. Variation of stability with elevation may be a factor in the stable categories.

σ_y values for large power plant plumes given by the TVA data (c) are similar in magnitude to those of Pasquill–Gifford and Brookhaven for the larger x range covered in Figure 8. There is less similarity for σ_z. For smaller x the power plant plume tends to have relatively larger σ_y and σ_z. This may be caused by large initial plume expansion which does not occur for the smaller effluent sources of the Brookhaven and Pasquill–Gifford-type experiments.

The effect of increased turbulence due to urban configuration is evident in the St. Louis data (d). Compared with open country σ's tend to be larger by about one category or more for intermediate x. σ_y tends toward open country values for larger x. The relatively larger σ_y for smaller x is analyzed by McElroy (*52*). He shows this may be due to initial plume expansion caused by nearby buildings which affects the low-level effluent emission. This is similar in principle to the case of effluent emission into the wake of a building discussed in Section III,C,3,c.

Comparison of standard deviations of wind-direction fluctuations σ_θ included in Table III for St. Louis and Pasquill–Gifford data shows that this indicator of diffusion accounts to a considerable degree for the increased turbulence due to urban configuration. Categories with similar σ_θ have similar σ_y and σ_z.

Selection of stability categories for urban diffusion requires consideration of effects of urban heating and roughness. Owing to urban heating an urban area will have greater instability than the surrounding open country areas. A neutral surface layer may continue into nighttime even though the surrounding country has a ground-based stable layer. These features are discussed in Chapter 8. If distance x is sufficiently small that the plume is wholly within the urban surface layer, it will diffuse in a manner appropriate to the stability and surface configuration of that layer.

The limited data for mountainous regions (e) shown in Figure 8 suggest that there is some regularity in plume diffusion. No firm conclusions should be drawn for other topography. Horizontal diffusion shown by the σ_y curve falls between that of level country and of urban regions. The σ_z data (see above text) is less definitive. Hinds (54) includes other forms of diffusion data.

While the above mountainous data may be the basis for obtaining crude estimates of plume concentrations, they do not locate plume center-line. It cannot be assumed that the plume will continue moving in a fixed direction. The plume centerline will be distorted depending on specific topographical features.

2. Multiple Stacks

While multiple stack arrangements are common for industrial installations, the effect of such stack configurations on plume behavior has had only limited exploration. Available means for evaluating multiple stacks are based on field studies of existing stack installations, generally power plants. Since such plumes are usually of the buoyant type, applicability of multiple stack corrections to jets is an open question.

Two stacks at sufficient separation distance will obviously have no mutual interaction. Concentration at a given downwind location will be a simple addition of the concentrations caused by the individual stacks. If two identical stacks were brought close together with negligible separation distance, and if there were no mutual interaction, plume concentration at a given location would be twice that from one stack. In terms of the various diffusion equations in Section III,C,1,a all terms including h_e would be the same except Q, which would be doubled. Concentration χ at a given location would, therefore, be twice as large. If the two adjacent stacks were viewed as one combined source, having heat emission Q_h twice that of one stack, then by the various plume rise equations of Section III,B, Δh and thereby h_e in the diffusion equations would increase. For maximum ground-level concentration by Equation (70), an increase in h_e will reduce concentration. The net result of doubling Q and increasing h_e will depend on relative magnitudes of the various quantities involved. Experience has shown that proximity of stacks tends to reduce ground-level concentrations compared with those obtained with simple super-position of individual stacks. At some elevated locations, concentrations could increase due to change in plume centerline location.

Montgomery et al. (41) include a correction factor for multiple stacks of approximately same height based on TVA power plant experience. It is expressed by the following equation applicable to ground-level concentrations:

$$\chi_N = \chi_1 N^{4/5} \tag{87}$$

where N is the number of stacks up to a total of 10. χ_1 is ground-level concentration due to one stack alone without the influence of the other stacks. χ_N is the ground-level concentration with N stacks operating simultaneously. For example, with three stacks χ_3 is $2.4\chi_1$ compared with $3\chi_1$, if mutual effects were omitted.

Briggs (56) has developed an "enhancement factor" E_N for multiple stack buoyant plume rise on the basis of TVA power plant data. This factor includes the effects of stack spacing as well as number of stacks. It is applied to plume rise Δh Equations (22) and (47). Enhancement of plume rise is obtained by the equation:

$$\Delta h_N = E_N \, \Delta h_1 \tag{88}$$

where Δh_1 is plume rise of a single stack without any effect of nearby stacks; Δh_N is rise of the same stack as one of a group of N enhanced by the presence of other stacks; and E_N is given in terms of a spacing factor S, and number of stacks N, as follows:

$$E_N = [(N + S)/(1 + S)]^{1/3} \tag{89}$$

For a line of N stacks with center spacing d_c of adjacent stacks:

$$S = 6[(N - 1)d_c/N^{1/3} \, \Delta h_1]^{3/2} \tag{90}$$

Briggs suggests that for a group of stacks not in a line, a conservative result is obtained by replacing $(N - 1)d_c$ with the maximum width of the group.

Briggs' (56) analysis includes the effect of multiple sources on plume rise in calm conditions and various other characteristics of multiple source plumes.

3. Nonuniform Atmospheric Conditions

Nonuniformities in atmospheric characteristics and surface features produce plume configurations and concentration distributions markedly different from those of the uniform type given by the equations of Section III,C,1. They may cause local concentrations higher than otherwise occur. Analytical methods are not as accurate as for uniform atmospheres over flat regions, but concentration estimates can be made with various procedures given below.

a. RESTRICTED DIFFUSION CAUSED BY INVERSION LAYERS. A neutral or unstable surface layer capped with a stable or inversion layer creates a situation in which plume diffusion is largely confined within the surface layer. The principal forms are fumigation (inversion breakup) and trapping (limited mixing layer). These are shown schematically in Figures 1, 9, and 10. Fumigation occurs during the morning hours following nighttime stable conditions. As the rising sun heats the surface, the resulting neutral or unstable surface layer increases in elevation until it reaches stack top. Up to this point the plume has been emitting into a stable layer where plume diffusion is low and concentrations are high. Little or none of the effluent has reached ground level. As the surface layer envelopes the plume, it is diffused downward and may produce relatively high ground-level concentration in the narrow region below the originally stable plume. This occurs for short periods generally less than an hour. This condition was studied in the early work of Hewson (57).

Trapping is caused by a similar restriction on upward diffusion but it is not related to morning breakup of the nocturnal inversion layer. It usually persists for longer periods and occurs at various times depending on the cause of its formation. It occurs with change of surface conditions from stable to neutral or unstable in the streamwise direction. This may be caused by airflow between land and a large body of water or between rural and urban areas. It may also be caused by passage of large-scale air masses of differing thermal characteristics or by the normal existence of an upper elevation stable layer. Stack height plays a minor role in reduction of ground-level concentration. A stack effluent emitted into the trapped layer at any elevation is so dispersed vertically that its initial elevation has little effect on its concentration distribution except in regions close to the stack. Pooler's (58) analysis of trapping and fumigation for power plant stacks showed they could be critically important. Carpenter et al. (38) found in their field experiments on Tennessee Valley Authority power plant plumes that with increasing size the nonuniform conditions of trapping and fumigation have become relatively more critical than the uniform coning condition. They report that for modern high stacks maximum ground-level concentrations caused by fumigating and trapping conditions were higher than those of the coning plume by factors of 2.4 and 3.3, respectively.

i. *Analysis.* The following equation for plume concentration distribution has application to some phases of both trapping and fumigation. It is assumed that horizontal distribution is of the Gaussian type, but that vertical distribution is constant at given x and y. It is also assumed that no effluent enters the inversion layer.

$$\chi_{i(x,y)} = \frac{Q}{\sqrt{2\pi}\, u\sigma_{yi} h_i} \exp\left[-\frac{1}{2}\left(\frac{y}{\sigma_{yi}}\right)^2 \right] \tag{91}$$

where h_i is the elevation of the inversion layer and σ_{yi} is representative of horizontal diffusion in the layer below the inversion. Maximum concentration is obtained on the vertical centerline plane ($y = 0$) with the following equation:

$$\chi_{i(x,y)} = \frac{Q}{\sqrt{2\pi}\, u\sigma_{yi}h_i} \tag{92}$$

In the discussions which follow, the subscript i is replaced with f and t for fumigation and trapping, respectively.

Trapping. In the trapping (limited mixing layer) configuration shown in Figure 9, the plume has initial rise and diffusion of the type appropriate to stability conditions of the trapped layer. The first effect of the inversion layer occurs when the upper plume edge reaches the inversion layer elevation h_t. The following analysis of plume development with downwind distance is that given by Turner (*48*). It is assumed that the distance x_{ta} where the plume first reaches the inversion layer is sufficiently far that it can be treated as a ground-level source given by Equations (72)–(74). Contact of the plume upper edge with the inversion layer is at the finite plume edge given by Equation (78) ($h_e = 0$). This yields the following equation which relates h_t and vertical standard deviation of concentration

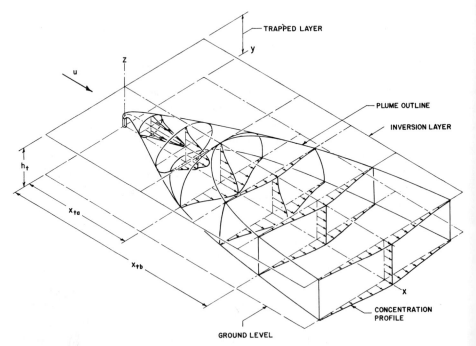

Figure 9. Plume concentration profiles under the trapping condition.

distribution $\sigma_{z,\text{ta}}$:

$$h_t = 2.15\sigma_{z,\text{ta}} \tag{93}$$

The corresponding downwind distance x_{ta} is found from σ_z vs x graphs of Figure 8b for neutral or unstable conditions of the trapped layer. The equations of Section III,C,1,a may also be used. Beyond x_{ta} the inversion layer causes downward diffusion which has little effect on ground-level concentration until the edge of the reflected plume reaches ground level at distance x_{tb}. At this location and beyond, the vertical distribution is substantially constant as given by Equations (91) and (92). Determination of x_{tb} requires use of an effective $\sigma_{z,\text{tb}}$ which, in Equation (73), gives the same ground-level concentrations as Equation (91). The following relation is obtained by equating Equations (73) and (91):

$$h_t = 1.25\sigma_{z,\text{tb}} \tag{94}$$

x_{tb} can be found by the procedure used for x_{ta} above. Turner (*48*) suggests x_{tb} be taken as $2x_{\text{ta}}$. If Equations (93) and (94) are substituted into one of the σ_z vs x equations of Section III,C,1,b for neutral conditions, x_{tb} is found to be approximately $2x_{\text{ta}}$. This confirms the Turner approximation.

Between x_{ta} and x_{tb} some form of interpolation on σ_z or concentration must be used. Turner (*48*) suggests the use of a linear graphical interpolation on log–log coordinates for ground-level concentration.

Horizontal standard deviation σ_y is selected to represent the stability of the trapped layer. Carpenter *et al.* (*38*) modify σ_y to give the trapping form σ_{yt} for large power plants:

$$\sigma_{yt} = \sigma_y + [(h_t/1.1) - 2.15\sigma_z]0.47 \tag{95}$$

They find that maximum ground-level concentrations occur at distances of 5–10 km. Montgomery *et al.* (*41*) use a value of 3 km in applying their nomograms for the trapping condition.

Fumigation. Analysis of fumigation (inversion breakup) is based on determination of the time t_f for the originally stable plume to diffuse to ground level as it is enveloped by the increasing nonstable surface layer. Downward diffusion begins when the surface layer reaches stack top. It continues as the layer increases until it reaches a maximum when the layer elevation reaches top edge of plume. Plume configuration is shown in Figure 10. Elevation of plume top h_f is the sum of effective plume centerline elevation h_e plus the vertical half-width of plume $2.15\sigma_{z,f}$ given by the following equations:

$$h_f = h_e + 2.15\sigma_{z,f} \tag{96a}$$

$$= h_s + \Delta h + 2.15\sigma_{z,f} \tag{96b}$$

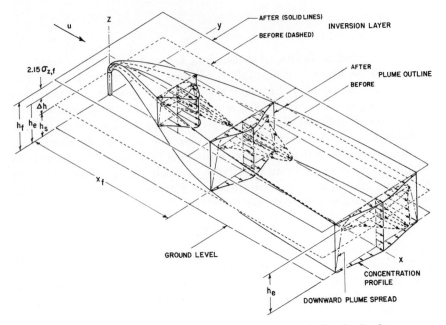

Figure 10. Plume concentration profiles before and after fumigation.

h_e is found for the originally stable layer into which the plume has emitted before being enveloped by the increasing surface layer. Carpenter *et al.* (*38*) modify Equations (96a) and (96b) with a multiplying factor of 1.1 to account for increase of the plume top elevation during transition.

Downwind distance x_f for maximum concentration is simply the product of time t_f by wind velocity u. Turner (*48*) gives two equations for determining t_f for different input data. The following form developed by Pooler (*58*) is based on potential temperature gradient $\partial\theta/\partial z$ and solar heating rate:

$$t_f = \frac{\rho c_p}{2q_{hf}} \frac{\partial \theta}{\partial z} (h_f - h_s)(h_f + h_s) \qquad (97)$$

q_{hf} is net rate of sensible heating of an air column by solar radiation, heat $\times$ length^{-2} $\times$ time^{-1}. Pooler (*58*) includes a value of 67 cal m^{-2} second^{-1} for q_{hf}. The second equation is derived from one given by Hewson (*57*) which has its origin in an expression developed by Taylor (*59*). It has the following form:

$$t_f = (h_f{}^2 - h_s{}^2)/4K_d \qquad (98)$$

K_d is eddy diffusivity for heat, length^2 $\times$ time^{-1}. Hewson (*57*) suggests a value of 3 m^2 second^{-1}.

A basically similar equation for t_f is used by Carpenter et al. (38), but they express it in terms of eddy conductivity K_c, heat $\times$ length^{-1} $\times$ temp^{-1} $\times$ time^{-1}, as follows:

$$t_f = \rho c_p [(h_f{}^2 - h_s{}^2)/4K_c] \qquad (99)$$

Carpenter et al. (38) treat K_c as a function of potential temperature gradient $\partial\theta/\partial z$ with a graph. The following empirical equation represents the graphical form:

$$K_c 10^3 = \exp[-0.99(\partial\theta/\partial z) + 3.22] \qquad (100)$$

K_c is in units of cal m^{-1} °C^{-1} second^{-1} and $\partial\theta/\partial z$ in °C/100 m.

The plume is originally in a stable layer and has σ_y appropriate to the layer. As the nonstable layer envelopes the plume, downward diffusion into that layer is greater than in the stable layer, and σ_y increases to produce $\sigma_{y,f}$ characteristic of the inversion breakup. Turner (48) treats $\sigma_{y,f}$ as produced by a downward angular spread of 15° at each plume edge as shown in Figure 10. This starts at effective plume elevation h_e. The effect of plume spread on standard deviation is given by the following equation:

$$\sigma_{y,f} = \sigma_y + h_e/8 \qquad (101)$$

Carpenter et al. (38) find from their power plant field studies the following equation to represent increase in horizontal diffusion:

$$\sigma_{y,f} = \sigma_y + 0.47 h_e \qquad (102)$$

With values of $\sigma_{y,f}$ determined by above methods, concentrations for distances $x \geq x_f$ are found with Equations (91) and (92).

b. LARGE THERMAL INSTABILITY AND THE LOOPING PLUME. High solar heating of ground surfaces with low winds produces large convective eddies of air motion. When these are large compared with plume dimensions, the looping plume is produced as shown in Figure 1. Concentrated segments of the plume may reach ground level and cause high effluent concentrations for very short time periods. Between these periods, concentrations are low or zero. Concentration variation with time at a given location will show wide fluctuations. Time average values may be a small fraction of peak values representative of the looping plume. For the longer sampling times given with the σ_y and σ_z data of Figure 8, concentrations are obtained with the various diffusion equations using the most unstable categories. Other sampling times treated by the method of Section III,D do not extend to the extremely short times representing the looping plume. Short-period concentration measurements needed for the looping plume are difficult to obtain.

The looping plume seems to play a lesser role in air pollution problems. Carpenter *et al.* (*38*) find it of lesser importance for large power plant stacks than those for other conditions which give concentrations of longer duration. Lesser attention has been given to techniques of analysis. The method of Bierly and Hewson (*60*) treats the plume as originating from a ground-level source but at a distance equal to the straight line distance s from stack top to ground-level point of concentration determination. They use a ground-level point source equation equivalent to Equation (74) but expressed in terms of Sutton's diffusion parameters. They give parameter values of 3-minute and 1-hour averages. Substitution of their 3-minute values yields the following result:

$$\chi = (14Q)/(us^{1.8}) \tag{103}$$

u is in m second^{-1} and s in meters. χ and Q are also in meter units appropriate to pollutant quantities used. Bierly and Hewson also recommend that u not exceed 4 m sec^{-1}. Their parameters for 1 hour reduces Equation (103) values by a factor of about $\frac{1}{4}$.

c. EFFLUENT EMISSION IN WAKES OF BUILDINGS. Effluents emitted from short stacks or flush vents on top of or close to a building may be partly or wholly immersed in the turbulent wake or cavity region. As shown schematically in Figure 2, effluent diffusion is greater than would otherwise occur in the absence of a building. The increased downward diffusion will usually cause increased ground-level concentrations in near downwind locations. The plume centerline, if it can be identified, will be highly distorted. Pollutant concentrations may occur where there would be none in the absence of a building. Some portions of the effluent in the cavity region may travel in upwind directions near the sides and top of the building. On the other hand, average concentrations at larger downwind distances may be lower due to plume diffusion over the wake cross section which will be larger than that of a simple plume. Concentrations at a given point will have short-period time-dependent fluctuations due to turbulent variations in motion.

Analytical techniques give only rough estimates of concentration in these highly nonuniform regions. Reliable results can only be obtained by experimental methods employing field experiments or scale model wind or water tunnel and tank experiments. Extrapolation of data from one building configuration to that for another may lead to erroneous results. Patterns of air motion are sensitive to small changes in building configuration and wind direction. Presence of nearby buildings introduces other variables. The following discussions include methods for presenting experimental data and estimating concentrations.

Plumes from stacks of sufficient height and plume rise will remain clear of the building wake and cavity. They will diffuse substantially as if the building were not present. A long-standing rule of thumb in the electric power industry for avoidance of building effects, sometimes termed downwash, is that the stack height be $2\frac{1}{2}$ times that of the associated building. This rule obviously omits such variations as are caused by differing building configurations but seems to be a generally valid criterion for a plume to be clear of adverse building effects. While it is desirable to have sufficient stack height to clear the wake region, there may ·be cases where pollutant concentration in the effluent is so low that plume entrainment by the adverse motions can be tolerated, thus reducing stack costs.

A plume located near the boundary of the wake region may be only partially entrained. The degree of entrainment is quite sensitive to change of plume elevation. A plume with a given momentum and/or buoyancy may have sufficient rise variation with wind speed to change from a clear to an entrained condition within a short time period. This may also occur with wind-direction changes for asymmetric configurations since wake and cavity configurations also change. Plume rise techniques of Section III,B do not account for the effects of nonuniform flow fields of wake regions.

Various forms of diffusion equations have been used in analysis and presentation of wake diffusion concentrations. Conversion of data from one form to another is not always possible. The following equation or its variations is often used for presentation of concentration data. Concentration χ at a given wake location is given in terms of the nondimensional concentration coefficient K_w:

$$\chi = K_w Q / A u \qquad (104)$$

χ and emission rate Q have the same definitions as in the diffusion equations of Section III,C,1. Speed u of the approaching wind should be defined with the K_w data. In absence of a defined speed, a mean value in the layer of wake height may be used. Concentration prediction techniques given below assume a uniform wind speed. K_w may, in addition to location, depend on the various variables which affect wake-free plume diffusion plus those caused by the building. The latter include building configuration and wind direction. Another variable is that of heat emission from building surfaces which is not accounted for in the following methods. For some industrial processes this could be sufficient to affect wake characteristics. It is likely to reduce ground-level concentrations because of upward convective motions.

Reference area A brings building size into Equation (104). It is usually defined as the projected area of building outline on a vertical plane normal to the wind direction, i.e., the area seen as viewed along a line parallel to the approaching wind. This is a crude indication of wake size and is variable with wind direction for all but cylindrical-type shapes. A constant A is obtained with projected area for a selected direction such as wind normal to the longer building dimension as used by Halitsky (*16*). In some presentations, A is replaced with l^2 where l is a characteristic length. It is evident that the precise definition of A or l must be used with given K_w data.

Typical K_w values (based on projected area normal to wind) are suggested by data such as found by Halitsky in Slade (*13*), by Meroney and Symes (*61*) from wind tunnel experiments on reactor containment-vessel-type shapes, and by Dickson *et al.* (*62*) from field experiments at a reactor complex. These are for leak-type emissions which have little or no plume rise. Maximum ground level K_w are of the order of 1.0 near the end of the close wake at downwind distances in the range of 2 to 5 building widths or heights. This distance is at or slightly beyond the end of cavity region. At closer distances, K_w may increase to large values as the effluent source is approached. It differs widely with configuration in the close region dominated by cavity flow. The three-dimensional K_w distribution shown by the wind tunnel experiments of Halitsky in Slade (*13*), treats a cylindrical vessel with a hemispherical top. Effluent source locations were varied from the top of the reactor shell to ground level at upwind, side, and downwind positions. They show a large variation in K_w with spatial location in the near wake region. Halitsky found that maximum concentration, wherever it occurred laterally, varied approximately as the inverse square of distance from the reactor shell surface. For elevated source locations, ground-level K_w values show lesser rate of reduction with distance. Meroney and Symes (*61*) wind tunnel experiments covered various tapered cone frustrums. They found wide variation of ground-level concentration in the close wake region with distance for various tapers. Their graphical data tended to follow a power law form with varying exponents up to about 6 diameters downwind. Their K_w values decreased approximately as $x^{-0.8}$ for cylindrical shapes and had a slight increase with x for the most highly tapered shape. All values tended toward a value of 1.0 at about 6 diameters downwind. It is impossible to predict concentration in the close wake region to even an order of magnitude for varying configurations without specific experimental information. There is, however, a tendency to approach a K_w value of the order of 1.0 at or shortly beyond the cavity-type close wake region.

The basis for the occurrence of K_w values of the order of 1.0 at the end of the close wake region is readily visualized. It must first be assumed that the effluent has become uniformly distributed throughout the wake cross section due to strong mixing in the cavity region. This is represented by the uniform concentration profile at x_c in Figure 11. None of the experimental data show a uniform concentration profile, but this appears as the best approximation when specific shapes are not involved. The wake is viewed as a duct having cross-section area A_w and average velocity u_w which gives volume flow $A_w u_w$. Concentration χ for emission rate Q distributed uniformly over wake cross section A_w is given by

$$\chi = Q/(A_w u_w) \tag{105}$$

The wake area behind buildings of cubical-type shapes is approximately twice A. Wake velocities are less than approach velocities by as much as $\frac{1}{2}$. With A_w twice A and u_w one-half u, χ becomes $Q/(Au)$ which requires K_w value of 1.0 in Equation (104).

The following equation for estimating concentration in close regions, as given by Gifford (47) and others, is similar in form to Equation (105):

$$\chi = Q/(cAu) \tag{106}$$

The uniform concentration distribution represented by this equation is shown schematically in Figure 11 at x_c. The suggested range of c values

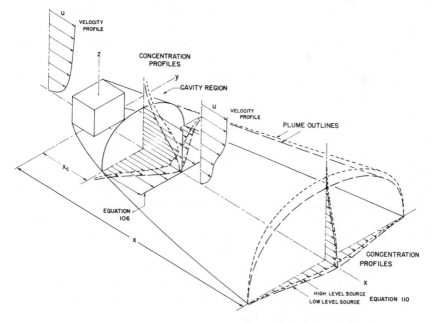

Figure 11. Stack effluent diffusion in the wake of a building.

is from 0.5 to 2 with the higher values being for blunt shapes which produce larger wakes. It is at least implied that c accounts for wake area, but Equation (106) does not account for wake velocities other than u values. Some correlations with field data suggest c values less than 0.5 which seem small on the basis of wake area alone. Such low c values could account for lower wake velocity as well as peak concentrations in non-uniform distributions. Large experimentally determined values of c may be due to partial plume entrainment in the wake; c then becomes another diffusion parameter similar in significance to K_w in Equation (104).

For increasing distance into the near wake region beyond the cavity, diffusion becomes more regular but still difficult to accurately predict. The widely varying close wake K_w values of Meroney and Symes (61) discussed above for various cone tapers tend to merge into a similar curve in the near wake continuing out to 30 diameters, the range covered. K_w varies approximately as $x^{-0.95}$. Dickson et al. (62) found with their field experiments on a reactor containment vessel complex that the average variation was as $x^{-0.6}$.

With increasing distance, the wake flow merges with the background of the approaching stream, and diffusion becomes dominated by pre-existing turbulence. The concentration variation with distance for a ground-level point source in the open atmosphere can be found with Equation (74). With σ_y and σ_z for the neutral Pasquill–Gifford category, its variation is approximately $x^{-1.8}$. Culkowski (63) compared measured ground-level concentrations for various rooftop emissions with those calculated for a simple ground-level point source. The buildings had a crosswind dimension considerably larger than height. With increasing distance, concentrations tend to merge with those of the simple ground-level source with neutral or slightly unstable diffusion. The distance for merging of values was of the order of 10 building heights although it differed for various experiments. This may be on the low side for other applications. Dickson et al. (62) found that aerodynamic effects as measured by turbulence characteristics tend to disappear at a distance which is of the order of 20 times the height of the reactor containment building. There were other buildings nearby which may have had an effect. However, it cannot be concluded that concentration distribution has reached that of a point source for this distance since initial diffusion caused by the building is not eliminated by return to free stream turbulence.

Gifford (47) gives the following equation [Eq. (107)] which combines the diluting action of the building as given by Equation (106), and that of the ambient atmosphere for a ground-level point source as given by Equation (74) for centerline concentrations:

$$\chi = Q/(\pi\sigma_y\sigma_z + cA)u \tag{107}$$

For close distances, σ_y and σ_z are small and χ is near that of Equation (106), thus accounting for initial diffusion. At large distances, the $\pi\sigma_y\sigma_z$ term dominates, and χ values approach those of the simple ground-level source.

Gifford presents in Slade (13) another diffusion equation employing modified standard deviations $\sigma_{y,w}$ and $\sigma_{z,w}$ to account for building induced diffusion and free atmospheric diffusion, total diffusion factors:

$$\sigma_{y,w} = (\sigma_y^2 + cA/\pi)^{1/2} \tag{108}$$

$$\sigma_{z,w} = (\sigma_z^2 + cA/\pi)^{1/2} \tag{109}$$

These are used in the following elevated point source equation identical with Equation (68) except for use of modified σ's. It gives ground-level concentrations:

$$\chi_{(x,y,0)} = \frac{Q}{\pi\sigma_{y,w}\sigma_{z,w}u} \exp\left[-\frac{1}{2}\left(\frac{y}{\sigma_{y,w}}\right)^2\right] \exp\left[-\frac{1}{2}\left(\frac{h_e}{\sigma_{z,w}}\right)^2\right] \tag{110}$$

Effective plume height h_e may be considerably different from that found with the plume rise equations of Section III,B because of the nonuniform wake motions. Halitsky's graphs in Slade (7) include vertical centerline plane concentration isopleths for top release of leak-type emission. If the line of maximum concentrations is taken as the plume centerline, plume elevation decreases with distance down to ground level in a distance of several building diameters. Use of outlet elevation as plume height for a nonbuoyant plume may give excessively high h_e. Equation (110) is equivalent to Equation (107) for $y = 0$, $h_e = 0$, and $\sigma_y = \sigma_z$. For $\sigma_y \neq \sigma_z$, the denominator of Equation (110) becomes larger than that of Equation (107).

Equation (110) with its exponential term widens the plume (at 0.1 of centerline concentration) over that for uniform distribution of Equation (106). This distribution is shown schematically in Figure 11 at x_c, but only the constant concentration distribution plume outline is shown. With more nearly uniform turbulence distribution at larger distances the Gaussian distribution of Equation (110) becomes more realistic. This shown at x in Figure 11.

Partial entrainment of plumes in building wakes has been observed in field experiments, but results do not seem to have been used to form partial entrainment analyses. Three-dimensional concentration and flow field measurements are rarely available as needed for such treatment. Munn and Cole (64) obtained ground-level concentrations for an effluent from a short stack extending 10 ft above the roof of a 60-ft-high heating plant building. Tracer material was introduced into the stack effluent for concentration measurements. Measured concentrations were of an order of magnitude less than those given by the preceding type equations which assume full entrainment. Lack of concentration measurements at elevations makes it impossible to assess the degree of entrainment. The nearest distance of measurement was of the order of ten times building height, well beyond the cavity region. Their data show a trend for increase in concentration with wind speed in contrast with the inverse ratio for the various wake diffusion equations (except where h_e is treated as variable with u). Lawson (65) in his discussion of the Munn and Cole (64) data interprets part of the results, including all below 12-ft sec^{-1} wind speed, to be cases where the plume "breaks clean" of the upper edge of the building wake in close regions. His mathematical model deduces effective plume height which decreases with wind speed. In the lower speed range these values are well above the building wake. This suggests that only a small portion of the stack effluent is entrained. Lawson does not interpret plume height in terms of momentum and buoyancy contributions. Effluent emission speed was relatively low, suggesting a low momentum rise, but the high effluent temperatures may produce a large buoyancy rise. Since most of momentum rise occurs in the shorter distances, it would be the more effective part of total rise for clearing the close wake and cavity region. Lawson's interpretative results indicate that regardless of low emission speed there was ample rise to initially clear the near wake for most cases tested. Once clearing the wake, buoyance rise may have kept much of the plume free of entrainment at larger distances.

Cases of partial entrainment were included in the experimental studies of Davies and Moore (66). Their field, air, and water tunnel experiments dealt with short rooftop stacks. Plume concentration profiles based on optical density measurements show that plume centerline remains above stack top, but there is some lowering at the larger distances where downward diffusion is greater than upward. This is in contrast with Halitsky's wind tunnel data in Slade (13) with a nonrising effluent (complete entrainment) emitted at the containment vessel top. His plume height decreased rapidly below emission elevation. Thus the estimation of an effective plume height h_e for Equation (110) is difficult in the absence of specific experimental data. Its inclusion, however crudely estimated, is likely to give better estimates of ground-level concentrations.

The experiments of Munn and Cole (64) and Davies and Moore (66) show the significance of plume rise in partial entrainment based on a parameter related to plume rise. Both series of experiments show that little or no ground-level concentrations were found when the ratio of effluent emission speed to wind speed V_s/u exceeds a certain value characteristic of the particular situation. Values varied between 2 and 5. This ratio appears directly in momentum rise as in Equation (63) but may also indicate buoyancy rise when V_s is proportional to mass emission rate. Davies and Moore (66) include in their scale-model experiments a case where V_s was increased without increase in mass emission (presumably by reduction in stack outlet area) which increased momentum rise only. This gave a favorable effect which would be achieved at considerable increase in fan power. Any estimates for plume rise as used in Equation (110) should include momentum rise.

Halitsky (67) presents a method for estimating the effect of partial plume entrainment in the near region where strong cavity mixing occurs. His method includes simplified mathematical models of plume concentration profiles and partial entrainment removal of the plume effluent into the cavity and wake regions. Plume rise is treated with a momentum rise equation, and the effect of flow curvature over the building top is included. Data of the type obtained from experiments must be used for location of the wake boundary.

The irregular turbulent motion in a wake produces short-period high concentrations. These may be much higher than those found for the longer periods representative of wake concentrations determined by preceding methods. This is similar in principle to the looping plume discussed in Section III,C,3,b. Effect of sampling time is inherent in the selection of σ_y and σ_z values for use in Equations (107)–(109). When taken from data as given in Figure 8, they show only time effects for the approaching airstream. Treatment of shorter period concentrations by sampling time for wake diffusion is discussed briefly in Section III,D. Another approach is given in the following paragraph.

In the method of Bierly and Hewson (60) it is assumed that the instantaneous plume has much the same characteristics as one produced by a ground-level source [Eq. (74)] in unobstructed flow for near-neutral conditions. This plume has time-varying locations within the wake region. It is similar to that shown in Figure 12 for the free atmosphere plume of short sampling time within the wider plume of longer time. The path is distorted to fit the instantaneous flow conditions within the wake region. Highest concentrations at ground level will occur for the shortest distance from the stack top to the ground-level location of concentration determination. The concept is identical with that for the looping plume of the preceding section. Equation (103) applies here with s as the shortest distance. The 1-hour sampling period also suggested by Bierly and Hewson may be calculated by the appropriate parameters which give lower concentration. Values obtained by this instantaneous plume method may be quite high. They apply to those very short periods when the plume is shifting past a given point.

Estimation of effluent concentration within a complex of closely spaced buildings as in an urban area is obviously even more difficult than for the problems discussed in the preceding paragraphs. With the stack top well above the rooftops of a complex of buildings of similar height, the simple plume diffusion techniques of preceding sections for uniformly rough surfaces give concentration estimates. Diffusion parameters are selected as appropriate to urban geometry. Some values are given in Figure 8. Such concentrations would apply to rooftop elevations. The effective ground-surface elevation is above the street level and may be near the rooftop for closely spaced buildings. It might be expected that rooftop concentrations would be transmitted to ground level between buildings. However, there may be horizontal components of air motion among buildings which cause an effective lateral diffusion. There is also a time lag in transmission of effluent between rooftop and street level. Change in wind direction may retard buildup or reduction of concentrations to instantaneous rooftop values.

Emission of an effluent near street level among closely spaced buildings presents a condition which cannot be treated with any of the preceding methods of plume analysis. Various urban studies such as those of Georgii et al. (19) and many others on multiple-source urban diffusion generally do not separate one source in a complex of many. A few complexities of air motion are shown in Figure 3. Diffusion of effluents will usually be much less among buildings than in open areas. Regions above streets parallel to wind direction may have high local velocities which cause much greater travel and diffusion out of the confined regions than in crosswind streets. Concentrations will be less. Motion in crosswind streets may produce unexpected concentration patterns. This is brought out by one of the wind tunnel experiments of Hoydysh and Chiu (20) for an effluent emitted at ground level at midblock location. At a given elevation, i.e., floor level, in the lower part of the building, concentrations are higher in midblock positions, as may be expected. At higher elevations, concentrations are more nearly constant or higher near building corners. The corner vortices shown in Figure 3 provide a mechanism for upward travel of effluent. Change in wind direction will alter this considerably. Another characteristic of urban configuration shown by the experiments is the favorable effect of nonuniform building heights. Compared with a uniform distribution of cubical-shaped buildings of constant height, increasing the height of one building reduces concentration in the nearby street regions. Likewise, removal of a building to create an open area is favorable as may be expected.

d. EFFECT OF TOPOGRAPHY ON PLUME DIFFUSION. The features of air-flow over regions of irregular topography discussed in Section II,B,3 introduce additional variables of plume travel and diffusion to the many discussed in the preceding sections. Accurate data on plume diffusion over complicated topography can be obtained only by experimental methods. Many field studies have been performed, but, with the infinite variations that exist in topographical features, they lack generality of application to new situations. The tremendous problems of instrumentation logistics discourage broad coverage beyond immediate needs of an experimental program. Spatial measurements of plume concentrations, velocity, and turbulence increase costs considerably and are usually not obtained for elevations comparable in size to the topographical features. Scale-model experiments in wind tunnels and water channels have been successful but are not as accurate as scale-model studies of diffusion around buildings. With the small linear scale of model needed to adequately cover a significant topographical region, scale effects become difficult to evaluate. Inclusion of nonneutral temperature structure makes modeling much more difficult. Modeling criteria are discussed in Section IV,B.

Except for differences in boundary configuration, the aerodynamics of flow over irregular topography is similar to that of buildings when thermally neutral conditions prevail. The cavity and wake flow shown in Figure 2 for a building has its counterpart in regions downstream of mountain ridges shown in Figures 4b, 4c, and 4d. Plume dimensions are, however, so much smaller in relation to size of topographical features that the plume may retain its identity for relatively longer distances before becoming widely diffused. The motion may be more regular and less diffusive because of smaller variations in boundary configuration as compared with those of building shapes. Except for the effects of plume rise, which are relatively small except for very low winds, the plume centerline tends to follow the local flow direction as shown by the various streamline patterns given in Figure 4. The unpredictable path that a plume can take over irregular topography on a large scale is shown by the wind tunnel experiments of Sakagami and Kato (68) for a mountainous region. Their three-dimensional concentration data determine the spatial position of the plume centerline as well as diffusion. At some locations, the centerline direction is as much as 60° from that of wind direction at higher elevations. The path elevation also varies but not to the degree of the surface variation below. One experiment shows the initial upstream travel of a low-level source located leeward of a ridge. This would be illustrated in the configuration of Figure 4d if the plume were emitted at low elevation between the ridges.

Addition of the effects of thermal stability and other thermally induced motions to the simpler neutral flow fields may radically alter plume travel and diffusion. This is especially true at lower elevations and in the valley-type configurations shown in Figure 5. These are often the cause of acute pollution episodes. Some topographical flow features are discussed in Section II,B,3. The reduced diffusion with stable gradients that normally occurs in open regions is supplemented with the confinement of topographical features. Stable conditions are common at night but may persist throughout the diurnal cycle and extend for days. The low or nonexistent winds aloft which often accompany these conditions prevent the usual plume travel and diffusion out of the region. If a plume has insufficient rise, effluent will remain within the region and accumulate. Large plumes with sufficient buoyancy may rise to elevations high enough to escape the restrictions to movement. The plume rise equations of Section III,B for calm conditions are a guide for estimating plume rise under these conditions.

Some analytical methods have been employed to treat certain simple topographical effects. They are special applications of various equations in preceding sections of this chapter. For limited distances over a relatively flat section of a larger complicated region, a plume behaves much like that over flat open areas. Small undulations in surface elevation as in Figure 4a have only secondary effects. At lower elevations, air movement tends to be parallel to the surface, and plume elevation can be treated as relative to local surface elevation. Plume from a stack upstream of a sharp rise in elevation as in Figures 4b and 4c will rise due to upward air movement but not as much as the increase in local surface elevation. Shown in the figure are the two-dimensional cases, as for a long constant elevation ridge normal to wind direction. With a three-dimensional shape such as a conical hill, rise will be less. This is further reduced with the addition of thermal stability which suppresses vertical movement. The plume tends to go around the hill rather than over it, more like the flow of water around an obstruction. A conservative technique of analysis omits the effect of upward movement due to increasing surface elevation. For concentration estimation on a ridge or plateau downstream at ridge elevation, plume diffusion is treated as if it occurs over an imaginary flat region of ridge top elevation but with actual stack top elevation. Stack height is, therefore, reduced by the difference in elevation between ground level at stack base and ridge top. Turner (48) suggests that when ridge elevation is at or above effective plume elevation, it be treated as a ground-level source to give highest expected concentrations.

Flow on the leeward side of a ridge with downslope configuration as shown in Figures 4c, 4d, 4e, and 4f may have radically differing patterns depending on stability conditions. The plume from a stack with sufficient height located on ridge top will, with neutral stability, travel out over the lee eddy and not reach the ground level until the downstream side of the eddy region. Some of the effluent may reach the leeward side of the ridge by reverse flow under the eddy. With the long travel distance before reaching the ground level, concentrations are likely to be less than for equivalent level region. If a stack is located at the downstream side of the eddy where the flow is vertically downward and divides as it meets the ground, the plume could be brought to ground level near the stack and cause local short-period high concentrations. This location of flow separation, a stagnation region, will shift around. With stable conditions, leeward flow is more nearly parallel to the local surface as shown in Figure 4e and 4f. The plateau approximation given above will apply since relative elevation of plume and local ground surface will be similar. Turner (48) recommends that the downslope side of a ridge be treated as flow parallel to local surface.

Berlyand (69) includes results of numerical solutions for various fundamental equations of motion and diffusion applied to hilly regions. His examples show a factor to be applied to stack height for level open regions to obtain equivalent concentrations when situated at various locations in hilly configurations. His graphs show that a stack located windward of a hill must be higher than for flat regions. His values suggest that the above assumption of an imaginary level surface upstream of a ridge for concentration calculations is conservative. They also show that location of a stack at hilltop or at the upstream edge of a plateau is favorable, requiring less stack height than over flat regions. His configuration does not have as sharp a ridge top as shown in Figure 4.

When wind direction is parallel to a steep ridge or bluff, the streamlines of flow may remain straight and parallel to the surface as over a level surface. A plume emitted in this region will, however, experience some suppression of horizontal diffusion by the presence of the bluff as if a vertical wall were there. Turner (48) treats this case with a modification of Equation (68) for elevated plume diffusion over a level region. The addition of a term to account for diffusion reflection by the adjacent bluff gives the following equation for which the x axis is parallel to the bluff:

$$\chi_{(x,y,0)} = \frac{Q}{\pi \sigma_y \sigma_z u} \left\{ \exp\left[-\frac{1}{2}\left(\frac{y}{\sigma_y}\right)^2 \right] \right.$$
$$\left. + \exp\left[-\frac{1}{2}\left(\frac{2(b-y)}{\sigma_y}\right)^2 \right] \right\} \left\{ \exp\left[-\frac{1}{2}\left(\frac{h_e}{\sigma_z}\right)^2 \right] \right\} \quad (111)$$

where b is the lateral distance of the bluff from the x axis. The positive

y axis must be in the direction of the bluff so that the term $b - y$ is less than b. Wind direction at a small angle to the bluff will induce the above type parallel flow at low elevations, particularly for stable conditions. This has some similarity to the valley flow of Figures 5a and 5b with the upstream valley wall removed. The magnitude of local velocity appropriate for diffusion equations may be quite different from that above the ridge top.

Steep sides of a narrow canyon restrict horizontal diffusion on both sides of the plume. The case of simple uniform flow parallel to the valley axis may be treated as one with an additional lateral restriction on diffusion to that covered above. This excludes cross-valley circulations caused by surface temperature variations shown in Figures 5c and 5d. The case of an inversion layer over an urban valley area is discussed below. For short downwind distances over a level valley floor the plume acts similar to that in open level regions. With increasing distance the plume edges will reach the valley sides and be reflected. The effect is similar to the situation caused by the inversion layer in the trapping condition of Section III,C,3,a but in the horizontal direction. This occurs when the distance from the plume centerline to a valley side is equal to $2.15\sigma_y$, similar to Equation (93) for reaching the trapping layer. With further increase in distance and sufficiently high valley walls, vertical diffusion will continue, but the effluent becomes uniformly distributed in the horizontal direction. The result is similar to the trapping condition of Figure 9 beyond x_{tb} with the y axis turned vertically. The vertical x,z plane of the figure becomes the ground plane. Turner (48) gives the following equation for ground-level concentration to include the effect of plume height.

$$\chi = \left(\frac{2}{\pi}\right)^{1/2} \frac{Q}{w\sigma_z u} \exp\left[-\frac{1}{2}\left(\frac{h_e}{\sigma_z}\right)^2\right] \tag{112}$$

where w is valley width.

The case of an urban region in a valley with a nocturnal inversion is a variation of the trapping condition in open regions or of the condition of unrestricted upward diffusion in a narrow valley discussed above. To the restriction of the inversion layer over the near-neutral urban surface layer is added that of the valley sides. Smith (24) reports results of nighttime field experiments and a method of analysis for the urban valley at Johnstown, Pennsylvania. Tracer diffusion experiments were made with a low-level source emitted in the urban region. Air movement in the valley was of the drainage type parallel to the valley axis. A surface inversion existed upstream of the urban region. A near-neutral layer of increasing depth formed over the urban region under the inversion layer. This caused a restriction on vertical diffusion before horizontal diffusion was sufficient to bring the plume edges to the valley walls. Analysis of diffusion was treated in three stages, the first two being of the trapping type described in Section III,C,3,a but with a modification of the second-stage equation. In the first stage, covering a distance of 2 km from the source equivalent to x_{tb} of Figure 9, ground-level concentrations were largely unaffected by the inversion layer above and can be treated with ground-level point source Equation (73). At the end of this stage (2 km) vertical distribution was essentially uniform up to the inversion layer at 100-m height but with a horizontal Gaussian distribution. The second phase, that beyond x_{tb} in Figure 9, covered distances between 2 and 5 km. Uniform vertical distribution continued but with increasing effect of valley walls. Smith's equation for this stage is equivalent to Equation (91) multiplied by a factor of 2. This equation will, at the end of the second stage, match the values obtained with the third-stage equation. In the third stage, occurring beyond 5 km where the urban layer was at least 140 m high, diffusion was handled with the "box" model equation which assumes uniform concentration distribution horizontally and vertically within the region bounded by the valley walls and the inversion layer.

$$\chi = Q/wh_i u \tag{113}$$

The valley width is given by w and h_i is the height of the urban layer under the inversion. The product wh_i is the cross-sectional area over which the plume effluent has become uniformly distributed.

Smith (24) found that the standard deviations of concentration distribution, σ_y and σ_z, for the first and second stages in the urban layer were equivalent to those between the C and D Pasquill–Gifford stability categories (between slightly unstable and neutral), generally closer to the D category. It is to be noted that wind speed u in the various equations is representative of the valley flow, but not necessarily that above the ridge tops which may be quite different, even calm, for drainage-type flow. Smith recommends that wind speed be evaluated at a height of at least 10 to 15 m to avoid localized effects of structures.

In the absence of field data or other guides, the various analytical methods for estimation of topographical effects require, in addition to local wind speed u, a selection of standard deviations of concentration distribution. The Vandenberg data for mountainous regions described in Section III,C,1,b and Figure 8 show σ_y and σ_z values higher than for level regions for comparable neutral conditions but with considerable scatter. Most of the values are for distances greater than 1 km (up to 10 km) and show the influence of ridges and canyons. For the more restrictive situations covered in the above analytical methods, these values may be on the high side. Any analysis should include values for open level regions at the appropriate stability conditions where they may lead to higher concentrations.

D. Plume Concentrations for Differing Sampling Times

1. Effect of Limited Sampling Times

Measurements of instantaneous effluent concentration at a given point in a plume will show variations with time due to the turbulent nature of diffusion. The degree of these variations vary widely depending on numerous variables. Atmospheric variables, surface configuration, plume elevation, relative location of plume, and point of concentration determination are among the more important factors which affect the time dependence of effluent concentration. Not included herein are such factors as changes in effluent physical and chemical properties with time and distance. In most air pollution applications the maximum instaneous concentration found in a time period of measurement is not the most important. It is also difficult to measure. Various items affected by air pollutants, plant and animal life, and materials, tend to respond to a form of integrated effect of concentration or an average over an increment of time during which the instantaneous value may show considerable variations. This is discussed in the chapters of Volume II.

From a record of instantaneously varying concentration versus time, the average concentration may be found for a selected time increment taken at an arbitrary time. As this time concentration is shifted, the average for the selected time increment may vary. One of these will be a "maximum average" concentration found within the longer time period record. This concentration is taken as representative of the selected time increment, the sampling time. As the selected time increment is increased, the maximum average tends to decrease. For plume diffusion in restricted flows, as in a wind tunnel, a value of time increment will be reached beyond which there will be no further reduction in maximum average concentration. It becomes statistically stationary. This concentration could be a reference value relative to which the maximum averages for the shorter time increments could be related. In the unrestricted atmosphere there is generally no such limiting time increment beyond which there is no effect of time on the maximum average value. This is caused by wind-direction variations which become increasingly important for longer times. Thus concentrations for various sampling times must be dealt with on a relative basis. The ratio of maximum average for a short time increment to the average of a longer period covering the short increment is sometimes referred to as the peak-to-mean or peak-to-average ratio. The longer period may be referred to as the sampling interval or duration of sampling. It may be the full length of time during which instantaneous or short-period values are obtained in a given experiment.

A physical representation showing the cause of the effect of sampling time is illustrated schematically and elementally in Figure 12. The shorter sampling or averaging time plume, as it is viewed at various times within the wider plume for a longer time increment, may have a meandering configuration. Three such plumes are drawn in the figure. The point x,y, at the time one of the shorter time plumes is at its location, will experience a peak concentration as shown. At other times it may have little or no concentration if only the shorter time plumes occurred. Average of the various short-time plumes over the longer time increment will produce the wider plume with the concentration profile shown. Point x,y will experience a much lower average or mean value.

Estimates of concentrations by the diffusion equations of Section III,C,1,a require standard deviations σ_y and σ_z that are dependent on sampling time. The values from field experiments given in Section III,C,1,b have relevant sampling times included, but these differ for various data sources. The common technique of correcting for the sampling time of an application, when it differs from that of field data, is one which is applied to concentrations rather than to σ_y or σ_z. Thus concentration calculations are made, first with the available σ_y and σ_z data, and then, if necessary, correction is made for sampling time of the application.

Various theoretical and empirical formulations of the effect of sampling time have been developed to find dependence of the ratio of concentration to ratio of sampling time. Results of these studies vary and depend on the

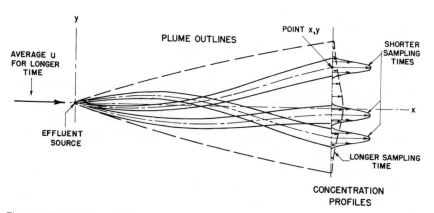

Figure 12. Shorter and longer sampling time plumes to show effect of time on concentrations.

assumptions made in setting up the theoretical model. Most show that the concentration ratio is given by a power law function of time ratio as follows:

$$\chi_2/\chi_1 = (t_2/t_1)^r \tag{114}$$

The exponent r has negative values which show decreasing concentration ratio with increasing time ratio. Because of the many possible variables which cannot all be included in a theoretical formulation, differing results are obtained for the value of the power law exponent. Heavy reliance must be placed on experimental data.

Table VI gives numerical values of concentration ratio by the above equation for various sampling times and exponents r as a guide to visualizing the effect of sampling time. They are given for various reference or base times t_1 covering the sampling times of the data in Section III,C,1,b. When field data for the effect of sampling time are treated on a peak-to-mean basis, χ_2 and χ_1 are peak and mean values, respectively. t_1 is the sampling interval (duration of sampling period) for the mean value from which a shorter time increment t_2 is selected to find the maximum average, the peak in this case. Such data would be represented in Table VI for $t_2/t_1 \leq 1$. This may be the range of sampling times covered in the experiments in which case extension to values of t_2 larger than t_1 is made on the assumption that the value of r found for shorter times still applies. In some presentations the inverse of t_2/t_1 is used which changes the sign of r.

Gifford (70) in his early paper on the effect of sampling time developed a fluctuating plume model [also covered in Slade (13)] which is the basis for the concept of Figure 12. The concentration distribution of an instantaneous plume is distributed to form the wider plume for longer sampling time by varying its centerline location with the Gaussian distribution. Gifford used various field data to arrive at numerical values. These cover sampling intervals that vary up to 140 minutes. He found that near the plume centerline, the effect of sampling time was less (r values were smaller negative) than for locations off-centerline. His data for the former were from ground-level source experiments; for the latter from elevated sources. His graphs (with some scatter of data points) include curves which show exponents of -0.2 and -0.75 for centerline and off-centerline locations, respectively, for limited downwind distances. He concludes that peak-to-mean ratios can be expected in the range of 1 to 5 for the centerline locations but may be as large as 50 to 100 at the ground for moderately high stacks of 50 to 100 m. Time ratios were as low as 0.005. With increasing downwind distance, the effect of off-centerline position becomes less, and centerline ratios are approached at distances around 5 km.

Table VI Concentration Ratios for Various Sampling Times and Power Law Exponents [Eq. (114)][a]

Sampling time	χ_2/χ_1							
	$r = -0.15$				$r = -0.20$			
1 min	1.18	1.41	1.67	1.85	1.25	1.58	1.97	2.27
3 min	1.00	1.20	1.41	1.57	1.00	1.27	1.58	1.82
10 min	0.85	1.00	1.18	1.31	0.80	1.00	1.25	1.43
30 min	0.71	0.85	1.00	1.11	0.63	0.80	1.00	1.15
1 hr	0.64	0.76	0.90	1.00	0.55	0.70	0.87	1.00
3 hr	0.54	0.65	0.76	0.85	0.44	0.56	0.70	0.80
24 hr	0.40	0.47	0.56	0.62	0.29	0.37	0.46	0.53

Sampling time	χ_2/χ_1							
	$r = -0.35$				$r = -0.50$			
1 min	1.47	2.24	3.29	4.19	1.73	3.16	5.48	7.75
3 min	1.00	1.52	2.24	2.85	1.00	1.83	3.16	4.47
10 min	0.68	1.00	1.47	1.87	0.58	1.00	1.73	2.45
30 min	0.45	0.68	1.00	1.27	0.32	0.58	1.00	1.41
1 hr	0.35	0.53	0.78	1.00	0.22	0.41	0.71	1.00
3 hr	0.24	0.36	0.53	0.68	0.13	0.24	0.41	0.58
24 hr	0.12	0.18	0.26	0.33	0.05	0.08	0.14	0.20

Sampling time	χ_2/χ_1				Time ratio			
	$r = -0.65$				t_2/t_1 (for all r)			
1 min	2.04	4.47	9.12	14.31	0.3	0.1	0.03	0.02
3 min	1.00	2.19	4.47	7.01	1.0	0.3	0.10	0.05
10 min	0.49	1.00	2.04	3.20	3.3	1.0	0.33	0.17
30 min	0.22	0.49	1.00	1.57	10.0	3.0	1.00	0.50
1 hr	0.14	0.31	0.64	1.00	20.0	6.0	2.00	1.00
3 hr	0.07	0.15	0.31	0.49	60.0	18.0	6.00	3.00
24 hr	0.02	0.04	0.08	0.13	480.0	144.0	48.00	24.00

[a] χ_1 is concentration at sampling time t_1, and χ_2 is concentration at sampling time t_2. t_1 is given on the line where $\chi_2/\chi_1 = 1.00$. t_2 is selected sampling time for which χ_2/χ_1 is given on the t_2 line in the same column where $\chi_2/\chi_1 = 1.00$.

Gifford's analysis shows that peak-to-mean ratios will approach a value of 1 at large downwind distances.

In his comparative study on the effect of sampling time on building wake diffusion, Hinds (71) includes results of field experiments for a ground-level source in unobstructed flow. With sampling intervals of 10 and 15 minutes he finds peak-to-mean values near the plume centerline larger than those in the earlier study of Gifford (70). In the time ratio range down to 0.005 his graphical presentation suggests r values of about -0.45 compared with Gifford's value of -0.20. It is to be noted that Gifford's data are from an earlier time and a number of data sources with various experimental conditions. Hinds includes other data in addition to those of his experiments to show the effect of off (horizontal) centerline position. Distance from centerline is expressed as a distance ratio to standard deviation of concentration distribution y/σ_y. These show that for a given time ratio the peak-to-mean ratio increases with off-centerline distance ratio. At large off-centerline distance ratios, Hinds finds r values larger (negative) than the -0.75 value of Gifford. The data are difficult to directly compare since Gifford did not analyze his in terms of the distance ratio. Furthermore, the off-centerline distances of Hinds are in the horizontal direction, while those of Gifford are in the vertical direction. It is noted that for the plume centerline at a given elevation and with the usual increase of σ_y and σ_z with distance, increasing the downwind distance effectively decreases off-centerline distance when expressed as a ratio to σ_y or σ_z. Thus the concentration ratio will decrease as observed by Gifford. The later study of Ramsdell and Hinds (72) covers in considerable detail various statistical properties of sampling time effects and off-centerline location. Their field data were obtained with a ground-level source and ground-level concentrations for sampling intervals between 10 and 20 minutes and with time ratios down to 0.03. Distances are up to 800 m. While the trends are similar to those given in Hinds (71), they find that the power law relation does not represent the data in the range of values covered. For time ratios close to 1 the slopes of their graphical data on log–log coordinates (the exponent r if this were approximated with the power law relation) tend to values of -1. These slopes decrease in magnitude (smaller negative) with decreasing time ratio, and vary with the off-centerline distance ratio y/σ_y. Values of concentration ratio at time ratios around 0.03 (limit of range covered) fall within the previous values but the slopes are different. The Ramsdell and Hinds (72) experiments cover a range of stabilities identified by Richardson number, but no dependence on stability is shown where off-centerline locations are in terms of distance ratio y/σ_y. The question arises as to whether there is a difference in effect of off-centerline location when it is obtained with a horizontal distance near the ground as compared with a vertical distance due to difference in elevation between the plume and ground level. The turbulence structure in the intervening regions would be different. Most of the data given in the following paragraphs are for ground-level concentrations from elevated sources.

The field studies of Singer *et al.* (*73*) [summarized in Slade (*13*)] with plumes from a 350-ft elevated source show the dependence of peak-to-mean concentrations on stability as expressed in terms of the Brookhaven gustiness categories (see Section III,C,1,b). The sampling interval was 100 minutes. Power law exponent r values taken from their graph vary from -0.6 for the B_2 category to -0.3 for the C category. Their supplementary studies (60-minute sampling interval) show that the effect of lightly and partly wooded areas reduces concentration ratios for the shorter sampling times less than approximately 3 minutes. The power law exponent for this range is approximately -0.15. Beyond the 3-minute range, the curves become those of the appropriate gustiness category. For heavily wooded areas the reduced r value extends throughout the 60-minute sampling interval. The effect of wooded areas is visualized as causing a dampening of short-period fluctuations of concentration due to irregular surface configuration. Singer *et al.* (*73*) include data from English and German authors for urban areas in which sampling intervals extend to 1 and 4 days, for which the power law exponents are approximately -0.15. Irregular surface configuration may play a role similar to that of wooded areas.

Turner (*48*) ("Workbook of Atmospheric Estimates") recommends that the r values be between -0.17 and -0.2 for sampling times less than 2 hours. He includes a concentration ratio tabulation for $r = -0.17$, which extends to 24 hours. These are applied to the Pasquill–Gifford data which are for 10-minute sampling time intervals (low-level sources).

The Task Group (*17*) (ASME Recommended Guide) lists r values of -0.65, -0.52, and -0.35 for very unstable, unstable, and neutral categories, respectively. They are to be used for estimating the effect of sampling times of less than 1 hour when using the σ_y and σ_z graphs of the Guide which are for 1 hour. These are closely similar to those of the Brookhaven B_2, B_1, and C categories, respectively.

Hino (*74*) reports a theoretical study and the results of field experiments on ground-level concentrations for thermoelectric power stations with high stacks located in Japan near the coast in the heavily industrialized zone of an urban area. Emphasis is on sampling times from 1 to 6 hours. The theoretical studies show that for the time range of interest, the power law exponent r is -0.5. The field data show this value to be generally valid for sampling times from 10 minutes to 5 hours. For times less than 10 minutes there was much scatter of data but the -0.2 value seems valid. Atmospheric stability had little effect except on the range of applicability of the -0.5 value. For unstable conditions, the -0.5 exponent was consistently obtained for a sampling interval of up to 5 or 6 hours, but with neutral conditions the value tended to be larger (negative) beyond 3 hours.

TVA data on σ_y and σ_z in Section III,C,1,b from Carpenter *et al.* (*38*) were obtained for large power plants in open country. In the application of similar data to a simplified technique for dispersion calculations, Montgomery *et al.* (*41*) indicate that the data give instantaneous concentrations. In their development of nomograms for 1-hour values, a factor of $\frac{1}{2}$ is applied to instantaneous concentrations. Assuming a power law relation and an instantaneous time of 1 minute, the exponent r would be -0.17; if the time were 3 minutes, r would be -0.23. In a further extension of the 1-hour concentrations out to 3 and 24 hours, Montgomery *et al.* (*41*) reduce their 1-hour concentrations by the factors $\frac{2}{3}$ and $\frac{1}{5}$, respectively. A single power law relation does not cover the various concentration reduction factors. The reduction from 1 to 3 hours gives $r = -0.37$, and from 1 to 24 hours $r = -0.51$. For the reduction between 3 and 24 hours, the r value is -0.58. Thus a progressive increase in effective r values (negative) is found with increasing sampling time.

The variation in estimates of the effect of sampling time on concentration obtainable with exponent values from the various above-quoted investigations show the complexity of the problem and potential inaccuracies of concentration estimates. Power law representation is useful for limited ranges of sampling time. There are some seemingly contradictory results; but, as more field data are obtained, the effect of other variables will be resolved. The above-quoted data serve as a guide to selecting the range of r values which may apply to various concentration calculations.

The effect of building wakes on dependence of concentration ratio with sampling time has not been investigated thoroughly enough to draw any definite conclusions. Studies such as those of Munn and Cole (*64*), Dickson *et al.* (*62*), and Hinds (*71, 75*) make contributions to this complex subject. As compared with diffusion in unobstructed flow, the effect of sampling time on concentration ratio may vary from none to substantial. Hinds (*71*) found little effect of a building on peak-to-mean ratio for a ground-level source located at the upwind side near the stagnation point. Concentrations were near a line of maximum concentration in the wake region, effectively plume centerline. On the other hand, a short stack with rising effluent on top of a building may give large variations in concentration ratios with sampling time combining the effects of atmospheric turbulence structure with variations in plume rise. Wake-type turbulence has more of the short-period fluctuations in motion. With increase in sampling time the larger period components of the unobstructed atmosphere will tend to dominate and to reduce the effect of building-induced turbulence; increase in downwind distance produces a similar effect.

2. Long-Period Concentrations and Application of Climatological Data

The preceding technique for estimating the effect of sampling time assumes that the time averages of the various variables can be determined. This becomes increasingly difficult as sampling time is increased, particularly for the direction of plume centerline, which is the average direction of the wind vector. Specific averages may not persist for sufficiently long periods so that the power law type of extrapolation can be extended indefinitely. Climatological characteristics covered in Chapter 12 become increasingly important. Consequently, the more important variables for plume diffusion (at a given location) are wind speed, direction, and atmospheric stability. The standard deviations of concentration distribution and plume rise may all depend on stability and wind speed. The wind direction enters the problem directly through the crosswind location of a point of concentration determination relative to plume centerline. For purposes of analysis the ranges of values for wind speed and stability for a given wind direction are divided into a suitable number of intervals or classes. The number of wind speed classes is N, and each class is designated by a number n which ranges from 1 through N. For example, if the classes are 1–2, 3–4, 5–6, 7–8, 9–12, and >12 m second^{-1}, $N = 6$ and $n = 1$ for class 1–2, $n = 2$ for class 3–4, etc. The number and sizes of classes may be dictated by the available climatological data. The range of stability is similarly divided into S number of classes with each class designated by a number s which ranges from 1 through S. Each wind speed class is given a representative wind speed u_n, such as the average of the class range; and each stability class is given a representative value of the stability variable used in the data such as $(\partial\theta/\partial z)_s$. Each wind speed class n will have S stability classes. For each pair of classes, concentration at a given location (for the wind direction of the group of classes) could be calculated by the diffusion equations given earlier, but this method is not feasible for handling the wind direction variable. For this purpose a different form of diffusion equation is used in which concentration is obtained with the assumption of uniform horizontal distribution. The y direction is taken along a circle of radius x; but, with the averaging process implicit in Equations (115) and (116) below, the y coordinate is no longer a variable. The complete circle of wind-direction angles (360°) is divided into B number of equal classes for which the wind direction representative of a class is designated θ. A common value for B is 16. This gives a sector angle covered by one class of $22\frac{1}{2}°$. Climatological data are often available for this division of wind directions. The equations given below are similar to those in Turner (*48*) and other references. Average concentration $\bar{\chi}$ for a wind direction class θ having class values of wind speed u_n and stability (designated by s) is found with the following equation. This gives ground-level concentration at downwind distance x and applies to all directions that fall in the class:

$$\bar{\chi}_{(x,\theta,n,s)} = \left(\frac{2}{\pi}\right)^{1/2} \frac{Q}{\sigma_z(n, s)u_n(2\pi x/B)} \exp\left[-\frac{1}{2}\left(\frac{h_e(n, s)}{\sigma_z(n, s)}\right)^2\right] \quad (115)$$

For the case of limited mixing layer or trapping (Section III,C,3,a) the following equation applies for distances equivalent to or beyond x_{tb} in Figure 9. This assumes uniform concentration distribution horizontally and vertically, i.e., the box model concept:

$$\bar{\chi}_{(x,\theta,n)} = Q/h_i u_n(2\pi x/B) \quad (116)$$

where h_i is height of the stable layer under which trapping occurs.

The summation of the contributions for a class wind direction θ and pairs of classes s and n can be expressed with the following double series:

$$\bar{\chi}_{(x,\theta)} = \sum_N \sum_S f(\theta, n, s) \, \bar{\chi}_{(x,\theta,n,s)} \quad (117)$$

where $f(\theta, n, s)$ is the frequency (fraction of time) of occurrence for each pair of classes s and n at the wind direction θ. If f is expressed in percent, the equation is multiplied by 0.01. With Equation (117) the average concentration for a given χ and θ may be found for the period of interest such as a month, season, or year.

In the case where frequency data on stability are not available, an approximate result is obtained by applying neutral stability to all wind-speed classes. This assumes that neutral stability represents a form of mean for the range of stabilities, unstable through stable. It must not be applied to time periods so short that an adequate range of stability occurrences is not covered. This can obviously lead to considerable inaccuracy where a nonneutral stability dominates the selected time period.

E. Particulate Plumes

Particulates in a stack effluent may, because of their differing physical properties, have motions different from the gaseous components. The greater mass density of particulates results in gravitational forces which cause a downward component of motion. The effect of gravitational forces is generally identified by free-fall (settling, terminal) velocity V_g of a particle in a motionless body of air. This depends on particle size, mass density, and configuration as well as mass density and viscosity of the ambient air. Free-fall velocities of particulates are given in Chapter 3, this volume with many references on particulate characteristics.

Fine particles may have such low free-fall velocities that their separate motions may be small compared with other motions of plume diffusion. They will then act the same as the gaseous components of a stack effluent, and the various diffusion equations of the preceding sections will apply. Particles less than 20 μm diameter generally fall in this category, but distance and time of plume travel affect the significance of particle fall velocity in a given application. For concentrations near ground level, deposition of particulates on ground surfaces removes them from the plume and reduces the ground-level reflection term in the diffusion equations. This is pointed out in Section III,C,1,a with the plume diffusion equations. It is to be noted that, while particulates may have negligible free-fall velocities, mass density of the effluent ρ_s (or an equivalent temperature) should include the contribution of particulates where it may affect plume rise. If, however, concentration χ is desired for the particulate component of the effluent only, the emission rate Q for the diffusion equations must only be that of the particulates.

Deposition rate ω of particulates on a horizontal surface is given by the following equation.

$$\omega = V_g \chi \tag{118}$$

ω has units of (particle property) $\times$ (area)$^{-1}$ $\times$ (time)$^{-1}$. The "property" may be mass or other selected property for which Q and χ are also evaluated. It is assumed that all particles are deposited when they reach the surface and are not reentrained into the airstream. Free-fall velocity V_g, as usually obtained for motionless air, is also assumed to apply, but local turbulence could have an effect.

Motions of particulates with significant free-fall velocities are extremely complex. They are subject to the aerodynamic forces of a turbulent medium as well as gravitational forces. The following elementary treatment of particulate plume diffusion assumes that the particles in a plume fall as if they were in nonturbulent air. The distance they fall is superimposed on the locations of a gaseous plume of equivalent effluent properties. Fall distance is simply the product of free-fall velocity and time. In terms of downwind distance of travel x and wind speed u, fall distance is xV_g/u. The particulate plume may be viewed as similar to a gaseous plume tilted downward through an angle having tangent of V_g/u. This can be introduced into the diffusion equations by replacing effective plume height h_e of a gaseous plume at given x with h_{ep}, effective plume height of a particulate plume, as shown by the following equation:

$$h_{ep}(x) = h_e(x) - [xV_g/u] \tag{119}$$

The above technique becomes increasingly inaccurate with increasing particle size since large particle motions are less effected by the diffusive motions of the atmosphere.

Equation (119) applies to one free-fall velocity. With a distribution of particle sizes, h_{ep} varies with particle free-fall velocity. Particulate free-fall velocities may be divided into classes for increments of velocity. Each is then treated as a separate particulate plume and results are superimposed. Bosanquet et al. (76) developed a particulate plume model which includes effects of turbulent motions. Deposition of particulates with various free-fall velocities can be readily determined with their graphical techniques. Soo (77) applies basic equations of diffusion in the treatment of particulates to obtain equations for concentrations and deposition of particulates from an elevated source. They include effect of partial reflection at the ground surface.

IV. Experimental Methods for Determination of Plume Characteristics

Theoretical formulations generally provide at least the foundations for various equations and methods of analysis relating to diffusion of stack effluents. There is a limit to their contributions owing to the complex nature of the various physical processes involved. Experimentally determined parameters are introduced in order to account for features of plume travel and diffusion which cannot be deduced solely from theoretical considerations. Some are of a fundamental nature, such as basic diffusion and entrainment parameters, which may be unrelated to the larger scale aspects of a particular problem. The latter require experimental data for a particular site or estimates based on data from similar sites. On a still larger scale are the effects of atmospheric motions occurring well beyond the region of plume diffusion which are treated with climatological data. This explains the need for experimental data ranging from the very general to the very specific.

A. Field Experiments

Many different experimental procedures have been used in field experiments on stack plume transport and diffusion. A number of these are described in the references of this chapter including Slade (13). Experimental problems relating to the material given herein are discussed in the following paragraphs.

With measurements of wind speed and temperature at a limited number of elevations, the problem of interpolation or extrapolation arises. Temperature profiles enter the diffusion equations indirectly through the selection of standard deviations of concentration distribution σ_y and σ_z. Wind-speed profiles also enter indirectly depending on the method of evaluating σ's. Actual speed at elevation directly enters in the diffusion and plume rise equations. It has a first-order effect on concentration estimations, therefore accurate evaluation of wind speed at elevation is important. With limited data, the velocity profile equations of Section II,A are useful for interpolation and extrapolation to other elevations.

Direct measurement of plume vertical concentration profiles requires sampling of effluent or placement of sensors *in situ*. Vertical profiles can be estimated from ground-level measurements based on the assumption of Gaussian distribution. An example is the application of Equation (74) for centerline concentration of a simple ground-level source. χ, Q, and σ_y can be obtained from ground-level measurements; σ_z for vertical distribution can then be calculated. Some of the field data quoted in preceding sections of this chapter include σ_z values obtained by this indirect manner.

Direct measurement of plume concentrations requires that the effluent have a measurable component. Since dilution of the effluent is very high, concentration-measuring instruments must be capable of detecting extremely low concentrations. Many field studies have used existing stack effluent components such as sulfur dioxide. Serious measurement problems exist in urban–industrialized areas where background concentration of the component may interfere with plume measurements. Tracer materials, gaseous and particulate, have been introduced into stack effluents to provide a clearly detectable component independent of background concentrations. Fine fluorescent particulates have been widely used. Many research studies on plume diffusion, apart from a specific stack, have been made with point source-type effluents placed at desired elevations. These usually do not include effect of plume rise. They have the advantage of freedom in selection of suitable effluent material.

Methods for indirect measurement of plume profile characteristics are many, and new ones are being developed. Photographic and other optical methods have been used for decades to obtain geometric features of visible plumes such as are shown for plume types in Figure 1. Hot plumes have been photographed by infrared photography. Determination of plume profile characteristics is based on the assumption of Gaussian distribution and a definition of the plume outline such as given by Equations (75) and (76). Remote sensing of plume cross-section characteristics from a ground-level location has obvious advantages.

The visualization of flow patterns around buildings to obtain wake characteristics as shown schematically in Figure 2 is useful in the design of short stacks so as to avoid adverse wake effects. Smoke bombs give smoke streamers of sufficiently long duration to be observed and photographed. They are placed at or above rooftop and other strategic locations to show features of wake flow.

Turbulence measurements are found to be closely correlated with standard deviations of concentration distribution σ_y and σ_z. There are good theoretical foundations for the existence of such relationships. A rapidly responding bivane instrument will provide wind vector angular fluctuations in azimuth and elevation from which standard deviations of these variables can be obtained. Singer et al. (55), in their simplified method of estimating diffusion parameters, give equations for obtaining approximate values of σ_y and σ_z from standard deviations of azimuth and elevation angles of the wind vector.

B. Scale-Model Experiments

Scale-model experiments on stack effluent plumes have been conducted using air and water as the fluid media, the most common being the air medium employed in wind tunnels. However, water tanks (water at rest) and water tunnels have also been used. In water tanks, relative motion of the ambient fluid and the model stack is obtained by drawing the model through the water. With the introduction of visible materials, water has certain advantages of flow visualization. Fluid density variations can be handled by varying the degree of salinity. Buoyant plumes have been simulated by inverting the stack and using fluids heavier than water. Many fundamental studies on entrainment parameters and rise characteristics of jets and buoyant plumes have been made in water, however, for lengthy test programs and extensive variation of test parameters, wind tunnels appear to have the advantage.

Fundamental to the design and application of scale-model experiments is the determination of modeling criteria or scale factors by which the scale-model variables are related to their atmospheric counterparts. In phenomena as complex as those occurring in the atmosphere, many variables may affect plume transport and diffusion. To form manageable criteria a selection must be made which includes a limited number most important to the phenomena being modeled.

Briggs' (28) form of plume rise equations in Section III,B,1 is used in the following presentation. Four equations cover jets and buoyant plumes at neutral and stable conditions. They are rewritten to show plume rise and downwind distance nondimensionalized with stack radius. They are all given in terms of stack effluent density ρ_s and emission velocity V_s to put them in comparable forms. The rise of a buoyant plume for the earlier stages of plume rise, $x < x^*$, is obtained from Equations (21) and (22) as follows:

$$\frac{\Delta h(x)}{r_\mathrm{s}} = C_1 \left(\frac{(1 - \rho_\mathrm{s}/\rho)(V_\mathrm{s}/u)}{u^2/gr_\mathrm{s}} \right)^{1/3} \left(\frac{x}{r_\mathrm{s}} \right)^{2/3} \tag{120}$$

For a jet in neutral atmosphere, from Equation (61):

$$\frac{\Delta h(x)}{r_\mathrm{s}} = 2.3 \left(\frac{\rho_\mathrm{s} V_\mathrm{s}^2}{\rho u^2} \right)^{1/3} \left(\frac{x}{r_\mathrm{s}} \right)^{1/3} \tag{121}$$

A condition usually required for proper model simulation of the full-scale counterpart is that they be geometrically similar. Exceptions are found in the use of distorted linear scales to compensate for other effects mentioned below. To obtain geometric similarity of the model boundary configurations for the various buildings, stacks, and topographical features, all linear dimensions of the full-scale configuration are reduced by the same factor, the linear scale (= model reference length/full-scale reference length). In the above equations, r_s may be replaced with any selected reference length l which may be stack or building height.

For geometric similarity of plume path, the terms $\Delta h/r_\mathrm{s}$ and x/r_s in Equations (120) and (121) must be the same in the model and in full scale. This situation is obtained by requiring the terms in the first parentheses to be invariant with linear scale. These become the modeling criteria for the respective cases of buoyant plumes and jets. When buoyant plumes (produced with flowing gases, not just heat Q_h alone) are emitted sufficiently close to buildings to be influenced by the adjacent flow field, both momentum and buoyancy characteristics may be important. In that case, the criteria obtained from both Equations (120) and (121) are applied simultaneously. This is accomplished by using the following three nondimensional groups as the modeling criteria. r_s is replaced with reference length l:

$$\text{Velocity ratio, } R_\mathrm{v} = V_\mathrm{s}/u \tag{122}$$

$$\text{Density ratio} = \rho_\mathrm{s}/\rho \tag{123}$$

$$\text{Froude number, Fr}' = u^2/gl \tag{124}$$

The above form of Froude number occurs in hydrodynamic scale modeling. It determines the velocity scale. This requires model test velocities to be in proportion to the square root of the linear scale. The above criteria have been used by Strom and Halitsky (78) in many scale-model wind tunnel studies on industrial-type stack effluents.

When only the buoyancy of a plume need be represented, the modeling criterion is the group of terms in the first parentheses of Equation (120). A modified Froude number is defined as follows for application to buoyant plumes.

$$\text{Modified Froude number, } \mathrm{Fr} = \frac{u^2}{gl(1 - \rho_s/\rho)} \qquad (125)$$

The following ratio expresses the buoyant plume criterion:

$$\text{Buoyant plume factor} = \frac{R_v}{F_r} = \frac{V_s gl(1 - \rho_s/\rho)}{u^3} \qquad (126)$$

In terms of heat emission rate, Q_h:

$$\text{Buoyant plume factor} = \frac{gQ_h}{\pi c_p T \rho r_s u^3} \qquad (127)$$

With increasing downwind distance, buoyancy rise becomes increasingly important. For many applications, simulation of buoyancy rise alone may be sufficient. The relative importance of momentum and buoyancy rise with distance is discussed by Hewett et al. (33) and Fay et al. (35).

For pure jets, the term in the first parentheses of Equation (121) is the criterion:

$$\text{Jet factor} = \rho_s V_s{}^2/\rho u^2 \qquad (128)$$

With stable atmospheres, an additional criterion is needed to account for the effect of varying atmospheric density with elevation. This is obtained with further application of Briggs' plume rise equations. Equation (47) for final rise of buoyant plumes is as follows when written in non-dimensional form and with terms arranged for comparison with Equation (120):

$$\frac{\Delta h}{r_s} = C_2 \left(\frac{(1 - \rho_s/\rho)(V_s/u)}{(u^2/gr_s)} \bigg/ \frac{(r_s/T)(\partial\theta/\partial z)}{(u^2/gr_s)} \right)^{1/3} \qquad (129)$$

For jets, Equation (64) is written as follows:

$$\frac{\Delta h}{r_s} = 1.5 \left(\frac{\rho_s V_s{}^2}{\rho u^2} \bigg/ \frac{(r_s/T)(\partial\theta/\partial z)}{(u^2/gr_s)} \right)^{1/6} \qquad (130)$$

Thus, for stable atmospheres, there is added another criterion in addition to those from Equations (120) and (121). It is the following variation of the gradient Richardson number:

$$\mathrm{Ri}''' = \frac{(l/T)(\partial\theta/\partial z)}{(u^2/gl)} \qquad (131)$$

Other forms are given by Equations (7) and (11). This criterion and its variations have been used by Sundaram *et al.* *(79)* and others in modeling the atmospheric surface layer together with other considerations of turbulence structure. For the case of combined momentum and buoyancy for which Fr' [Eq. (124)] is a criterion, the denominator of Ri''' [Eq. (131)] is already determined. This leaves the numerator as a criterion for temperature gradient as follows:

$$\text{Temperature gradient factor} = (l/T)(\partial\theta/\partial z) \qquad (132)$$

The inclusion of temperature gradient as a feature in the modeling of plume diffusion introduces difficult experimental problems in equipment design and operation. Most applications for a single or small group of stacks do not require it. Nonneutral gradients occur at low wind velocities for which the plume rise may be sufficiently high that ground-level concentrations are small. At higher winds, the temperature gradient tends to the neutral (adiabatic) value due to the strong influence of mechanical turbulence. This is shown in Table IV for the selection of stability categories with the Pasquill–Gifford diffusion data. As atmospheric wind velocity approaches 6 m second^{-1}, the neutral D category becomes the plume diffusion condition for nearly all thermal conditions.

The above plume rise equations include the effect of turbulent diffusion caused by interaction of the plume with the ambient atmosphere. The constants C_1 and C_2 as given in Section III,B,1 are based on a variety of experimental data from model size to full scale. Briggs *(36)* points out that the wind tunnel scale-model plume values for C_1 and C_2 obtained by Hewett *et al.* *(33)* are much like those obtained in full-scale measurements. With increasing downwind distance, the role played by ambient atmospheric turbulence becomes increasingly important. Additional requirements must be considered for this phase of plume diffusion.

Many of the air pollution problems treated with scale-model experiments involve atmospheric turbulence and flow patterns caused by buildings and topography since these are the most difficult to handle with analytical methods. The above criteria for plume modeling determine the model testing velocity [Eq. (124) in the case of combined buoyancy and momentum] or at least the range of velocities when the practical aspects of model plume simulation are considered. These velocities are generally much lower than the atmospheric velocities they represent. This introduces the problem of the effect of model size and velocity on the turbulence and flow patterns significant to plume transport and diffusion. Fluid viscosity may have an important effect. Forces due to viscosity are always present in fluid phenomena where there are velocity gradients across streamlines. When effects of viscosity are present, they can usually be expressed by some function of Reynolds number, Re:

$$\text{Reynolds number, Re} = \rho u l / \mu \qquad (133)$$

where μ is fluid dynamic viscosity. Fortunately, many airflow and plume diffusion characteristics have a weak dependence on Reynolds number, otherwise scale models on plume diffusion would be impossible. The vast body of knowledge on the aerodynamics of blunt shapes with sharp corners, such as represented by buildings, shows that flow patterns and wake turbulence are largely independent of Re except at very low values. This lower limit varies with shapes. The turbulent wake generally originates at an upstream sharp edge, as shown in Figure 2, where the flow separates from the surface. With smooth and curved surfaces, the point of flow separation may change with Reynolds number in the low range due to the change in the local surface boundary layer from the turbulent to the laminar type. This may cause a marked change in wake configuration. This effect may be minimized by artificially roughening the surface.

Local wake characteristics at the downstream side of a cylindrical stack may change with Reynolds number in the range of model test velocities. While the wake size is small and may have little effect on the other model flow characteristics, it may be important near the stack top where there could be excessive downwash of stack effluent into the stack wake. This is not likely to occur if the effluent flow rate meets the criteria of Section III,B,4 for avoidance of effluent downwash.

Scale-model experiments on stack effluent transport and diffusion over buildings and irregular topography have given good results for linear scales as small as 1/1000. The more irregular the configuration the smaller the linear scale which may safely be used. While the larger features of plume diffusion may accurately be reproduced at the low Reynolds number of a model, some important details may not be. Close to the surface of a model, particularly a smooth surface, there may be a local laminar boundary layer where flow characteristics do not properly represent their atmospheric counterpart. Sampling of a plume effluent close to a smooth surface may give erroneous results. As the linear scale of a model is reduced to accommodate large flat regions, as in urban modeling, the presence of a laminar layer creates a difficult modeling problem. Chaudhry and Cermak (80) treat this problem by distorting the vertical linear scale in relation to the horizontal linear scale. Such small model scales, which are encountered in multiple-source diffusion, are not likely to be encountered with plume diffusion for a single or small group of stacks.

The finite dimensions of a wind tunnel test section, or any modeling facility, will obviously introduce restrictions on flow simulation that do not exist in the open atmosphere. The lateral restrictions caused by the walls and ceiling of a wind tunnel test section prevent the formation of the large-scale components of atmospheric turbulence and wind-direction variations. Thus the concentration data from a scale-model experiment are representative of short sampling times. In the streamwise direction there is the problem of developing, in the limited length of the modeling region, a boundary layer of velocity, turbulence, and temperature profiles representative of the unlimited atmosphere upwind of the plume diffusion region. In cases where irregular topographical features play a major role, as in mountainous areas, a significant portion of the upwind region should be included in the model. A rough rule for cases of similar heights of elements (valley floor to mountain top) is that the length of the upwind region be of the order of 10 times the height of element. In some cases this rule may be conservative, such as when one topographical feature dominates the upwind configuration. Modeling techniques for the introduction of thermal characteristics with topography are not well established.

The case of flat upwind regions requires a different approach. This problem was studied by Jensen and Franck (*10*) with wind tunnel and field experiments on the wake characteristics of buildings. They found that for neutral conditions the full scale wake characteristics were simulated in the model most accurately when the velocity profile as given with Equation (2) has the proper roughness length z_0. This requirement may be expressed as a ratio with reference length l to form another modeling criterion as follows:

$$\text{Roughness length factor} = l/z_0 \qquad (134)$$

This factor should have the same value in the model as in full scale. Thus, the effects of surface configuration in flat upwind regions can be represented with a surface roughness which is geometrically dissimilar from that of full scale, provided Equation (134) is satisfied. This does not account for discrete shapes such as nearby buildings whose wake characteristics have not been diffused into the background flow before reaching the plume diffusion region. In the absence of specific field data on z_0 values, Table I serves as a guide.

The velocity profile for a nonneutral gradient over flat surfaces is given by Equation (5) in which the Monin–Obukhov stability length L appears in the z/L term. This requires an evaluation of heat transfer at the air–surface interface. Gradient forms of the Richardson number [Eqs. (7), (11), and (131)] are, with some restrictions, related to z/L through Equations (9) and (10). A minimum of criteria for modeling the atmospheric surface layer are the roughness length factor [Eq. (134)] and some form of the gradient Richardson number. This subject has been studied extensively by Cermak (*81*) and his associates at Colorado State University. Snyder (*82*) and Cermak (*81*) derive various criteria, beyond those shown above, from consideration of the fundamental equations for conservation of mass, momentum, and energy. They include Coriolis acceleration due to the earth's rotation.

Experimental means for developing a flat surface boundary layer of sufficient height to encompass the region of plume diffusion has been the subject of much study. At the beginning of a wind tunnel test section, the boundary layer height is small. With a long test section as used by Cermak (*81*), the layer height develops in a "natural" manner with profiles appropriate to the intervening surface boundary conditions of temperature and roughness. Thick boundary layers can be developed "artificially" in a relatively short distance by various inducing devices. This subject has been investigated by Counihan (*83*), Sundaram et al. (*79*), and others. Sundaram et al. (*79*) and Cermak (*81*) have excellent reference lists on the subject.

REFERENCES

1. O. G. Sutton, "Micrometeorology." McGraw-Hill, New York, New York, 1953.
2. E. J. Plate, "Aerodynamic Characteristics of Atmospheric Boundary Layers," AEC Crit. Rev. Ser., TID-25465. U.S. At. Energy Comm., Washington, D.C., 1971.
3. A. G. Davenport, *in* "Wind Effects on Buildings and Structures," Symp. No. 16. Dept. Sci. Ind. Res., Nat. Phys. Lab., HM Stationery Office, London, England, 1965.
4. C. H. B. Priestley, "Turbulent Transfer in the Lower Atmosphere." Univ. of Chicago Press, Chicago, Illinois, 1959.
5. J. L. Lumley and H. A. Panofsky, "The Structure of Atmospheric Turbulence." Wiley, New York, New York, 1964.
6. R. E. Munn, "Descriptive Micrometeorology." Academic Press, New York, 1966.
7. L. R. Orlenko, *in* "Building Climatology," Tech. Note No. 109. World Meteorol. Organ., Geneva, Switzerland, 1970.
8. P. M. Jones, M. A. B. de Larrinaga, and C. B. Wilson, *Atmos. Environ.* **5**, 89 (1971).
9. D. H. Slade, *J. Appl. Meteorol.* **8**, 293 (1969).
10. M. Jensen and N. Franck, "Model-Scale Tests in Turbulent Wind," Part I. Danish Technical Press, Copenhagen, Denmark, 1963.
11. H. Lettau, *J. Appl. Meteorol.* **8**, 828 (1969).
12. H. Lettau, *in* "Multiple-Source Urban Diffusion Models" (A. C. Stern, ed.), Air Pollut. Contr. Office Publ. No. AP-86. United States Environmental Protection Agency, Research Triangle Park, North Carolina, 1970.
13. D. H. Slade, ed., "Meteorology and Atomic Energy 1968," TID-24190. U.S. At. Energy Comm., Oak Ridge, Tennessee, 1968.

14. G. A. De Marrais, *J. Meteorol.* **16**, 181 (1959).
15. O. G. Sutton, *Quart. J. Roy. Meteorol. Soc.* **73**, 426 (1947).
16. J. Halitsky, *Trans. Amer. Soc. Heat., Refrig., Air.-Cond. Eng.* **69**, 464 (1963).
17. Task Group for Second Edition, "Recommended Guide for the Prediction of the Dispersion of Airborne Effluents," Second edition. Amer. Soc. Mech. Eng., New York, New York, 1968.
18. B. H. Evans, "Natural Air Flow Around Buildings," Res. Rep. No. 59. Texas Eng. Exp. Sta., College Station, Texas, 1957.
19. H. W. Georgii, E. Busch, and E. Weber, "Investigation of the Temporal and Spatial Distribution of the Immission Concentration of Carbon Monoxide in Frankfurt/Main" ("Untersuchung uber die zeitliche und raumliche Verteiling der Immisions-Konzentration des Kohlen-monoxid in Frankfurt am Main"), Tr. 0477, APTIC 10550. Reports of the Institute for Meteorology and Geophysics of the University of Frankfurt am Main, Federal Republic of Germany, 1967.
20. W. G. Hoydysh and H. H. Chiu, "An Experimental and Theoretical Investigation of the Dispersion of Carbon Monoxide in the Urban Complex," Urban Technol. Conf., 1971, Pap. No. 71-523. Amer. Inst. Aeronaut. Astronaut., New York, New York, 1971.
21. R. S. Scorer, "Natural Aerodynamics." Pergamon, Oxford, England, 1958.
22. H. Blenk and H. Trienes, *in* "Grundlagen der Land-technik" (Herausgeben von Prof. Dr.-Ing. W. Kloth, ed.), No. 8. VDI Verlag, Düsseldorf, Federal Republic of Germany, 1956.
23. K. Yabuki and S. Suzuki, *Bull. Univ. Osaka Prefect., Ser. B.* **19**, 51–193 (1967).
24. D. B. Smith, *J. Air Pollut. Contr. Ass.* **18**, 600 (1968).
25. F. Defant, *in* "Compendium of Meteorology" (T. F. Malone, ed.), pp. 655–672. Amer. Meteorol. Soc., Boston, Massachusetts, 1951.
26. B. Davison and P. K. Rao, *Int. J. Air Water Pollut.* **7**, 907 (1963).
27. J. Z. Holland, "Meteorology Survey of the Oak Ridge Area," ORO-99. U.S. At. Energy Comm., Oak Ridge ,Tennessee, 1953.
28. G. A. Briggs, "Plume Rise," AEC Crit. Rev. Ser., TID-25075. U.S. At. Energy Comm., Div. Tech. Inform. Ext., Oak Ridge, Tennessee, 1969.
29. J. E. Carson and H. Moses, *J. Air. Pollut. Contr. Ass.* **19**, 862 (1969).
30. H. Moses and M. R. Kraimer, *J. Air Pollut. Contr. Ass.* **22**, 621 (1972).
31. S. B. Carpenter, J. M. Leavitt, F. W. Thomas, J. A. Frizzola, and M. E. Smith, *J. Air Pollut. Contr. Ass.* **18**, 458 (1968).
32. F. W. Thomas, S. B. Carpenter, and W. C. Colbaugh, *J. Air Pollut. Contr. Ass.* **20**, 170 (1970).
33. T. A. Hewett, J. A. Fay, and D. P. Hoult, *Atmos. Environ.* **5**, 767 (1971).
34. B. Bringfelt, *Atmos. Environ.* **3**, 609 (1969).
35. J. A. Fay, M. Escudier, and D. P. Hoult, *J. Air Pollut. Contr. Ass.* **20**, 391 (1970).
36. G. A. Briggs, *Atmos. Environ.* **6**, 507 (1972).
37. B. Bringfelt, *Atmos. Environ.* **2**, 575 (1968).
38. S. B. Carpenter, T. L. Montgomery, J. M. Leavitt, W. C. Colbaugh, and F. W. Thomas, *J. Air Pollut. Contr. Ass.* **21**, 491 (1971).
39. T. L. Montgomery, S. B. Carpenter, W. C. Colbaugh, and F. W. Thomas, *J. Air Pollut. Contr. Ass.* **22**, 779 (1972).
40. H. Moses and J. E. Carson, *J. Air Pollut. Contr. Ass.* **18**, 454 (1968).
41. T. L. Mongomery, W. B. Norris, F. W. Thomas, and S. B. Carpenter, *J. Air Pollut. Contr. Ass.* **23**, 388 (1973).

42. K. G. Brummage, chairman, "The Calculation of Atmospheric Dispersion from a Stack." Stichting CONCAWE, The Hague, The Netherlands, 1966.
43. J. Halitsky, *Int. J. Air Water Pollut.* **10**, 821 (1966).
44. H. E. Cramer, Proc. Nat. Conf. Appl. Meteorol., 1st, 1957 C-33. Amer. Meteorol. Soc., Boston, Massachusetts.
45. J. S. Hay and F. Pasquill, *Advan. Geophys.* **6**, 345–365 (1959).
46. F. Pasquill, *Meteorol. Mag.* **90**, 33 (1961).
47. F. A. Gifford, Jr., *Nucl. Safety* **2**, 47 (1961).
48. D. B. Turner, "Workbook of Atmospheric Dispersion Estimates," Office of Air Programs Publ. No. AP-26. United States Environmental Protection Agency, Research Triangle Park, North Carolina, 1970.
49. M. E. Smith and I. A. Singer, *J. Appl. Meteorol.* **5**, 631 (1966).
50. I. A. Singer and M. E. Smith, *Int. J. Air Water Pollut.* **10**, 125 (1966).
51. I. A. Singer and M. E. Smith, *J. Meteorol.* **10**, 121 (1953).
51a. M. E. Smith and T. T. Frankenberg, *J. Air Pollut. Contr. Ass.* **25**, 595 (1975).
52. J. L. McElroy, *J. Appl. Meteorol.* **8**, 19 (1969).
53. F. Pooler, Jr., *J. Air Pollut. Contr. Ass.* **11**, 677 (1966).
54. W. T. Hinds, *Atmos. Environ.* **4**, 107 (1970).
55. I. A. Singer, J. A. Frizzola, and M. E. Smith, *J. Air Pollut. Contr. Ass.* **16**, 594 (1966).
56. G. A. Briggs, "Plume Rise from Multiple Sources," ATDL Contrib. No. 91. Atmospheric Turbulence and Diffusion Laboratory, Nat. Oceanogr. Atmos. Admin., United States Dept. of Commerce, Oak Ridge, Tennessee, 1974.
57. E. W. Hewson, *Quart. J. Roy. Meteorol. Soc.* **71**, 266 (1945).
58. F. Pooler, Jr., *U.S., Pub. Health Serv., Publ.* **999-AP-16** (1965).
59. G. I. Taylor, *Phil. Trans. Roy. Soc. London, Ser. A* **215**, 1 (1915).
60. E. W. Bierly and E. W. Hewson, *J. Appl. Meteorol.* **1**, 383 (1962).
61. R. N. Meroney and C. R. Symes, *in* "Entrainment of Gases by Buildings of Rounded Geometry," pp. 132–135. Conf. on Air Poll. Meteorology, Raleigh, North Carolina, April 5–7, 1971. Amer. Meteorol. Soc., Boston, Massachusetts.
62. C. R. Dickson, G. E. Start, and E. H. Markee, Jr., *in* "Aerodynamic Effects of the EBR-II Containment Vessel Complex on Effluent Concentration" (C. A. Mawson, ed.), U.S. At. Energy Comm. Meteorol. Inform. Meet., 1967, AECL-2787, pp. 87–104. At. Energy Can. Ltd., Chalk River, Ontario, Canada, 1967.
63. W. M. Culkowski, *Nucl. Safety* **8**, 257 (1967).
64. R. E. Munn and A. F. W. Cole, *Atmos. Environ.* **1**, 33 (1967).
65. T. V. Lawson, *Atmos. Environ.* **1** (Discussion), 177 (1967).
66. P. O. A. L. Davies and D. J. Moore, *Int. J. Air Water Pollut.* **8**, 515 (1964).
67. J. Halitsky, *Amer. Ind. Hyg. Ass., J.* **26**, 106 (1965).
68. J. Sakagami and M. Kato, "Effects of Complicated Topography on Diffusion—Wind Tunnel Experiments," Natur. Sci. Rep. No 19, p. 1. Ochanomizu University, Tokyo, Japan, 1968.
69. M. E. Berlyand, *Atmos. Environ.* **6**, 379 (1972).
70. F. Gifford, *Int. J. Air Water Pollut.* **3**, 253 (1960).
71. W. T. Hinds, *Atmos. Environ.* **3**, 145 (1969).
72. J. V. Ramsdell, Jr. and W. T. Hinds, *Atmos. Environ.* **5**, 483 (1971).
73. I. A. Singer, K. Imai, and R. G. Del Campo, *J. Air. Pollut. Contr. Ass.* **13**, 40 (1963).
74. M. Hino, *Atmos. Environ.* **2**, 149 (1968).
75. W. T. Hinds, *in* "On the Variation of Concentration in Plumes and Building

Wakes" (C. A. Mawson, ed.), U.S. At. Energy Comm. Meteorol. Inform. Meet., 1967, AECL-2787, pp. 105–129. At. Energy Can. Ltd., Chalk River, Ontario, Canada 1967.

76. C. H. Bosanquet, W. F. Carey, and E. M. Halton, *Proc. Inst. Mech. Eng. (London)* **162**, 355 (1950).
77. S. L. Soo, *Atmos. Environ.* **5**, 283 (1971).
78. G. H. Strom and J. Halitsky, *Trans. ASME* **77**, 789 (1955).
79. T. R. Sundaram, G. R. Ludwig, and G. T. Skinner, *AIAA J.* **10**, 743 (1972).
80. F. H. Caudhry and J. E. Cermak, *in* "Simulation of Flow and Diffusion over an Urban Complex," Conf. Air Pollut. Meteorol., Raleigh, North Carolina, April 5–7, 1971, pp. 126–131. Amer. Meteorol. Soc., Boston, Massachusetts.
81. J. E. Cermak, *AIAA J.* **9**, 1746 (1971).
82. W. H. Snyder, *Boundary-Layer Meteorol.* **3**, 113 (1972).
83. J. Counihan, *Atmos. Environ.* **3**, 197 (1969).

10

Urban Air Quality Simulation Modeling

Warren B. Johnson, Ralph C. Sklarew, and D. Bruce Turner

I. The Nature of Urban Air Quality Simulation Models and Their Role

A. The General Need

An urban air quality simulation (AQS) model is a numerical technique or methodology, based upon physical principles, for estimating pollutant concentrations in space and time as a function of the emissions distribution and the attendant meteorological and geophysical conditions. AQS models thus serve as tools to provide objective answers to the many "what if . . .?" questions that are regularly faced by decision makers in the fields of air pollution control, transportation planning, and land-use planning.

The intent of this chapter is to discuss substantive technical areas, issues, and problems rather than to review all of the past work in this field. For the latter purpose, several excellent documents are available for reference, such as Stern (*1*), Wanta (*2*), Moses (*3*), Seinfeld *et al.* (*4*), Eschenroeder and Martinez (*5*), Hoffert (*6*), and Fan and Horie (*7*).

One approach to air pollution control is simply to control every source to the maximum extent possible using "best available" technology. While effective, this approach is not usually cost effective in capital or resources. Maximum control across the board also generally involves maximum expense. In view of the rapidly increasing costs required as finer degrees of control are achieved, as well as the mounting shortages of resources such as "clean" fuels involved in pollution control, it is clear that increasing attention must be paid to the cost effectiveness of control plans. Thus the problem becomes one of how to allocate available resources to produce the greatest improvement in air quality. In this connection, questions such as the following are typical:

What emissions reductions are needed to meet ambient air quality standards?

What are the relative contributions to ground-level concentrations from residential area sources and from tall stacks?

Where should a new emissions source be sited?

What air quality improvements can be achieved through staggering of working hours and thus leveling of commuter traffic loads?

What will be the change in ozone concentration if the emissions of its precursor, nitrogen oxide, are reduced by 50%?

What will the air quality be like tomorrow?

Until recently AQS models have not been applied extensively. There are several reasons for this:

Detailed knowledge about many of the basic meteorological and chemical processes operating on pollutants in the lower atmosphere is still incomplete.

Fitting knowledge of these processes into an economical and easily used computational framework is difficult to achieve without excessive loss of realism.

There is a shortage of suitable data that are sufficiently complete for model evaluation purposes, with the result that few models have been sufficiently validated.

There is a shortage of adequately compiled data to run models.

Many otherwise useful models have never been converted from research purposes to a practical applications format, due partly to the difficulty in identifying the diverse requirements of users.

Many potential using agencies have neither the staff nor the computational resources to take advantage of AQS models. There also appears to be an understandable disinclination on the part of some agencies to attempt to use sophisticated mathematical techniques which they may not fully understand or in which they may lack confidence.

Despite the existence of these problem areas, much progress has been made through efforts toward achieving better utilization of models. It should also be recognized that, while the state of the art in AQS modeling is imperfect, the application of properly evaluated "best available" assessment techniques is legitimate and necessary. Thus the question has increasingly become not *whether* models should be applied, but *how* and *when*.

B. Alternative Approaches

1. Proportional Scaling Methods

Estimates of the effects on air quality of changes in emissions have commonly been based upon proportional scaling methods known as "rollback methods." In the simplest version of this method, concentrations are assumed to be linearly related to emissions, so that a desired reduction in concentrations can be accomplished by a corresponding across-the-board reduction in emissions. Thus, to calculate the degree of improvement in air quality needed for attainment of a national standard, the following equation is used:

$$R = 100(A - C)/(A - B) \tag{1}$$

where

- R is the percent reduction needed
- A is the existing air quality at the location having the highest measured or estimated concentration in the region
- B is the background concentration
- C is the national standard

Such an approach may not be cost effective because it does not account for any of the factors—such as source height, source location, meteorology, or topography—which cause different sources to contribute in differing proportions to ground-level concentrations at a given location. Thus, noncontributors are required to control as much as contributors. Rollback cannot be used for predicting tomorrow's air quality, since meteorology does not enter into the technique. Furthermore, the rollback technique is not appropriate for reactive pollutants because of the nonlinearities inherent in the chemical transformations. Thus while sometimes useful for initial estimates, rollback calculations even if applied carefully can give misleading or wrong results. Implementation of a control strategy based upon rollback will likely be unnecessarily expensive.

Horie and Overton (8) have analyzed the characteristics of rollback equations with regard to the effect of the statistical distributions of pollutant concentrations. They found that the higher the percentile values of concentrations considered, the greater the uncertainty in the emissions reduction ratios calculated by the rollback technique.

A recent comprehensive review of rollback modeling by de Nevers and Morris (9) serves to put the role of this type of modeling in perspective. These authors derive four modified forms of the rollback equation which extend the basic technique to include multiple source types, different stack heights, different source-receptor distances, and wind-direction frequencies. These modified rollback models may find a useful role as intermediate tools between the basic rollback formula and the more advanced and complex diffusion models.

2. Empirical Techniques

A number of methods for calculating concentrations have been devised based upon detailed analyses of actual measured concentrations. Typically, such pollutant data are correlated or otherwise related to measured meteorological conditions or other relevant variables. One of the best examples of this approach is the "Tabulation Prediction" scheme developed by Moses (10) on the basis of data from a number of air monitoring stations in Chicago, Illinois. In this method SO_2 concentration frequency distributions for each station are tabulated for each combination of classes of the meteorological variables. Hence, given a predicted meteorological condition, concentrations are "predicted" simply by looking them up in a set of tables.

Perhaps the most widely used empirical technique has been that devised by Larsen (11). Based upon detailed analyses of measured pollutant concentrations in a number of cities, Larsen found that the concentration frequency distributions generally could be reasonably represented by log-normal distributions, usually with a different standard geometric deviation for each city, for each pollutant, and for each concentration averaging time. Given a measured concentration at a certain averaging time, Larsen's technique has greatly facilitated the calculation of the concentration to be expected at a different averaging time. [See also Kornreich (12) and Chapters 9 and 12, Vol. III.]

Both Larsen (11) and Stern (13) have discussed the properties of frequency distributions of air quality data as a function of averaging time and of the so-called "arrowhead charts" that display them. Stern has proposed the use of such charts for predictive purposes by means of an analysis that would remove meteorological effects from the air quality data and leave the component due to source factors. If the latter could be separated into source types, then the effects of alternative control strategies could be tested.

The chief limitations of all such empirical techniques is that they are only strictly valid for the same city and for emissions and other conditions corresponding to those occurring during the period of measurement of the concentrations used in the air quality analysis. Thus it is difficult to generalize with these methods.

3. Fluid Models

In the case of detailed physical settings involving complex natural or man-made terrain features, numerical simulations of pollutant dispersion must generally be rather drastically idealized in order to avoid excessive computational requirements. An experimental or empirical approach is often the only feasible means of dealing with these complex situations. Since atmospheric experiments are costly, laboratory simulation using scaled-down models in wind tunnels or water channels is often the best approach, even though simplifications are often still necessary.

Fluid models have the advantage that the scale-model geometry, as well as flow speeds and the other essential variables, can be easily changed and controlled. Concentrations resulting from simulated pollutant releases within the physical model and measured at various locations can be converted to equivalent atmospheric concentrations through the use of appropriate scaling relationships. The application of wind tunnel modeling to air flow and diffusion over an urban complex has been discussed by Chandry and Cermak (14) and Cermak and Arya (14a).

Perhaps the most significant area of concern in fluid modeling is that of ensuring that the scaling (or similarity) criteria used are suitable to permit a realistic simulation of the atmosphere. A detailed treatment of this subject has been published by Snyder (15), who shows that five nondimensional parameters plus a set of nondimensional boundary conditions must be matched in the model and the prototype (atmosphere). With regard to scaling, Snyder says:

> For correct modeling, certain nondimensional parameters in the prototype must be duplicated in the model. Almost invariably, duplication of these nondimensional parameters is impractical or impossible. Hence, a decision must be made as to which parameters are dominant. The less important ones must be ignored. This decision will generally depend upon the scale in which the investigator is interested. For example, when studying the upper air flow above a city, the waffle-like topography may be treated as surface roughness. The heat-island effect may be modeled by using a heated plate. If the city is large enough, Coriolis forces may be important. If, however, the interest is in dispersion in the immediate vicinity of buildings, the topography cannot be treated as surface roughness. The heat-island effect would require a detailed distribution of heat sources, and Coriolis forces could be ignored because the aerodynamic effects of the flow around the buildings would dominate.

In general, it is less difficult to achieve validity in fluid model simulations as the scale of the prototype atmospheric situation is decreased. However, mesoscale simulations covering special cases for rather large areas have been attempted [e.g., Cermak and Peterka (16)].

A unique technique in fluid modeling developed by the French National Meteorological Institute [Facy *et al.* (17)] uses laminar water flow in a channel to represent transport wind, and an acid-solution release system to simulate emissions from point sources. The scale model is coated with a chemical which turns various colors corresponding to the various levels of acidity to which it becomes exposed. This color pattern can then be interpreted in terms of ground-level pollutant concentrations.

C. Basic Model Applications

1. Planning (Strategic) Applications

Control agencies need objective techniques that can simulate, *with the aid of available historical data*, the pollutant concentration distribution changes that are expected to result from simulated changes in the emissions distribution. There are two widely differing spatial scales of interest for this application: the urban scale (1 to 25 km) and the local scale (10 to 1000 m).

Urban-scale AQS models are in use for both quasi-stable (CO, SO_2, particulates) and reactive (NO_2, O_3) pollutants from both stationary and mobile sources. The time scales of interest correspond to the averaging times specified for air quality standards, and basically range from $\frac{1}{2}$ hour to 1 year. Most models have been designed to calculate only annual average concentrations. However, models that calculate short-term ($\frac{1}{2}$ to 1 hour) concentrations, as well as concentration frequency distributions at any receptor using climatological records, have been under active development and evaluation.

Local-scale planning AQS models are mainly needed for analysis of transportation-related pollutant sources, particularly near congested downtown streets, highways, airports, and such high-traffic areas as shopping centers and sports complexes. The emphasis here is usually on the shorter time scales.

Strategic AQS models play only one part in the general process leading to decisions on air pollution control (Fig. 1). Significant roles also exist for control costs models; health, biological, and materials effects models; and social, economic, and political impact models. Interactions between pollution in the atmosphere and that in other media must also be considered. Finally, an objective technique or model is needed to appropriately assign weights to each of these factors, and thus to furnish guidance to the decision maker.

2. Predictive (Tactical) Applications

Control *tactics* are normally intermittent in nature and are intended to temporarily reduce concentrations occurring under high-pollution episodes. Real-time tactical prediction models for this application differ in several ways from strategic planning models. The most basic difference

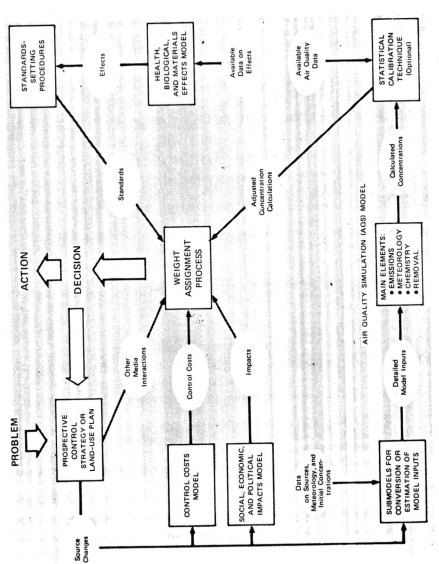

Figure 1. Elements and functional diagram of the air pollution control planning procedure.

lies in the type of model inputs required, in that the types of strategic models used for planning purposes generally use historical (climatological) records for meteorological data input, in contrast to the requirements of predictive (tactical) models that require forecast emission and meteorological data as input. The accuracy of tactical model prediction is directly dependent upon the accuracy of the forecast values for the input data. Although a detailed discussion of urban-scale meteorological models with predictive capabilities is beyond the scope of this chapter, such models clearly are needed for furnishing detailed meteorological inputs to predictive AQS models. In turn, the accuracy of predictive meteorological model is dependent upon the quality of their input information, the mesoscale, or synoptic scale meteorological forecasts.

Some of the more substantial efforts in the field of urban-scale meteorological modeling have been those of Pandolfo *et al.* (*18, 19*); Knox (*19a*) and his colleagues at Lawrence Livermore Laboratory; and the modeling group at the United States Environmental Protection Agency (*20, 21*). Models are being developed by this latter group to provide a three-dimensional dynamic prediction of the wind and temperature fields over an area of about $30 \times 30 \times 3$ km, with a resolution of $1 \times 1 \times 0.2$ km.

There are several other differences in the design of predictive AQS models compared with that of planning models. Predictive models must include more detailed treatment of the source types that lend themselves most readily to control on an intermittent basis, such as point sources, compared to those which are difficult to control on this basis, such as areal type sources. Predictive models also should be designed to optimize accuracy in predicting *changes* in concentration levels, since current air quality measurements will generally be available for reference. Simplicity is important so that computation time is minimized. Finally, to facilitate usage, output predictions should be provided in a graphical format, as well as in a form such that individual source contributions to predicted concentrations can be readily identified. Horie and Svendsgaad (*22*) and Horie (*22a*) examined the predictive performance of four types of AQS models and found only slight or no improvement over persistence, indicating that much work remains to be done in this area.

D. General Composition of Models

1. Submodels for Furnishing Required Model Inputs

The component parts of AQS models and their relationships are illustrated in the center box at the bottom of Figure 1. The need for the input submodels arises because few of the input variables for the main elements—emissions, meteorological transport and diffusion, chemical transformations, and removal—are directly observed or routinely available in sufficient detail to satisfy the requirements of most models. For example, usually only the horizontal winds near the surface are known for perhaps 5 minutes out of each hour. Such surface wind measurements are never available at more than a few widely dispersed stations, and sometimes only at a single airport station outside the city of interest. Direct measurements of atmospheric stability and turbulence are quite uncommon. In lieu of actual measured data, most researchers have resorted to objective submodels (or skilled analysis) to convert, estimate, or interpolate from the available data the detailed fields of input variables that AQS models require.

2. Main Model Elements

a. EMISSIONS.* In an emissions submodel, emissions estimates must be developed from such basic information as traffic distribution, industrial process rates, power plant loads, fuel types, home heating degree-days, and the like. Normally, empirically derived emission factors in the form of emissions per unit of input, output, or other relevant activity are applied to these basic data to estimate total emission strengths. Temporal distributions of emissions are difficult to estimate because of the shortage of such information in the basic data. Pollutant concentrations calculated by AQS models obviously can never be any more accurate than the emissions estimates, unless a compensating adjustment is included in the model. Unfortunately, field evaluation of the accuracy of emissions estimates is almost impossible, except by testing individual sources one by one.

b. METEOROLOGICAL TRANSPORT AND DIFFUSION. The meteorological elements of AQS models are generally rather simple in their treatment of advection and diffusion. Common practice is to use either the quasi-empirical Gaussian diffusion assumption or a version of the equation of mass conservation. A simplified treatment of the meteorology is possible since the transport wind field is considered given by the input submodel on the basis of available data or forecast values. More realistic treatments of the meteorology become feasible as comprehensive urban meteorological models which include the conservation equations of momentum and energy as well as that of mass become available. Until then, improvements in the meteorological elements of AQS models should be possible

---

*See also Chapter 16, Vol. III.

as better data are obtained on the variation of diffusion parameters as a function of meteorological conditions and city structure.

c. CHEMICAL TRANSFORMATIONS. The treatment of chemical transformations of pollutants is a relatively new inclusion in AQS models, ranging in sophistication from the treatment of SO_2 transformation as an empirical exponential decay factor in a Gaussian-type diffusion equation to simulation of the photochemical transformation processes by coupled differential equations. The chemical processes involved in pollutant transformations are exceedingly complex, and are a source of considerable uncertainty in AQS models. Major areas of concern include the gas-to-solid transformations, such as photochemical generation of aerosols and conversion of SO_2 to sulfates [Eschenroeder et al. (*23, 24*)], the effects of turbulent mixing rates on chemical reaction rates [Donaldson and Hilst (*25, 26*)], and even the basic chemical reactions [Hecht et al. (*27*), Roth et al. (*27a*)].

d. REMOVAL PROCESSES. AQS models generally neglect the potentially important processes of pollutant removal from the atmosphere, either through chemical reactions at the surface, dry deposition, or precipitation scavenging. In this area, knowledge is especially scarce; little is known about surface removal rates for any of the major air pollutants, although the limited studies performed indicate that surface uptake can be significant [Abeles et al. (*28*)]. Even carbon monoxide, which is considered to be chemically quasi-inert, has been found to be taken up at substantial rates by fungi normally present in soil [Inman et al. (*29*), Inman and Ingersoll (*30*), Ingersoll et al. (*31*)]. Inclusion of pollutant removal in models would permit the assessment of effects on surface vegetation or materials, as well as improve the accuracy of calculated concentrations through proper account of pollutant losses.

3. Calibration of Model Outputs

Figure 1 shows provision for the calibration of AQS models. Such calibration procedures generally involve a statistical comparison between calculated and measured concentrations. On this basis suitable adjustment factors can be incorporated into the model to give better agreement with measurements. Model calibration is frequently necessary because of the rather large uncertainties involved in the emissions estimates. However, such procedures are not on the whole desirable for a model meant to be generally applicable, since recalibration may be required for each city for which the model is used. Calibrations of this sort using air quality data are a poor substitute for a true model evaluation or validation process, which involves examination of the performance of the individual portions of a model with the aid of detailed data from specially designed experiments (see Section VII).

II. Basic Numerical Modeling Approaches

A. Gaussian Diffusion Models

By far the most frequently used approach to AQS modeling has been the Gaussian diffusion formulation. This approach stems from the fact that the well-known normal, or Gaussian, distribution function provides a fundamental solution to the classic Fickian diffusion equation. In the Gaussian plume model, the crosswind plume concentration distributions are taken to be Gaussian in form. This has been partially substantiated through field experimentation for typical meteorological conditions and for averaging times of 1 hour or longer. In the strict sense, the Gaussian diffusion model is valid only for long diffusion times and for homogeneous, stationary conditions. However, this type of model has been found to give useful results for many practical applications. A detailed discussion of the Gaussian diffusion model is given by Gifford (*31a*).

In its basic form, for a continuous point source at height h and a receptor at ground level ($z = 0$), the Gaussian diffusion formulation is given by

$$\chi(x,y) = \frac{Q}{\pi \sigma_y \sigma_z u} \exp\left[-\left(\frac{y^2}{2\sigma_y^2} + \frac{h^2}{2\sigma_z^2}\right)\right] \qquad (2)$$

where χ is the time-averaged ground-level concentration at horizontal coordinates (x,y), Q the continuous source strength, u the time-averaged magnitude of the wind velocity in the x direction (considered constant and unidirectional over the height interval between the source and the ground), and σ_y and σ_z are the standard deviations of the concentration distribution in the y and z directions. These latter are usually taken to be functions of downwind distance x in the forms

$$\sigma_y = ax^b \qquad (3)$$

and

$$\sigma_z = cx^d \qquad (4)$$

where the constants a, b, c, and d depend upon atmospheric stability and are derived empirically.

The point-source formula can be integrated in one or two dimensions to give line- or area-integral representations. The basic concept of vertical diffusion from a line source according to the Gaussian formulation

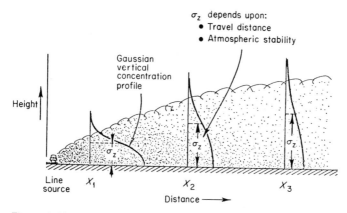

Figure 2. Vertical diffusion according to the Gaussian model (48).

is illustrated in Figure 2 (see Section V). Generally both point- and area-source versions of the model are used. In practice, the contributions of all sources to the concentration at a given receptor are usually calculated separately and then added to give the total concentration. This superposition capability adds flexibility and is an important advantage of the Gaussian technique, an addition to its inherent simplicity, ease of use, and short computation times.

The basic disadvantages of the Gaussian plume approach are

(a) Concentrations are not time-dependent in the usual sense (the approach is "quasi-steady-state" in that the input variables are updated, say, once per hour)

(b) Spatial variability in the meteorological parameters are difficult to incorporate

(c) Difficulties are encountered when the wind is light and ill defined

(d) The approach cannot be used for reactive or secondary pollutants, because in these cases superposition of individual source contributions at a receptor is not valid

The Gaussian "puff" model [Roberts et al. (32), Shieh et al. (33)], which tracks individual pollutant plume elements as they move along wind trajectories and diffuse in Gaussian fashion, was designed to overcome these problems, with the exception of (d) above. However, its large computational requirements have not encouraged extensive applications of the puff model.

B. Physical Basis of the Mass Conservation Approach

In addition to the Gaussian type of formulation for an air pollution model, another fundamental approach is based upon the equation of mass conservation. There are two main classes of models which fall into this category: the Eulerian (multibox) models and the Lagrangian (moving-cell) models. These approaches have the following common characteristics:

(a) All are adaptable to photoreactive pollutants
(b) All are time dependent
(c) Diffusion is treated by K theory (concentration flux proportional to gradient) in all models that include diffusion
(d) Each of these formulations can be placed on a firm mathematical footing

1. Eulerian (Multibox) Models

Whereas the Gaussian model was initially developed for point source emissions (plumes), the multibox model has been developed as an urban or regional model to specifically treat area distributed emissions. In the multibox model, the atmosphere over a region is divided into a grid of boxes, and the flow of pollutants into and out of each box is calculated. The boxes are usually defined by rectangles on the ground and the height of an inversion or arbitrary upper boundary. In the simplest form of this model, the change in concentration ΔC_{ij} in a box (i,j) during a time Δt is calculated from the net flow of pollutant through the sides, plus the pollutant emissions into the box, divided by the box volume:

$$\Delta C_{ij} = [(F_{i-j/2} - F_{i+j/2}) + (F_{ij-1/2} - F_{ij+1/2}) + S_{ij}\,\Delta t]/V_{ij} \qquad (5)$$

where the pollutant emissions are denoted by S_{ij}, the box volume V_{ij}, and the pollutant fluxes $F_{i+j/2}$ through the sides between boxes (i,j) and $(i+1,j)$ is defined by

$$F_{i+j/2} = C_{ij}A_{i+j/2}u_{i+j/2}\,\Delta t \qquad (6)$$

with the side area $A_{i+j/2}$ and the velocity through the side $u_{i+j/2}$. Thus the simplest type of multibox model neglects horizontal dispersion and assumes instantaneous vertical mixing up to an assumed lid. This type of model will be discussed further in the more general comments on Eulerian grid models later in this section. The multibox model calculates concentrations of pollutants throughout a region at one time, and not, as is done in Gaussian models, source by source (or receptor by receptor). Multibox models of the simple type have been used by Ulbrick (*34*) for computations in Los Angeles, California, by Reiquam (*35*) in the Willamette Valley, Oregon, and by MacCracken *et al.* (*36*) in the San Francisco Bay area, California.

2. Lagrangian (Moving-Cell) Models

In the moving-cell formulation, attention is focused on the polluted air itself in a Lagrangian or moving frame. The technique considers a "parcel" or column of air that ingests pollutant emissions as it passes over the various sources. The horizontal extent of the moving column has been assumed by various researchers to be as small as 1 m² or as large as many square kilometers, the size of an urban complex of emissions. The concentrations calculated by this model normally depend on a vertical dimension which may be specified as the distance from ground level to an inversion base, or some other height related to the vertical extent of mixing. As the mixing depth changes (due, say, to convection or convergence), the cell volume increases and the concentration decreases. In other versions of the moving-cell model, the concentrations are taken to be independent of inversion height and so depend only on emissions.

The equations to be solved in this model are designed to determine the change in position, Δx and Δy, of a cell moving with local velocity u, v during a time period, Δt:

$$\Delta x = u \, \Delta t \tag{7}$$

$$\Delta y = v \, \Delta t \tag{8}$$

and the change in cell concentration due to source emissions is given by

$$\Delta C = S(x,y) \, \Delta t \tag{9}$$

This approach neglects horizontal dispersion, and the results can be sensitive to cell position in regions of divergent horizontal winds. Again, detailed discussions of the advantages and shortcomings of the model are deferred until the mathematical foundations of this approach are presented later in the context of the Lagrangian grid model. Moving-cell models have been used by Wayne (*37*), Behar (*38*), and Eschenroeder and Martinez (*5*) to investigate Los Angeles, California, photochemical smog, and by Leahey (*39*) for modeling SO_2 in New York, New York. Eschenroeder and Martinez's model includes treatment of vertical diffusion through the use of several vertically stacked cells.

C. *Mathematical Foundations of the Mass Conservation Approach*

1. Equation of Mass Conservation

A proper beginning is with the equation expressing conservation of mass for a pollutant whose concentration is sufficiently small as not to perturb ambient atmospheric flow. For this pollutant, in a single dimension for simplicity, the time rate of change of the instantaneous concentration C is due to motion with the ambient flow and to molecular diffusion D as expressed by the first and second terms on the right side of Equation (10):

$$\frac{\partial C}{\partial t} = -\frac{\partial}{\partial x}(uC) + D\frac{\partial^2 C}{\partial x} \tag{10}$$

The concentration and the velocity u can be divided into mean (denoted by overbars) and fluctuations about the mean* (denoted by primes):

$$C = \bar{C} + C' \tag{11}$$

$$u = \bar{u} + u' \tag{12}$$

When these are substituted back into the basic equation and the resultant averaged, the following is obtained:

$$\frac{\partial \bar{C}}{\partial t} = -\frac{\partial}{\partial x}(\bar{u}\bar{C}) - \frac{\partial}{\partial x}\overline{u'C'} + D\frac{\partial^2}{\partial x^2}(\bar{C}) \tag{13}$$

That is, the time rate of change in mean pollutant concentration is caused by transport of mean concentration by the mean wind, transport of concentration fluctuations by correlated velocity fluctuations, and molecular diffusion. The second term on the right-hand side represents the turbulent diffusion. Since this term contains fluctuation terms for which we usually have little or no data, it is frequently modeled in terms of mean quantities. By analogy with molecular diffusion or by other assumptions, the hypothesis is established that this term may be approximated by diffusion of mean concentration

$$\overline{u'C'} = -K\frac{\partial \bar{C}}{\partial x} \tag{14}$$

where the turbulent diffusivity is introduced (in three-dimensional generality, K is a tensor quantity). In the atmosphere, turbulent diffusion is much greater than molecular diffusion, so the mass conservation equation can be rewritten as follows:

$$\frac{\partial \bar{C}}{\partial t} = -\frac{\partial}{\partial x}(\bar{u}\bar{C}) + \frac{\partial}{\partial x}\left(K\frac{\partial \bar{C}}{\partial x}\right) \tag{15}$$

This is the basic equation of "K theory." Its validity for a given application and values for the turbulent diffusivity must be based upon atmo-

---

* Subtleties arise dependent upon the type of mean—spatial, temporal, or ensemble—but for the purpose of simplicity, these are neglected in this discussion.

spheric observations. Other models for the turbulent transport term have received recent attention and some success; the interested reader is referred to Donaldson (40).

2. Finite Differencing Methods

The rest of this section will treat methods for solving "K-theory," Equation (15). For simplicity the turbulent diffusivity will be considered constant and the overbars will be suppressed on mean values. The effect of pollutant emissions can be included by adding a source rate S to the right-hand side, giving the basic diffusion equation:

$$\frac{\partial C}{\partial t} = -\frac{\partial}{\partial x}(uC) + K\frac{\partial^2 C}{\partial x^2} + S \tag{16}$$

For a constant wind and diffusivity and a continuous point source, Equation (16) can be solved analytically to give the Gaussian plume formula [Eq. (2)].

A straightforward method of solution of Equation (16) is by finite differencing. The differential operators are replaced by differences over finite distances. These approximations are valid as the distances approach zero, but in actual practice are usually taken to be as large as a few miles. Finite-difference methods are classified by the order (the number of terms kept in a Taylor series expansion of the differential), by the interpretation of results as volume averages (flux form) or point values, and by the time centering of the right-hand side of the equation (implicit—all terms on the right-hand side are taken at the advanced time, explicit—all terms at the present time, or an admixture of each). As an example, consider a simple finite differencing—first order, flux form, and explicit. The basic equation, simplified to one spatial dimension, becomes

$$\frac{C_i^{n+1} - C_i^n}{\Delta t} = \frac{u_{i+1/2}^n C_{i+1/2}^n - u_{i-1/2}^n C_{i-1/2}^n}{\Delta x}$$

$$+ K\frac{C_{i+1}^n - 2C_i^n + C_{i-1}^n}{\Delta x^2} + S_i^n \tag{17}$$

where superscripts denote times $(t_{n+1} = t_n + \Delta t)$, and subscripts are used to identify positions, with $i + \frac{1}{2}$ used to indicate values at the cell boundary between i and $i + 1$. [The latter could be formed by averaging,

$$C_{i+1/2} = \tfrac{1}{2}(C_i + C_{i+1}) \tag{18}$$

but the algorithm is unstable, and a better choice is the upstream concentration.] Solving Equation (17) for the concentration at the advanced time t_{n+1}, in terms of concentration (and velocities) known at the nth time, t_n, gives

$$C_i^{n+1} = C_i^n + \Delta t \left(\frac{u_{i-1/2}^n C_{i-1/2}^n - u_{i+1/2}^n C_{i+1/2}^n}{\Delta x} \right.$$
$$\left. + K \frac{C_{i+1}^n - 2C_i^n + C_{i-1}^n}{\Delta x^2} + S_i^n \right) \quad (19)$$

If the turbulent diffusion term is dropped, this is the same equation as that used in the multibox model. That is, a multibox model is not more than a first-order, explicit, finite-difference solution to the diffusion equation with the diffusion term dropped, which usually considers only two horizontal directions and assumes vertical homogeneity. Thus inaccuracies in the box model are threefold:

(a) Complete vertical mixing is assumed
(b) Diffusion between boxes is dropped
(c) First-order differencing is used

3. Solution Techniques

The error ϵ in approximating the above equation without diffusion by truncating a Taylor expansion at first order, in the case of the upstream choice for $C_{i+1/2}$ and constant velocity u, can be calculated from

$$\epsilon = \frac{u \, \Delta x}{2} \left(1 - \frac{u \, \Delta t}{\Delta x} \right) \frac{\partial^2 C}{\partial x^2} \quad (20)$$

This is the inaccuracy in the right-hand side of the finite difference approximation. It is a diffusion-type term, has an associated diffusivity

$$D_n = \frac{u \, \Delta x}{2} \left(1 - \frac{u \, \Delta t}{\Delta x} \right) \quad (21)$$

and is called numerical or truncation diffusion. For $u \, \Delta t/\Delta x > 1$, ϵ is negative and errors would tend to propagate in an expanding fashion. In regional modeling, $\Delta x \sim 1$ km is reasonable; then $D_n \approx 5 \times 10^3$ m²/ second, for 10-m/second winds. This is larger than typical atmospheric turbulent diffusivities. Thus, in a first-order, explicit solution, the numerical errors in approximating the advection term can mask the atmospheric dispersion being modeled. Both higher-order differencing and time centering can reduce the numerical error so it is no longer a diffusion-type term. Another difficulty is introduced by using higher-order differencing:

sharp gradients can cause negative concentrations. The effect is similar to fitting data points in the vicinity of a discontinuity by a high-order polynomial; extraneous ripples occur. The other alternative, centering in time, requires the solution of a set of linear equations, or inversion of a matrix, as was done by Lantz (*41*). Another technique was introduced by Egan and Mahoney (*42*) that focuses on the pollutant distribution and advects the various moments of the distribution instead of the pollutants. This also minimizes the truncation diffusion problem. All of these solution techniques are Eulerian in nature; they are based on a fixed coordinate system or grid of cells on which the finite differences are calculated.

The most important errors introduced by an Eulerian formulation stem from approximating the advection term, but this term merely describes translation of the solution with local wind speed, i.e.,

$$C(x,t) = C(x - ut, 0)$$

Thus, by changing variables to a coordinate system moving with the local wind velocity, the basic diffusion equation [Eq. (16)] can be simplified to

$$\frac{dC}{dt} = K \frac{\partial^2 C}{\partial x} + S \tag{22}$$

where the Lagrangian time derivative is represented by

$$\frac{dC}{dt} = \frac{\partial C}{\partial t} + u \frac{\partial C}{\partial x} \tag{23}$$

The main difficulty of this approach is in actual application to realistic three- or even two-dimensional flows. The coordinate system is distorted by the *local* wind velocities which can have considerable shear. When the coordinate system becomes distorted, input and output of data, and subsidiary calculations (such as diffusion) become extremely difficult or impossible.

Instead of interpreting Equation (22) to apply to a continuous space or grid of cells, a single cell or volume of air can be considered. The equation then relates the time history of pollution concentration in the cell as the cell moves along the air trajectory. From this interpretation, the models based on this formulation have been called trajectory models. Data from which three-dimensional wind fields can be constructed are virtually nonexistent, so trajectory models have been based on horizontal wind fields. For a single cell, the turbulent diffusion term must be approximated from external data or dropped (the approach taken in most moving cell models). When multiple cells are used to directly calculate

the diffusion terms, simplifications are usually invoked to restrict or eliminate wind shear and the resulting grid distortion. In this manner models
can be envisioned which are developed around a column of cells, a crosswind line or plane of cells, or a downwind line or plane of cells. Such
modeling approaches have been discussed by Eschenroeder and Martinez
(*43*).

The approximations necessary to solve the Lagrangian formulation
limit its applicability and the accuracy of the solution. Principle approximations involve smoothing of the wind field and neglect of one or more
components of diffusive transport. Interpretation and use of the resulting
solution can also lead to difficulties, since concentrations are only known
as a function of time at a point, along a line, or within a plane which
moves in time. The position corresponding to the calculated concentration
is computed by integration of the calculated motion of the air parcel
during the incremental time period. This formulation has been most useful
when confined to source or receptor-oriented calculations, since in these
cases one terminus of the trajectory is specified.

4. Particle-in-Cell and Other Methods

The difficulties with solution of the Lagrangian and Eulerian formulations have led to development of other methods for solution of the diffusion equation. In particular, the particle-in-cell method was modified to
include the effects of diffusion by Sklarew *et al.* (*44*). In this method
the mass of pollutant is separated into individual elements, and centroids
of these discrete masses ("particles") are tracked. The advection and
diffusion terms are combined by definition of a diffusion velocity u_d where

$$u_d = -K \frac{\partial C/\partial x}{C} \qquad (24)$$

which can be interpreted as the effective velocity of diffusive transport.
Then the total velocity effective in transporting pollutants is

$$U = u + u_d \qquad (25)$$

and the diffusion equation becomes

$$\frac{\partial C}{\partial t} = -\frac{\partial}{\partial x}(UC) \qquad (26)$$

Hence, the local pollutant mass is being transported by the local total
effective velocity. A simple means to simulate this is for the mathematical
particles, into which the pollutant mass has been discretized as previously

mentioned, to move according to

$$\frac{\partial x}{\partial t} = U \qquad (27)$$

That is, the particles are individually like Lagrangian cells which move in accordance with the total effective velocity. Eulerian cells are used to define mass averages (concentrations) based upon the number of particles in each cell at a given time. This method eliminates the truncation diffusion of first-order Eulerian methods and the distorted grids of Langrangian methods. It also permits effective graphical output, as illustrated in Figures 3 and 4.

Other methods have been suggested for the calculation of advection and turbulent diffusion of atmospheric pollutants, but in most cases have not received extensive application. These include spectral techniques, more general orthogonal expansions, even more general eigenfunction expansions and similarity solutions [e.g., Friedlander and Seinfeld (45), Gifford and Hanna (46)]. The latter offer the possibility of stochastic solutions. Each of these methods has been successful in closely related problems.

D. Computer Requirements

To quantify the computer requirements of various types of numerical models, three typical applications are hypothesized:

(a) Simulation of regional pollutant concentrations for an inert pollutant every hour for 24 hours at 100 locations

(b) Evaluation of the reduction in inert pollutant concentrations at 10 high-pollution areas ("hot spots") after application of a short-term control strategy

(c) Determination of the best site of 10 candidates for minimum air pollution impact from a new facility

The values presented in Table I are estimates based upon the experience of various investigators. The estimates assume the use of a large third-generation computer, and are more appropriate as relative rather than absolute measures of computer requirements. The table illustrates that each of the three main types of models has computational advantages for specific applications. Eulerian models are computationally more efficient for regional simulation, Lagrangian for receptor-oriented "hot spot" reduction, and particle-in-cell for source-oriented siting. (The reason for the latter is perhaps not readily apparent, but occurs because the particles trace the emissions from each source.)

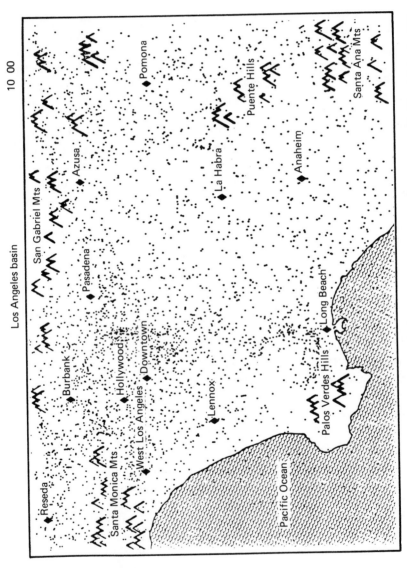

Figure 3. Plot of particle positions at 10:00 a.m. in the area of the Los Angeles (California) basin (86).

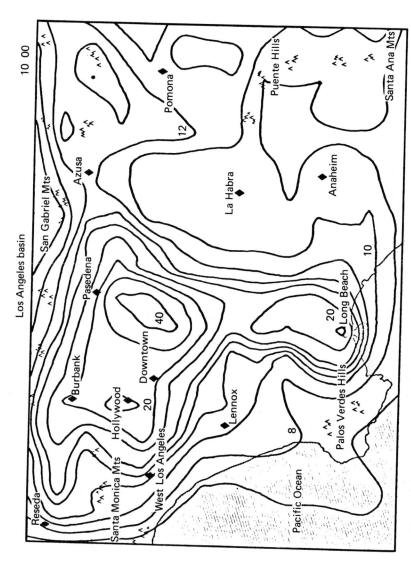

Figure 4. Isopleths for CO at 10:00 a.m. in the area of the Los Angeles (California) basin (86).

Table I   Estimated Computer Requirements for Various Model Types

| | | Time (CPU,[a] minutes) | | |
| | Memory | (a) | (b) | (c) |
Model type	(10³ words)	Regional	Strategy	Siting
Eulerian				
Multibox	2	2	1	20
Finite difference				
(Three-dimensional)				
Explicit—first order	50	15	7	150
higher order	50	20	10	200
Implicit—first order	50	30	15	300
higher order	100	40	20	400
Lagrangian				
Moving cell	<1	5	<1	<1
Column of cells	<1	20	2	2
Plane of cells				
Parallel to wind	1	30	2	2
Perpendicular to wind	1	30	10	10
Particle in cell	150	30	15	3

[a] CPU: Central Processor Unit.

III. Inherent Problems in Air Quality Simulation Modeling

Each commonly used numerical technique introduced in the previous section has difficulties in numerics or in applications. The choice of technique should be based on minimizing errors and on ease of application for specific cases.

In addition to computational limitations and finite-differencing problems, other areas of difficulty such as those associated with boundary conditions and the spatial scale of calculations face AQS modelers.

A. Boundary Conditions

The solution of the mass conservation equation requires specification of the initial ($t = 0$) pollutant concentrations throughout the solution volume, plus the concentrations on the boundary surfaces for all times. The simplified moving-cell approach, however, requires only one set of concentrations at the starting time. Although most models are fairly sensitive to the pollution levels specified for the initial time and for the air flowing across the boundaries into the solution volume, on the outflow boundaries simple approximations will usually suffice.

Ideally a model should be designed to cover a sufficient volume so that the flow through the boundaries will not reach the area of interest during the time period covered by the calculations. Unfortunately, this is usually impracticable, since the volume so enclosed would be up to ten times larger than that of the area of interest, and computing times would be excessive. In addition, the lack of available data to specify initial conditions becomes even more severe as the solution volume increases in size.

An alternative method for regional modeling is to start with pollutant levels equivalent to natural background concentrations, and then to calculate over a time period sufficiently long to reach a steady state. This approach is necessary for evaluating a control strategy in a region with recirculating air flow, such as in the Los Angeles, California basin, where pollutants can have an extended residence time on the order of days.

B. *Spatial Resolution and Compatibility with Available Data*

Another of the inherent difficulties of modeling lies in the treatment of subgrid or local scale effects. Figure 5 illustrates the diverse range of spatial scales that must be considered, as well as the time scales of interest. The limits of the latter are reasonably well defined as $\frac{1}{2}$ hour to 1 year, in accordance with air quality standards. However, the standards make no reference to the spatial scale.

Figure 5 shows the wide range of concentration averaging area covered by two types of models: a regional SO_2 model for Connecticut developed

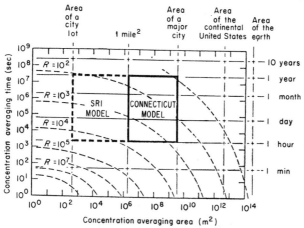

Figure 5. Time and space scales of typical model types (*47*). The dashed curves represent estimates of the ratio (*R*) of maximum to minimum concentrations expected for a given averaging time and area.

by Hilst (*47*), and an urban microscale CO model developed at Stanford Research Institute (SRI) by Johnson, Ludwig, and Dabberdt (*48, 49*). In addition, estimates are shown of the value of R, the ratio of maximum to minimum concentrations, as a function of temporal and spatial averaging. As indicated, the concentration variability, as represented by R, increases as the scale of space and time decrease, and thus becomes more difficult to model. The choice of spatial averaging scale for the design for a given model must be based upon its intended use.

In the case of CO, which is almost totally emitted from ground-level sources, the Stanford Research Institute studies (*49, 50*) showed that hourly concentrations at a height of 3 m frequently varied by a factor of 2 or 3 from one side of a street to the other, because of the aerodynamic flow patterns set up by surrounding buildings (see Section IV,A). The results of these studies emphasize the fact that "local air quality," as measured by an air monitoring station, varies drastically over distances as short as a few meters for a pollutant emitted near the surface, such as CO.

Similar findings have been reported by Ott and Eliassen (*50*) who carried out an extensive field study in San Jose, California, with the aid of walk-around bag samples, to determine the spatial variation of CO concentration. They found that concentrations significantly higher than those recorded at the Bay Area (California) Air Pollution Control District air monitoring station (located 2 miles from downtown San Jose, California) prevail on the sidewalks of downtown streets, and that these showed little correlation with the values recorded at the air monitoring station. In addition, there was evidence that relocation of the station by only 60 m could remove it from the influence of local traffic and reduce long-term average CO concentrations by nearly one-half.

Since, because of computation time constraints, all urban models calculate concentrations over a finite averaging area, typically 1 to 2 km on a side, there is a basic incompatibility with point measurements that makes model evaluation difficult. Achieving spatial compatibility either requires some form of spatial average measurement, or else a microscale analytical or empirical model that can reduce the averaging scale of the calculation. Such models could also be used to adjust fixed-point air quality measurements to a common datum on the basis of the siting characteristics of the individual stations and the magnitudes of nearby emissions sources.

IV. Transportation-Related (Small-Scale) Modeling

Air pollution has important impacts on the local scale as well as on the urban or regional scales. Pollutants from transportation-related sources, such as automobiles, are emitted near large numbers of people. Thus, source-receptor distances are short, and concentrations can be high.

A. Street Canyon Models

Street effects on air pollution in congested areas of cities have great importance for two reasons. First, such effects contribute substantially to those concentrations to which large parts of the population are exposed. Second, street effects must be considered in order to use existing data to verify the performance of urban models, since most available observations are taken near streets in downtown areas where local effects are likely to be significant. Models which do not include the effects of micro-scale diffusion will normally undercalculate concentrations in comparison with those measured at streetside stations. For example, the model used by Ott et al. (51) gave average concentrations that amounted to 36% of the streetside measurements.

Pollutant plumes emitted from vehicles within a street canyon surrounded by buildings are small in size compared with the nearby structures. Hence aerodynamic effects are important [McCormick (52)]. Experiments by Georgii et al. (53), Johnson et al. (48), and Dabberdt et al. (49) have shown that the CO concentrations at street level near the leeward sides of buildings are normally considerably higher than those near the windward sides, implying a helical cross-street circulation component near the surface in the opposite direction from the roof-level wind. This effect is illustrated (in two dimensions) in Figure 6.

Based upon these experimental results, Johnson et al. (48) developed a simple model to calculate the contribution of local emissions in a street to streetside CO concentrations. The equations* developed for ΔC_L and ΔC_W, the respective concentration components on the leeward and windward sides of the streetside buildings, are

$$\Delta C_L = \frac{(0.1)KNS^{-0.75}}{(U + 0.5)[(x^2 + z^2)^{1/2} + 2]} \tag{28}$$

and

$$\Delta C_W = \frac{0.1KNS^{-0.75}}{W(U + 0.5)} \tag{29}$$

where N is the traffic flow (vehicles/hour), S the average vehicle speed (miles/hour), U the wind speed at roof level (m/second), W the width of the street (m), x and z the horizontal distance and the height (m) of the

---

* For details on the development of these equations, the interested reader may refer to the original paper (48).

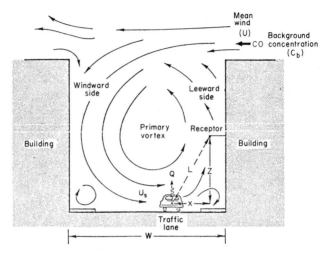

Figure 6. Schematic of cross-street air circulation in a street canyon surrounded by buildings (*48*).

receptor relative to the center of the traffic lane (Fig. 6), and K a dimensionless constant obtained from a least-squares fit to the measured concentrations.

These equations are taken to apply for wind directions in the two quadrants bisected by a line perpendicular to the street. When the wind direction is such that neither a leeward nor a windward case is appropriate, an intermediate concentration component ΔC_I, which applies on both sides of the street, is found by averaging the results of the previous two equations:

$$\Delta C_I = \tfrac{1}{2}(\Delta C_L + \Delta C_W) \tag{30}$$

The total concentration C is found by adding the concentration component from the street canyon model to the urban "background" concentration C_b calculated by an urban-scale model:

$$C = C_b + \Delta C \tag{31}$$

This simple model was evaluated in conjunction with the Stanford Research Institute CO model (see Section V,B) in extensive field tests in two cities, San Jose, California, and St. Louis, Missouri. From these field data a value of $K = 7$ was empirically derived [see Equations (28) and (29)]. A sample comparison between calculated and measured concentrations at one station on 2 days for three receptor heights is presented in Figure 7. The performance of the model was found to be best for mid-block stations. For intersections surrounded by tall buildings, it was concluded that the resulting complex flow and concentration patterns are

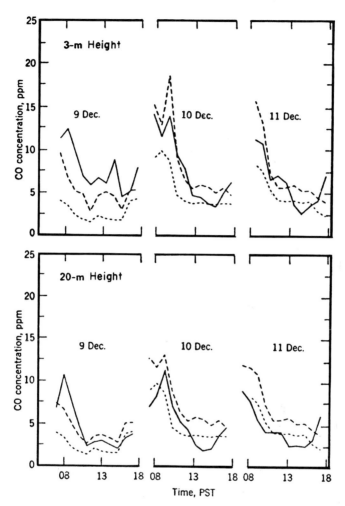

Figure 7. Calculated and observed CO concentrations at two heights for 3 days in 1970 at a special streetside station in San Jose, California (48). Solid line: observed; short dashed line: calculated urban background concentration (C_b) only; long dashed line: calculated total concentration ($= C_b +$ local street contribution, ΔC).

probably best treated by fluid (scale) models, rather than by analytical or numerical approaches.

B. Highway Models

The factors influencing air pollution levels in the vicinity of a major highway in an area where the surrounding structures are too short to cause significant recirculatory flow can be qualititatively described in three phases:

(a) Pollutant generation and emissions into the air surrounding the vehicle

(b) Transport and reactions during the first few minutes

(c) Transport and reactions during the next few hours

Phase (a) starts with pollutant generation by combustion and evaporation leakage. The generation rate per unit distance is normally a function of vehicle speed and weight, among other factors. In a stream of moving traffic, the vehicles also induce turbulent zones that dilute the exhaust and form a rather uniformly mixed "cell." This mixing cell is normally arbitrarily taken to be approximately twice the dimensions of the vehicle cross section, and can be used as the effective emission source.

In phase (b), the local winds and atmospheric and vehicle-induced turbulence transport and disperse the pollutants. Generally, the greater the distance from the highway, the lower the concentrations. Important phenomena occurring on this local scale include the fallout of large particles, interactions with terrain, and the establishment of chemical equilibrium among NO, NO_2, and O_3.

At some distance downwind (approximately 1 km), the effect of a single highway is negligible—this defines phase (c) or the mesoscale. Concentrations on the mesoscale are the sum of emissions from many highways and other sources. It is a background concentration that is superposed (for inert pollutants) on the microscale sources. On the mesoscale, the mean winds and mixing height assume more importance. Here, also, the photochemical reactions occur that produce smog.

This discussion will be restricted to the phenomena occurring in phase (b), and to an inert pollutant, such as CO. The shortest concentration averaging time in the present specification of the United States ambient air quality standards for CO is 1 hour. For this averaging time, knowledge of the fine structure of the emissions is largely irrelevant, and what is of prime concern is the mean hourly rate of emission of carbon monoxide per unit length of the highway line source. Thus in estimating the effective emission rate, the movement of the sources can be ignored, as can the highly variable nature of the emissions from moment to moment at any given location on the highway. This rationale would also apply to a $\frac{1}{2}$-hour averaging time.

Many of the techniques previously discussed can be applied to highway modeling. However, horizontal concentration gradients can be large, so that trajectory models may not be completely appropriate, although Eschenroeder (*23*) has reported using a vertical Lagrangian plane in this context. The two types of highway models that have received the greatest attention are those based on a Gaussian plume or an Eulerian solution in two or three dimensions.

For an infinite length line source, the Gaussian diffusion model reduces to

$$C = \left(\frac{2}{\pi}\right)^{1/2} \frac{Q \exp[-\tfrac{1}{2}(H/\sigma_z)^2]}{u\sigma_z \sin \phi} \tag{32}$$

where C is the ground-level pollutant concentration at distance x downwind from the highway, Q the line source strength, ϕ the angle between the wind direction and the highway, and H the effective height of emission of the line source. This equation has been suggested by Turner (54) to estimate concentrations from highways. If d is the perpendicular distance from the highway, the downwind distance x is given by

$$x = d/\sin \phi \tag{33}$$

The effects of the turbulent wake behind the moving vehicles can be incorporated by assuming that an initial nonzero value of σ_z (on the order of 2 or 3 m) occurs at $x = 0$. The point-source Gaussian plume equation can be used directly for finite length highways or for small values of ϕ. The highway is approximated by a series of point sources; as the number of point sources is increased the result converges to an effective line source. The principle researchers developing and applying Gaussian highway models include Beaton et al. (55), Sklarew et al. (56), Lissaman (57), and Johnson (57a). Lissaman has also used a non-Gaussian analytical solution incorporating the effects of ground roughness and stability.

Eulerian highway models have been developed to specifically treat the complex flows that occur over these short distance scales. Many highways are elevated or depressed from the surrounding terrain which thus perturbs the local wind flow. An Eulerian model can solve the diffusion equation with a different wind velocity specified for each grid cell. The cells falling within the mixing zone then contain a source term, with the sum of the source terms equal to the highway emissions. The model is run to steady-state with specified winds and emissions.

This approach has been followed by Danard et al. (58), Sklarew et al. (59), and Egan and Lavery (60). It is in the specification of the winds and diffusivities to be used in the model that each of the approaches differ. Sklarew has used a modified potential-flow fit to wind measurements as they are available. Egan and Danard chose to enter the flow patterns manually. The turbulent diffusivity used by Sklarew was based on local wind shear, roughness, and stability. In the work of Egan and of Danard, it is chosen by manual intervention. Lantz has fit the diffusivities to give concentrations similar to the Gaussian plume results. A more fundamental approach would be to undertake the solution of the Navier–Stokes equations describing the wind field. Notable in this regard is the work by Hotchkiss and Hirt (*61*).

Though many highway models have been developed, typically to analyze the air quality impact of new roadways, there has been less effort devoted to the verification of the model results. This is mainly due to the shortage of field data. A very limited validation analysis of a number of highway models using a small data sample (six highway-wind cases) has been conducted by one of the authors (RCS). Three of the cases were for at-grade (level with the terrain) roadway sections, and three for depressed sections. The results of this analysis were lumped by type of model, Gaussian or grid. One measure of accuracy is the correlation of model calculations and measured concentrations at various positions near the roadway. Gaussian models had correlation coefficients of 0.6 to 0.9 for at-grade and of 0.4 to 0.6 for depressed sections. Grid models are considerably better with 0.9 for at-grade and 0.8 to 0.9 for depressed sections. The root-mean-square (RMS) difference between model results and measurements (averaging 5 ppm) were about 2 ppm CO for the Gaussian models and about 1 ppm for the grid models.

One of the greatest difficulties with highway modeling involves the specification of emissions. Emissions from vehicles, unlike those from most stationary sources, are from highly variable combustion processes. Changes in emissions occur due to changes in operating mode. In addition, large differences in emissions occur between vehicles in similar operating modes. Modeling to date has focused only on average emission characteristics, although, in principle, it should be possible to incorporate some of the statistics of emissions variations.

C. Airport Models

A difficult task in air quality simulation modeling is to estimate the contributions from various sources to air quality in the vicinity of airports. The need is to determine air quality levels for averaging times from 1 hour up to 24 hours at a large number of locations, both on the airport itself and in the surrounding community. In addition to determining the total concentration from all sources, it is necessary to determine how much of the total concentration comes from each source type. Because of the highly intermittent nature of some of the emissions, and the large number of source configurations—area sources, elevated point sources, point sources in the vicinity of structures, horizontal line sources, and slant-path line sources—it is difficult to formulate submodels for each that have adequate reliability.

Pollutants of importance in airport modeling include suspended particulates, CO, NO_x, and various hydrocarbons. The latter two, because of their complex reactivities, are considerably more difficult to model than the more stable CO and particulates.

Sources of these pollutants inside the airport boundaries include aircraft, service vehicles, fueling operations, incinerators, heating plants, cars, taxis, buses, and trucks. Sources outside the airport include industrial, commercial, residential, and motor vehicle emissions. Emissions from aircraft are highly variable, depending upon both aircraft type and mode of operation. Additional complexities arise because the aircraft move from place to place, and operate in a single mode only for relatively short intervals of time (minutes). The velocity of emissions released from aircraft engines also serves to cause some initial diffusion of the pollutants.

The aerodynamic flow patterns about terminal structures serve to transport and diffuse the emissions from idling aircraft at the gate, and from service and fueling operations. It is important that airport models have the capability for calculating local concentrations in the vicinity of the terminal, because of the large number of persons exposed to pollutants there.

One of the first airport diffusion models, developed by Platt *et al.* (*62*), included a detailed aircraft pollutant emissions module. In this brief initial study, a number of simplifications were made, such as using the continuous point-source Gaussian model to simulate diffusion from all sources. Line sources were considered by using a series of points, and large segments of the urban area surrounding the airport were considered as single point sources. The complexities of intermittent and highly variable emissions as well as that of moving aircraft were avoided by determining emissions and the resulting air quality on an hourly average basis. Estimates of air quality were made with this model using emission estimates for four major airports. The results of studies to validate and refine this model have been reported by Thayer (*63*).

V. Air Quality Simulation Models for Quasi-Inert Pollutants

A. Sulfur Dioxide and Particulate Models

1. Long-Term Models

It is not surprising that initial efforts in air pollution modeling were with sulfur dioxide and particulates, since these pollutants long have been recognized as troublesome pollutants in many locations.

Although the concept of simulating urban air pollutant transport and dispersion was suggested by Frenkiel (*64*), the general lack of urban inventories of pollutant emissions hindered investigators from testing models. An emissions inventory of sulfur dioxide for Nashville, Tennessee, enabled Pooler (*65*) to apply a model to determine monthly concentrations at a large number of points. Comparison of these estimates with monthly measurements from a network of stations showed the feasibility of long-term models. One-half of the observed concentrations were between 80 and 125% of predicted values for the 5-month test period. An example of a comparison of observed and predicted concentrations for 1 month are shown in Figure 8.

Later, Martin (*66*) in collaboration with Tikvart formulated a longterm model (seasons to decades) similar to that of Pooler. It has become known as the Air Quality Display Model (AQDM) as reported by the

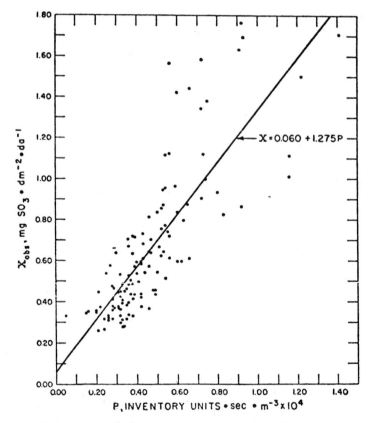

Figure 8. Comparison of observed (X_{obs}) and predicted (P) mean SO_2 concentrations for Nashville, Tennessee, December 1958 (*65*).

TRW Systems Group (67), and has been applied extensively to the estimation of sufur dioxide and particulate matter from point and area sources for a number of urban areas. Required meteorological input consists of a joint frequency distribution of wind direction, wind speed, and stability, referred to as stability wind rose.* The model uses the Holland plume-rise formula for determining the effective heights of point sources (see Chapter 9). The concentrations from area sources are determined by using the virtual-point-sources method; i.e., the concentration is determined by assuming dispersion from a point placed upwind of the area source of interest. Most of the evaluations conducted to date indicate that the AQDM tends to overestimate concentrations. Attempts have been made to overcome this by model calibrations involving the use of available air quality measurements.

More recently Zimmerman and Busse, using suggestions of Calder (68), have programmed a long-term model, the Climatological Dispersion Model (CDM), and applied it to St. Louis, Missouri [Zimmerman (69)]; Ankara, Turkey [Zimmerman (70)]; and the New York City, New York area [Turner et al. (71)]. In the case of the New York area, results from the AQDM were available for comparison. The CDM uses plume-rise estimates from Briggs (72) and calculates concentrations from area sources using the narrow-plume hypothesis [Calder (68) and Hanna (73)]. This model also assumes an increase in wind speed with height dependent on stability. Several versions of the simple model of Gifford and Hanna (74) were also tested on the same data. These comparisons indicated that the CDM yielded estimates nearer the measurements than the AQDM and the simpler models, but the simpler models were a surprisingly close second. It is recommended that the performance of such simplified models be used as a baseline upon which to judge results achieved with any newly formulated models.

2. Short-Term Models

Although the long-term models are especially useful for planning purposes in assessing the problems due to area sources, there is also an important need for short-term models to calculate concentrations for time periods of $\frac{1}{2}$ hour to 1 day and to produce predicted frequency distributions of these concentrations through iterative calculations. Such short-term models are particularly helpful in examining the contribution of point

---

* This type of data is available for most United States National Weather Service stations from the National Climatic Center in Asheville, North Carolina using their "STAR" Program.

sources to urban pollution. Because of the normal variability in wind directions over the long term, point sources are generally more likely to violate short-time period air quality standards than annual standards.

Using the same Nashville, Tennessee emissions inventory mentioned previously, Turner (*75*) estimated 2- and 24-hour sulfur dioxide concentrations for a number of randomly selected days. Although this model was not as successful as the long-term models, with 70% of the 24-hour estimates within a factor of 2 of observed values, the results were useful and pointed to the need for properly estimating the hour-to-hour and day-to-day variations of emissions.

Clarke (*76*) used a receptor-oriented approach with emissions partitioned in polar-coordinate fashion about the receptor. This allows better spatial resolution of emission sources near the receptor, which have a large effect on concentrations. For SO_2 and NO_x, Clarke's model gave reasonably good correlations between 24-hour estimates and measurements, ranging from 0.67 to 0.81. The principal disadvantage of this model is that the emission inventory must be reorganized for each receptor.

Koogler *et al.* (*77*) using a model similar to that of Turner, calculated 24-hourly SO_2 concentrations for Jacksonville, Florida. Although modeled concentrations in this area were quite low, estimates compared well with measurements, with 95% of the estimates within 1 pphm of the measurements.

All of the above short-term models use a single wind direction and speed for each time period. Hilst (*47*) developed a regional model for the State of Connecticut that took account of the variations in wind across the state. An example of the concentration field estimated from this model is shown in Figure 9.

Through an integration of the Gaussian puff equation, Shieh *et al.* (*33*) derived a model which they applied to a number of periods in New York City, New York. Their model tests resulted in mean observed and predicted concentrations of 0.19 and 0.18 ppm, respectively, with a standard error of estimate of 0.09 ppm.

In their model for Chicago, Illinois, Roberts *et al.* (*32*) also used a Gaussian puff formulation in order to handle the situation where winds are light and nonuniform in space and time. From their tests, they concluded that a steady-state plume model produces results nearly as good as the puff model and requires fewer calculations. However, because of the lack of light wind cases available for the evaluation, the capabilities of the puff model may not have been adequately tested. Calculated and observed concentrations for five Chicago, Illinois locations for 2 days are shown in Figure 10.

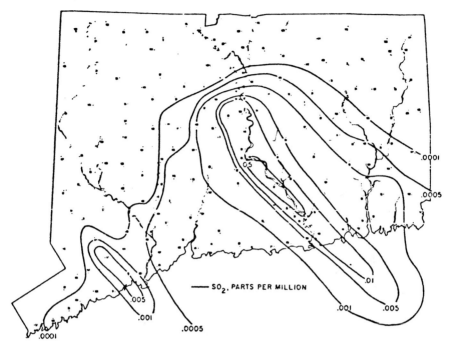

Figure 9. Ground-level SO₂ concentrations (ppm) over Connecticut at midnight in January with a northwest wind (47).

Gifford (78) has evaluated the performance of several simple models with available data used to test other models. The estimates from the simple models often are as good as those of the more complex models.

Fortak (79) applied a short-term model to the city of Bremen, Federal Republic of Germany, to calculate SO₂ concentrations at four locations. He also calculated cumulative frequency distributions of the concentrations and compared these with measurements. Concentration estimates at all frequencies were within a factor of 2 of observed values, with many much closer. Monthly average estimates and measurements were nearly the same at three of the four stations.

A generalized urban air pollution model, based on numerical integration of the concentration equation, has been developed by Shir and Shieh (79a). A new method is used to estimate the turbulent diffusivity and atmospheric stability. The model was used to study SO₂ distributions in the St. Louis, Missouri metropolitan area during 25 consecutive days in February 1965. The computed results agree favorably with experimental measurements for both long-term and short-term average concentration.

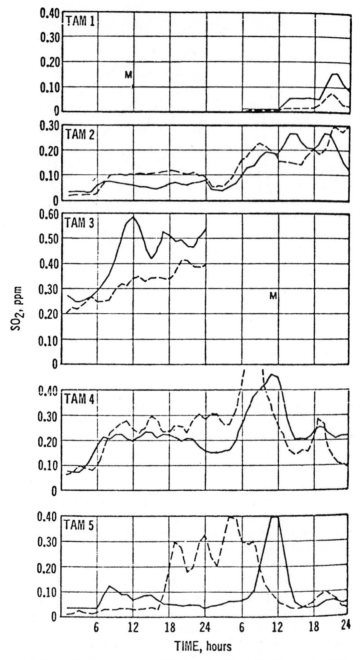

Figure 10. Hourly values of calculated and observed SO₂ concentrations at five Chicago, Illinois telemetered air monitoring (TAM) stations on January 17–18, 1967 (*32*). Solid line: observed; dashed line: estimated; M: missing.

Features of a few selected examples of short-term models are summarized in Table II. Most of the Gaussian plume models perform quite well when there is a well-defined wind across the urban area, approximating a quasi-steady-state situation. However, these conditions seldom prevail during the worst periods of air pollution. Typically, adverse pollution episodes occur under stagnant meterological situations, when the air movement is governed principally by local effects such as urban and topographical influences which can cause recirculating flow patterns. To handle these situations, future work must rely heavily on non-steady-state meteorological models which can simulate phenomena such as the urban heat-island effect, valley flow, and land–water circulations.

B. Carbon Monoxide Models

On the basis of the receptor-oriented technique developed by Clarke (76), Ott et al. (51) estimated future carbon monoxide concentrations for Washington, D.C. This study required assumptions about future changes in emissions from the current estimates. Decreases in carbon monoxide concentration from 15 to 29% at four sites were calculated for 1985 on the basis of expected traffic increases from 1964 to 1985 in combination with an anticipated 50% reduction in emissions from each vehicle. The model was found to substantially undercalculate concentrations in comparison with measured values at streetside air monitoring stations, probably because sources near the stations were lumped into large area sources.

Johnson et al. (48, 48a) used a Gaussian-type diffusion model similar in some respects to that of Clarke, in that emissions were considered in considerable spatial detail near the receptor, but with less spatial resolution at greater distances away. Emissions estimates in this model are generated from traffic information consisting of vehicle volume and average vehicle speed on each of a large number of roadway links. Wind speed, wind direction, stability class, and mixing height, considered as averages for the urban area, serve as meteorological inputs. The model can be applied in two ways: (a) to calculate hour-by-hour concentrations at one or more receptors, and (b) to calculate frequency distributions of concentrations at any receptor on the basis of long-term sequences of meteorological data. Figure 11 depicts examples of carbon monoxide concentration distributions calculated by this model on two grid sizes (48–49).

After initial comparisons with measured streetside concentrations in several urban areas, it was found that the measured concentrations were higher than those calculated, similar to the results of the previous study by Ott et al. (51). After the addition of a submodel to account for this

Table II Selected Examples of Short-Term Models for Quasi-Inert Pollutants

Investigators	Model type	Pollutants	Source inputs	Meteorological inputs	Outputs	Special features	Areas tested	Model performance
Clarke (76)	Integrated Gaussian, partitioned area sources	SO_2, NO_x	Polar coordinate area inventory about receptor, plus point sources	Wind speed, direction, stability	Hourly concentrations at each receptor	Simple to use	Cincinnati, Ohio Washington, D.C.	Correlations from 0.61 to 0.81 for 24-hour values
Hilst (47)	Steady-state Gaussian plume following wind trajectories	SO_2	Point, area, background	Wind field, stability	Hourly concentrations at each receptor	Regional scale	State of Connecticut	—
Fortak (79)	Gaussian plume	SO_2	Industrial point; industrial area, space-heating area	Frequency distribution of wind speed direction, stability	Frequency distributions of concentration at each receptor	Statistical output	Bremen, Federal Republic of Germany	Frequency distributions compare well at half of receptors
Roberts et al. (32)	Integrated Gaussian puff	SO_2	Point, area	Wind field, stability, mixing height	Hourly concentrations at each receptor	Can handle light winds	Chicago, Illinois	66% of 24-hour average SO_2 within 0.05 ppm of observed
Johnson et al. (48)	Integrated Gaussian, partitioned area sources	CO	Polar coordinate area inventory of roadway links about receptor, plus background	Wind speed, direction, stability, mixing height	Hourly concentrations or frequency distributions	Can handle microscale (street canyon) effects	Five cities in the United States	80% of calculated hourly values within 3 ppm of observed values
Gifford and Hanna (74)	Integrated Gaussian, partitioned area sources	SO_2, CO, particulates	Point, area	Wind speed, direction	Hourly concentrations at each receptor	Simple to use	Several cities in the United States	Two-thirds of predictions within ± 50% of measurements

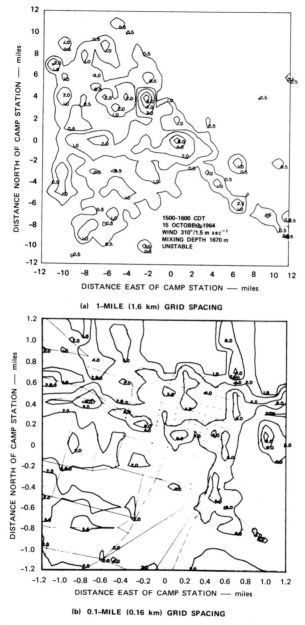

(a) 1-MILE (1.6 km) GRID SPACING

(b) 0.1-MILE (0.16 km) GRID SPACING

Figure 11. Computer-generated plots of calculated carbon monoxide concentration patterns in St. Louis, Missouri for two grid sizes (48–49). (a) One mile (1.6 km) grid spacing; (b) 0.1 mile (0.16 km) grid spacing.

"street-canyon" effect (see Section IV,A), correlation coefficients between calculations and measurements for hourly values ranged from 0.6 to 0.7. Approximately 80% of the calculated values were within 3 ppm of the measured hourly concentrations, which ranged as high as 16 ppm.

VI. Air Quality Simulation Models for Photochemical Pollutants*

A. Background

The modeling of air pollutants that participate in the reactions leading to photochemical smog has been a major undertaking in the air pollution modeling community. The main efforts, however, have each been of limited scope and duration. Three such models have been partially developed to explain certain features of observed data or to provide an operational tool for implementation planning: Each is formulated differently and contains different assumptions and approximations. A three-dimensional Eulerian model has been developed by Roth *et al.* (*80*) and Reynolds *et al.* (*81*), a single-moving-cell model by Weisburd *et al.* (*82*) and Wayne *et al.* (*83*), and a Lagrangian column-of-cells model by Eschenroeder and Martinez (*4*) and Eschenroeder *et al.* (*85*). A fourth photochemical model has been developed by Sklarew *et al.* (*86*) incidental to demonstrating the particle-in-cell technique. In addition, Hanna (*80a*) has developed a simplified approach utilizing dimensional analysis.

A controlled evaluation of the first three of the models mentioned above was undertaken in 1972. Since all the models have been developed for Los Angeles, California, this was the test area. Six days from the 1969 smog season were chosen for the evaluation, which is discussed further later in the section.

Each of the photochemical models contains input-conditioning submodels for emissions, initial and boundary concentrations, winds, and diffusivities. Computational submodels are used for advection, diffusion, and chemical simulation (Table III). The common emission inventory used included CO, NO_x, and hydrocarbons (both reactive and unreactive) from mobile and stationary sources within each cell of a 2-mile grid extending 25 cells on a side. The initial (and boundary) concentrations and the winds (and diffusivities) were interpolated, extrapolated, or estimated from available data. For example, no ambient hydrocarbon concentration measurements were available over the ocean (or even over most of the Los Angeles, California basin) and the vertical temperature structure was monitored only twice on some of the days. Parts of the

---

* See also Chapter 6.

Table III   Submodels in a Photochemical Air Quality Simulation Model

Input	Input-conditioning submodels	Calculation submodels
Measured winds	Full wind field or local wind trajectory	Advective transport and diffusion
Measured inversion heights	Inversion height as a function of position and time; vertical diffusivities	Vertical transport and diffusion
Traffic patterns	Mobile source emissions as a function of position and time	
Mobile source emission factors		Source additions
Stationary source emission factors	Emissions as a function of time	
Chemical reaction chamber experiments	Chemical mechanism and reaction rates	Concentration changes due to reactions
Insolation	Ultraviolet flux for photochemical reactions	
Concentration observations	Initial and boundary concentrations	Initial and boundary conditions

emissions inventory and the advection and diffusion submodels were evaluated by comparison of predictions with CO measurements, since carbon monoxide is essentially inert on these time scales, is similarly advected by winds and dispersed by turbulent diffusion, and thus serves as a check of these submodels. The chemistry simulation submodel was evaluated against data from chemical reaction chambers. Finally, each model taken as a whole was evaluated against available ambient air observations of NO, NO_2, hydrocarbon, and (mainly) O_3 concentrations.

B. Features of Photochemical Air Quality Simulation Models

In the derivation of the diffusion equation (Section II), the instantaneous pollutant concentration was divided into mean concentration and fluctuation about the mean, and the equation was written for the mean concentration. If the effects of reactions between pollutants were included in the original equation for conservation of mass, the derived equations would contain products of concentrations, such as $\bar{C}_i\bar{C}_j$ and $\overline{C_iC_j}$, where the subscripts refer to different chemical species. The latter example is the covariance of fluctuations of the two species. Since this type of term is not known in terms of mean concentrations, it has been dropped in the three models. The diffusion equation is modified only to the extent that the source term is generalized to include net changes in concentration due to chemical reactions. The net changes are calculated using the mean concentrations in the chemical kinetics equations. That is, a reaction such as

$$A + B \rightarrow C + D$$

with rate constant k, would result in a rate of change of each species equal to $kC_A C_B$, removal for A and B, and production of C and D.

Each of the three photochemical models has a chemical simulation submodel in which the coupled nonlinear differential equations are solved numerically for small increments of simulated time, on the order of a few minutes, corresponding to the time step used for advection and source emissions. The methods of numerical solution include multistep Adams *(87)*, Crank-Nicolson *(88)*, and use of Padé approximates *(89)*. The advantages of a multistep implicit method, such as has been developed by Gear *(90)* may be outweighed by the additional memory requirements in an Eulerian approach, but will probably be adopted in trajectory approaches.

Of course, the basic differences in the three chemical submodels are the reaction mechanisms used; Table IV gives an example of one mechanism consisting of 16 steps. All three proposed mechanisms have the same features of (a) an inorganic loop that pushes NO, NO_2, and O_3 toward equilibrium, (b) an oxidative attack on a hydrocarbon-producing radical, the radicals converting NO to NO_2, and (c) a final removal mechanism. The possibilities of branching are included in the first two steps to increase the O_3 yield.

The single moving-cell model by its simplicity in all other aspects permits the use of a more complex 33-step mechanism. In the mechanism for the moving-cell model, the focus is on elementary reactions so any variations in concentrations will produce the proper scaling. It is not possible to detail all of the elementary reactions for the many hydrocarbons in the atmosphere so this mechanism treats propylene and a second hypothetical hydrocarbon of lesser reactivity. The other two mechanisms combine or lump reactions and species into typical ones. This "lumped parameter" approach started simply and has increased in complexity to the present 15 or 16 reaction steps required to simulate observed data. Both lumped-parameter mechanisms are very similar.

The chemical mechanisms were evaluated against data for experiments in reaction chambers covering four hydrocarbon systems (propylene and ethane, toluene, toluene and *n*-butane, and dilute auto exhaust). The various parameters—rate constants, reactivities, dummy hydrocarbon and branching ratios—for which values are available in the literature were fixed or set within the experimental range. Parameters with unknown values were varied to obtain the best fit to the chamber results.

Table IV   Sixteen-Step Chemical Mechanism Used in Lagrangian Column Model (85)

Reaction	Experimental rate constant values[a]
1. $h\nu + NO_2 \rightarrow NO + O$	$2.67(-1)$ min^{-1}
1a. $O + O_2 + M \rightarrow O_3 + M$	$1.32(-5)$ ppm^{-2} min^{-1}
2. $NO + O_3 \rightarrow NO_2 + O_2$	$2.2(+1)$ to $4.4(+1)$ ppm^{-1} min^{-1}
3. $O + HC \rightarrow (b_1)RO_2$	b
4. $OH + HC \rightarrow (b_2)RO_2$	b
5. $O_3 + HC \rightarrow (b_3)RO_2$	b
6. $RO_2 + NO \rightarrow NO_2 + (y)OH$	
7. $RO_2 + NO_2 \rightarrow PAN$	
8. $OH + NO \rightarrow HONO$	$1.5(+3)$ ppm^{-1} min^{-1}
9. $OH + NO_2 \rightarrow HNO_3$	$3.0(+3)$ ppm^{-1} min^{-1}
10. $h\nu + HONO \rightarrow OH + NO$	
11. $NO + NO_2 \xrightarrow{\text{H}_2\text{O}} 2HONO$	
12. $NO_2 + O_3 \rightarrow NO_3 + O_2$	$5(-2)$ to $1.25(-1)$ ppm^{-1} min^{-1}
13. $NO_3 + NO_2 \rightarrow N_2O_5$	$4.5(+3)$ ppm^{-1} min^{-1}
14. $N_2O_5 \rightarrow NO_3 + NO_2$	$1.4(+1)$ min^{-1}
15. $N_2O_5 + H_2O \rightarrow 2HNO_3$	$2.5(-3)$ ppm^{-1} min^{-1} [c]
16. $NO_2 + \text{particulates} \rightarrow \text{products}$	

[a] The number in parentheses denotes the power of ten by which the coefficient must be multiplied, e.g., $2.67(-1) = 2.67 \times 10^{-1}$.

[b] Experimental values for these rate constants are often known for particular hydrocarbons.

[c] For the validation process, k_{15} was converted to a pseudo-first-order rate constant by lumping water vapor content of air at 50% relative humidity into k_{15} since the smog chamber experiments were conducted at this level of humidity. The resulting rate constant is 60.5 min^{-1}.

In this way, results such as those presented in Figure 12 were obtained. The general experimental observations are reproduced, but this is hardly a true test of the mechanisms since the data are incorporated in the selection of parameters.

One evaluation of the three models is their capability to simulate CO concentrations. To provide further tuning of the models, three of the days were used for "calibration." The "calibration" included adjustments in the input-conditioning methodologies used to provide the full set of required input data from more limited observations. For example, in the Eulerian model, the observed winds were used to generate by manual analysis a full wind field throughout the three-dimensional grid; and, in the Lagrangian column model, the temporal and spatial variations of the inversion height between the sparse observations were adjusted to yield consistent CO levels. The other 3 days were then simulated using

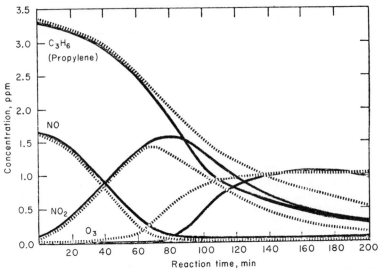

Figure 12. Time profiles of calculated contaminant concentrations compared with smog chamber data. Solid line: experimental data; dotted line: predicted values (80).

the semiobjective techniques developed. In the moving-cell model, an interpolation technique was used throughout and so all 6 days were "hands-off" calculations. All models were initialized with the early morning concentrations and the simulations were continued into the afternoon hours.

In the trajectory models, the calculated concentrations correspond to the average air quality in a volume moving across the area under consideration. To compare trajectory results with data taken at fixed monitoring stations, the station data must be spatially interpolated to the trajectory or the trajectory must pass near a station. In the Eulerian model, the cells closest to a station may be used to interpolate a calculated value to compare to the station observation. In either of these types of models, the calculated concentrations correspond to spatially averaged air qual-

Table V Statistical Comparison of Model Calculations and Observations of CO

Model	Number of data points	Correlation coefficient	RMS error (ppm)	Regression line ($y = ax + b$)	
				a	b
Moving cell	61	0.82	3.0	0.91	−0.70
Lagrangian column	149	0.80	3.7	1.01	−0.37
Eulerian	514	0.79	3.4	1.09	0.68

Table VI   Statistical Comparison of Model Calculations and Observations of O₃

Model	Number of data points	Correlation coefficient	RMS error (ppm)	Regression line (y = ax + b)	
				a	b
Moving cell	63	0.49	6.9	0.46	−7.6
Lagrangian column	151	0.92	2.1	0.84	2.3
Eulerian	574	0.69	6.3	0.76	3.9

ity, while the station monitors the local concentrations at a point. Since the station data thus can be severely influenced by local emissions, this poses a special problem in using these data for model validation purposes, as was discussed in Section III. These limitations in the available data should be kept in mind during the following discussion.

C. Results of Model Evaluation Tests

The results of statistical comparisons of concentrations calculated by the three models with observed concentrations of CO and O₃ are presented in Tables V and VI. Two of the models show better performance with

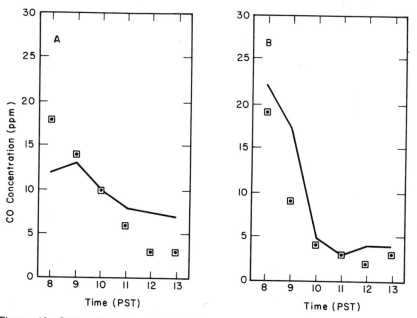

Figure 13. Comparison of observed with calculated CO concentrations from the moving-cell model (83). A: Downtown Los Angeles, California, 9/29/69; B: Whittier, California, 10/30/69. ▣: measured; solid line: predicted.

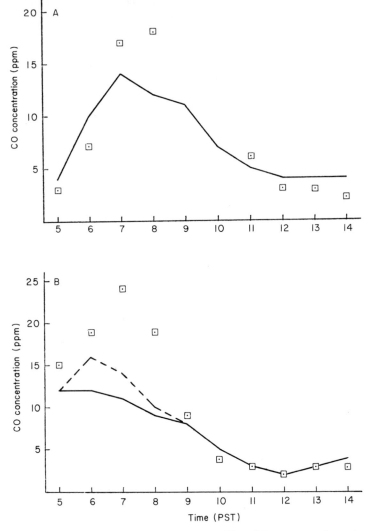

Figure 14. Comparison of observed with calculated CO concentrations from the Eulerian model (*81*). A: Downtown Los Angeles, California, 9/29/69; B: Whittier, California, 10/30/69. ⊡: measured; solid line: predicted; dashed line: represents a correction for local effects calculated by a submodel for the subgrid scale.

CO than with O_3, as might be expected due to the difficulty in simulating the complex photochemical process leading to the formation of O_3.

Comparisons of calculated and observed concentrations as a function of time of day are also illustrative, as shown by Figures 13–15 for CO, and Figures 16–18 for O_3. The time histories for the moving-cell and the

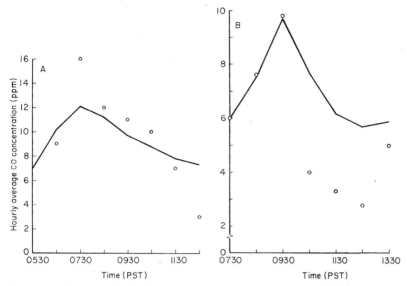

Figure 15. Comparison of observed with calculated CO concentrations from the Lagrangian column model (85). A: Along an air trajectory from central Los Angeles, California to El Monte, California, 9/29/69; B: along an air trajectory from Pasadena, California to Orange County, California, 10/30/69. Solid line: model results; ○: interpolated station data.

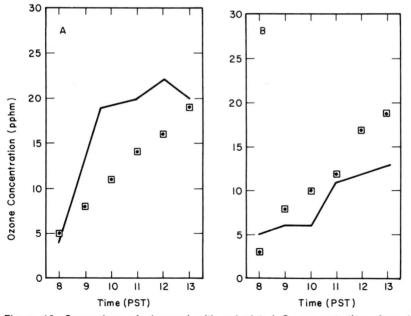

Figure 16. Comparison of observed with calculated O₃ concentrations from the moving-cell model (83). A: Downtown Los Angeles, California, 10/29/69; B: Azusa, California, 10/29/69. ⊡: measured; solid line: predicted.

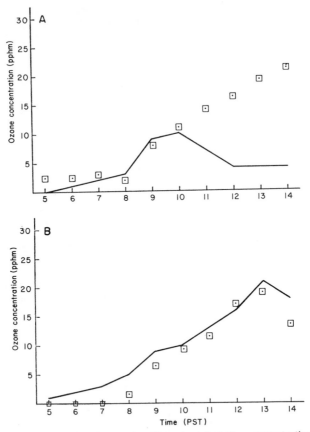

Figure 17. Comparison of observed with calculated O_3 concentrations from the Eulerian model (81). A: Downtown Los Angeles, California, 10/29/69; B: Azusa, California, 10/29/69. ⊡: measured; solid line: predicted.

Eulerian models are for the same two stations, while the Lagrangian column-of-cells model results are compared with interpolated observations along air trajectories. (For example, the results in Figure 18 reflect concentrations along a trajectory that started at the coast near Los Angeles, California International Airport and ended 10 hours later near Anaheim, California.) In general, the models reproduce much of the temporal distribution of the observed concentrations, but at later times for O_3 the calculated values and observations tend to diverge. This is even more apparent in the results for NO_2, for which observations show a much more rapid removal in the afternoon than was accounted for in the simulations.

Another and perhaps more interesting test of the model is the simula-

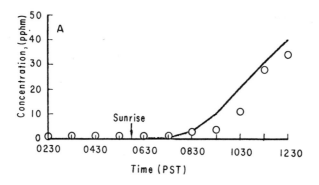

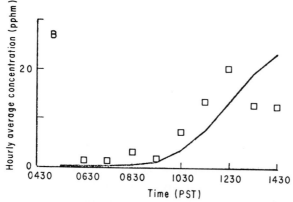

Figure 18. Comparison of observed with calculated O_3 concentrations from the Lagrangian column model (*85*). A: Along an air trajectory from Redondo Beach, California to Anaheim, California, 9/29/69; B: Along an air trajectory from central Los Angeles, California to Van Nuys, California, 11/4/69. Solid line: model results; ○ and □: interpolated station data.

tion of the temporal and spatial distribution of maximum concentrations. The Eulerian model produces results easily analyzed in this way. For September 29, 1969 (a "hands-on" day), the Eulerian model generally calculated the O_3 maxima within 1 hour of occurrence and within 20% of the magnitude, as shown in Table VII. When the differences were over 1 hour or 20% the cause could usually be traced to a large gradient in O_3 levels near the station.

D. Areas Needing Further Work

Although the results from this first controlled test of three photchemi-cal models indicate that much progress has been made, there are still

Table VII O_3 Maxima Calculated by the Eulerian Model on September 29, 1969

	Observed values		Calculated values	
Station (California)	Time	Concentration (pphm)	Time	Concentration (pphm)
Reseda	1200	20	1100	22
Burbank	1200	23	1100	23
Pasadena	1100	40	1100	40
Azusa	1300	39	1200	40
El Monte	1200	25	1300	40
Downtown Los Angeles	1000	12	1000	7
West Los Angeles	1100	14	1100	10
Lennox	1100	10	1200	6
Whittier	1100	32	1000	27
La Habra	1200	42	1100	38
Long Beach	1100	7	1300	6
Anaheim	1200	40	1200	38
Santa Ana	1200	21	1200	17

serious shortcomings in the state of the art. As one illustration, in the Lagrangian column model the NO_x emissions flux is reduced by 75% before any calculations are made, whereas the other two models obtain similar correlation with observations using the total NO_x emissions. At the time of writing this chapter, sensitivity analyses had yet to be performed to determine the requirements on emissions and meteorological input data, on the detail in the chemical mechanism, or on spatial resolution. The list of future in-depth research topics is a long one, including the following:

Chemical simulation
　Hydrocarbon breakdown by reactivity
　Aerosol formation
　Reactions on surfaces and in solution, including pollutant removal
　Non-Los Angeles, California applications, especially with SO_2 present
　Treatment of localized reactive emissions

Meteorological simulation
　Delineation of applicability of Lagrangian approach
　Objective analysis techniques to develop complete wind field
　Objective techniques to develop diffusivities
　Specification of proper ultraviolet flux
　Large-point-source modeling

Modeling concepts
  Scale of resolution required for regional models
  Subgrid effects, such as turbulence effects on chemical reactions
  Local-scale submodels for comparison of model results to point
  measurement

Model applications
  Implementation planning
  Non-Los Angeles, California regions

It is perhaps on this last topic that there has been the least progress. With all the weaknesses in photochemical (and other air pollution) models, they still represent an organized framework encompassing our best knowledge of the behavior of pollutants in the atmosphere. Especially with regard to photochemical smog, previous nonmodeling approaches can result in the wrong decision. In describing the application of photochemical models to the evaluation of implementation plans, Sklarew (*91*) discusses a simulation of the effects of reducing downtown traffic. Instead of reducing O_3, the model calculates *higher* O_3 levels—due to the reduction of NO to react with the O_3—which is counter to conclusions reached by rollback-type arguments.

VII. Model Validation and Application

The structural or developmental aspects of AQS models are only a part of the total picture, since a developed model does no good until it is used, and before a model is used it must be evaluated and validated.

A. Model Validation

An important factor that has discouraged extensive model application has been a general lack of confidence in their capabilities. This follows from the fact that few models have been adequately evaluated and validated,* because of the general lack of suitable data. To gather more data requires field programs, which are expensive.

A contributing element has been the frequent absence of model performance goals to serve as standards for validation. Setting such goals is

---

* In this discussion the terms "evaluate" and "validate" are used interchangeably, and taken to have almost the same meaning. Strictly speaking, however, a model must be *evaluated* (dictionary: "appraised carefully") as a necessary step *before* it can be *validated* (dictionary: "confirmed").

not a straightforward matter, since they must necessarily reflect the intended use of the model, as well as the capabilities of the users, and must also involve a number of compromises among such things as accuracy, resolution, and computation time. In view of the state of the art of modeling, the best policy is to be as liberal in the specifications as the intended model application will permit, rather than to strive for some desirable, but possibly unattainable, goal.

In the absence of resources for special data-gathering programs, modelers have had to rely upon available data for checking their models. Most efforts have been confined to comparisons of the model calculations with observed air pollutant concentrations. This procedure, which is basically a *calibration* (see Section I,E), can help improve the performance of a model for a specific application. However, the adjustments that are made do not usually contribute to confidence in the generality of the model, unless the latter can be proved through the use of data from several cities. Indeed, it is quite possible for a model to have defective, but compensating, elements, which overall may give adequate agreement with air quality observations for the "calibration" city, but nowhere else.

In contrast, a true evaluation and validation program must involve the evaluation of individual model elements, on the basis that the whole can be no better than the weakest component part. As previously discussed, almost all models consist of several elements or submodels. Some of these submodels estimate detailed model inputs on the basis of available gross data. In order to validate such submodels, special experiments are necessary to furnish data by which comparisons between submodel estimates and measured values of the same variables can be made.

Validation of the main portions of a model is best accomplished through special experiments designed to isolate and obtain data on the specific phenomena being simulated. For example, validation of a meteorological submodel for simulating three-dimensional flow patterns in complex settings might involve such experimental techniques as trajectory tracing with constant-level balloons, or wind-field mapping with an instrumented aircraft. Validation of a chemical transformations submodel might involve laboratory chamber experiments, or atmospheric experiments designed to minimize the effects of meteorological diffusion and other such complications.*

---

* Dr. F. Pasquill has summed up the preceding discussion rather well in the form of a verse:

> Calibration may be consoling
> But is a practice we shouldn't condone,
> And for validation to have virtue
> The physics and chemistry must be known.

The United States Environmental Protection Agency (EPA), recognizing the paucity of available data for model validation, has sponsored a large, multiyear field program in the St. Louis, Missouri area called the Regional Air Pollution Study (RAPS). One of the principal objectives of this study has been the refinement and validation of AQS models for general use in determining least-cost air pollution abatement strategies [Pooler (92)]. RAPS has involved a network of some 25 fixed aerometric and meteorological ground stations, along with instrumented aircraft, mobile stations, and other facilities to support special experiments.

B. Model Application

In order to encourage and facilitate the use of models, EPA has provided a library of models in a central computer, located at Research Triangle Park, North Carolina, that users can access and run remotely. This system is called the User's Network for Applied Modeling of Air Pollution (UNAMAP). The basic advantages of UNAMAP are

(a) The users always have access to a set of models reflecting the latest state of the art

(b) The users do not have to program any models

(c) The users need only minimum modeling expertise

(d) EPA can efficiently carry out its responsibility within the U.S. for maintaining and periodically updating the models and model inventories, and for providing a message service to users concerning UNAMAP changes

(e) There should eventually be a manageable number of consistent models in use, which should facilitate review of control plans and the like

Several models are now in storage in the UNAMAP system, and others will be added upon completion of development and validation activities.

REFERENCES

1. A. C. Stern, *Clear Air (Aust. N.Z.)* **4**, (1970).
2. R. C. Wanta, *in* "Air Pollution" (A. C. Stern, ed.), 2nd ed., Vol. 1, p. 187. Academic Press, New York, New York, 1968.
3. H. Moses, "Mathematical Urban Air Pollution Models," ANL/ES-RPY-001. Argonne Nat. Lab., Argonne, Illinois, 1969.
4. J. H. Seinfeld, P. M. Roth, and S. D. Reynolds, *Advan. Chem. Ser.* **113**, 58 (1972).

5. A. Q. Eschenroeder and J. R. Martinez, *Advan. Chem. Ser.* **113**, 101 (1973).

6. M. I. Hoffert, *AĨAA J.* **10**, 377–387 (1972).

7. L. T. Fan and Y. Horie, *Crit. Rev. Environ. Contr.* pp. 431–457 (1971).

8. Y. Horie and J. H. Overton, *in* "Proceedings of the Symposium on the Statistical Aspects of Air Quality Data" (L. D. Kornreich, ed.), EPA-650/4-74-038. United States Environmental Protection Agency, Research Triangle Park, North Carolina, 1974.

9. N. de Nevers and R. Morris, *66th Annu. Meet. Air Pollut. Contr. Ass., 1973* Pap. No. 73-139 (1973).

10. H. Moses, *in* "Multiple-Source Urban Diffusion Models" (A. C. Stern, ed.), Air Pollut. Contr. Office Publ. No. AP-86, pp. 14-13. United States Environmental Protection Agency, Research Triangle Park, North Carolina, 1970.

11. R. I. Larsen, *J. Air Pollut. Contr. Ass.* **19**, 24–30 (1969) ; "A Mathematical Model for Relating Air Quality Measurements to Air Quality Standards," Office of Air Programs Publ. No. AP-89. United States Environmental Protection Agency, Research Triangle Park, North Carolina, 1971.

12. L. D. Kornreich, ed., "Proceedings of the Symposium on the Statistical Aspects of Air Quality Data," EPA-650/4-74-038. United States Environmental Protection Agency, Research Triangle Park, North Carolina, 1974.

13. A. C. Stern, *in* "Multiple-Source Urban Diffusion Models" (A. C. Stern, ed.), Air Pollut. Contr. Office Publ. No. AP-86, Chapter 13. United States Environmental Protection Agency, Research Triangle Park, North Carolina, 1970.

14. F. H. Chandry and J. E. Cermak, *in* "Proceedings of Conference on Air Pollution Meteorology," p. 126. Amer. Meteorol. Soc., Boston, Massachusetts, 1971.

14a. J. E. Cermak and S. P. S. Arya, *Boundary-Layer Meteorol.* **1**, 40–60 (1970).

15. W. H. Snyder, *Boundary-Layer Meteorol.* **3**, 113–134 (1972).

16. J. E. Cermak and J. Peterka, "Simulation of Wind Fields over Point Arguello, California, by Wind Tunnel Flow over a Topographic Model," Final Rep., Contract No. N123(61756)34361A(PMR). Colorado State University, Fort Collins, Colorado, 1966.

17. L. Facy, P. Brichambaut, A. Doury, and R. Quinio, *Rev. Meteorol.* **69**, No. 71 (1963).

18. J. P. Pandolfo, M. A. Atwater, and G. E. Anderson, "Prediction by Numerical Models of Transport and Diffusion in an Urban Boundary Layer," Final Rep., Contract No. CPA 70-62. Center for the Environment and Man, Inc., Hartford, Connecticut, 1971.

19. J. P. Pandolfo and C. A. Jacobs, "Tests of an Urban Meteorological-Pollutant Model Using CO Validation Data in the Los Angeles Metropolitan Area," Vol. 1, No. R4-73-025a. United States Environmental Protection Agency, Research Triangle Park, North Carolina, 1973.

19a. J. B. Knox, *J. Air Pollut. Contr. Ass.* **24**, 660–664 (1974).

20. D. G. Fox and F. Pooler, *in* "Proceedings of the Fourth Meeting of the Expert Panel on Air Pollution Modeling" (K. L. Calder, ed.), Rep. No. 30, p. XIII-1. NATO Committee on the Challenges of Modern Society, 1973.

21. D. G. Fox, *67th Annu. Meet. Air Pollut. Contr. Ass., Denver, Colorado, 1974* Pap. No. 74-271 (1974).

22. Y. Horie and D. Svendsgaad, "Development and Critical Evaluation of Air Pollution Episode Prediction Schemes," 2nd Conf. Numerical Prediction. Amer. Meteorol. Soc., Monterey, California, 1973.

22a. Y. Horie, *67th Annu. Meet. Air Pollut. Contr. Ass., 1974* Pap. No. 74-269 (1974).

23. A. Q. Eschenroeder, J. R. Martinez, and R. Nordsieck, *in* "Proceedings of the

2nd Summer Computer Simulation Conference," Library of Congress Catalog Card No. 73-76709, p. 1013. Simulation Council, Inc., La Jolla, California, 1972.

24. V. A. Mohnen and E. G. Walther, *67th Annu. Meet. Air Pollut. Contr. Ass., 1974* Pap. No. 74-150 (1974).

25. C. du P. Donaldson and G. R. Hilst, *Environ. Sci. Technol.* **6**, 812–816 (1972).

26. C. du P. Donaldson and G. R. Hilst, in "Proceedings of the 1972 Heat Transfer and Fluid Mechanics Institute" (R. B. Landis and G. J. Hordemann, eds.), p. 353. Stanford Univ. Press, Stanford, California, 1972.

27. T. A. Hecht, J. H. Seinfeld, and M. C. Dodge, *Environ. Science and Tech.* **8**, 327–339 (1974).

27a. P. M. Roth, T. A. Hecht, and J. H. Seinfeld, "Existing Needs in the Experimental and Observational Study of Atmospheric Chemical Reactions," EPA Report R4-73-031. United States Environmental Protection Agency, Research Triangle Park, North Carolina, 1973.

28. F. B. Abeles, L. E. Craker, L. E. Forrence, and G. R. Leather, *Science* **173**, 914–916 (1971).

29. R. E. Inman, R. B. Ingersoll, and E. A. Levy, *Science* **172**, 1229–1231 (1971).

30. R. E. Inman and R. B. Ingersoll, *J. Air Pollut. Contr. Ass.* **21**, 646–647 (1971).

31. R. B. Ingersoll, R. E. Inman, and W. R. Fisher, *Tellus* **26**, 151–159 (1974).

31a. F. A. Gifford, in "Meteorology and Atomic Energy" (D. H. Slade, ed.), Nat. Tech. Inform. Serv. No. TID-24190, Chapter 3, p. 65. U.S. At. Energy Comm., Oak Ridge, Tennessee, 1968.

32. J. J. Roberts, E. J. Croke, and A. S. Kennedy, in "Multiple-Source Urban Diffusion Models" (A. C. Stern ed.), Air Pollut. Contr. Office Publ. No. AP-86, p. 6-1. United States Environmental Protection Agency, Research Triangle Park, North Carolina, 1970.

33. L. J. Shieh, B. Davidson, and J. P. Friend, in "Multiple-Source Urban Diffusion Models" (A. C. Stern, ed.), Air Pollut. Contr. Office Publ. No. AP-86, p. 10-1. United States Environmental Protection Agency, Research Triangle Park, North Carolina, 1970.

34. E. A. Ulbrick, *Socio-Econ. Plan. Sci.* **1**, 423 (1968).

35. H. Reiquam, *Atmos. Environ.* **4**, 233 (1970).

36. M. C. MacCracken, T. V. Crawford, K. R. Peterson, and J. B. Knox, "Development of a Multibox Air-Pollution Model and Initial Verification for the San Francisco Bay Area," Rep. No. UCRL-733348. Lawrence Livermore Laboratory, Livermore, California, 1971.

37. L. G. Wayne, R. Danchick, M. Weisburd, A. Kokin, and A. Stein, *J. Air Pollut. Contr. Ass.* **21**, 334 (1971).

38. J. V. Behar, "Simulation Model of Air Pollution Photochemistry," Project Clean Air Res. Rep. No. 4. Statewide Air Pollution Research Center, University of California, Riverside, California, 1970.

39. D. M. Leahey, *J. Air Pollut. Contr. Ass.* **22**, 548 (1972).

40. C. du P. Donaldson, "Construction of a Dynamic Model of the Production of Atmospheric and the Dispersal of Atmospheric Pollutants," Workshop in Micrometeorology. Amer. Meteorol. Soc., Boston, Massachusetts, 1972.

41. R. B. Lantz, K. H. Coats, C. V. Kloepter, Air Pollution Turbulence Diffusion Symp. Amer. Meteorol. Soc., Boston, Massachusetts, 1971.

42. B. A. Egan and J. R. Mahoney, *J. Appl. Meteorol.* **11**, 312 (1972).

43. A. Q. Eschenroeder and J. R. Martinez, "Mathematical Modeling of Photochemical Smog," 8th Aerosp. Sci. Meet., Pap. No. 70-116. Amer. Inst. Aeronaut. Astronaut., New York, New York, 1970.

44. R. C. Sklarew, A. J. Fabrick, and J. E. Prayer, *J. Air Pollut. Contr. Ass.* 22, 865 (1972).

45. S. K. Friedlander and J. H. Seinfeld, *Environ. Sci. Technol.* 3, 11 (1969).

46. F. A. Gifford and S. R. Hanna, *Atmos. Environ.* 6, 131–136 (1972).

47. G. R. Hilst, *in* "IBM Scientific Computing Symposium on Water and Air Resource Management," pp. 251–274. International Business Machines Corp., White Plains, New York, 1968.

48. W. B. Johnson, F. L. Ludwig, W. F. Dabberdt, and R. J. Allen, *J. Air Pollut. Contr. Ass.* 23, 490–498 (1973).

48a. W. B. Johnson, F. L. Ludwig, and A. E. Moon, *in* "Multiple-Source Urban Diffusion Models" (A. C. Stern, ed.), Air Pollut. Contr. Office Publ. No. AP-86, Chapter 5. United States Environmental Protection Agency, Research Triangle Park, North Carolina, 1970.

49. W. F. Dabberdt, F. L. Ludwig, and W. B. Johnson, *Atmos. Environ.* 7, 603–618 (1973).

50. W. Ott and R. Eliassen, *J. Air Pollut. Contr. Ass.* 23, 685–690 (1973).

51. W. Ott, J. F. Clarke, and G. Ozolins, "Calculating Future Carbon Monoxide Emissions and Concentrations from Urban Traffic Data," Nat. Air Pollut. Contr. Admin. Publ. No. 999-AP-41. United States Environmental Protection Agency, Research Triangle Park, North Carolina, 1967.

52. R. A. McCormick, *Phil. Trans. Roy. Soc. London, Ser. A* 269, 515–526 (1971).

53. H. W. Georgii, E. Busch, and E. Weber, "Investigation of the Temporal and Spatial Distribution of the Immission Concentration of Carbon Monoxide in Frankfurt/Main," Rep. No. 11 of the Institute for Meteorology and Geophysics, University of Frankfurt am Main (Translation No. 0477. United States Nat. Air Pollut. Contr. Admin., Research Triangle Park, North Carolina), 1967.

54. D. B. Turner, "Workbook of Atmospheric Dispersion Estimates," No. AP-26. United States Environmental Protection Agency, Research Triangle Park, North Carolina, 1970.

55. J. L. Beaton, A. J. Ranzieri, E. C. Shirley, and J. B. Skog, "Mathematical Approach to Estimating Highway Impact on Air Quality," Air Quality Manual IV, Rep. No. FHWA-RD-72-36. United States Federal Highway Administration, Washington, D.C., 1972.

56. R. C. Sklarew, D. B. Turner, and J. R. Zimmerman, "Modeling Air Pollution Near Highways." Conference Proc., VDI-Kommission Reinhaltung der Luft, Dusseldorf, Germany, 1972.

57. P. B. S. Lissaman, *66th Annu. Meet. Air Pollut. Contr. Ass., 1973* (1973).

57a. W. B. Johnson, *in* "Proceedings of the Symposium on Atmospheric Diffusion and Air Pollution," pp. 261–266. Amer. Meteorol. Soc., Boston, Massachusetts, 1974.

58. M. B. Danard, R. S. Koneru, and P. R. Slawson, *Air Pollut. Turbulence Diffusion Symp.*, p. 152. Amer. Meteorol. Soc., Boston, Massachusetts, 1971.

59. R. C. Sklarew, A. J. Fabrick, and J. E. Prager, "Atmospheric Simulation Modeling of Motor Vehicle Emissions in the Vicinity of Roadways," Proc. 2nd Summer Simulation Conf. Simulation Council, Inc., La Jolla, California, 1972.

60. B. A. Egan and T. F. Lavery, "Highway Designs and Air Pollution Potential," Proc. 3rd Urban Technol. Conf., 1973. Amer. Inst. Aeronaut. Astronaut., Boston, Massachusetts, 1973.

61. R. S. Hotchkiss and ‘ C. W. Hirt, "Particulate Transport in Highly Distorted Three-Dimensional Flow Fields," Proc. 2nd Summer Simulation Conf. Simulation Council, Inc., La Jolla, California, 1972.

62. M. Platt, R. C. Baker, E. K. Bastress, K. M. Chang, and R. D. Siegel, "The Potential Impact of Aircraft Emissions upon Air Quality," Final Rep. Contract 68-02-0085, Northern Research and Engineering Corporation Report 1167-1. Cambridge, Massachusetts, 1971.

63. S. D. Thayer, *in* "Proceedings of the Symposium on Atmospheric Diffusion and Air Pollution," pp. 368–375. Amer. Meteorol. Soc., Boston, Massachusetts, 1974.

64. F. N. Frenkiel, *in* "Smithsonian Report for 1956," pp. 269–299. Smithsonian Institution, Washington, D.C., 1956.

65. F. Pooler, Jr., *Int. J. Air Water Pollut.* 4, 199–211 (1961).

66. D. O. Martin, *J. Air Pollut. Contr. Ass.* 21, 16–19 (1971).

67. TRW Systems Group, "Air Quality Display Model," Nat. Air Pollut. Contr. Admin. Contract PH-22-68-60 (Nat. Techn. Inform. Serv. No. PB 189-194). United States Environmental Protection Agency, Washigton, D.C., 1969.

68. K. L. Calder, *in* "Proceedings of the Second Meeting of the Expert Panel on Air Pollution Modeling," Chapter 1. NATO Committee on the Challenges of Modern Society, 1971.

69. J. R. Zimmerman, *in* "Proceedings of the Third Meeting of the Expert Panel on Air Pollution Modeling," Chapter 4. NATO Committee on the Challenges of Modern Society, 1972.

70. J. R. Zimmerman, *in* "Proceedings of the Second Meeting of the Expert Panel on Air Pollution Modeling," Chapter 15. NATO Committee on the Challenges of Modern Society, 1971.

71. D. B. Turner, J. R. Zimmerman, and A. D. Busse, *in* "Proceedings of the Third Meeting of the Expert Panel on Air Pollution Modeling," Chapter 8. NATO Committee on the Challenges of Modern Society, 1972.

72. G. A. Briggs, "Plume Rise," Nat. Tech. Inform. Serv. No. TID-25075. U.S. At. Energy Commission, Oak Ridge, Tennessee, 1969.

73. S. R. Hanna, *J. Air Pollut. Contr. Ass.,* 21, 774–777 (1971).

74. F. A. Gifford and S. R. Hanna, *Atmos. Environ.* 7, 131–136 (1973).

75. D. B. Turner, *J. Appl. Meteorol.* 3, 83–91 (1964).

76. J. F. Clarke, *J. Air Pollut. Contr. Ass.* 14, 347–352 (1964).

77. J. B. Koogler, R. S. Sholtes, A. L. Danis, and C. I. Harding, *J. Air Pollut. Contr. Ass.* 17, 211–214 (1967).

78. F. A. Gifford, *in* "Proceedings of the 4th Meeting of the Expert Panel on Air Pollution Modeling" (K. L. Calder, ed.), Chapter 16. NATO Committee on the Challenges of Modern Society, 1973.

79. H. G. Fortak, *in* "Multiple-Source Urban Diffusion Models" (A. C. Stern, ed.), Air Pollut. Contr. Office Publ. No. AP-86, Chapter 9. United States Environmental Protection Agency, Research Triangle Park, North Carolina, 1970.

79a. C. C. Shir and L. J. Shieh, *J. Appl. Meteorol.* 13, 185–204 (1974).

80. P. M. Roth, S. D. Reynolds, P. J. W. Roberts, and J. H. Seinfeld, "Development of a Simulation Model for Estimating Ground Level Concentrations of Photochemical Pollutants," Rep. 71-SAI-21. Systems Applications, Inc., Beverly Hills, California, 1971.

81. S. D. Reynolds, M. K. Liu, T. A. Hecht, P. M. Roth, and J. H. Seinfeld, "Further Development and Validation of a Simulation Model for Estimating Ground Level Concentrations of Photochemical Pollutants," Rep. No. EPA-R4-73-030a to f. United States Environmental Protection Agency, Research Triangle Park, North Carolina, 1973.

82. M. Weisburd, L. G. Wayne, R. Danchick, A. Kokin, and A. Stein, "Development of a Simulation Model for Estimating Ground Level Concentrations of Photo-

chemical Pollutants," Final Rep. TM-(L), 4673/000/00. Systems Development Corp., Santa Monica, California, 1971.

83. L. G. Wayne, A. Kokin, and M. I. Wiesburd, "Controlled Evaluation of the Reactive Environmental Simulation Model (REM)," Rep. No. EPA-R4-73-013. United States Environmental Protection Agency, Research Triangle Park, North Carolina, 1973.

84. A. Q. Eschenroeder and J. R. Martinez, "Concepts and Applications of Photochemical Smog Models," Technical Memo 151b. General Research Corporation, Santa Barbara, California, 1971.

85. A. Q. Eschenroeder, J. R. Martinez, and R. A. Nordsieck, "Evaluation of a Diffusion Model for Photochemical Smog Simulation," CR-1-273. General Research Corporation, Santa Barbara, California, 1972.

86. R. C. Sklarew, A. J. Fabrick, and J. E. Prager, "A Particle-in-Cell Method for Numerical Solution of the Atmospheric Diffusion Equation and Applications to Air Pollution Problems," Final report (NTIS No. PB-209-290). Systems, Science and Software, Inc., La Jolla, California (1971).

86a. S. R. Hanna, *Atmos. Environ.* **7**, 803–817 (1973).

87. C. W. Gear, "Numerical Initial Value Problems in Ordinary Differential Equations," p. 111. Prentice-Hall, Englewood Cliffs, New Jersey, 1971.

88. D. Greenspan, "Discrete Numerical Methods in Physics and Engineering," p. 134. Academic Press, New York, New York, 1974.

89. R. S. Varga, *J. Math Physics* **40**, 220 (1961).

90. C. W. Gear, *Commun. ACM* **14**, 176 (1971).

91. R. C. Sklarew, *J. Air Pollut. Contr. Ass.* (in press).

92. F. Pooler, *J. Air Pollut. Contr. Ass.* **24**, 228–231 (1974).

11

Meteorological Measurements

E. Wendell Hewson

I. Introduction

As air pollution problems become both more widespread and more complex the need for carefully planned programs of appropriate atmospheric measurements becomes more acute. Computer modeling of physical and chemical processes in polluted atmospheres can produce much relevant information at moderate cost, but such models must be tested against measurements taken in the atmosphere in order that still more adequate models can be generated. Maximum progress is made when models and measurements leap frog over each other in an orderly manner to produce new syntheses of significant knowledge. The previous chapter (Chapter 10) presents meteorological modeling of air pollution. The present chapter describes many of the meteorological measurements and the instruments needed to obtain them which are so essential for an effective interaction of models and measurements.

The analysis of air pollution in its various aspects is increasingly being undertaken in a systems context. The problem in its larger outlines is often considered as a three-component system: source, atmospheric dispersion, and receptor. Unless such a systems approach is adopted, an incomplete view of any particular air pollution problem is often the result. The evaluation of the various functions of the atmosphere in relation to atmospheric dispersion of airborne contaminants is also being undertaken more and more on a systems basis. These dispersing atmospheric characteristics must be first sensed, then transmitted, next processed, then subjected to rigorous quality control, and finally classified. Meteorological measuring systems and their various components are described in subsequent sections of this chapter. A complete description of systems engineering in its many aspects is available (1). There is also a discussion of weather data systems in general (2) and an application of the systems approach to the development of specifications for automatic weather stations (3).

A number of comprehensive treatises on meteorological measurements have been published (4–10), and there is a comprehensive bibliography of published papers on the subject (11). A useful discussion of practical considerations in instrument design is available (12). A critical evaluation of meteorological instrumentation for use in air pollution surveys is most valuable (13).

II. Primary Meteorological Measurements

There are certain meteorological measurements which experience has shown to be necessary for the evaluation of virtually all problems in air pollution. These primary observations are those of wind direction, wind speed, wind turbulence (*14*), and a related quantity known as the mixing height. There are other measurements which are often but not always important; these will be referred to as secondary meteorological measurements.

The measurement of these primary quantities leads directly into a brief consideration of the dynamic response characteristics of meteorological sensors such as those for wind and temperature (*15–18*). There are two distinct types of response characteristics which correspond to two different mathematical systems: the first-order system and the second-order system.

The first-order system is a nonoscillatory system represented by one of the rotation anemometers or resistance thermometers mentioned below. The response characteristics of such a system are completely defined by a single quantity, the time constant; for sensors whose response to step changes in input is exponential in nature, the time constant is the time required for the sensor to indicate 63.2% of the total change.

The second-order system is a damped oscillatory system such as the wind vane whose dynamic response characteristics are completely specified by two quantities: its damping ratio and its undamped natural frequency or wavelength. The damping ratio is the ratio of the amplitude of any one of a series of damped oscillations to that of the following one at the same phase; the natural frequency is that at which a system will oscillate freely in the absence of external forces.

The complete specifications of meteorological sensing and recording systems should include values of these and related quantities. Unfortunately, not all manufacturers supply dynamic response data.

A. Wind Direction

Wind direction measurements are conveniently discussed under the following headings: surface wind direction and wind direction aloft.

1. Surface Wind Direction

There are four main types of wind vane for measuring wind direction (*4, 9–11, 19*).

a. FLAT PLATE VANE. In this type, the sensing element which governs the azimuth angle of a vertical shaft specifying the wind direction is a vertical plate mounted at one end of a horizontal rod. There is a counterweight at the other end of the horizontal rod. The rod is fastened to the vertical shaft. Wind pressure acting on the flat plate keeps the counterweight heading into the wind.

In the flat plate vane shown in Figure 1, as the wind direction changes, the vane responds, and the vertical shift turns in low torque bearings, causing a wear-free air-gap capacitive transducer to phase shift a 1-kHz sinusoidal signal relative to a 1-kHz reference signal by an amount directly proportional to the wind direction azimuth. The output information is contained in phase-shifted frequencies for accurate transmission to a translator, which is not shown. In another often-used system, the

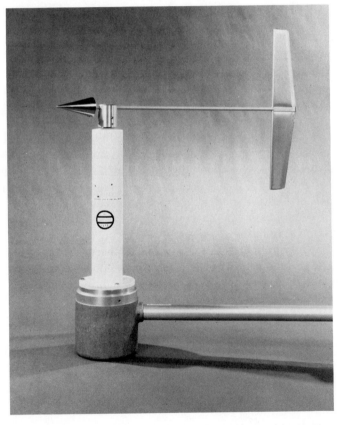

Figure 1. Rapid response flat plate wind vane. (Courtesy of Teledyne Geotech, Garland, Texas.)

vane is coupled directly to the shaft of a synchro generator, which is wired to a synchro receiver in a recorder used to position a direction pen (*19*).

b. SPLAYED VANE. In this type, two flat plates joined at a small angle at one end of a horizontal rod act as the wind direction sensor.

c. AIRFOIL VANE. The vane of this instrument has an airfoil cross section, with the span often being three or four times the chord.

d. MULTIPLE ELEMENT VANE. This vane appears in a variety of responding surfaces. It may consist of two or more plates or airfoil sections arranged in various combinations. These vanes are sometimes used when very rapid response to wind direction changes is required. It should be noted that double element wind vanes which have their plates at angles >10° with the vane axis may refuse to point into the wind, having instead two symmetric equilibrium positions at angles up to 16° with respect to the wind direction (*20*).

e. ANALYSIS OF VANE ACTION. Electrical circuitry is the preferred technique for indicating or recording wind direction (*6*, *19*). Theoretical and wind tunnel analyses of vane action have been made (*6*, *15*, *20–24*). The dynamic response characteristics of a variety of commercial wind vanes have been studied (*15*).

f. RUNNING-AVERAGE ANEMOGRAPH. It is often advantageous to be able to obtain directly a recording of the running average of the wind speed or direction or both. There is no problem in obtaining a running average of the wind speed, except for very long time constants. In averaging the wind direction, however, a difficulty arises from the discontinuity 360°–0° as the wind direction fluctuates around north. An anemograph which automatically produces the running averages of both wind speed and direction has been developed (*25*).

2. Wind Direction Aloft

The wind direction at the height of a plume from one or more stacks may be important in some analyses and investigations. All of the available methods involve considerable time and effort (*5*, *7*, *8*, *10*, *26*).

a. PILOT BALLOONS (PIBALS). The direction of the wind aloft may be determined by tracking by means of one or more theodolites a small balloon inflated with hydrogen or helium (7, 26). The average wind direction in successive height intervals is obtained by triangulation techniques. Expandable neoprene balloons are commonly used. Those with an initial diameter of 1 m, or a little more, operate in the subcritical range of Reynolds numbers, and are generally satisfactory; larger balloons are in the supercritical range and experience random lateral motions which obscure all but the gross features of the wind distribution (27). Specially roughened large superpressure spheres such as the Jimsphere are much less unstable (28, 29) but usually such larger balloons are not required for the analysis of ordinary air pollution problems.

The pilot balloon technique of determining winds aloft is of limited value with low cloud, fog, or smoke, in which the balloon may be quickly lost.

b. TETROONS. The tetroon is a constant volume Mylar balloon in the shape of a tetrahedron which is ballasted to a zero lift condition and carried by the wind in an almost horizontal direction (30). A pillow-shaped Mylar balloon has also worked well (31), and zero-lift balloons have been produced by filling pilot balloons with a mixture of two gases, one more dense and the other less dense than air (32). As the mixture of gases leaks from the balloon, the loss of the heavier gas causes its weight to decrease. Thus an appropriate mixture of gases will keep the balloon in a zero-lift or balanced condition for lengthy periods of time. Such a balloon, if accurately tracked, acts as a sensor for wind direction at any lower level of the atmosphere (33). Because of the large areal coverage of such a horizontally floating balloon, it is an excellent indicator of wind trajectories near cities (34, 35), over shorelines, and in valleys where complicated patterns of airflow are commonly observed. Tracking over substantial distances usually requires the use of a radar installation, perhaps with a transponder suspended from the tetroon. Such equipment is often expensive to obtain, operate, and maintain. Two-theodolite tracking is less demanding but greatly limits the horizontal range.

c. KITE BALLOONS. A kite balloon is an elongated captive balloon with fins at one end; it is tethered in such a way that it behaves as an ordinary captive spherical balloon in light winds and as a kite in stronger winds, and thus maintains altitude under both conditions. The wind direction at the height of the balloon may be determined continuously or at intervals by suitable measurements at the ground of the azimuth angle of the horizontal projection of the tethering cable. The balloon cannot be left untended for long periods since it must be refilled at intervals to replace the inflating gas lost by slow leakage; another danger is that a sudden wind squall may carry the kite balloon away before it can be reeled in to safety.

d. RADIO AND RADAR. Wind directions aloft may be obtained under all weather conditions, including low cloud, fog, or smoke by the use of radio or radar theodolites (*7, 8, 36*). With the former, a small radio transmitter is carried aloft by a freely rising balloon and is tracked. With the latter system, pulses of electrical energy emitted by the radar are reflected back to it by a target carried by the free balloon; the system measures distance to the target as well as azimuth and elevation angles and is thus inherently more accurate than the radio direction finder. Operation and maintenance costs for both radio and radar theodolites are so high that their use in air pollution investigations is not justified except in most unusual circumstances.

e. SMOKE TRAILS. Highly detailed information on wind directions aloft may be obtained by determining at intervals the position in space of smoke trails released above the surface as by a rising rocket (*37, 38*). If detailed wind directions aloft are required, the technique has much to recommend it. On the other hand, observations in fog and smoke, or at night, are not possible with present methods; and more operating personnel are required than for simpler methods such as single theodolite pilot balloon measurements.

B. Wind Speed

The meteorological analysis of air pollution problems often requires wind speed measurements both near the surface and aloft, such as at and above the top of a chimney.

1. Surface Wind Speed

Instruments of a mechanical nature for the measurement of wind speed are generally referred to as anemometers. Of these the rotation anemometers are most generally satisfactory for air pollution studies (*4, 6, 8, 9*).

a. ROTATION ANEMOMETERS. The windmill or propeller anemometer consists of a circular array of blades, usually not less than three, mounted at one end of a horizontal shaft; the blades are free to rotate in response to wind forces acting on them. A wind vane is attached to the opposite end of the shaft. The horizontal shaft is connected to a vertical shaft which, being mounted in bearings, turns as the varying wind direction acts on the vane. The blade assembly is thus kept heading into the wind at all times by the vane. The rate of rotation of the blades is proportional to the wind speed. In one well-established design the blades drive a dc generator which produces a voltage directly proportional to wind speed (*39*). A similar type of instrument of more recent design is shown in Figure 2. The response characteristics of instruments of this type have been analyzed in detail (*15, 16, 24*). Such instruments may be used to measure wind direction as well as wind speed, both of which may be indicated on dials or a suitable recorder.

Figure 2. Windmill or propellor anemometer with wind vane. (Courtesy of G. C. Gill and R. M. Young Company, Traverse City, Michigan.)

The cup anemometer, the second type of rotation anemometer, consists of three conical cups attached, by horizontal arms 120° apart, to a freely rotating vertical shaft mounted in bearings (40–43). The cups rotate in a horizontal plane at a rate proportional to the wind speed. The cup anemometer has been developed in a highly self-contained and portable model, illustrated in Figure 3. In such devices the measuring element is a permanent magnet fixed on the vertical axis supporting the three cups. Electrical current induced in a concentrically mounted aluminum drum causes the drum to move against the bridling action of a spiral spring—the rotational displacement of a pointer attached to the drum being then proportional to the rate of rotation of the cups, and hence to

Figure 3. The Deuta hand anemometer. (Courtesy of Epic, Inc., New York, New York.)

the wind speed. This device is particularly valuable for air pollution surveys where spot checks of wind speeds at various locations are required.

If more fully representative wind speeds are required, dial registering or chart recording equipment is indicated. In one type of cup anemometer the vertical shaft carrying the cups drives a dc generator whose output is indicated on a dial instrument or on a chart-type recorder, to give instantaneous wind speed. One type of instrument, known as the "Staggered Six," employs two three-cup assemblies, one above the other, with each of the six rods supporting the cups making an angle of 60° with the adjacent rods, as illustrated in Figure 4. This staggered 60° spacing of cups is effective in smoothing out the ripples in speed registering and recording that are due to the 120° spacing of most cup assemblies. The instrument contains an air-gap capacitive transducer which detects the rotation of the anemometer shaft without contact or friction. The transducer rotates to produce an amplitude-modulated signal which is demodulated to provide a sinusoidal output voltage, the frequency of which is 40 times that of the shaft rotation. Such sensitive and ripple-free operation may be important where accurate measurements of the horizontal structure of the wind turbulence are required for air pollution investigations.

A second kind of cup anemometer is the wind-run or totalizing type in which the passage of each 1/10 or 1/60 mile of wind and each 1 mile of wind past the instrument is counted, either by a counter that must be read at intervals or by a recording counter. A detailed sketch of a counter-equipped wind-run or totalizing cup anemometer is presented in Figure 5. The average wind speed during any period is obtained by dividing the difference in counter or recording readings at the end and beginning of the period by the time elapsed. Since in air pollution studies the average wind speed is most often required, the wind-run or totalizing cup anemometer is generally preferred.

Both types of rotation anemometers may be adversely affected by freezing rain forming a layer of ice on the blades or cups. Severe freezing rain may stop the rotation altogether, but this rarely occurs. Other types of anemometers are affected equally or more by precipitation.

b. PRESSURE-TUBE ANEMOMETERS. This instrument consists of a horizontal pitot tube which is kept heading into the wind by a vane. A static tube surrounds the upper part of the vertical shaft that supports the pitot tube and vane. Pipes from the pitot and static heads serve to transmit pressure differentials to the recorder, which linearizes the response of the sensor either by a float of special design or by a system of springs (4,

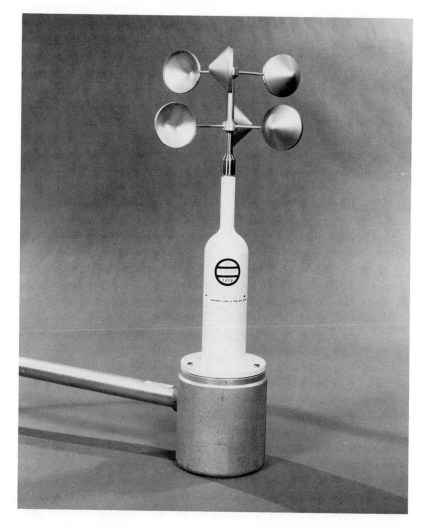

Figure 4. The "Staggered Six" precision cup anemometer. (Courtesy of Teledyne Geotech, Garland, Texas.)

6, 9). A pressure-tube anemometer was used effectively in the air pollution investigations in the Columbia River Valley near Trail, British Columbia (*44*).

Freezing rain and snow may obstruct the entrances to the pitot and static heads. Heating equipment may be installed to prevent ice and snow accumulation if necessary. If subfreezing temperatures are anticipated,

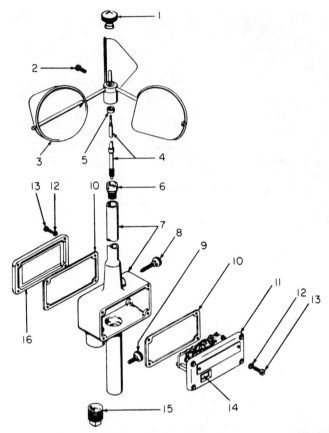

Figure 5. An exploded view of a wind-run or totalizing cup anemometer showing the component parts: (1) rotor retaining nut; (2) hub setscrew; (3) rotor assembly; (4) spindle; (5) upper ball bearing; (6) ball-bearing housing; (7) anemometer case; (8) spindle retaining screw; (9) pintle thumb screw; (10) gasket; (11) movement plate; (12) lock-washer; (13) plate screw; (14) odometer dial; (15) ¾-in. pipe plug; (16) rear cover plate. (Courtesy of the Environmental Science Division, The Bendix Corporation, Baltimore, Maryland.)

an antifreeze solution of the correct specific gravity must be used in the float chamber. The main advantage of this instrument is the fact that an electrical power supply is not required, since the recording pen is actuated by air pressure differentials; the rotating drum holding the chart is usually driven by clockwork.

c. Hot-Wire Anemometers. In this instrument the temperature of an electrically heated element which is cooled by the passing wind is recorded by suitable means. Several types have been developed for use in the atmosphere (*4, 45, 46*). Sometimes a hot thermistor is used instead of a hot wire (*47*).

The principle of the compensated heated thermocouple type of hot-wire anemometer (*48*), as used in the Hastings air meter, is illustrated in Figure 6. Alternating current heats the fine thermocouple wires extending between and supported by studs A, D, and C. The butt-welded thermojunctions halfway between AD, BD, and CD are supported only by fine wires, 0.08 mm in diameter, which meet to form each thermojunction. Since the electrical resistance of the wire between A and D is equal to that between B and D and since the secondary of the transformer is center-tapped, only the thermoelectric direct current passes through the millivoltmeter G. The thermocouples between A and D and between B and D are in parallel with reference to meter G, which therefore registers the thermoelectric voltage appropriate for a single thermocouple. A third but unheated thermocouple between C and D in the meter circuit eliminates the influence of temperature fluctuations in the wind; by wiring this thermocouple as a source of bucking voltage, any fluctuations in the thermoelectric output of the heated wires caused by fluctuating air temperatures are exactly canceled out by the counterelectromotive forces developed in this unheated thermocouple. Laboratory tests have shown perfect compensation even for temperature fluctuations with an amplitude as great at 27°C. The time constants range from 0.22 to 0.08 seconds for wind speeds of from 1 to 8 m/second.

The fine wire sensor must be carefully handled and the whole apparatus requires considerable attention while in use. The instrument is not suitable for foggy conditions when small water droplets may impinge on the fine hot wire. In heavily polluted atmospheres enough particulate matter may collect on the sensing hot wire to change significantly its response to passing wind. Such an instrument should be employed in air pollution studies only if the other types mentioned above have been proved inadequate for one reason or another.

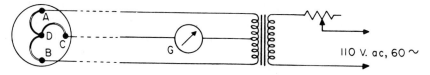

Figure 6. Circuit diagram illustrating the principle of the compensated heated thermocouple type of hot-wire anemometer used in the Hastings air meter. Heated thermojunctions are located between supports A and D and between supports B and D; the unheated compensating thermojunction is suspended between supports C and D. Millivoltmeter G measures the direct thermoelectric force and hence the wind speed (*48*).

2. Wind Speed Aloft

The methods described for obtaining wind direction aloft also give wind speed at higher elevations. This earlier section and the scientific works referred to in it should be consulted for information on methods of measuring wind speed aloft. Wind speeds without wind directions aloft were determined in a study of the wind velocity profile over an urban area (49). Wind speeds aloft were measured by means of two totalizing cup anemometers suspended from the cable of a kite balloon, the upper instrument being maintained at a height of about 1000 ft.

C. Wind Turbulence

An entirely satisfactory definition of wind turbulence is difficult to achieve. However, for present purposes, turbulence in the atmosphere may be defined as highly irregular motion of the wind. It is this highly irregular wind motion which is the primary diffusing agency in the atmosphere; its measurement or estimation is therefore nearly always of basic importance in the analysis of air pollution problems. Since turbulent wind flow with a pronounced vertical component is so effective in promoting the upward diffusion of air contaminants from their surface sources, it is often desirable to measure this vertical component as well as one or both of the horizontal components. In some circumstances, however, it is sufficient to measure one horizontal component.

The vertical component of wind turbulence is closely related to the degree of atmospheric stability existing. If the temperature decreases rapidly with height, the air is unstable and vertical turbulence and mixing are pronounced. If the air is isothermal, the vertical mixing is much less, and if the temperature increases upward, in an inversion layer, the air is very stable and vertical turbulence and mixing are almost entirely suppressed. The rate of decrease of temperature with height, known as the temperature lapse rate, may thus be used to estimate the degree of mixing in the vertical. The highly important vertical component of wind turbulence may thus be measured directly by sensitive wind instruments or may be estimated indirectly from temperature lapse rate measurements. The various means of making direct measurements of wind turbulence will be described first.

1. Direct Measurements of Wind Turbulence

Rapid response wind instruments of low inertia are needed for accurate measurements of the rapid fluctuations of wind speed or direction which

Figure 7. A two-channel sigma computer for calculating the standard deviation of wind direction fluctuations. (Courtesy of Climet Instruments Co., Sunnyvale, California.)

are the direct manifestation of wind turbulence. Many sensitive wind vanes or anemometers meet these requirements. If it is necessary to use a high inertia instrument, it is possible to make allowance for the influence of the high inertia in making observations of a turbulent wind field (*50*). Some of the methods by which the records made by low inertia instruments may be analyzed to yield meaningful turbulence information are presented elsewhere (*4, 51*).

a. WIND VANES. A number of rapid response wind vanes suitable for use in measuring the horizontal crosswind component of turbulence are available, such as that shown in Figure 1. A low moment of inertia favors rapid response. At the same time the dynamics of the system should be such as to permit the vane to follow, with little or no overshooting, the horizontal turbulent elements which are of such importance in the lateral dispersion of air pollutants (*6, 15, 24, 52*). Wind vanes subject to substantial overshooting should not be used for turbulence measurements. A wind vane especially designed for turbulence measurements in air pollution investigations has proved satisfactory in extensive field use (*44*).

It has been found convenient to express the horizontal crosswind diffusion coefficients in terms of the standard deviation σ of the azimuth fluctuations of appropriate wind vanes. A number of devices for determining σ directly from the electrical output of a suitable wind vane have been developed (*53, 54*). Such a device has come to be known as a sigma meter or a sigma computer. A commercially available unit is illustrated in Figure 7. A simpler method of estimating σ has been developed (*55*) and its use facilitated by employing electronic computing techniques to give σ directly (*56*).

b. BIVANES. A wind vane which is constructed so as to measure wind-direction fluctuations in vertical planes as well as those in horizontal planes is known as a bidirectional vane or, more simply, as a bivane. In one type, synchro generators rotate with the vertical and horizontal mounting shafts of the instrument; the synchro receivers actuate recording instruments which, by separate traces, give the azimuth and elevation angles of the fluctuating wind velocity vector (57). In another design microtorque potentiometers are used instead of synchro generators and receivers (58, 59). A bivane of this type is shown at the left in Figure 8; both microtorque potentiometers are located in the cylindrical housing at the base of the instrument. Changes in azimuth angle of the bivane are transmitted to one potentiometer by gearing. Changes of elevation angle are transmitted to the other potentiometer by a bead chain which passes over a pulley in the upper cylindrical housing and under an equal pulley in the lower housing. The design shown at the left in Figure 8 includes a mill or propellor anemometer at the upwind end of the shaft, so that this combination permits the measurement of the full three-dimensional turbulent structure of the wind. An illustration of yet another similar instrument is available elsewhere (24). A method of machine reduction of data from such an anemometer-bivane is available (60). One way of using bivane data in atmospheric diffusion studies is presented in (61).

Another type of bivane uses a simple counting mechanism to indicate turbulence which makes it especially suitable for air pollution investigations (62, 63). The standard deviation σ of the vertical and horizontal wind-direction fluctuations as measured by such a bivane may be estimated with sufficient accuracy for most air pollution investigations (55). A simple counting bivane of this type has given satisfactory service for a number of years at the plants of the American Smelting and Refining Company near Salt Lake City, Utah.

c. ANEMOMETERS. Turbulent winds are characterized by fluctuations in wind speed, and their horizontal components of turbulence may thus be evaluated by means of a suitable recording anemometer. The recording anemometers described earlier, in Section II,B,1,a, are suitable for this purpose; the only exception is the wind-run or totalizing cup anemometer which, because it counts cup assembly rotations rather than giving instantaneous wind speed, records only the long-period wind fluctuations caused by very large eddies and makes no record of the short-period fluctuations which are due to the passage of smaller eddies.

The rotation anemometers are, with the exception noted above, generally satisfactory for turbulence measurements in most air pollution investigations. The heavier rotation anemometers, with a large moment of inertia, accelerate more rapidly in an increasing wind than they deceler-

Figure 8. Wind turbulence measurements by bivane (*left*) and by *u-v-w* ane-mometer (*right*). (Courtesy of G. C. Gill and R. M. Young Company, Traverse City, Michigan.)

ate in a decreasing wind. These and other errors have been analyzed in detail (*61, 64–69*), and in other broader investigations (*15, 16, 24*). Precision cup anemometers with low friction and a small moment of inertia, in which errors are small, are available (*70–72*). A number of present-day three-cup anemometers, as well as that shown in Figure 4, are acceptable as wind turbulence devices.

There are, at times, advantages in being able to measure wind components and their turbulent fluctuations directly in an orthogonal system of coordinates. An instrument which does this is termed a *u-v-w* or tri-axial anemometer, the *u* and *v* representing the horizontal components of the wind velocity, and *w* the vertical component (*73–75*). This instrument

Figure 9. Close up of *u-v-w* anemometer. (Courtesy of G. C. Gill and R. M. Young Company, Traverse City, Michigan.)

is shown at the right in Figure 8 and in Figure 9. Helicoidal mills or propellors are mounted at the ends of the three orthogonal shafts and sense continuously the three components of the wind velocity and their turbulent fluctuations.

Another type of fast response instrument is the drag sphere anemometer (*76*). In this, strain gauges sense the horizontal component of drag force exerted by the wind on the sphere. A relatively small sphere must be used in order to ensure that the Reynolds number is always below its critical value. Unless this condition is met, the drag coefficient of the sphere is not constant, and the instrument no longer delivers meaningful information. For a 6-in. smooth sphere the drag coefficient begins to decrease at wind speeds of about 40–60 mph. This type of instrument has still to undergo additional field testing in the use for which it was designed, viz, the wind loading of structures.

The sensors of another fast response instrument known as the anemoclinometer are mounted in a sphere but the principle of operation is different (77).

A rapid response hot-wire anemometer has been designed to provide accurate measurements of the vertical components of wind turbulence (78). Two such anemometers properly positioned will give a continuous record of both upward and downward components of wind turbulence. The hot-wire anemometer has proved its value in research on the structure of turbulence. Only an elaborate and sophisticated air pollution investigation would employ such anemometers.

d. Gust Accelerometers. This instrument provides a direct measure of turbulent wind acceleration, positive and negative, for any specified time interval. The sensor is a horizontal aluminum wheel around whose periphery are equally spaced, vertical curved blades, each of the same size and shape; this wheel is attached to the top of a vertical shaft mounted in bearings (4, 44). The rotation of this shaft in response to wind forces is bridled and linearized by sets of steel springs located at the bottom of the constrained vertical shaft. This constraint prevents the bladed wheel from turning freely like a cup anemometer, but allows it to rotate in one direction as the wind speed increases and in the opposite direction as the wind speed decreases. Maximum rotation of the blades is limited by stops to one complete revolution. A make or break in an electrical circuit occurs when the wind speed changes by a specified amount. The average gust acceleration during a given time interval is then proportional to the number of makes and breaks during that interval.

A prime advantage of this instrument is that it gives simply and directly a numerical measure of turbulence. The gust accelerometer was adopted as a standard instrument for turbulence measurements by the Trail Smelter Arbitral Tribunal in drawing up its code of regulations for the maximum possible emission of SO_2 from the plant of the Consolidated Mining & Smelting Company of Canada, Ltd., at Trail, British Columbia (44). A gust accelerometer has operated for years with a minimum of maintenance at Sudbury, Ontario, near the Copper Cliff plants of the International Nickel Company.

e. Tetroons. The tetroon as a meteorological sensor has been described in Section II,A,2,b. Tetroon tracking may also be employed to measure wind turbulence. The use of single tetroon flights to estimate the characteristics of horizontal and vertical components of atmospheric turbulence has been explored (79, 80). Tetroons released in succession from a specified location have been used to estimate the lateral turbulence in relation to the lateral diffusion of a contaminant from a continuous point source (81). It should be noted that the sensor, viz, the tetroon, travels with the wind rather than being fixed in space as an anemometer, wind vane, or bivane is, and may therefore be expected to give a more meaningful representation of the turbulence which is actually effective in diffusing a puff or a plume. As pointed out earlier, however, the tetroon technique, as practiced to date, requires elaborate equipment and skilled supporting and operating personnel, and is therefore unlikely to be used in air pollution surveys of the more routine kind.

f. Aircraft-Mounted Turbulence Sensors. The mobility of aircraft, both in the horizontal and vertical, makes them attractive mounts for turbulence sensors. There have been two approaches to measuring atmospheric turbulence from aircraft.

Counting accelerometers have been used to specify the motions of the aircraft caused by atmospheric turbulence (82). The relations between aircraft response and atmospheric turbulence have been summarized (83) and the sampling errors of such measurements analyzed (84).

Mill or propellor anemometers mounted on aircraft may also provide a measure of turbulence (85). Such an instrument is illustrated in Figure 10. It gives the intensity of horizontal turbulence in the inertial subrange as a number which is proportional to the one-third power of the energy dissipation rate ϵ. This is a fundamental meteorological quantity which expresses the rate of conversion of the kinetic energy of turbulence into heat energy. Since the inertial subrange is a range of eddies having dimensions from a few centimeters up to several hundred meters, and since these eddies appear to be the primary turbulent diffusing agency in the lower atmosphere, the energy dissipation rate is a good indicator of the atmospheric diffusing capability by eddies within the inertial subrange.

Turbulence measurements by the mill or propellor instruments are independent both of the flying speed and the response characteristics of the aircraft on which they are mounted, which is not true for the aircraft-mounted counting accelerometers. The mill or propellor instrument thus has several advantages which suggest that it will be used increasingly in air pollution investigations.

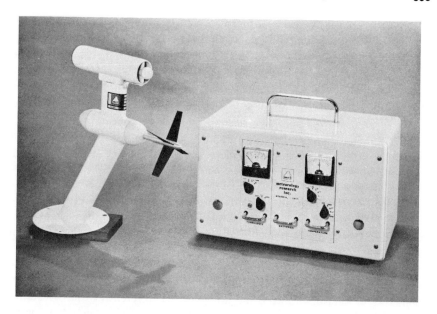

Figure 10. Vortex thermometer (*above left*) and wind turbulence indicator (*below left*) with indicating meters (*right*) for aircraft mounting. (Courtesy of Meteorology Research, Inc., Altadena, California.)

g. BALLOON-BORNE TURBULENCE SENSORS. A turbulence probe for mounting on the tethering cable of large captive balloons has been developed and tested (*86*). The basic probe consists of an eight-cup anemometer, a wind inclination meter, and a platinum wire resistance thermometer.

2. Indirect Measurements of Wind Turbulence

At times it is advisable, or even essential, to estimate the degree of turbulence by indirect means rather than by direct measurements of the types described above. If vertical stability prevails, the vertical component of eddy motion is suppressed and upward diffusion is limited. With instability, vertical motion is promoted and active air churning and mixing occurs. The atmosphere is unstable when the temperature decreases rapidly with height, and stable when it increases with height. It is in neutral equilibrium, i.e., neither stable nor unstable, when there is a small rate of temperature decrease with height. Thus measurements of the vertical temperature gradient may be, and often are, used to estimate indirectly the wind turbulence in the lower atmosphere.

a. TEMPERATURE LAPSE RATES. The rate of temperature decrease with increasing height is known as the temperature lapse rate. There are a number of methods of measuring lapse rates, including thermoelectric methods (87). When the temperature sensors are located at the heights between which the lapse rate is being determined so that each unit senses the temperature of the ambient air, then the method may be referred to as *in situ air temperature sensing*. When the measuring equipment is at a remote location and the temperatures of appropriate volumes of air are determined by indirect means, then the method is conveniently described as *remote air temperature sensing*.

i. *In situ air temperature sensing.* Among the *in situ* air temperature sensing techniques used in air pollution investigations, tower measurements find a prominent place. Since air temperature differences are often small over the height of the tower, it is sometimes necessary to use differential sensors, such as differential thermocouples, to achieve acceptable accuracy, although carefully calibrated and maintained resistance thermometers or thermistors may be adequate. If differential thermocouples are installed at several heights on a tower, a stepping switch may be used to take the output of each pair of thermocouples in succession to a sensitive recording potentiometer. Matched sets of thermistor sensors are being used increasingly with satisfactory results. Even with completely functioning and well-calibrated sensors, anomalous lapse rates are sometimes found in the early morning near the surface (88).

Wiresonde measurements of lapse rate are less often taken in air pollution analyses (59). In this technique a temperature sensor is carried aloft by a kite balloon and measurements made at appropriate height intervals as the kite balloon ascends. The output from the sensor is transmitted to the surface by wire or by a radio link. In one type, the sensor is a ceramic rod whose electrical resistance varies with temperature (89). Another type uses a thermistor (90). The electrical resistance of the sensing element at each elevation is measured by a Wheatstone bridge at the surface. Heights are computed by multiplying the footage of cable paid out by the sine of the elevation angle of the balloon. Neglecting the curvature of the cable results in an error of less than 2% under ordinary operating conditions. Lapse rate measurements up to a height of 500 m are conveniently made by wiresonde when a relatively small number of measurements is required, or when measurements are required at a number of locations. A wiresonde is expensive and time consuming to operate for a long series of measurements, and it should not be used where the cable might drag over overhead power lines.

Dropsonde measurements of lapse rate are made by a suitable temperature sensor which is lowered by a parachute. The unit may be carried aloft by an airplane or by a small cold propellant rocket (*91*). Sensor output signals are usually telemetered to a receiving unit. Over heavily forested or otherwise inaccessible terrain, however, where the probability of recovery is low in comparison with that over open country, the convenience of this instrument tends to be offset by the cost of replacement sondes.

Another type of rocket sounding permits the measurement of lapse rate and other meteorological variables without the penalty of frequent losses of the equipment over rough terrain (*92*). As Figure 11 illustrates the spent rocket is lowered by parachutes and drawn back to the launch site by a nylon thread which pulls in the rocket just as if it were a kite. Two rockets have been developed: in one the measurements are recorded on

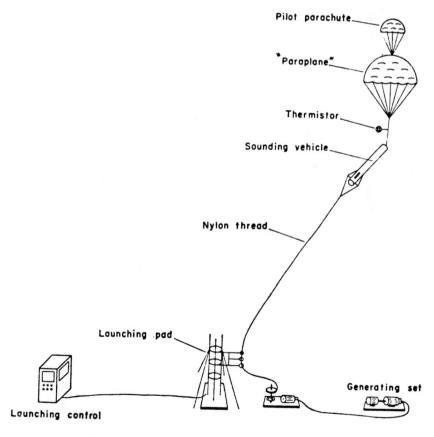

Figure 11. Recovery phase for sounding rocket (*92*).

board; in the other they are transmitted to ground by radio telemetry at 400 MHz.

Manned airplane measurements are often very convenient and permit serial soundings over wide areas. Some special duty airplane installations may be quite elaborate (*93*). In Figure 10 a vortex thermometer for manned airplane lapse rate measurements is shown mounted above the turbulence sensor. Such a thermometer is designed so as to minimize adiabatic heating effects which might lead to spurious high temperature readings.

Drone model airplane measurements of lapse rate are an interesting possibility at large distances from airports and other concentrated areas of air traffic. Remote control systems for model aircraft have improved to the point where such lapse rate measurements are entirely feasible. The method is unlikely to be acceptable from the safety point of view near cities and airports.

Radiosonde or rawinsonde measurements of lapse rates are customarily made at greater heights. In the radiosonde an assembly of pressure, temperature, and humidity sensors is carried aloft by a freely rising balloon (*9*). When radio or radar theodolites are used for wind speed and direction measurement the device is referred to as a rawinsonde (*36*). A radiosonde assembly of sensors carried aloft by a captive balloon may have substantial advantages for certain air pollution studies over the ordinary wiresonde (*94*).

ii. *Remote air temperature sensing.* There are two primary techniques for remote lapse rate sensing: electromagnetic sounding and acoustic sounding. A treatment covering both methods which is at the same time complete and comprehensive is available (*95*). Several more concise introductions have also been published (*96, 97*). Electromagnetic sounding may involve either radio or optical wavelengths and may be active or passive in nature. Active probing involves the emission of electromagnetic radiation of known characteristics by suitable equipment and deduction of the air temperature from the nature of the signal received, often backscattered radiation. In passive probing the air temperature is deduced from the nature of the radiation emitted by one or more specified volumes of air.

Radio remote sensing has developed substantially in a number of ways. Active systems, such as those involving radar technology, are of limited usefulness for lapse rate measurements. The passive systems show substantial promise for air pollution investigations. An instrument known as the radiometric thermasonde has been extensively tested (*98*). The radiometric technique requires only the detection of electromagnetic energy, in this case the naturally occurring energy radiated from oxygen molecules in the atmosphere at a wavelength of 5 mm. The intensity of this energy is determined by the radiometric brightness temperature as a function of elevation angle which can be equated to ambient air temperature by appropriate mathematical inversion techniques. Figure 12 shows a chain of functional blocks representing the various components of the radiometric thermasonde; a detailed engineering description of this equipment is available (*99*).

In contrast to this technique involving multi-angle-scan of brightness temperatures of atmospheric O_2 at a single wavelength of 5 mm, there is the multi-spectral-frequency approach which involves inverting observations of brightness temperature as a function of frequency at a fixed elevation angle (*100*). There is evidence that a combination of angular-scan and multispectral input data leads to more reliable lapse rate information than is furnished by either system used separately (*100*). Both precipitation and low clouds having high water content interfere with radiometric lapse rate determinations. This limitation of the method is, however, of relatively little significance because the more severe air pollution episodes tend to occur under atmospheric stagnation conditions characterized by fair weather patterns. Air pollution disasters such as those which have occurred in England and the United States have often been accompanied by persistent fog. Relief from such severe stagnation conditions comes from changes in regional weather patterns which are predicted by computer or other methods rather than by measured lapse rate changes within or above the fog. Although there are some inaccuracies in the lapse rate values obtained, the testing conducted suggests that this type of equipment is satisfactory for the investigation of many air pollution problems.

Optical remote sensing employing an active system also shows considerable promise as a method for remote sensing of temperature lapse rates near the earth's surface (*101, 102*). The method utilizes the part of the laser backscatter arising from the Raman rotational spectrum of atmospheric N_2. The differences in the amplitudes of appropriately chosen portions of the Raman spectrum as a function of height become, in effect, a means of obtaining a temperature profile; temperatures are obtained to within 1°K up to 2 km with a 100-m height resolution. This height resolution would permit the determination of approximate lapse rate values only within the first few hundred meters above the surface which are so important for air pollution analyses. As with the radiometric thermasonde, Raman backscatter methods have limited utility when low cloud or fog are present.

Both the radiometric and the Raman methods of remote temperate

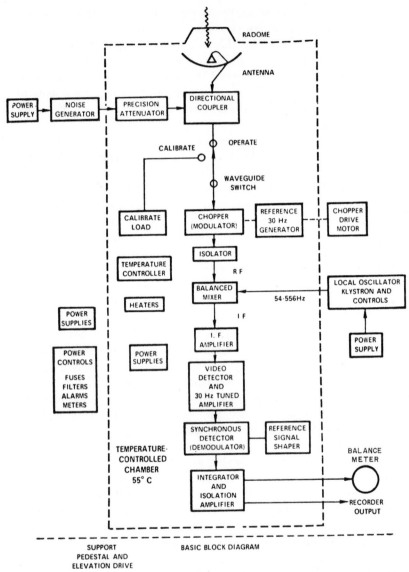

Figure 12. Basic block diagram of the radiometric thermasonde (*98*). RF, Radio frequency; IF, intermediate frequency.

sensing of the lower atmosphere hold promise of widespread utilization within the next few years. Both methods give low-level lapse rates which are subject to some error. Both lose their utility when fog or low cloud are present. The radiometric instrument has been extensively tested (*98*) whereas the Raman method has not yet been the object of such tests. Further development and testing will be necessary before the relative merits of the two systems can be determined. A tall meteorological tower provides more accurate and reliable lapse rate measurements than either system, but the remote sensing methods offer so many advantages in convenience and mobility that meteorological towers for air pollution studies are likely to become a thing of the past.

Theory suggests that it may be possible to utilize acoustic sounding for the determination of temperature lapse rates in the lower atmosphere. The necessary methods have not yet been developed, however (*96, 103*).

b. RADIATION.   The temperature lapse rate near the surface is often governed by the net radiational gain or loss of heat at the ground. After sunrise on a clear morning solar radiation causes the lapse rate, and hence the turbulence, to increase substantially. After sunset with clear skies loss of heat of the ground by terrestrial radiation causes the lapse rate and the associated turbulent mixing to decrease as a temperature inversion grows upward from the ground. An instrument which measures the net heating or cooling load directly at or near the surface thus supplies information which may be used to directly estimate the lapse rate and the associated turbulence patterns.

Such an instrument is the thermal or net radiometer (*104–109*), one type of which is illustrated in Figure 13. In this model, the radiation sensor assembly consists of a thin horizontal plate, a blower, and a nozzle for directing a steady flow of air over the upper and lower surfaces of the plate, thus eliminating the variable cooling effect of changes in the natural wind. The sensor consists of a thermopile made up of a series of thermocouples which are so located that one set of thermojunctions is in a plane adjacent and parallel to the upper surface of the plate and the other set of thermojunctions is in a plane adjacent and parallel to the lower surface of the plate. This plate is of a sandwich type construction, with the thermopile located between sheets of an electrical insulating material, and covered above and below with sheet aluminum, the outer surfaces of which have been blackened. Figure 14 is a cross-sectional sketch of a typical plate assembly.

Differential radiational heating or cooling of the upper and lower blackened surfaces establishes a vertical temperature gradient which leads to an electromotive force in the thermopile. This emf may be determined by a direct reading or a recording potentiometer.

The radiometer will not operate satisfactorily if rain, snow, fog, dew, or frost is deposited on the sensor. Exposure to heavy rain or hail or to atmospheres heavily polluted with particulates makes reblackening

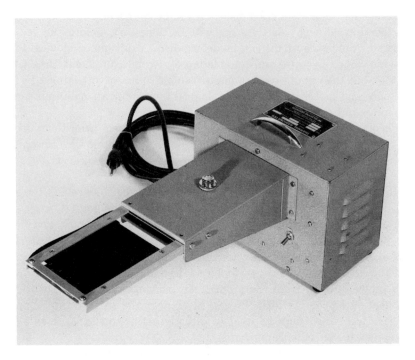

Figure 13. Thermal or net radiometer for measurements of total radiation. (Courtesy of Teledyne Geotech, Garland, Texas.)

and recalibration necessary. Light rain will not change the calibration and may help to maintain accuracy by washing away deposits of soot and dust. During one 6-month period of continuous operation, the calibration did not vary by more than ±5%.

Another instrument is the Commonwealth Scientific and Industrial Research Organization (Australia) (CSIRO) net radiometer (107, 108). In

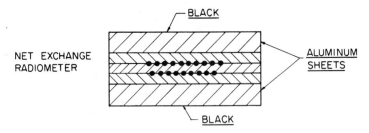

Figure 14. Cross section of typical thermal radiometer sensing element.

type S-1 the sensor element consists of a copper-constantan thermopile which is protected by polyethylene spheres, sealed with O rings for operation over a temperature range from −40° to 110°C. Inside and outside forced ventilation prevent fogging and improve transparency of windows. The time constant is 30 seconds for 98% of full scale and linearity is 1%.

Radiometer measurements are a less reliable indicator of turbulence than are temperature lapse rate measurements. Near large bodies of water and during seasons when the temperature of rapidly moving air masses is markedly greater or less than that of the underlying surface, radiometer measurements will have little or no relation to wind turbulence.

c. SMOKE PHOTOGRAPHY. Another indirect technique for estimating wind turbulence is by determining photographically the rate of growth of smoke both vertically and horizontally. The method is especially appropriate for air pollution studies, since it utilizes directly the diffusion resulting from wind turbulence. There are several possible smoke configurations to be considered, including point and vertical line sources.

The point source method has been well developed both theoretically and experimentally (100–116). In one method of analyzing the photographs of smoke plumes for a continuous point source the ratios of plume dimensions only are required, not absolute magnitudes; this minimizes difficulties which might arise when photographing the plume against backgrounds affording limited contrast. Time exposures are often desirable to give the average dimensions of the plume. Smoke plume photographs are particularly valuable in special cases, as in studying air pollution meteorology at shorelines, when unusual patterns of turbulence are found (114). The method has a number of advantages but fails when the visibility is low, owing to fog or low cloud, or to contaminants from other sources.

The vertical line source method has been used mainly for obtaining wind directions and speeds aloft (38). Photographs of an instantaneous vertical line source of smoke released from a small cold-propellant rocket can also be analyzed to determine in the vertical the atmospheric turbulence and diffusion that exist aloft. Further development of this technique could well be undertaken.

D. Mixing Height

The fourth of the primary meteorological measurements is the mixing height. It is usually related to one or more of the first three: wind direction, wind speed, and wind turbulence. In certain circumstances it may be related to all three.

1. Definition

The mixing height may be defined as that height above the earth's surface to which released pollutants will extend, primarily through the action of atmospheric turbulence. The phrase mixing depth has sometimes been used in the past for this quantity, but it is best confined to use with bodies of water such as lakes, seas, and oceans. Since the oceans are a sink for much atmospheric pollution there is the possibility of much confusion unless mixing height is used for the atmosphere and mixing depth for the ocean.

2. Direct Determinations of Mixing Height

The definition of the mixing height provides an indication as to how a direct determination may be made. Since the vertical mixing of pollutants is accomplished by the vertical component of atmospheric turbulence, it follows that the mixing layer is characterized by unstable air and turbulence with a marked vertical component which is surmounted by a stable or inversion layer of air with very little vertical turbulence. Such a direct determination may be accomplished by an aircraft-mounted turbulence sensor such as is described in Section II,C,1,f. As the aircraft ascends through the mixing layer, pronounced vertical turbulence is detected. The intensity of this turbulence decreases greatly as the aircraft ascends into the stable or inversion layer aloft. The mixing height is the height at which this pronounced decrease in turbulence occurs.

Turbulence sensors carried aloft by captive balloons or on tall towers would serve the same purpose.

3. Indirect Determinations of Mixing Height

There are several indirect methods of estimating the mixing height. These may be classified as either temperature lapse rate methods or remote sensing methods.

a. TEMPERATURE LAPSE RATE METHODS. These methods have been commonly employed in determining mixing heights and involve the use of elementary meteorological concepts. These are perhaps best illustrated by considering a typical diurnal variation of the mixing height. The process of atmospheric mixing by means of turbulence tends to cause the actual lapse rate to approach the dry adiabatic lapse rate of 10°C/km: the more intense the turbulence and hence mixing, the closer the approach to this value. Thus the mixing layer is characterized by a temperature lapse rate close to 10°C/km. Now let us consider a typical daily variation of the mixing height over open countryside commencing at sunrise on a clear day (Fig. 15). During the night loss of heat to space by long-wave terrestrial radiation has caused the ground to cool. This in turn has caused a cooling of the air above and thus a surface-based inversion layer. Hence, at sunrise the mixing height is zero. After sunrise the solar radiation reaching the ground increases and in due course becomes equal to the net terrestrial radiation loss at the ground. Thereafter the ground temperature commences to rise with further solar heating. Heat from the ground is transported upward by turbulence leading to an initially thin mixing layer in which the lapse rate may be many times greater than the dry adiabatic lapse rate because of the proximity of the ground. As ground heating proceeds during the morning, the height of the top of the mixed layer, i.e., the mixing height, becomes progressively greater and reaches its maximum values, perhaps 1 or 1.5 km, by early afternoon when the maximum surface temperature is attained. As the surface temperature decreases the surface air becomes progressively more stable, the energy supply for the turbulence aloft is cut off, and the mixing height decreases to zero.

Over a city, solar energy stored in buildings and pavements in summer and space heating of buildings in winter may cause a mixed layer to persist until well into the night, or even until sunrise for a very large city.

Methods for ascertaining the mixing height based on temperature lapse rates may be inferred from the above discussion. If radiosonde or aircraft

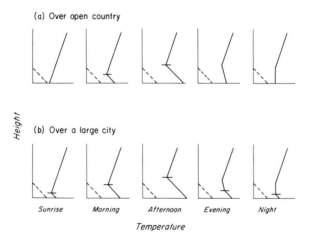

Figure 15. Typical diurnal variations of the mixing height (a) over open country; (b) over a large city.

lapse rate values are available at intervals, the top of the mixing layer characterized by the dry adiabatic lapse rate as shown in Figure 15 is readily determined. If twice-daily radiosondes only—for early morning and early afternoon, for example—are available, then intermediate values between these times are readily estimated. Mixing heights may also be forecast. If one wishes to forecast the noon mixing height, for example, the early morning radiosonde values are plotted on a temperature–height or other appropriate graph. The noon surface temperature is then forecast and plotted on the graph, and a line representing the dry adiabatic lapse rate of 10°C/km is drawn through this noontime temperature. The intersection of this line with the temperature–height graph of the early morning radiosonde then specifies the predicted noon mixing height. Similarly the maximum mixing height for the afternoon may be predicted using the forecast afternoon maximum temperature.

b. ELECTROMAGNETIC REMOTE SENSING METHODS. Active or passive electromagnetic remote sensing methods used to estimate lapse rates and hence mixing heights have been described in Section II,c,2,a,ii. Another technique is to utilize laser radar, often referred to as lidar, to detect the top of the aerosol layer which so often coincides with the mixing layer (117–120). As illustrated in Figure 16, in its simplest and most common form, the laser radar is aligned in a monostatic configuration, i.e., with transmitter and receiver located at the same place, and with the optical axes of the transmitter and receiver parallel. The laser emits a short, highly collimated light pulse that propagates through the atmosphere and, at any instant, illuminates a well-defined volume often containing suspended particles. If present, these aerosols will backscatter a small amount of the laser radiation which is collected by a suitable receiving telescope. If there is a well-defined mixing layer present with its upper boundary marked by a pronounced decrease of aerosol concentration with height, the backscatter from the pulsed energy emitted decreases greatly as the pulse leaves the surface aerosol layer and enters the relatively clean air above, thus locating the mixing height. There are several factors to be considered when evaluating the usefulness of this method.

First, there must be aerosols present in the mixing layer to scatter back the energy. Although there doubtless are air masses so clean that little backscatter occurs, pollution dispersion is then so active that atmospheric conditions are of little concern. Even in the absence of pronounced industrial pollution there are usually enough natural aerosols carried up from the surface to permit one to locate the mixing height with satisfactory accuracy.

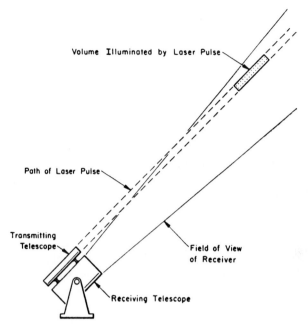

Figure 16. Schematic representation of a monostatic laser radar (*120*).

Second, when winds aloft are light for long periods of time as under conditions of prolonged atmospheric stagnation, serious errors are possible unless special care is taken. The reason for such possible errors may be seen by reference to Figure 15. Consider the afternoon situation over open country when the aerosols extend to their maximum height for the day. With light winds aloft the aerosols near the top of the mixing layer will not be carried far during the subsequent hours nor will they diffuse much vertically, being in stable air. Thus, although the mixing height has become effectively zero during the following evening, night, and sunrise periods, there will be laser backscatter from these aerosols aloft with a sharp discontinuity at their top just as if the mixing layer had persisted throughout the night. The aerosols may still be there later in the morning, suggesting a much greater mixing height than is actually present. It is clear that appropriate techniques must be developed to ensure that false mixing heights are identified as such.

c. Acoustic Remote Sensing Methods. Acoustic probing offers much promise as a method for determining mixing heights (*96, 97, 103*). The technique involves emitting a pulse upward and recording the acoustic energy scattered back to the surface from temperature inhomogeneities aloft and the time-of-flight of the sound pulse. The time-of-flight is directly proportional to the height of the inhomogeneities. Temperature fluctuations in the mixing layer are small, but tend to become more pronounced at the base of a stable layer extending above a mixing layer. Variations in the lapse rate within a mixing layer may result in temperature fluctuations which are picked up by an acoustic echo-sounder, so that the record must be interpreted carefully if the mixing height is to be determined accurately.

Several possible limitations of the acoustic probe should be recognized (*121*). The first is limited range, which has been estimated to be 1.5 km for sound pulses of 1 kHz. A second limitation may result from the strength of the interaction of the acoustic waves and the atmospheric irregularities: earlier theory assumes single scattering, but for highly turbulent conditions, or for longer paths or higher frequencies a multiple scatter theory may be required. A third possible problem may arise from sounds generated by adverse weather conditions such as strong winds,

Figure 17. Acoustic radar for remote sensing. (Courtesy of AeroVironment Inc., Pasadena, California.)

rain, hail, and thunderstorms. Air pollution is not likely to be of concern, however, under such meteorological conditions. Acoustic echo-sounders are also likely to be of limited value where there is pronounced man-made noise, as near airports.

An acoustic echo-sounder known as an acoustic radar is illustrated in Figure 17. It is a monostatic unit in which sound source and receiver are located at the same point. A burst of 1600-Hz sound of 50, 100, or 200 msec is emitted from a speaker and directed vertically upward by a 4-ft-diam. parabolic reflector. Backscattered sound is received by the same reflector-pickup unit and recorded on a time–height graph by a time-phased cycling stylus. The type of record obtained may be seen in Figure 17. An acoustic enclosure is required for the antenna to shield it from ambient noise.

E. Mesoscale Wind Fields

Mesoscale meteorological phenomena may be defined roughly as those having horizontal scales in the range 1 to 100 km. Such phenomena include: land and sea breezes; mountain, valley, and slope winds; urban heat-island winds; etc. These are significant factors in determining the

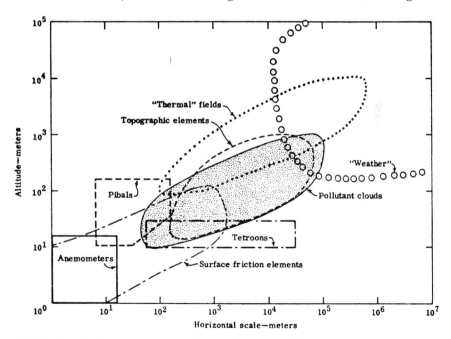

Figure 18. Scale ranges of significance for air pollution problems and various measuring devices (122).

wind fields which are often important in larger scale air pollutin problems. The measurement of these mesoscale wind fields which are often required for computer modeling of air pollution situations presents special problems which cannot be explored here (*122*). Figure 18 shows how some of the wind measurement techniques that we have discussed, such as anemometers, tetroons, and pilot balloons (pibals), are related to relevant factors including the size scales of: pollutant clouds; surface friction elements; topographic elements; "thermal" fields which generate land and sea breezes; urban heat-island winds, etc.; and the much larger "weather" systems such as the cyclones and anticylones of middle latitudes. Persons faced with problems involving mesoscale wind fields, especially in computer modeling, should consult the original source (*122*).

III. Secondary Meteorological Measurements

Although wind in its various manifestations is of primary importance in the evaluation of air pollution problems, there are a number of other meterological elements which play secondary roles. Among these are visibility, turbidity, humidity, precipitation, and solar radiation. It should be emphasized that, in certain circumstances, these normally secondary elements may become primary ones.

A. Visibility and Turbidity*

Meteorological visibility is defined as the greatest distance at which a black object of suitable dimensions can be seen and recognized against the horizon sky, or, in the case of night observations, could be seen and recognized if the general illumination were raised to the normal daylight level. It should be emphasized that the criterion of recognizing an object should be used—not merely the seeing of the object without recognizing what it is. Lower visibilities are expressed in meters or yards, or in fractions of a mile; higher visibilities in kilometers or miles. The term may express the visibility in a single direction, or the prevailing visibility, based on all directions. Observations of visibility should be made without the aid of binoculars, telescopes, or theodolites (*4*). Visibility is a quantity that may give some indication of air particulate matter, but the use of visibility observations for such a purpose presents a number of problems (*123–126*).

* See also Chapter 12.

Atmospheric turbidity refers to any condition of the atmosphere which reduces its transparency to radiation, and especially to visible radiation. Ordinarily, turbidity is considered in relation to a cloud-free portion of the atmosphere that owes its reduced transparency to particulates such as smoke, dust, and haze, and to scintillation effects (*127*). Thus it follows that, in general, the greater the turbidity, the smaller the visibility, and vice versa. Turbidity measurements have found some use in air pollution studies (*128*).

1. Visibility Objects

The exact method of choosing visibility objects around an observing station varies from one national meteorological service to another, although general guidelines have been established by the World Meteorological Organization (*5*). If visibility observations are planned for an air pollution study, the system of visibility objects to be used should be chosen in conference with representatives of the appropriate national weather service.

a. PLAN OF VISIBILITY OBJECTS. Personnel manning a station which expects to take visibility observations should prepare a plan of visibility objects to be used for the program, showing their distances and bearings from the observation point. The plan should include objects suitable for determining the visibility by night as well as by day.

b. DAYTIME VISIBILITY OBJECTS. Suitable objects at as many distances as possible should be selected for daytime observations. Objects should be black or nearly so and should subtend an angle greater than 0.5°, but less than 5°.

c. NIGHT VISIBILITY OBJECTS. The most suitable objects for night observations are unfocused lights of moderate intensity at known distances, and the silhouettes of hills and mountains against the sky. The brilliance of stars near the horizon may also be a useful indication. The relations between day and night visibility have been analyzed in detail (*4*).

d. ESTIMATIONS OF VISIBILITY. If suitable distant objects are not available, visibilities must be estimated by noting the clearness with which the farthest available object stands out. When such an object is seen in sharp outline and relief, the visibility is much greater than the distance of the object; a blurred or indistinct object indicates that the visibility is approximately equal to or slightly less than the distance of the object.

2. Transmissometers

The transmissometer measures the transmission of light by a portion of the atmosphere. The instrument consists primarily of an intense source of carefully focused light and, at a suitable distance—500 ft is a convenient value—a photocell receiver on which the transmitted light from the source is focused (*129*). A typical installation is sketched in Figure 19. If necessary, the indicating and recording apparatus may be located as far as 10 miles from the sensing equipment (*130*). The output of the photocell varies with the amount and size of obscuring particulate matter between source and receiver.

There are other related types of instruments. In the scattered light meter it is the light which is scattered by the obscuring particulates which is measured, rather than the transmitted light (*131*). Although the instrument is more compact than the conventional transmissometer, it may give less representative readings because the path length sampled is less.

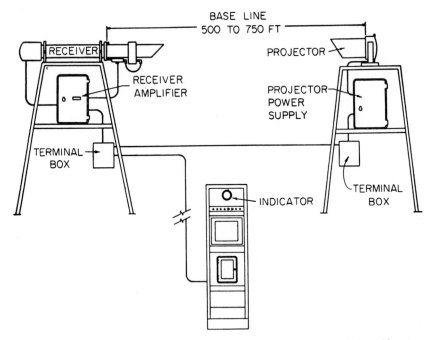

Figure 19. Schematic diagram of a typical transmissometer installation. (Courtesy of the National Oceanic and Atmospheric Administration, Washington, D.C.)

The greatest usefulness of a transmissometer is in giving a continuous record of the absence or presence in significant amounts of particulate contaminants along a specified path, e.g., between a possible source of pollution and a sensitive area such as a residential district. The general utility of the transmissometer is limited by two facts. First, unlike a human observer, it can estimate the visibility in one direction only. Second, standard transmissometers are designed for airport runway applications which emphasize the highly restricted visibilities which may hinder airport operations; such instruments are therefore likely to be relatively inaccurate in measuring the moderate and high visibilities which are often important in specifying the magnitude of an air pollution problem. For some air pollution investigations, however, the advantage of continuous recording may outweigh such disadvantages.

3. Telephotometers

The telephotometer operates on the principle that a perfectly black target viewed from a distance appears gray or white due to light scattered by the haze between the observer and the target. One convenient model operates as follows (*132*). A photometer or photocell and amplifier alternately views the target and the horizon sky by means of a long focus lens. A motor-driven diaphragm with two different apertures produces successive images of the target and the horizon sky on the photocell, the output of which is taken to a recorder.

A telephotometer operates satisfactorily with a much shorter light path than a transmissometer. It also requires a clear view of the horizon sky. In heavily built-up areas where the horizon sky is not visible, the transmissometer is therefore the preferred instrument, provided that a sufficiently long unobstructed path length is available.

4. Sun Photometers

The sun photometer estimates the attenuation of the direct solar beam by atmospheric particulates. In one type, when the instrument is pointed at the sun, the radiation enters a lens opening in whose focal plane is a diaphragm which limits the field of view to about 1°. The light is then diffused by a ground-glass plate and passes through a filter combination which transmits a nearly monochromatic beam of wavelength about 0.5 μm. This radiation falls on a selenium photocell whose output is indicated by a microammeter (*128, 133*). The instrument provides an estimate of the particulate loading of the lowest kilometer or so of the atmosphere under clear sky conditions, but not with cloudy skies, or at night.

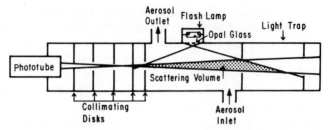

Figure 20. Schematic diagram of an integrating nephelometer. The vertical dimension is exaggerated by a factor of about 3 in this sketch. (Reproduced by permission from S. S. Butcher and R. J. Charlson, "An Introduction to Air Chemistry." Academic Press, New York, New York.)

5. Nephelometers

Another instrument capable of providing measurements of atmospheric visibility is the integrating nephelometer, the basic elements of which are illustrated in Figure 20 (*134, 135*). An air sample is drawn continuously through a chamber where it is illuminated by a pulsed flash lamp. The scattered light is detected by a photomultiplier tube looking at the illuminated air sample. The signal produced by the photomultiplier is averaged and compared with a reference voltage. Scattering from clean, filtered air and from Freon-12 is used as a calibration reference. A more convenient calibration technique is to bring a scattering medium of known properties—a piece of opal glass—into the narrow cone which is illuminated by the flash lamp; this procedure simulates an aerosol with a scattering coefficient which can be readily calculated (*136*). Sources of error in the integrating nephelometer have been identified (*137, 138*).

The output of the integrating nephelometer may be recorded on a chart to give visibility in miles from 0.3 miles to infinity. This method suffers from the limitations of all those which involve sampling at one point. Results are reliable when the aerosol concentration in the atmosphere is uniform. When there is substantial inhomogeneity in aerosol loading more meaningful results are obtained by using an instrument such as the transmissometer which samples over a long path length.

Other instruments have been devised for measuring the optical properties of the atmosphere (*6, 124, 139, 140*), but their use has been limited.

B. Humidity

There is a good deal of information available on the measurement of atmospheric humidity, much of it of recent origin. This ranges from monographs on the subject (*141*) to a very comprehensive bibliography covering the period from 1665 to 1962 (*142*). Also available is an encyclopedic presentation of principles and methods of measuring humidity in gases (*143*). The following list of major section headings illustrates the scope of this work: psychrometry; dew-point hygrometry; electric hygrometry; spectroscopic hygrometry; coulometric hygrometry; and miscellaneous methods. These represent very rich resources in reference materials for those who face special problems in the measurement of atmospheric humidity in relation to an air pollution problem. Only a few of the most important methods can be mentioned here.

Of the various means by which atmospheric humidity may be expressed, the relative humidity is most frequently used in air pollution studies. The degree of corrosion by SO_2, for example, is closely related to the relative humidity. The relative humidity of the atmosphere will also have an important bearing on the length of the visible plume from a stack whose effluent is warm and has a high water vapor content; as the plume cools by mixing with the ambient atmosphere, some of this water vapor will condense and become visible as small droplets which persist much longer in an atmosphere with high relative humidity than in one with a low relative humidity.

1. Wet and Dry Bulb Hygrometer or Psychrometer

The sensing elements in the psychrometer are two identical thermometers, one with its temperature-sensing element covered with a piece of muslin which is moistened by a suitable means. In saturated air the two thermometers give the same reading; in unsaturated air evaporation from the wet bulb occurs, lowering its temperature. The relative humidity is then obtained from tables.

The muslin covering the wet bulb must be wet. In one type of psychrometer water is carried from a container along a wick to the muslin, thus keeping it wet; a motor-driven fan causes the steady flow of air, past the wet bulb, which is necessary to ensure a steady-state, accurate reading (*144*). The sling psychrometer consists of two thermometers mounted side by side on a frame attached by a swivel connector to a handle; the bulb of one is covered by tubular gauze. The gauze-covered bulb is dipped in water, and the framework and two thermometers are whirled steadily (*144*). The instrument must be whirled until a steady state of evaporational cooling leads to a wet bulb reading which is constant with time. If the whirling continues too long, all the water in the gauze will evaporate and the indicated reading will rise to that of the dry bulb thermometer. A hand-aspirated psychrometer employing a rubber hand pump for ventilation of the wet bulb is also available (*144*).

2. Hair Hygrometer

The sensor in this instrument is a group of human hairs kept extended to their full length by light tension applied at one end of the group. Human hair elongates slightly as the relative humidity increases and shortens slightly as it decreases. These changes in length with varying humidity are magnified and read directly in the indicating instrument, or recorded by a pen making a trace on a moving chart (144). For temperatures between 0° and 30°C, and for relative humidities between 20 and 80%, a good hair hygrometer will, if subjected to an abrupt change in relative humidity, register 90% of the true change in about 3 minutes. At lower temperatures the rate of response is much slower. Full descriptions of the operational characteristics, maintenance requirements, etc., for both psychrometers and hair hygrometers are available (4–6, 9).

3. Infrared Hygrometer

The infrared hygrometer is a more accurate and also a more complex instrument; it gives mass of water vapor per unit volume of air by measuring the relative absorption of infrared radiation over a 1-m path or less by comparing the absorption by the 1.37-μm water vapor absorption band with that by a 1.24-μm reference band, the absorption at the latter wavelength being effectively zero. The details are shown schematically in Figure 21. The beam from an infrared lamp is chopped by a sector wheel driven at 900 rpm and consisting of four 1.37-μm filters and four 1.24-μm filters positioned alternately in the wheel. Detection of the chopped infrared beam is accomplished by a lead sulfide photocell and amplifier operating a self-balancing null system whereby the energy in the absorption band is kept equal at all times to the energy in the reference band. Balance is maintained by automatically varying the temperature of the lamp supplying the infrared energy. The temperature of the lamp is a measure of the water vapor in the sensing path. An index of the lamp temperature is obtained by a monitor photocell which drives a remote self-balancing recording potentiometer (144, 145).

The prime advantage of the method is that no phase change or absorption of water vapor is involved and, consequently, there are no complications at subfreezing temperatures: there is no decrease in sensitivity at low temperatures and low vapor concentrations. Smoke in small or medium concentrations has no appreciable influence on the readings, but at very high concentrations errors are caused by differential scattering effects. Polluted atmospheres containing gaseous contaminants having

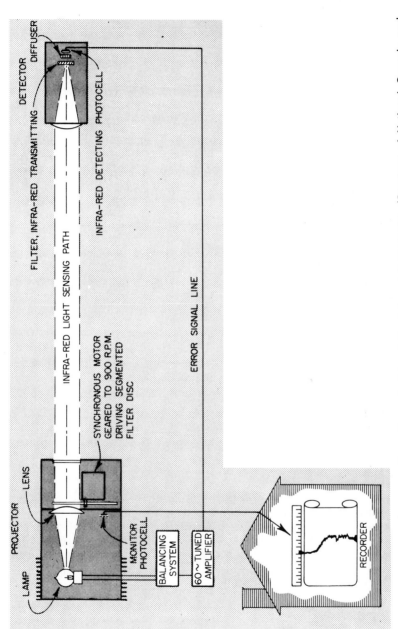

Figure 21. Schematic diagram illustrating principle of the infrared hygrometer. (Courtesy of National Oceanic and Atmospheric Administration, Washington, D.C.)

absorption bands at 1.24 or 1.37 μm, or both, may produce erroneous readings.

Many other types of hygrometers have been devised (6, 8, 9, 13, 143).

C. Precipitation

Rainfall is one of the natural cleansing processes which are active in the atmosphere: it cleanses by washing out water-soluble contaminant gases and particulate matter having dimensions greater than 1 μm. The washout by precipitation of airborne radioactive wastes is a matter of particular concern. [For discussions of special problems in precipitation measurement such as accuracy, aerodynamic effects, measurement of snowfall, and evaporational losses, see (5, 6, 9, 146, 147); analyses of these special problems in precipitation measurement made in the U.S.S.R. have been published (148).]

1. Simple Rain Gauge

The basic parts of a simple rain gauge are a funnel supported with its mouth horizontal, and a collecting vessel located below it. To prevent rain from splashing in and out, the top of the funnel should consist of a cylinder with its axis vertical, and the slope of the funnel wall should be steep, at least 45°. The outlet of the funnel should be small, in order to minimize evaporational losses.

To permit accurate measurement of precipitation amounts, the inside cross-sectional area of the collector is often one-tenth of that of the mouth of the funnel, giving a magnification of 10 in measuring the depth of the collected water. The collector may be a glass vessel with suitable units of depth or volume engraved on it for direct readings, or it may be of metal, and a dipstick may be used to measure the amount of precipitation.

2. Weighing Rain Gauge

This instrument, the simplest of the recording rain gauges, weighs the rain, hail, sleet, or snow which is funneled into a bucket having a 12-in. rainfall capacity. The weighing mechanism converts the weight of precipitation to its equivalent in millimeters or inches, and actuates a pen arm which traces a record on a moving chart. The weighing rain gauge is particularly suitable for use in climatic regions having cold winters, where the funnel is removed to allow free entry of snow into the bucket, and a salt antifreeze solution is placed in the bucket to prevent damage by freezing and to permit rapid emptying. Fresh antifreeze must be added after each emptying. For subtropical regions and areas with low rainfall another type of weighing instrument has been developed (*149*).

3. Siphoning or Float Rain Gauge

Instead of weighing the precipitation collected, this gauge senses the amount which has fallen up to any given time by a light, hollow float in the vessel below the funnel; the vertical movement of the float as the level of the water rises is converted, by a suitable mechanism, into the movement of a pen on a chart. When the pen reaches the top of the chart scale the water in the float chamber is automatically siphoned out into a larger collecting can at the base of the instrument. As this happens, the recording pen descends rapidly to the bottom of the chart scale. If

Figure 22. Automatic siphoning or float rain gauge of the Hellmann type. (Courtesy of Wilh. Lambrecht KG, Göttingen, Germany.)

rain continues the pen will commence to rise again. A rain gauge of this type is illustrated in Figure 22.

If winter temperatures fall below freezing, thermal insulation and an electrical heater are incorporated to prevent damage by freezing to the float and its chamber. Only a minimum amount of heating should be supplied in order to prevent excessive water losses by evaporation. If winter temperatures are very low, it is difficult to prevent freezing without causing excessive evaporation. Under these conditions, the weighing rain gauge is a better choice.

4. Tipping Bucket Rain Gauge

This rain gauge registers precipitation by counting small increments of rain collected. The rain leaving the funnel runs into a container divided into two equal compartments by a partition, as illustrated in Figure 23. The empty container has been designed to balance in unstable equilibrium about a horizontal axis, and in its normal position is tilted as shown, with one side or the other resting against a stop. When a specified small amount of rain, perhaps 0.01 in. or 0.1 mm, has drained from the funnel into the upper compartment, the bucket tilts the opposite way, so that the compartment containing the rain comes to rest against the stop on the opposite side, the rain empties out, and the other compartment commences to fill. The amount and intensity of the rain are given by the number and rate of bucket movements (5, 6, 9).

As with the siphoning or float gauge, thermal insulation and electrical heating may 'be used in climates with moderate winters to melt snow and prevent freezing. If this is done, however, the heating rate must be

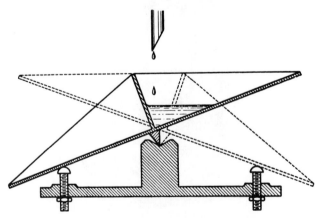

Figure 23. Principle of tipping bucket rain gauge. (Reproduced from Fig. 90 in M.O. 577 "Handbook of Meteorological Measurements Part I" by permission of the Controller of Her Majesty's Stationery Office.)

kept at a low value to minimize evaporational losses. If negligible water losses by evaporation cannot be attained, it is better to use a weighing instrument.

5. Radar Measurements

Radar has proved to be a powerful tool in the analysis of precipitation patterns and of precipitating clouds (7, 150–153). Hourly rates of precipitation can usually be calculated to within a factor of 2 by assuming that the reflectivity factor varies as a power of the rainfall rate; satisfactory surface rainfall measurements in heavy rain can be made by the reflectivity method only at radar wavelengths of about 5.5 cm, and longer, with beam widths up to 2° (153). Advanced methods of processing radar precipitation data are beccming available (154).

Weather radar technology will be used increasingly in air pollution studies in which washout by rain is an important factor. Weather radar will be especially valuable for precipitation washout studies over areas where it is difficult if not impossible to maintain regular rain gauges, such as over the Great Lakes.

D. Solar Radiation

In an earlier section the thermal radiometer was mentioned as an appropriate instrument for assisting in making a rough estimate of turbulence and diffusion. Solar radiation is important in another way: it has been shown to cause highly significant photochemical reactions in smog (155–157). It may therefore be necessary in some air pollution investigations to make direct measurement of the amount of solar radiation.

Solar radiation in its many aspects has been fully described in several monographs (158, 159). The various instruments which may be used to measure the radiant energy received from the sun are known as actinometers. In one of the simpler types of actinometers, two bimetalic strips lie side by side under a glass dome. The blackened strip is heated more by the sun than the reflecting strip, leading to a differential bending of the strips, which actuates the arm holding the recording pen.

A more accurate instrument of the actinometer family is the Eppley pyrheliometer, in which the sensor consists of two concentric silver rings, the inner one covered with lampblack and the outer one with magnesium oxide. The temperature difference between the two rings is indicated by the output of a thermopile of either 10 or 50 junctions, which may be carried to a millivolt recorder. The sensor is mounted at the center of a 3-in. sealed glass bulb containing dry air.

There are other instruments having similar capabilities for measuring

Figure 24. Solarimeter for measurements of solar radiation, direct and diffuse. (Courtesy of Kipp & Zonen, Delft, Holland.)

solar radiation (*5, 160, 161*). One such instrument is the solarimeter illustrated in Figure 24. The sensing element consisting of a Moll thermopile is made up of 14 manganin–constantan thermocouples and is protected from weather influences by two concentric spherical glass domes. Condensation within the hemispheres is prevented by the use of silica gel as a drying agent. A matching recording millivoltmeter is available. There are also chemical actinometers (*162*), of which the *o*-nitrobenzaldehyde actinometer has some special advantages for use in photochemical air pollution studies (*163*). The relative accuracy of most of the various types of radiometers has been analyzed (*5*), including four inexpensive models (*164*).

IV. Instrument Siting, Mounting, and Protection

The proper placing of instruments is necessary if representative values of the various meteorological elements are to be obtained and if trouble-free operation is to be achieved.

A. Principles of Instrument Siting

The cardinal principle of instrument siting is that the sensor should be located in such a position and in such a manner that it yields values of the variable which are representative of the atmosphere for the area of interest.

B. Fixed Instruments

1. Fixed Instruments near the Surface

The various wind instruments are the most important group of meteorological instruments for air pollution investigations. Since the same criteria apply in the proper siting of instruments for measuring wind direction, wind speed, and wind turbulence, no differentiation will be made in the treatment below.

a. WIND INSTRUMENTS. The standard exposure of wind instruments over level, open terrain is 10 m above the ground. Open terrain is defined as an area where the distance between the anemometer and any obstruction is at least 10 times the height of the obstruction. Where a standard exposure is not available, the wind instruments are often mounted at some height that is greater than 10 m by an amount depending on the extent, height, and distance of the obstructions; the wind at this height should represent as closely as possible that which would occur at 10 m in the absence of such obstructions. The concept of an "effective height" of the instruments in such situations has been introduced; effective height is the height over open level terrain in the vicinity of the wind sensor which, it is estimated, would have the same mean wind characteristics as those actually recorded by the existing sensor with its various obstructions (6). Thus the effective height above a given point on the ground might be different for wind vanes, anemometers, and bivanes, and might be a function of wind direction and wind speed as well. Unfortunately, the concept gives no firm guidance in determining the effective height, which in the final analysis is a matter of individual judgment.

The difficult problem of the proper exposure of wind sensors is discussed in detail in other publications (5, 6, 9).

b. RADIATION INSTRUMENTS. Correct exposure requires (1) that the site should provide an uninterrupted view of the sun at all times of the year, throughout the whole period when it is above the horizon; and (2) that the instrument be mounted securely so that the receiving surface is truly horizontal. The sensor must not be exposed to reflected solar radiation nor to artificial radiation sources. The flat roof of a building may be a desirable location for radiation instruments (5, 9).

c. VISIBILITY INSTRUMENTS. Since most visibility data used in air pollution analyses have been visual observations of systems of objects, and since very few specially planned and designed programs of instrumental observations for air pollution surveys have been undertaken, there is little experience available on which to base sound advice on the siting of visibility instruments. As mentioned earlier, in built-up areas the transmissometer has advantages over the telephotometer when a long enough unobstructed light path between source and receiver is available. Another obvious guiding principle is that the visibility should be measured in areas and directions, and at heights of primary concern for the particular air pollution problem under consideration.

d. HUMIDITY INSTRUMENTS. Hygrometers without artificial ventilation are usually located about 5 ft above the ground in instrument shelters with louvered sides which ensure a free flow of air past the sensor, and, at the same time, provide protection from heating by solar radiation. With artificial ventilation, which should be at the rate of at least 4 m/second, shielding from radiation must be provided. Instruments should also be located away from buildings and other objects which may have temperatures appreciably higher or lower than that of the air at the same height over open ground. A roof location should be avoided in favor of an open ground site if the latter is available.

The measurement of humidity is regions having substantial air pollution requires special precautions. For example, since ammonia is very destructive to the human hair filaments in the hair hygrometer, this instrument should not be used in the immediate vicinity of industrial plants which discharge ammonia into the atmosphere (5).

e. PRECIPITATION INSTRUMENTS. The optimum location of rain gauges presents a difficult problem because the catch tends to be related to the local structure of the wind. Local eddies induced by the gauge itself usually decrease the amount of water collected. On the other hand, eddies caused by the local topography and nearby objects may either increase or decrease the catch, depending on particular circumstances. A study in which 30 sites above 8000 ft in Utah have been classified according to gauge exposure provides valuable information (165).

Wherever possible, the rain gauge should be exposed with its mouth horizontal over level ground, and surrounding objects should not be closer than a distance equal to four times their height. Subject to this limitation, however, a site that is sheltered from the full force of the wind should be chosen. Slope and roof sites are generally unsatisfactory and should be avoided. The mouth of the gauge should be high enough to prevent rain from splashing in, but no higher. The whole question of minimizing errors due to exposure has been discussed (5). A method which is sometimes effective in reducing wind errors is to provide a wind shield for the gauge; many types of shield have been proposed (*5, 6, 9, 166*).

f. REMOTE INSTALLATIONS. For certain types of air pollution investigations it is necessary to measure wind speed and direction at remote inaccessible locations. Measurements of other meteorological variables are also sometimes needed. The sensors alone may be in a remote location with the indicating or recording equipment close at hand, or both sensors and recorders may be inaccessible (*167*).

Remote wind sensors, with the recorder nearby, have been developed. One model consists of an anemometer with transmitter and indicator units connected by a two-conductor cable, there being practically no limit to the length of cable that can be used (*168*). The sensing unit employed is a cup anemometer of the wind-run or totalizing type. The equipment therefore cannot be used to obtain smaller scale turbulence data.

There are a number of automatic weather stations designed so that the sensors, including wind speed and direction sensors, may be installed at remote locations; the information is telemetered to conveniently located receiving and recording equipment (*169–171*).

Remote wind sensors and recorders have also become widely available. In the Woelfle type of mechanical wind recorder, the sensors are a wind-run or totalizing cup anemometer and a two-element wind vane assembly. The installation is completely self-contained. The chart must be changed and the clock rewound at least once a month. A more versatile mechanical weather station (illustrated in Fig. 25) which records wind direction, wind run (speed), temperature, and rainfall may also be obtained (*172*). This equipment will operate as long as 2 months without requiring attention, and is available in either English- or metric-unit models. Elaborate installations have been designed for remote island locations (*173*).

A battery-operated type of wind speed and direction recording assembly has been designed to function without attention for periods up to 6–12 months (*174, 175*).

2. Fixed Instruments Mounted on Towers

Tower installations of meteorological instruments and associated read-out equipment are becoming increasingly complex (*176, 177*). Simpler tower systems have generally sufficed for air pollution surveys and investigations. For such purposes the two important categories of instruments are those for measuring wind and those for measuring temperature lapse rate.

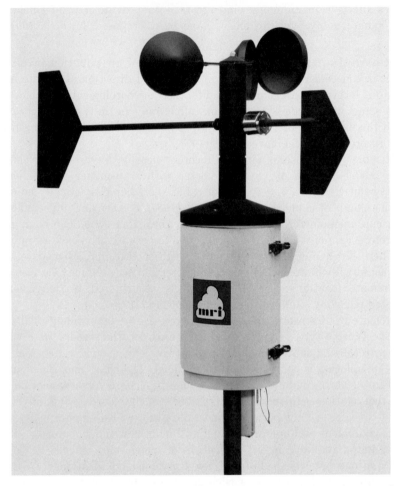

Figure 25. Mechanical weather station. (Courtesy of Meteorological Research, Inc., Altadena California.)

a. WIND INSTRUMENTS. Tower wind sensors should be mounted at the end of horizontal booms extending outward a minimum distance equal to the maximum horizontal dimension of the tower at that height, if large errors in wind measurements are to be avoided (*178–184*). Figure 26 summarizes the results of a comprehensive wind tunnel study carried out to determine the nature and magnitude of the errors that occur with various boom lengths (*181*). The figure shows that, with the boom mounted as indicated on a relatively open tower, wind speed and direction are accurate to ±5% over a direction range of 180° if the length R of the boom is equal to the maximum dimension D of the tower but that

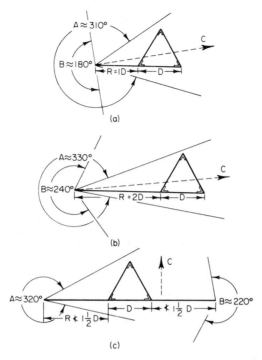

(a)

(b)

(c)

Figure 26. Limits of error of tower wind speed and direction measurements for various wind directions in relation to boom length and boom orientation with respect to tower: A and B represent arcs for approaching winds for which speeds are true within 10% and directions are true within 10° (arc A), and for which speeds are true within 5% and directions true within 5° (arc B); the broken arrow C points toward the wind direction of minimum concern with regard to air pollution, or the wind direction of lowest frequency of occurrence; and the maximum horizontal dimension of the tower is D. Boom length is (a) equal to D; (b) equal to 2D; and (c) not less than 1.5D (181).

the direction range for ±5% accuracy increases from 180° to 240° if the boom is lengthened from D to 2D. For a boom length of not less than 1.5D, the corresponding value is 220°. The direction ranges for ±10% accuracy are correspondingly larger. The orientation of the booms should be determined by such factors as the bearing from the tower of particular areas of concern. For example, if the influence of atmospheric conditions on pollution concentrations in a residential area lying to the north of an industrial plant emitting the contaminant is being studied, the booms should extend southward from the tower.

A logarithmic spacing of wind sensors in the vertical on a tower is often advantageous, e.g., at 10, 20, 40, and 80 m. Provision for bringing the instruments near the tower for servicing, as by using a telescoping boom, must be made. For various methods of mounting wind instruments, see Lettau and Davidson (*185*).

Wind instruments have also been mounted on booms extending horizontally from a stack. Because of the disturbance of the wind field by the stack, however, such instruments will give accurate measurements for wind directions from half the compass only. Wind tunnel studies lead to the same result (*181*). For example, instruments on a boom having a length three times the outside diameter of the stack at that height and extending westward from the stack will give accuracies of ±10% for wind speed and ±5% for wind direction only for winds from the south, west, north, and intermediate directions. If information on winds from all directions is required, two wind sensors at the end of oppositely pointing booms may be employed. The accuracy of speed and direction measurements is substantially improved by mounting the topmost set of wind sensors above the top of the stack—by an amount equal to the outside radius of the stack or more—but corrosion of the instruments by the effluent may be severe (*181*).

b. TEMPERATURE LAPSE RATE MEASUREMENTS. The temperature sensors should be mounted on horizontal booms, as in the case of the wind instru-

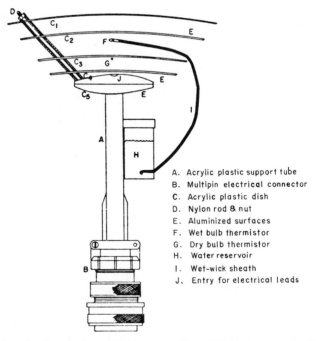

A. Acrylic plastic support tube
B. Multipin electrical connector
C. Acrylic plastic dish
D. Nylon rod & nut
E. Aluminized surfaces
F. Wet bulb thermistor
G. Dry bulb thermistor
H. Water reservoir
I. Wet-wick sheath
J. Entry for electrical leads

Figure 27. Detail of radiation shield construction. Shield is approximately 12 in. high (*187*).

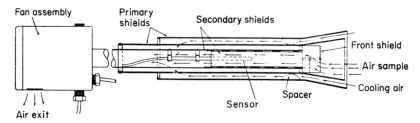

Figure 28. Schematic diagram of artificially ventilated temperature lapse rate sensor for tower or other similar installation. (Courtesy of Climet Instruments Company, Sunnyvale, California.)

ments, in order to avoid measuring the temperature of air locally heated or cooled by the tower structure.

The temperature sensors must also be shielded from radiational heating and cooling, and from precipitation. In one type of shield, protection is afforded by one or more horizontal or curved plates with special thermal characteristics located above and below the sensor (*186, 187*). An effective design is shown in Figure 27. The wind provides natural airflow and ventilation of the temperature sensing element. In a second type, artificial ventilation is provided by drawing air by an exhaust fan through concentric cylindrical shields, as illustrated in Figure 28. This aspirated shield is designed to limit errors from radiation to 0.2°F or less when exposed to radiation levels of 1.6 cal cm^{-2} min^{-1}. Unless special circumstances dictate otherwise, the unit should be mounted horizontally and, in the northern hemisphere, with its open end pointing toward the north in order to minimize penetration of solar radiation into the shield at sunrise and sunset.

Mounting temperature lapse rate sensors on booms attached to a stack is not recommended; even small differential errors due to the proximity of a warm stack may lead to serious errors in temperature lapse rate measurements.

C. Movable Instruments

It is often desirable to move the meteorological instruments from one location to various other ones in the course of an air pollution survey.

1. Movable Surface Instruments

There is a wide range of usefulness of mobile meteorological stations; a number of such stations have been built and used in air pollution investigations, usually in conjunction with sampling programs also conducted from the mobile station (*188*).

a. Design of Mobile Meteorological Stations. Many of these stations consist of a truck chassis with a body of the delivery van type. When in use, all meteorological instruments must be located so as to be independent of disturbing influences caused by the presence of the truck itself. In particular, if wind instruments are mounted on the truck, when in use they should be at a minimum height above the ground, which is twice the height of the highest part of the truck, and preferably higher. The most satisfactory height is the standard height of 10 m above the ground.

The interior layout of a rather complete mobile meteorological station mounted on a truck chassis is shown in Figure 29. Tower sections are carried. When observations are to be made, these are assembled and erected nearby to form a guyed tower of any specified height from 12 to 75 ft, or even higher if required. Such a tower provides an excellent mounting base for instruments for measuring wind, radiation, vertical wind profile, and vertical temperature profile, i.e., temperature lapse rate. The recording equipment for these is located in the working area of the truck, as shown in Figure 29. Radio facsimile equipment permits reception of the latest analyzed weather maps to allow assessment of developing meteorological systems which may influence the local observational program. There is also provision for meteorological measurements aloft by pilot balloon, wiresonde, and radiosonde.

b. Use of Mobile Meteorological Stations. Stations of this type are especially useful for air pollution surveys of an area where there is sufficient local difference in topography to make observations at a single fixed station unrepresentative; they are also useful for brief surveys of a number of areas. Care must be exercised, however, to be certain that the atmospheric characteristics being measured are the primary ones required. For example, the mobile station should not be set up in a pronounced hollow unless the limited air movement and stability conditions in the hollow are believed to have particular significance for the air pollution problem being studied. A site less sheltered from the main wind currents of the region is probably preferable. The exposure criteria which must be obeyed to ensure representative measurements for fixed instruments apply with equal force for movable surface instruments.

2. Movable Airborne Instruments

There are a number of ways of obtaining meteorological measurements by sensors which are carried aloft.

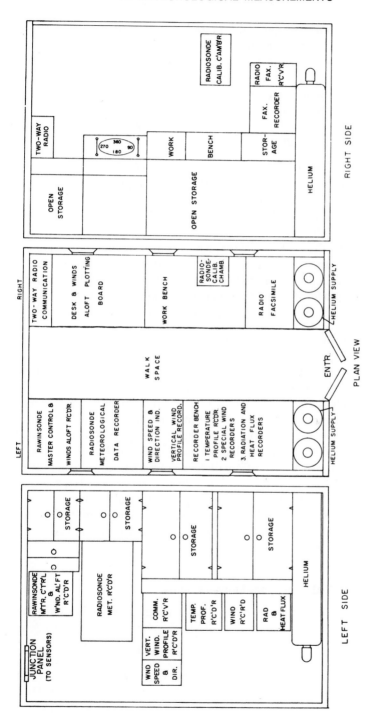

Figure 29. Interior layout of mobile meteorological station. Definitions of abbreviations used are: M'T'R: master; C'T'R'L: control; W'ND: wind; AL'FT: aloft; R'C'D'R: recorder; MET: meteorological; DIR: direction; R'C'V'R: receiver; COMM: communications; VERT: vertical; PROF: profile; R'C'R'D: recorder; RAD: radiation; IND: indicator; RECORD: recorder; CALIB: calibration; C'AM'B'R: chamber; FAX: facsimile.

a. RAWINSONDE. The site for radio or radar wind equipment should be on high ground with the horizon as free from obstructions as possible. Metal objects such as corrugated steel roofs of farm buildings or radio towers within the horizon will cause confusion of signals, but a single isolated object a few miles away may prove to be useful as a reference marker. A symmetrical hill with lower ground nearby and surrounded at greater distances by hills rising gently, at a slope of 1° to 2°, affords a good site; it eliminates ground echoes beyond short range (5,7).

b. WIRESONDE. There are restrictions on the use of wiresondes. A wiresonde should not be operated near overhead power lines because of the hazard both to personnel and equipment. Government regulations prohibit the flying of kite balloons and associated equipment at heights where they may be dangerous to low-flying aircraft unless special authorization has been given. The government agency having discretionary jurisdiction over air space will vary from country to country.* Such authorization is usually for limited geographic areas and for limited periods of time. Application should be made well before the planned time for wiresonde use to permit completion of the necessary formalities.

c. AIRPLANE, AIRSHIP, AND HELICOPTER. These various types of aircraft have been used primarily for atmospheric sampling of contaminants (189–191), but measurements of lapse rate have on occasion been made simultaneously with the sampling (44), or with other duties such as smoke plume photography (114). Detailed discussions of individual meteorological instruments mounted on airplanes and their proper exposure (7–9), as well as an elaborate airborne installation whose sensor outputs were recorded on a multichannel oscillograph (93), are available. A special study of the wakes of small helicopters has shown that these do not, in general, extend more than 200 ft below the craft (192), suggesting that helicopters, with their hovering capability, could advantageously be more widely used in air pollution investigations, including the meteorological phases. Helicopters have played a major role in a New York City survey (193).

V. Processing of Meteorological Measurements

In recent years new and versatile methods of data processing have become available. As with many new products, however, unit costs tend

---

* In the United States such authority is vested in the Federal Aviation Administration. For detailed information on restrictions in the United States and procedures for applying for authorization, contact Regional Air Space Sub-Committee, c/o Federal Aviation Administration, U.S. Department of Transportation, Washington, D.C.

to be high until volume sales and mass production methods interact to lower costs. Because of these earlier high costs, automatic data processing is just beginning to be utilized in the analysis of the results of air pollution investigations; many studies still rely heavily on manual methods.

A. Manual Methods

Although for convenience the present treatment discusses data processing in two main categories only—manual methods and automated methods—in reality there are a number of combinations of these two methods, and one or two important examples are discussed below.

1. Manual Methods without Mechanical Aids

In cases where the number of observations to be processed is small, it may be economical to complete the task using manual skills alone. For example, it may be required to obtain hourly average values of wind speed. If the wind sensor is a wind-run or totalizing cup anemometer, the required quantity is proportional to the difference in cumulative counts at successive hourly intervals, and is thus very simply determined. On the other hand, if wind speed is obtained continuously by a recording millivoltmeter actuated by the output of a small dc generator which is attached coaxially to a cup anemometer, determining an hourly average wind speed is more complicated. What must be done is, in effect, to place a horizontal line through a 1-hour portion of the wind speed trace in such a position that equal areas lie above and below the horizontal line. The position of this line on the wind speed scale of the chart paper then gives the average wind speed for that hour. The process is then repeated hour by hour for the whole chart.

If wind turbulence statistics are required, several options are available. For highly accurate information by manual methods it is necessary to take records of wind speed or direction with fast drive speeds in order that the recorder traces are clear, well separated, and not blurred by superposition of pen traces. Individual values must then be read off at appropriate time intervals and the standard deviations calculated by the standard root-mean-square methods. Before going to all this trouble, one should make certain that the response characteristics of both the wind sensor and the recorder are such as to justify the time and effort (*15*). For most air pollution purposes, however, more approximate methods will suffice. With slow chart speeds it is not difficult to estimate a range of the chart trace for each hourly interval. From this number it is possible to obtain an approximate value of the standard deviation of the record by the method recommended in Markee (*194*) or by similar techniques.

2. Manual Methods with Mechanical Aids

Mechanical aids have been devised for the purpose of simplifying and expediting the abstraction of values from strip chart records. In such devices the chart to be abstracted is initially placed on a drum at the left of a rectangular working surface. The strip chart is moved manually from left to right across the working surface by rotating a take-up drum on the right to which it is attached. Transparent straight-edge plastics with appropriate scales and cross hairs engraved on them are hinged so as to remain always horizontal or vertical as they are moved up and down or sideways. The purpose of these is to facilitate the estimation of areas and the assigning of numerical values.

Experience with such mechanical aids has been variable. Some personnel find them of distinct value whereas others find they can abstract data just as rapidly or even more rapidly without the encumbrance of moving the take-up drum and the straight-edge scales. The choice depends on the type of strip chart used, the time interval employed, and the nature of the quantities to be determined.

3. Manual Methods with Electronic Aids

It is possible to combine manual and automated methods, and this is often done advantageously. For example, wind records of instantaneous values may be made on strip charts advancing at a suitable rate. These values are read off at chosen time intervals and then punched on cards. With the data on cards it is then possible to process it in an almost unlimited number of ways with the aid of a high speed electronic digital computer: averages for periods of minutes, hours, or days may be obtained; standard deviations of wind-direction records may be calculated; power spectrum analyses of the energy in various scales of wind turbulence may be computed; etc.

There are digitizers available which combine mechanical, electrical, and electronic devices. In such point-by-point digitizing equipment the operator centers movable cross hairs over the chart trace with cranks or hand wheels, enters the coordinates, and then cranks the cross hairs to a new position. The position of the cross hairs is sensed by potentiometers, and an analog voltage is generated when the coordinates are entered. The generated voltage is processed through an analog-to-digital converter and recorded on a punch card. The punch cards are usually processed through a small computer to enter the digital coordinates in usable form on magnetic tape. Finally the tape is processed by a large computer, and data analyses of the types mentioned above are performed.

More efficient and rapid digitizers are also available. For example, one relatively low-cost and flexible digitizing system with output to punched cards can be expanded as need arises without making obsolete any of the initial components (*195*). In this digitizer commands are entered by teletype keyboard and cursor control buttons. This unit includes the following components: a 48 in. $\times$ 60 in. digitizer with x–y display and events counter; a computer with 4-K memory; a teletype with paper tape punch and reader; a keypunch interface; and all necessary software and a number of options available.

Such methods and devices have a good deal to recommend them for use until such times as fully automated methods can be employed. The cost of such digitizers tends to go up with their efficiency. It is therefore important to make certain that it will not be more economical in the long run to go directly to fully automated equipment.

B. Automated Methods

Automation in the processing of meteorological data ranges from such relatively simple processes as the electronic computing of the standard deviation of wind directions (*53, 54*) by such devices as the two-channel sigma computer illustrated in Figure 7, to very complex, versatile, and expensive installations. The cost effectiveness is an important aspect of any system, large or small (*196*).

1. Automatic Strip Chart Recorders

Strip chart recorders come in many sizes and forms: some have narrow charts and others have wide ones; some produce a more or less smooth continuous trace whereas operations recorders draw only steplike graphs. Recorders designed to sample the outputs from a number of sensors in succession, through the operation of a stepping switch, such as those from temperature lapse rate sensors on a tower, make their record in a series of disconnected dots. Some strip charts have a rectilinear coordinate system whereas others have a curvilinear system because the recording pen pivots on a fixed bearing. Red, green, blue, brown, black, or some other color of ink may be used; a hot stylus may cut a visible trace in a waxed chart paper; or a needle may mark a trace as it presses against a plastic-coated pressure-sensitive chart paper, as in Ref. (*172*). These differences are spelled out in order to emphasize the difficulties of designing an automatic strip chart digitizer which is capable of handling a wide variety of strip charts.

A highly automatic strip chart digitizer (*197*) that optically scans the entire chart at preselected intervals is, nonetheless, available. The digitized data can be recorded on magnetic tape for later computer processing or they can be read directly into the computer; punch cards and paper tapes are eliminated.

Other methods for automatically digitizing strip chart traces have been proposed, such as microfilming the strip charts and using automatic microfilm scanners in conjunction with computers to permit economical processing of the data.

The relatively high cost of such equipment prohibits its use for all but a very small number of air pollution investigations.

2. Data Loggers

Data loggers is the name given to equipment which records electrical inputs directly in digital form, usually by punch paper tape, magnetic tape, or by printer.

a. DATA LOGGING ON PAPER TAPE. Most of the early data loggers developed for handling the electrical signals from meteorological sensors have used punched paper tape for the system output, usually with printed read-out as well (*176, 198–200*). All the systems are capable of handling the electrical outputs from numerous meteorological sensors, two of the systems being capable of recording data from 49 analog and pulse generating sensors (*199, 200*). One of the systems has been installed in a mobile trailer van (*200*). It is not possible in this brief account to describe the various features of the several systems.

Punch paper tape has one clear-cut advantage over its chief competitor, magnetic tape. With punch paper tape it is readily possible to detect malfunctions in the equipment by visual inspection of the tape to determine whether or not data are being recorded and whether these data have reasonable values. Visual inspection of magnetic tape reveals nothing. Similar information about possible malfunctioning can be obtained only by breaking the record and making test playback runs.

b. DATA LOGGING ON MAGNETIC TAPE. This type of tape has not been used as frequently as punch paper tape for meteorological data loggers. Although the art of magnetic tape recording has developed to a high degree so that both analog and digital signals are readily handled, there are other drawbacks in addition to the one just mentioned above. For

Figure 30. A compact magnetic tape data logger. (Courtesy of Teledyne Geotech, Garland, Texas.)

example, magnetic tape must be protected carefully from contamination: if even small amounts of dust are allowed to fall on it, the information recorded on the tape may not be recoverable.

A magnetic tape logger was installed on the United States Coast and Geodetic Survey ship *Pioneer* and used to record meteorological data during the 1964 International Indian Ocean Expedition (*201*). The tape recorder in this system is an incremental magnetic tape recorder using ½-in.-wide tape. The 24 data-channel inputs are cycled in succession; a total of 8.48 in. of tape is required for 3.20 hours of continuous recording. By this method the data are efficiently stored in a format compatible for data processing by automatic computer.

A compact magnetic tape data logger is shown in Figure 30. This unit is designed to multiplex analog inputs, convert the values to binary coded decimal data, and output these binary coded decimal data for recording on an incremental tape transport in a computer-compatible format. There are 8 channels for analog inputs, with optional expansion to 32 in increments of 8 (*202*).

c. DATA LOGGING BY PRINTER.   It is sometimes sufficient to use a printer to record the information in data logging. For example, such a system has been employed for logging time-integrated meteorological data (*203*). The printer employed has six digit wheels: all six are used for time, four digits are used for radiation, four for wind speed, and only three are used for temperature.

d. ELECTRONIC DATA PROCESSING EQUIPMENT. After the meteorological data have been logged, next comes the step of data processing. Data processing may be accomplished in many different ways. The resources of the Meteorology Computing Laboratory of the University of Michigan will be described briefly as an example of a modern and versatile computing and data processing facility.

The installation consists of three medium-sized analog computers, a small general purpose digital computer, and an analog/digital linkage, so that all of the computers can be operated together as a hybrid system.

The primary functions of the digital computer are acquisition and reduction of field data and research in meteorological simulation. The three analog computers are versatile and of sufficient capacity to be useful for many analog data processing tasks, as well as for simulation of atmospheric models such as those used in the study of atmospheric diffusion and micrometeorology.

3. On-Line Computers

At the present time there are on-line computers, analog or digital, which are in operation or planned for the near future. Some of these are single purpose units, but general purpose computers will find increasing use.

a. ON-LINE ANALOG COMPUTERS. The sigma computer (53, 54) described earlier and illustrated in Figure 7 is a small single purpose on-line analog computer. Another example is provided by the meteorologically operated stack control system installed at the Enrico Fermi Atomic Power Plant near Monroe, Michigan (204). The wind speed and wind direction signals from a mill or propellor anemometer and wind vane assembly are recorded and simultaneously transmitted to a small analog computer for processing. The running mean of the wind speed and the standard deviation of the wind direction for selected averaging times are obtained from the computer as well as CAUTION and STOP signals based on certain preset limits of these parameters. The valve leading to the stack is physically opened or closed, depending upon the computer output. Provisions are made to contain the radioactive waste gases should the stack be closed.

In the future more versatile on-line analog computers will be used for a wider range of purposes in the processing of the meteorological measurements needed for the assessment of air pollution conditions, actual or potential.

b. On-Line Digital Computers. One example of a mobile recording and data processing system uses a digital computer with a 16-K memory (205, 206); it provides 96 channels for data sampling. The mobility of equipment of this type and the versatility of present-day digital computers, even the relatively small ones, suggest that both mobile and fixed data processing systems incorporating on-line digital computers will be employed more and more for atmospheric diffusion analyses.

More complex systems are coming into use, such as the data management system developed by the National Center for Atmospheric Research (207). It consists of two completely independent computer systems interconnected by a single digital communications link. These minicomputers are 16-bit machines featuring 1200-nsec memory cycle time, 16-K memory, and multiply and divide hardware. Although this system was developed for aircraft installation, it would function well in air pollution investigations.

VI. Quality Control of Meteorological Measurements

The design of a program of meteorological measurements should incorporate, as an integral part of the plan, a carefully worked-out program of quality control to ensure that the final data obtained measure up to the highest standards of accuracy and reliability. Unfortunately, it sometimes happens that the provisions for quality control are insufficient to permit a measurement program to achieve its full potential: initial calibration may be inadequate or completely lacking; poor servicing and maintenance of equipment may lead to large gaps in the data; lack of routine detailed inspection of chart records may allow unsuspected errors to accumulate; etc. A regularly scheduled program of data validation will pay high dividends in reliable measurements and observations (208, 209). Some of the things to be kept in mind in the planning, establishment, and execution of a quality control program for meteorological measurements are set forth briefly below.

A. Optimal Response Characteristics

The concept of quality control should be kept in mind even in the initial planning of a meteorological installation. Many wind vanes overshoot badly in a turbulent wind, so that turbulence statistics from such a sensor may often be in error; choice of a wind vane having optimal characteristics for turbulence measurements will greatly reduce this problem (15, 16, 24, 208). There are other examples, such as that of lapse rate measurements. More representative values of average lapse rates will be obtained in a simple manner with a time constant of 1 minute than with one of 1 second because the rapid fluctuations of air temperature are of no direct concern in air pollution surveys. It is also often important to choose a recorder whose response time is compatible with that of the sensor (208).

With an informed and discriminating initial choice of sensors and recorders which have optimal response characteristics for the type of survey planned, a good foundation for a sound quality control program has been laid. Even the most carefully conceived and executed plan of quality control in the later stages will not yield good data unless such a solid foundation has been laid initially.

B. Calibration of the Measurement System

Between receipt of equipment from the manufacturer and its incorporation into an operational program of routine measurements, it is important to calibrate the equipment, both as individual components, and, later, as an integrated measuring, recording, and data processing system.

1. Calibration of the Components

Most meteorological equipment is not, and cannot be, as carefully engineered and tested as a present-day electrical refrigerator, which may run for 20 years after being plugged in without an instant's attention from a service man. Meteorological equipment may not arrive in the same condition it left the factory: a number of detrimental things may have happened in transit despite careful packing.

If the manufacturer has not supplied a calibration curve or equivalent information, such a calibration should be carried out. Special facilities such as temperature calibration chambers and calibration wind tunnels are invaluable; perhaps it will be possible to obtain the use of such facilities belonging to others for the necessary periods of time, if they are otherwise unavailable. Or it may be quicker and more economical to hire a commercial testing agency to conduct the required calibrations.

2. Calibration of the System

It is valuable to calibrate the system both in the laboratory and as installed in the field.

a. System Calibration in the Laboratory. It is sometimes possible to assemble all the units in the laboratory and to check the whole system by subjecting the sensors to controlled conditions which simulate closely the actual conditions as they occur in the outside atmosphere. Such laboratory calibration is especially valuable if the system is a complex one, incorporating, for example, data processing equipment. Even though the individual components operate satisfactorily, malfunctions may occur when they are assembled into a system. It is easier to locate the trouble under laboratory than under field conditions.

b. System Calibration in the Field. A thoroughgoing check and calibration of the system installed in the field, and undertaken before measurements are taken on a routine basis, is a prime necessity. The field tests, especially if undertaken during and after wet weather, may reveal flaws in the system which were not even hinted at by the system calibration in the laboratory.

Exact calibration methods cannot be specified in advance because of the great variety in the possible components which may comprise the system. A substantial amount of information on the calibration of meteorological measuring systems is available, including examples of commonly occurring problems (*208, 210*).

C. Routine Maintenance and Servicing

If the various calibrations recommended above have been carried out completely, the operation of the system has had an auspicious beginning. The next task is to maintain the initial high quality of the data output by instituting routine maintenance and servicing of the equipment. Preventive maintenance is the key to success. The following general advice should form the basis for such a program (*208, 210*).

For reliable observations over periods of months the recorder should be checked daily at a specified time to ensure proper operation, and to place time marks on the chart roll. Daily maintenance should include a check for proper inking, proper indication of the time, and general system operation. Whenever chart rolls are changed, the operator should place enough data on the starting end of the roll to distinguish it positively from any other charts that might be used in the system complex. For instance, the wind direction chart at one level on a tower might be identified as follows: "Wind direction, 256-ft level, Charlevoix, on 0803 EST, Feb. 4/73, John Doe." A similar entry placed on the end of the roll thus completely identifies the chart records.

Generally such instrument systems should be thoroughly checked at about quarter-yearly intervals. This should include routine checks on the basic sensors, oiling and servicing, where appropriate, and full maintenance and servicing of the recording system. Some inking systems require only very occasional cleaning of the pen points and the ink wells, e.g., at quarterly intervals. Other systems will require thorough monthly flushing of the ink wells and weekly cleaning of the pens for consistent fine-line traces. A careful maintenance and servicing routine can yield good records 99% of the time, whereas moderately careless servicing may yield less than 50%.

If the system includes a data logger, equally thorough maintenance and servicing routines must be established.

Of the various meteorological instruments which have been described, probably the ones requiring closest surveillance are wind turbulence instruments, hygrometers, and actinometers. Detailed studies of the management of the last two have been published (5), and a manual on the maintenance of these and a number of other meteorological instruments has been prepared (211).

D. Regular Recalibration

Present-day recording systems usually require full calibrations only once a year, if check calibrations are made periodically. One might check a multipoint temperature recorder by immersing a temperature sensor in a well-stirred bath of carbon tetrachloride at bimonthly periods. Carbon tetrachloride combines the desirable features of high thermal conductivity and low electrical conductivity and because of its high vapor pressure it evaporates rapidly from the sensor and its support and shield after removal of the bath and is therefore less likely to cause corrosion than water. The temperature of the bath would be measured by a mercury or alcohol thermometer whose calibration was known. If the instrument were still within the previous limits of error of the system, full calibration would not be required. The frequency with which a system must be recalibrated depends a great deal on the degree of pollution of the air in which it operates. In highly corrosive atmospheres it may be essential to thoroughly overhaul and recondition the equipment and then recalibrate it more frequently than once a year.

If an instrument system has been in prolonged operation, the system should be calibrated before it is adjusted or serviced. The calibration then applies to the readings that were taken during previous operation. For future use of the recording system, the basic sensor and the recorder should be carefully checked before the second calibration is made. For null-balance potentiometer recorders, one should check the freedom of operation of the writing system, the absence of backlash in the writing pen, the absence of end play in the chart-drive roller, performance of the servo-drive system (as shown by the pen returning to within $\pm 1/100$ in. when deflected to right or left); condition of the battery, if any; and operation, adjustment, and lubrication of all other moving parts of the system. Galvanometer recorders require fewer adjustments but should be oiled and checked for proper adjustment before a calibration run (208).

E. Detailed Periodic Inspection of Chart Records

It may be thought superfluous to mention such an apparently obvious point as the routine inspection of recorder or data logger output. It is surprising, however, to find how often personnel will remove a wind direction chart from its take-up cylinder without any wonderment at seeing a perfectly straight line on the chart signifying that the wind has been blowing from the northeast without even fluctuating in direction by as much as a degree for the past 2 weeks. Shafts do seize in their bearings at times, and are especially prone to do so in the corrosive atmospheres in which air pollution surveys are often conducted.

Regular and frequent inspection of chart or other types of records should be at the heart of a program of quality control if severe losses of data are to be avoided. If the personnel who inspect charts are not technically trained, they should be instructed by a competent person and given a checklist of the most common symptoms of malfunctioning equipment. In this way even unskilled persons who are seriously interested in their work and the success of the undertaking will be able to detect most of the aberrations which at times overtake even the best equipment. Such attention to detail may spell the difference between success and failure.

VII. Instrument Requirements

There is a very wide range in the types of air pollution surveys made and in their needs for meteorological measurement systems. It is convenient to classify the requirements as follows: minimum, intermediate, and maximum. One insistent question which comes up, especially when a survey covers a substantial area, such as a city, a county, or a state, is how many meteorological instruments are needed, and with what spacing over the area of concern. The topography of areas varies so widely, however, that it is not yet possible to establish sound guidelines. A study of the relative effectiveness of various grouping of wind instruments over the city of Nashville, Tennessee, provides a good background for reaching similar decisions for other cities (*212*). The study suggests that the discriminating judgment of the skilled and experienced meteorologist is still the best source for obtaining a sound decision.

Whenever possible, sampling for atmospheric contaminants should be conducted simultaneously with a program of meteorological measurements; such a combined program is virtually a necessity even in a minimum investigation.

A. Minimum Instrumentation

The minimum requirement will depend on the type of source and on the topography.

1. Single Source

a. LEVEL AND UNIFORM TERRAIN. The minimum here is a recording wind vane and a totalizing cup anemometer at a height of 10 m, with a turbulence indicator such as a recording wind vane or a gust accelerometer at a greater height, perhaps 30 m. The lower instruments are provided to permit comparison with long period records from instruments in the area at the standard height that may be available. The higher instrument should be located at a height as nearly as possible equal to the average height of the plume. In general, a small tower is required to mount the instruments at these heights. For details of a similar program, see Walke *et al.* (*115*).

b. COMPLEX TERRAIN. In complex terrain, more than one of the level terrain installations may be required. In a valley, instruments on the valley slopes may be needed (*44*). At a shoreline location, there should be an installation of the level terrain type as near the shoreline as possible (*84*). If there is a small low flat island within a mile or so of shore, a similar installation on it would be most helpful, since meteorological conditions there may be very different from those at the shoreline for offshore or alongshore winds. Such differences will be especially important for a deep lake with several population centers on its shores.

2. Multiple or Area Source

An industrial city represents a multiple, or area, source.

a. LEVEL AND UNIFORM TERRAIN. If buildings are high in the business section, direct measurements of turbulence and wind speed, as by a gust accelerometer attached to a cable between buildings and well above street level, would be valuable. A tower, instrumented as for a single source over level terrain, should be used on a relatively unobstructed site in a neighborhood with low buildings.

b. Complex Terrain. With irregular topography, the level terrain minima above should be met, with valley-side instrumentation in addition for a city in a valley. A shoreline city should have additional wind instruments at a height of 10 m, and at several distances from the shore, to determine the distance of penetration of sea or lake breezes in the late spring, summer, and early autumn.

B. Intermediate Instrumentation

The instrumentation requirements will depend, to a degree, on whether pollution from a single source or from a multiple or area source is being analyzed.

1. Single Source

In the great majority of cases, the single source of pollution will be an industrial stack. The key facility required for a program of meteorological measurements is an adequate tower on which to mount instruments. If the industrial plant is situated on a level plain, and if a television tower happens to be located on the same plain, even a number of miles away, the tower would serve as an excellent mount. Atmospheric conditions at the tower could safely be taken as representative of those at the plant except, perhaps, in the amounts of shower precipitation. Such meteorological measurements have been and are being made on TV towers. Television stations in the United States have an obligation to provide a certain amount of public service without cost. Some stations have served the public by permitting their towers to be used for meteorological measurements needed to assess and control air pollution problems. A wider use of more TV towers may confidently be anticipated in the future. Since only the topmost radiating antenna is electrically charged, the lower portion of the tower structure is as safe to use as a tower designed and erected for meteorological purposes.

If there is no television tower nearby, the height of the tower to be erected must be related to the average height of the plume from the stack. If possible, the top of the tower should extend to that height. For purposes of illustration we may think of a tower 80 m high, with wind vanes and totalizing cup anemometers at 10, 20, 40, and 80 m; shielded temperature sensors at 2, 20, 40, and 80 m; and a single gust accelerometer, bivane, or u-v-w anemometer at the top. Alternately, the wind instruments may be mounted on the stack itself, in the manner described earlier. Since lapse rate measurements near a stack are not likely to be reliable, direct turbulence measurements at the top of the stack would be relied upon for information on diffusion conditions, or additional direct turbulence indicators could be installed at lower levels, if necessary. A sigma computer for data processing would be desirable.

A rain gauge should be installed and perhaps a hygrometer as well. If significant photochemical reactions are suspected in a stack effluent, an actinometer should also be installed. With a valley or shoreline plant location, additional wind instruments may be set up on the valley sides or near the shore, and wiresonde lapse rate measurements may be made using a kite balloon at places and times of particular interest.

2. Multiple or Area Source

The same considerations hold for a multiple or area source such as an industrial city. If available, one or more TV towers should be used, the number depending on the location of major sources of pollution in relation to local topographic features. Unless the terrain is very flat and uniform, a study of air pollution meteorology for a city requires a wider horizontal deployment of sensors than does a point source study. Complicated systems of valley or shoreline winds or both may exist and may need detailed investigation. Wiresonde studies of the prevalence of low-level inversions may be indicated.

It should be emphasized again that a full sampling program for atmospheric contaminants must be conducted simultaneously with the meteorological measurements. Either set of data is difficult to interpret without the other. In the past, sampling programs for air quality have received greater emphasis than adequate meteorological studies, but the situation is improving in this respect, and atmospheric conditions are receiving more attention. In difficult situations, it is becoming more customary to introduce artificial tracer materials into the air and to follow their movements by air trajectory studies and sampling programs, in order to chart diffusion patterns more accurately.

If investigations of the above scope are contemplated, the services of one or more experienced air pollution meteorologists should be obtained, to ensure that maximum results are achieved for the substantial investment required.

C. Maximum Instrumentation

It is impossible to make very specific recommendations for a problem so complex and so serious as to require maximum meteorological instrumentation. The range of possible combinations of atmospheric conditions, local topography, and location of sources is so great that it would be fruitless to attempt to classify them. Virtually all the instrumental resources described earlier in this chapter may be required, including wiresondes, rocketsondes, rawinsondes, aircraft observations, remote sensing systems, mobile meteorological stations, meteorological data loggers, and extensive sampling, both of indigenous contaminants and of artificial tracers.

Finally, the whole meteorological program should be under the immediate direction of individuals with the highest competence in air pollution meteorology and closely coordinated with the other phases of the total investigation.

A number of the concepts developed in this chapter have been expressed in other terms in a survey of meteorological instrumentation for air pollution studies, including a brief look at probable future developments (213). For those faced with the task of making a selection of instruments for a survey, there is a valuable list of suppliers, both domestic and foreign, of meteorological instruments of the type needed for air pollution investigations (214).

ACKNOWLEDGMENTS

The writer wishes to acknowledge his indebtedness to two of his associates, Fred V. Brock and Gerald C. Gill, for helpful criticism and advice. He has drawn heavily on both their experience and their publications in preparing the present chapter.

REFERENCES

1. R. E. Machol, W. P. Tanner, Jr., and S. N. Alexander, eds., "System Engineering Handbook." McGraw-Hill, New York, New York, 1965.
2. R. Perley, *Bull. Amer. Meteorol. Soc.* **45**, 740 (1964).
3. M. E. Ringenbach, *Meteorol. Monogr.* **11**, 334 (1970).
4. E. W. Hewson, *in* "Encyclopedia of Instrumentation for Industrial Hygiene" (C. D. Yaffe, D. H. Byers, and A. D. Hosey, eds.), p. 521. Univ. of Michigan Institute of Industrial Health, Ann Arbor, Michigan, 1956; also H. Moses, *in* "Meteorology and Atomic Energy 1968" (D. H. Slade, ed.), p. 257. U.S. At. Energy Comm., Div. Tech. Inform., Washington, D.C., 1968; also "Second Symposium on Meteorological Observations and Instruments." Amer. Meteorol. Soc., Boston, Massachusetts, 1972; also E. F. Bradley and O. T. Denmead, eds., "The Collection and Processing of Field Data." Wiley (Interscience), New York, New York, 1967.

5. "Guide to Meteorological Instrument and Observing Practices," 4th ed. World Meteorol. Organ., Geneva, Switzerland, 1971.
6. "Handbook of Meteorological Instruments, Part I: Instruments for Surface Observations," Meteorol. Office, M. O. 577. HM Stationery Office, London, England, 1956.
7. "Handbook of Meteorological Instruments, Part II: Instruments for Upper Air Observations," Meteorol. Office, M. O. 577. HM Stationery Office, London, England, 1961.
8. C. F. Campen, Jr. et al., eds., "Handbook of Geophysics," Chapter 20. Macmillan, New York, New York, 1960.
9. W. E. K. Middleton and A. F. Spilhaus, "Meteorological Instruments," 3rd ed. Univ. of Toronto Press, Toronto, Canada, 1953.
10. A. Perlat and M. Petit, "Mesures en Météorologie." Gauthier-Villars, Paris, France, 1961.
11. M. F. Haas, Meteorol. Abstr. Bibliogr. **10**, Suppl. 1, 2241–2415 (1959).
12. M. Thaller, Meteorol. Monogr. **11**, 211 (1970); also E. O. Doebelin, "Measurement Systems: Application and Design." McGraw-Hill, New York, New York, 1966.
13. G. C. Gill, H. Moses, and M. E. Smith, J. Air Pollut. Contr. Ass. **11**, 77 (1961).
14. R. A. McCormick, Bull. Amer. Meteorol. Soc. **41**, 175 (1960).
15. G. C. Gill, in "Proceedings of the First Canadian Conference on Micrometeorology" (R. E. Munn, ed.), Part I, p. 1. Department of Transport, Meteorol. Branch, Toronto, Ontario, 1967.
16. P. B. MacCready, Jr., Bull. Amer. Meteorol. Soc. **46**, 533 (1965).
17. P. B. MacCready, Jr., Meteorol. Monogr. **11**, 202 (1970).
18. G. C. Gill and P. L. Hexter, Bull. Amer. Meteorol. Soc. **53**, 846 (1972).
19. C. D. Yaffe, D. H. Byers, and A. D. Hosey, eds., "Encyclopedia of Instrumentation for Industrial Hygiene," pp. 567, 569, 579, 583, 608, 617, 634, and 669. University of Michigan Institute of Industrial Health, Ann Arbor, Michigan, 1956.
20. J. Wieringa and F. X. C. M. van Lindert, J. Appl. Meteorol. **10**, 137 (1971).
21. M. A. Garbell, J. Meteorol. **4**, 82 (1947).
22. H. P. Barthelt and G. H. Ruppersberg, Beitr. Phys. Atmos. **29**, 154 (1957); **31**, 262 (1959).
23. F. Weidenhammer, Beitr. Phys. Atmos. **33**, 123 (1960).
24. P. B. MacCready, Jr. and H. R. Jex, J. Appl. Meteorol. **3**, 182 (1964).
25. A. Longheto and A. Persano, Atmos. Environ. **2**, 77 (1968).
26. H. G. Müller, in "Handbuch der Aerologie" (W. Hesse, ed.), p. 587. Akad. Verlagsges., Leipzig, Germany, 1961.
27. P. B. MacCready, Jr., J. Appl. Meteorol. **4**, 504 (1965).
28. H. M. Morrow and R. M. Henry, J. Appl. Meteorol. **4**, 131 (1965).
29. J. R. Scoggins, J. Appl. Meteorol. **4**, 139 (1965).
30. N. J. Cherry, J. Appl. Meteorol. **10**, 982 (1971).
31. E. S. Mason, J. Appl. Meteorol. **7**, 512 (1968).
32. A. Longhetto, Atmos. Environ. **5**, 327 (1971).
33. J. K. Angell and D. H. Pack, J. Atmos. Sci. **19**, 87 (1962).
34. F. Pooler, Jr., J. Air Pollut. Contr. Ass. **16**, 677 (1966).
35. J. K. Angell, D. H. Pack, L. Machta, C. R. Dickson, and W. Hoecker, J. Appl. Meteorol. **11**, 451 (1972).
36. T. O. Haig and V. E. Lally, Bull. Amer. Meteorol. Soc. **39**, 401 (1958).
37. T. H. Cooke, Quart. J. Roy. Meteorol. Soc. **88**, 83 (1962).

38. G. C. Gill, E. W. Bierly, and J. N. Kerawalla, *J. Appl. Meteorol.* **2,** 457 (1963).
39. C. D. Yaffe, D. H. Byers, and A. D. Hosey, eds., "Encyclopedia of Instrumentation for Industrial Hygiene," pp. 579 and 583. University of Michigan Institute of Industrial Health, Ann Arbor, Michigan, 1956.
40. C. D. Yaffe, D. H. Byers, and A. D. Hosey, eds., "Encyclopedia of Instrumentation for Industrial Hygiene," pp. 569, 608, 617, 634, and 669. University of Michigan Institute of Industrial Health, Ann Arbor, Michigan, 1956.
41. I. Karmin, *Bull. Amer. Meteorol. Soc.* **40,** 473 (1959).
42. S. Ramachandran, *Quart. J. Roy. Meteorol. Soc.* **95,** 163 (1969).
43. S. Ramachandran, *Quart. J. Roy. Meteorol. Soc.* **96,** 115 (1970).
44. E. W. Hewson and G. C. Gill, *U.S., Bur. Mines, Bull.* **453,** 23 (1944).
45. L. J. Anderson, *Bull. Amer. Meteorol. Soc.* **40,** 49 (1959).
46. L. J. Fritschen and R. H. Shaw, *Bull. Amer. Meteorol. Soc.* **42,** 42 (1961).
47. H. Bardeau and R. Saporte, *J. Rech. Atmos.* **1,** 57 (1964).
48. G. C. Gill, *Bull. Amer. Meterol. Soc.* **35,** 69 (1954).
49. P. M. Jones, M. A. B. de Larrinaga, and C. B. Wilson, *Atmos. Environ.* **5,** 89 (1971).
50. R. L. Kagan, *Bull. Acad. Sci. USSR, Geophys. Ser.* No. 2, p. 175 (1964).
51. F. Pasquill, "Atmospheric Diffusion." Van Nostrand-Reinhold, Princeton, New Jersey, 1962.
52. J. W. Corcoran, "Theoretical Analysis of Wind Vanes." Beckman & Whitley Inc., Mountain View, California, 1962.
53. J. I. P. Jones and F. Pasquill, *Quart. J. Roy Meteorol. Soc.* **85,** 225 (1959).
54. F. V. Brock and D. J. Provine, *J. Appl. Meteorol.* **1,** 81 (1962).
55. E. K. Harris and R. A. McCormick, *J. Appl. Meteorol.* **2,** 804 (1963).
56. R. N. Sachdev and K. K. Rajan, *J. Appl. Meteorol.* **10,** 1331 (1971).
57. D. A. Mazzarella, *Bull. Amer. Meteorol. Soc.* **33,** 60 (1952).
58. G. C. Gill, *in* "Encyclopedia of Instrumentation for Industrial Hygiene" (C. D. Yaffe, D. H. Byers, and A. D. Hosey, eds.), p. 627. University of Michigan Institute of Industrial Health, Ann Arbor, Michigan, 1956.
59. K. J. Marsh, K. A. Bishop, and M. D. Foster, *Atmos. Environ.* **1,** 551 (1967).
60. F. V. Brock, *J. Appl. Meteorol.* **2,** 755 (1963).
61. A. W. Waldron, Jr., *J. Appl. Meteorol.* **2,** 740 (1963).
62. M. D. Thomas, *Proc. Nat. Air Pollut. Symp., 2nd, 1952* p. 16 (1952).
63. M. D. Thomas and J. O. Ivie, *J. Air Pollut. Contr. Ass.* **3,** 41 (1953).
64. S. P. Fergusson, *Harvard Meteorol. Stud.* No. 4 (1939).
65. G. E. W. Hartley, *Proc. Inst. Elec. Eng., Part 3* **98,** No. 64, 430 and 456 (1951).
66. G. B. Schubauer and G. H. Adams, *Nat. Bur. Stand. (U.S.), Rep.* **3245** (1954).
67. P. B. MacCready, Jr., *J. Appl. Meteorol.* **5,** 219 (1966).
68. P. Hyson, *J. Appl. Meteorol.* **11,** 843 (1972).
69. D. T. Acheson, *Meteorol. Monogr.* **11,** 252 (1970).
70. P. A. Sheppard, *J. Sci. Instrum.* **17,** 218 (1940).
71. E. L. Deacon, *J. Sci. Instrum.* **25,** 44 and 283 (1948).
72. C. D. Yaffe, D. H. Byers, and A. D. Hosey, eds., "Encyclopedia of Instrumentation for Industrial Hygiene," p. 569. University of Michigan Institute of Industrial Health, Ann Arbor, Michigan, 1956.
73. R. Drinkrow, *J. Appl. Meteorol.* **11,** 76 (1972).
74. R. R. Brook, *J. Appl. Meteorol.* **11,** 443 (1972).
75. T. W. Horst, *J. Appl. Meteorol.* **12,** 716 (1973).
76. W. H. Reed, III and J. W. Lynch, *J. Appl. Meteorol.* **2,** 412 (1963).

77. G. W. Thurtell, C. B. Tanner, and M. L. Wesely, *J. Appl. Meteorol.* **9**, 379 (1970).

78. R. J. Taylor, *J. Sci. Instrum.* **35**, 47 (1958).

79. J. K. Angell, *Mon. Weather Rev.* **90**, 263 (1962).

80. D. H. Pack, *Mon. Weather Rev.* **90**, 491 (1962).

81. J. K. Angell and D. H. Pack, *J. Appl. Meteorol.* **4**, 418 (1965).

82. J. Taylor, R. & M. "Reports and Memoranda," No. 2812, Aeronautical Research Council. HM Stationery Office, London, England, 1950.

83. J. K. Zbrozek, R. & M. "Reports and Memoranda," No. 3216, Aeronautical Research Council. HM Stationery Office, London, England, 1961.

84. N. I. Bullen, R. & M. "Reports and Memoranda," No. 3063, Aeronautical Research Council. HM Stationery Office, London, England, 1956.

85. P. B. MacCready, Jr., *J. Appl. Meteorol.* **3**, 439 (1964).

86. D. A. Haugen, "Atmospheric Technology," No. 2, p. 81. National Center for Atmospheric Research, Boulder, Colorado, 1973.

87. L. R. Struzer and A. P. Istomin, *Tr. Leningrad. Glavnaia Geofiz. Observ.* No. 129, p. 66 (1962).

88. G. A. De Marrais, *J. Appl. Meteorol.* **4**, 535 (1965).

89. L. J. Anderson, *Bull. Amer. Meteorol. Soc.* **28**, 356 (1947).

90. C. D. Yaffe, D. H. Byers, and A. D. Hosey, eds., "Encyclopedia of Instrumentation for Industrial Hygiene," p. 636. University of Michigan Institute of Industrial Health, Ann Arbor, Michigan, 1956.

91. "Cricketsonde Meteorological Rocket," Publ. EIR-392. Friez Instrum. Div., Bendix Corporation, Baltimore, Maryland, 1961.

92. E. Dubois and R. Bouscaren, *Atmos. Environ.* **2**, 83 (1968).

93. V. I. Skatskii and V. V. Shchelokov, *Bull. Acad. Sci. USSR, Geophys. Ser.* No. 8, p. 772 (1963).

94. G. A. Cleeves, T. J. Lemmons, and C. A. Clemons, *J. Air Pollut. Contr. Ass.* **16**, 207 (1966).

95. V. E. Derr, ed., "Remote Sensing of the Troposphere." National Oceanic and Atmospheric Administration and University of Colorado, Boulder, Colorado, 1972.

96. C. G. Little, *Bull. Amer. Meteorol. Soc.* **53**, 936 (1972).

97. C. G. Little, *in* "Atmospheric Technology, No. 2, p. 51. National Center for Atmospheric Research, Boulder, Colorado, 1973.

98. C. R. Hosler and T. J. Lemmons, *J. Appl. Meteorol.* **11**, 341 (1972).

99. W. D. Mount, A. C. Anway, C. V. Wick, and C. M. Maloy, "Development of Mark-I Radiometric and Simulation and Experimental Studies for Passively Probing Temperature Structure within the First Mile of the Atmosphere," Final Rep., Contract PH 22-68-22, SRRC-CR-70-6. Sperry Rand Research Center, Sudbury, Massachusetts, 1969.

100. J. B. Snider, *J. Appl. Meteorol.* **11**, 958 (1972).

101. R. G. Strauch, V. E. Derr, and R. E. Cupp, *Appl. Opt.* **10**, 2665–2669 (1971).

102. J. Cooney, *J. Appl. Meteorol.* **11**, 108 (1972).

103. C. G. Little, *in* "Remote Sensing of the Troposphere" (V. E. Derr, ed.), p. 19-1. National Oceanic and Atmospheric Administration and University of Colorado, Boulder, Colorado, 1972.

104. J. T. Gier and R. V. Dunkle, *Trans. Amer. Inst. Elec. Eng.* **70**, 339 (1951).

105. V. E. Suomi, M. Franssila, and N. F. Islitzer, *J. Meteorol.* **11**, 276 (1954).

106. J. MacDowall, *Meteorol. Mag.* **84**, 65 (1955).

107. J. P. Funk, *J. Sci. Instrum.* **36**, 267 (1959).
108. C.S.I.R.O. Net Radiometer, Middleton Instruments, 75–79 Crockford Street, Port Melbourne 3207, Australia.
109. C. D. Yaffe, D. H. Byers, and A. D. Hosey, eds., "Encyclopedia of Instrumentation for Industrial Hygiene," p. 565. University of Michigan Institute of Industrial Health, Ann Arbor, Michigan, 1956.
110. A. B. Kazanskii and A. S. Monin, *Bull. Acad. Sci. USSR Geophys. Ser.* No. 8, 1020 (1957).
111. F. Gifford, Jr., *Int. J. Air Poolut.* **2**, 42 (1959).
112. N. E. Bowne, *Bull. Amer. Meteorol. Soc.* **42**, 101 (1961).
113. A. R. Orban, J. D. Hummell, and G. G. Cocks, *J. Air Pollut. Contr. Ass.* **11**, 103 (1961).
114. E. W. Hewson, G. C. Gill, and G. J. Walke, "Smoke Plume Photography Study, Big Rock Point Nuclear Plant, Charlevoix, Michigan, 1963." Univ. Mich. Rep. No. 04015-3-P.
115. G. J. Walke, E. W. Hewson, and G. C. Gill, *Nucleonics* **23**, No. 2, 72 (1965).
116. U. Hogström, *Tellus* **16**, 205 (1964).
117. G. G. Goyer and R. Watson, *Bull. Amer. Meteorol. Soc.* **44**, 564 (1963).
118. P. M. Hamilton, *Air Water Pollut.* **10**, 427, (1966).
119. R. G. Strauch and A. Cohen, *in* "Remote Sensing of the Troposphere" (V. E. Derr, ed.), p. 23–1. National Oceanic and Atmospheric Administration and University of Colorado, Boulder, Colorado, 1972.
120. G. W. Grams and C. M. Wyman, *J. Appl. Meteorol.* **11**, 1108 (1972).
121. C. G. Little, *Meteorol. Monogr.* **11**, 397 (1970).
122. G. E. Anderson, *J. Appl. Meteorol.* **10**, 377 (1971).
123. E. W. Hewson, *in* "Compendium of Meteorology" (T. F. Malone, ed.), p. 1140. Amer. Meteorol. Soc., Boston, Massachusetts, 1951.
124. C. Steffens, *in* "Air Pollution Handbook" (P. L. Magill, F. R. Holden, and C. Ackley, eds.), Sect. 6. McGraw-Hill, New York, New York, 1956.
125. H. L. Green and W. R. Lane, "Particulate Clouds: Dusts, Smokes and Mists," 2nd ed. Spon, London, England, 1964.
126. E. W. Burt, *Amer. Ind. Hyg. Ass., J.* **22**, 102 (1961).
127. S. Fritz, *in* "Compendium of Meteorlogy" (T. F. Malone, ed.), p. 24. Amer. Meteorol. Soc., Boston, Massachusetts, 1951.
128. R. A. McCormick and D. M. Baulch, *J. Air Pollut. Contr. Ass.* **12**, 492 (1962).
129. V. D. Rockney, *Bull. Amer. Meteorol. Soc.* **40**, 554 (1959).
130. C. A. Douglas and L. L. Young, Tech. Div. Rep. No. 47. Civil Aeranaut. Admin., Washington, D.C., 1945.
131. G. H. Ruppersberg, *Beitr. Phys. Atmos.* **37**, 252 (1964).
132. C. D. Yaffe, D. H. Byers, and A. D. Hosey, eds., "Encyclopedia of Instrumentation for Industrial Hygiene," p. 572. University of Michigan Institute of Industrial Health, Ann Arbor, Michigan, 1956.
133. F. Volz, *Arch. Meteorol., Geophys. Bioklimatol., Ser. B.* **10**, 100 (1959).
134. R. J. Charlson, N. C. Ahlquist, H. Selridge, and P. B. MacCready, Jr., *J. Air Pollut. Contr. Ass.* **19**, 937 (1969).
135. S. S. Butcher and R. J. Charlson, "Introduction to Air Chemistry." Academic Press, New York, New York, 1972.
136. H. Horvath, *Atmos. Environ.* **7**, 521 (1973).
137. D. S. Ensor and A. P. Waggoner, *Atmos. Environ.* **4**, 481 (1970).
138. R. A. Rabinoff and B. J. Herman, *J. Appl. Meteorol.* **12**, 184 (1973).

139. W. E. K. Middleton, "Vision Through the Atmosphere." Univ. of Toronto Press, Toronto, Ontario, 1952.
140. J. S. Nader, G. C. Ortman, and M. T. Massey, *Amer. Ind. Hyg. Ass., J.* **22**, 42 (1961).
141. H. Spencer-Gregory and E. Rourke, "Hygrometry." Crosby Lockwood, London, England, 1957; also A. Pande, "Modern Hygrometry." Somaiya Publications Pvt. Ltd., Bombay, India, 1970.
142. T. Sinha, *Meteorol. Geoastrophys. Abstr.* **13**, 3622 (1962).
143. R. E. Ruskin, ed., "Humidity and Moisture: Measurement and Control in Science and Industry" (A. Wexler, ed.), Vol. 1. Van Nostrand-Reinhold, Princeton, New Jersey, 1965.
144. C. D. Yaffe, D. H. Byers, and A. D. Hosey, eds., "Encyclopedia of Instrumentation for Industrial Hygiene," pp. 583, 587, 593, 594, 602, 611, 614, 649, 657. University of Michigan Institute of Industrial Health, Ann Arbor, Michigan, 1956.
145. L. W. Foskett, N. B. Foster, W. R. Thickstun, and R. C. Wood, *Mon. Weather Rev.* **81**, 267 (1953).
146. E. R. C. Reynolds, *Meteorol. Mag.* **93**, 65 (1964).
147. T. Anderson, *Ark. Geofys.* **4**, 359 (1965).
148. N. P. Rusin, *Meteorol. Monogr.* **11**, 283 (1970).
149. E. Nothmann, *Bull. Amer. Meteorol. Soc.* **39**, 273 (1958).
150. L. J. Battan, "Radar Observation of the Atmosphere." Univ. of Chicago Press, Chicago, Illinois, 1973.
151. J. W. Wilson, *J. Appl. Meteorol.* **3**, 164 (1964).
152. G. B. Walker, L. S. Lamberth, and J. J. Stephens, *J. Appl. Meteorol.* **3**, 430 (1964).
153. D. Atlas, *Advan. Geophys.* **10**, 318 (1964).
154. K. E. Wilk and E. Kessler, *Meteorol. Monogr.* **11**, 315 (1970).
155. A. J. Haagen-Smit, *Science* **128**, 869 (1958).
156. N. A. Renzetti and G. J. Doyle, *J. Air Pollut. Contr. Ass.* **8**, 293 (1959).
157. P. A. Leighton, "Photochemistry of Air Pollution." Academic Press, New York, New York, 1961.
158. C. Perrin de Brichambaut, "Rayonnement soliare et échanges radiatifs naturels." Gauthier-Villars, Paris, France, 1963.
159. N. Robinson, "Solar Radiation." Elsevier, Amsterdam, Netherlands, 1964.
160. D. M. Gates, *Bull. Amer. Meteorol. Soc.* **46**, 539 (1965).
161. P. S. Hariharan, *Indian J. Meteorol. Geophys.* **12**, 619 (1961).
162. J. G. Calvert and J. N. Pitts, Jr., "Photochemistry." Wiley, New York, New York, 1965.
163. J. N. Pitts, Jr., J. M. Vernon, and J. K. S. Wan, *Air Water Pollut.* **9**, 595 (1965).
164. G. W. Smith, *Quart. J. Roy. Meteorol. Soc.* **98**, 855 (1972).
165. M. J. Brown and E. L. Peck, *J. Appl. Meteorol.* **1**, 203 (1962).
166. C. F. Brooks, *Int. Ass. Hydrol. Bull., Riga* No. 23 (1938).
167. N. E. Rider, *Meteorol. Monogr.* **11**, 405 (1970).
168. J. C. Bhattacharyya and S. Prakash, *Indian J. Meteorol. Geophys.* **15**, 277 (1964).
169. V. P. Petrov, A. M. Bogomolov, and E. N. Shadrina, Jr. *Leningrad. Glavnaia Geofiz. Observ.* No. 103, p. 10 (1960).
170. Automatic Weather Stations, *World Meteorol. Organ., Tech. Note* No. 52 (1963).
171. K. N. Manuilov, V. A. Usoltzev, and A. L. Zlatin, *Meteorol. Monogr.* **11**, 358 (1970).

172. "Mechanical Weather Station," Models—English Units: 1071, 1072, 1076, 1077; Metric Units: 1081, 1082, 1086, 1087. Meteorol. Res., Inc., Box 637, Altadena, California.

173. D. T. A. Langford, *Aust. Meteorol. Mag.* No. 43, 24 (1963).

174. C. J. Sumner, *J. Sci. Instrum.* **36**, 475 (1959).

175. C. J. Sumner, *Quart. J. Roy. Meteorol. Soc.* **91**, 364 (1965).

176. J. R. Gerhardt, W. S. Mitcham, and A. W. Straiton, *Proc. IRE* **50**, 2263 (1962).

177. R. I. Glass, Jr., *Bull. Amer. Meteorol. Soc.* **45**, 601 (1964).

178. H. Moses and H. G. Daubek, *Bull. Amer. Meteorol. Soc.* **42**, 190 (1961).

179. E. V. Borovenko *et al.*, *in* "Investigation of the Bottom 300-meter Layer of the Atmosphere" (N. L. Byzova, ed.). Izd. Akad. Nauk SSSR, Inst. Prikl. Geofiz., Moscow, USSR, 1963 (English transl., p. 83, 1965).

180. J. E. Cermak and J. D. Horn, *J. Geophys. Res.* **73**, 1869 (1968).

181. G. C. Gill, L. E. Olsson, J. Sela, and M. Suda, *Bull. Amer. Meteorol. Soc.* **48**, 665 (1967).

182. W. F. Dabberdt, *J. Appl. Meteorol.* **7**, 359 (1968).

183. Y. Izumi and M. L. Barad, *J. Appl. Meteorol.* **9**, 851 (1970).

184. G. C. Gill, *J. Appl. Meteorol.* **12**, 732 (1973).

185. H. H. Lettau and B. Davidson, eds., "Exploring the Atmosphere's First Mile," Vol. I. Pergamon, Oxford, England, 1957.

186. D. J. Portman, *in* "Exploring the Atmosphere's First Mile" (H. H. Lettau and B. Davidson, eds.), Vol. I, p. 159. Pergamon, Oxford, England, 1957.

187. R. Hadlock, W. R. Seguin, and M. Garstang, *J. Appl. Meteorol.* **11**, 393 (1972).

188. C. D. Yaffe, D. H. Byers, and A. D. Hosey, eds., "Encyclopedia of Instrumentation for Industrial Hygiene," pp. 671, 672, 674, and 676. University of Michigan Institute of Industrial Health, Ann Arbor, Michigan, 1956.

189. F. E. Gartrell and S. B. Carpenter, *J. Meteorol.* **12**, 215 (1955).

190. R. H. McQuain, J. M. Leavitt, R. C. Wanta, and W. W. Frisbie, *Proc. 51st Annu. Meet. Air Pollut. Contr. Ass., 1958* pp. 27–1 to 27–24.

191. Joint District, Federal and State Project for the Evaluation of Refinery Emissions, Interim Progr. Rep. Los Angeles County Air Pollution Control District, California, 1956.

192. D. L. Randall, *Bull. Amer. Meteorol. Soc.* **33**, 416 (1952).

193. B. Davidson, *J. Air Pollut. Contr. Ass.* **17**, 154 (1967).

194. E. H. Markee, Jr., *Mon. Weather Rev.* **91**, 83 (1963).

195. "Calmagraphic." Calma Company, 707 Kifer Road, Sunnyvale, California, 1973.

196. M. E. Ringenbach, *Meteorol. Monogr.* **11**, 334 (1970).

197. "Electroscanner Strip Chart Digitizer." United Gas Corporation, UGC Instrum. Div., Shreveport, Louisiana.

198. H. Moses and F. C. Kulhanek, *J. Appl. Meteorol.* **1**, 69 (1962).

199. L. J. Fritschen and C. H. M. van Bavel, *J. Appl. Meteorol.* **2**, 151 (1963).

200. V. J. Valli, "A Biometeorological Data Logging System for Agricultural Research." United States Dept. of Commerce, ESSA, Washington, D.C., 1966; also *Ga., Agr. Exp. Sta., Mimeogr. Ser.* [N.S.] No. 244 (1966).

201. D. J. Portman, *in* "International Indian Ocean Expedition—Meteorological Data Logged by the University of Michigan," Vol. 1, p. 129. United States Dept. of Commerce, ESSA, Washington, D.C., 1964.

202. "Operation and Maintenance Manual Data Conversion Unit, Model 36460." Teldyne Geotech, Garland, Texas, 1973.

203. D. N. Baker and S. G. Williams, *J. Appl. Meteorol.* **5**, 33 (1966).

204. G. C. Gill and E. W. Bierly, *J. Appl. Meteorol.* **2**, 431 (1963).

205. D. A. Haugen, *J. Appl. Meteorol.* **2,** 306 (1963).
206. J. C. Kaimal, D. A. Haugen, and J. T. Newman, *J. Appl. Meteorol.* **5,** 411 (1966).
207. T. M. Duncan, "Atmospheric Technology," No. 2, p. 47. National Center for Atmospheric Research, Boulder, Colorado, 1973.
208. G. C. Gill, *in* "Symposium on Environmental Measurements—Valid Data and Logical Interpretation," Pub. Health Serv. Publ. No. 99-AP-15. Robert A. Taft Sanit. Eng. Cent., Cincinnati, Ohio, 1964.
209. O. M. Essenwanger, *Meteorol. Monogr.* **11,** 141 (1970).
210. A. Mani, *Meteorol. Monogr.* **11,** 227 (1970).
211. "Pictorial Guide for the Maintenance of Meterological Instruments," Meteorol. Office, M.O. 725. HM Stationery Office, London, England, 1963.
212. R. H. Frederick, *Air Water Pollut.* **8,** 11 (1964).
213. P. B. MacCready, Jr., *in* "Air Pollution Instrumentation" (D. F. Adams, ed.), p. 4. Instrum. Soc. Amer., Pittsburgh, Pennsylvania, 1966.
214. "List of Meteorological Instrument Suppliers, March 1973." National Weather Service, Silver Spring, Maryland, 1973.

12

Air Pollution Climatology

Robert A. McCormick and George C. Holzworth

I. Introduction

Air pollution climatology is concerned with the aggregate of weather as it may effect the atmospheric concentrations of pollutants. It is based on documented records of pertinent meteorological elements and air quality measured at particular places during particular time periods. In a broad climatological sense, as well as for air pollution purposes, the locations and time periods of such records are very important since climate often varies from place to place and from time to time. Air pollution climatology is usually described in terms of statistical tables of wind, temperature, stability, sunshine, stagnation, diffusion parameters, atmospheric composition, etc.

A. Sources of Data

1. Meteorological Observations

Standard routine meteorological data are obtained by those National Weather Services at whose field stations regular and comprehensive surface observations of meteorological elements are made and recorded. Reference can be made to issuances of the World Meteorological Organization for lists of such services (1). In the United States there are approximately 300 stations manned by Weather Service personnel who make 24 surface observations at 1-hour intervals every day, as well as more frequently when warranted by weather changes. This basic network is supplemented by special purpose, secondary official networks, usually of limited scope with respect to completeness and frequency of their observational programs. Included in this category are observations made at fire-weather or fruit-frost stations, military posts, small airports, and at strategic locations by cooperative observers that are supervised by the Weather Service. It is not unusual to find excellent private records of observations made by "weather hobbyists" for their own enjoyment, and by industrial and educational organizations for operational, research, or documentary purposes. In addition, some air pollution control organizations routinely monitor the wind, and in some cases dry bulb and dewpoint temperatures, as well as air quality. In the more sophisticated networks, the measurements are made with automatic equipment and telemetered to a central location for real time display, processing, and archiving.

The number of stations at which upper air observations are made is considerably smaller than for surface observations. The latest official list of sites in the United States is given in Table I. The worldwide *scheduled* times for synoptic upper air observations are 0000 and 1200 Greenwich Meridian Time (GMT), although in the United States and probably throughout much of the world the balloon-borne instruments are commonly released about 45 minutes before the scheduled hour. At some places additional soundings are scheduled for 0600 and 1800 GMT but this practice varies throughout the world. In certain United States cities where air pollution is particularly troublesome special low-level soundings are made near sunrise and about noon on regular workdays and at other times during air pollution episodes. At these stations (see Table I) the balloons rise more slowly than in regular upper air soundings in order to reduce the effect of instrument sensor lag. The elements mea-

Table I   Rawinsonde Stations, United States, July 1973

Alabama	Idaho	North Carolina
Birmingham[a]	Boise	Cape Hatteras
Montgomery	Illinois	Greensboro
Alaska	Chicago[a]	North Dakota
Anchorage	Peoria	Bismarck
Annette	Salem	Ohio
Barrow	Kansas	Dayton
Barter Island	Dodge City	Oklahoma
Bethel	Topeka	Oklahoma City
Cold Bay	Louisiana	Oregon
Fairbanks	Boothville	Medford
King Salmon	Lake Charles	Salem
Kodiak	Shreveport	Pennsylvania
Kotzebue	Maine	Philadelphia[a]
McGrath	Caribou	Pittsburgh[b]
Nome	Portland	South Carolina
St. Paul Island	Massachusetts	Charleston
Yakutat	Chatham	South Dakota
Arizona	Michigan	Huron
Tucson	Flint	Rapid City
Winslow	Sault Ste. Marie	Tennessee
Arkansas	Minnesota	Nashville
Little Rock	International Falls	Texas
California	St. Cloud	Amarillo
El Monte[a]	Mississippi	Brownsville
Los Angeles[a]	Jackson	Del Rio
Oakland	Missouri	El Paso
San Diego	Monett	Forth Worth
San Nicholas Island	Montana	Houston[a]
(irregular)	Glasgow	Midland
Vandenberg	Great Falls	Victoria
Colorado	Nebraska	Utah
Denver	North Platte	Salt Lake City
Grand Junction	Omaha	Virginia
District of Columbia	Nevada	Wallops Island
Washington	Ely	Washington
Florida	Winnemucca	Quillayute
Key West	Yucca Flat	Spokane
Miami	New Mexico	West Virginia
Tampa	Albuquerque	Charleston[a]
Georgia	New York	Huntington
Athens	Albany	Wisconsin
Waycross	Buffalo	Green Bay
Hawaii	New York[b]	Wyoming
Hilo		Lander
Lihue		

[a] Low-level soundings only.   [b] Synoptic and low-level soundings.

sured aloft are temperature, pressure, relative humidity, and wind speed and direction.

A considerable number of "meteorological towers" are in routine service, operated by both private and public organizations, for purposes directly related to atmospheric diffusion and air pollution studies. They range in height from a few tens to a few hundreds of meters, and are generally instrumented at several elevations. While most of them are associated with research institutions and located in rural areas, some are situated within urban environments. Some television towers have also been used for this purpose. The National Meteorological Service is usually the best initial source of information on tower location, observations, and data availability.

2. Climatological Data

Current national practices in the publication of climatological data and summaries vary widely throughout the world. In the United States the basic routine source of climatological information is the Department of Commerce's *Local Climatological Data* (LCD). It is currently published for about 300 locations, including most large cities and, with rare exceptions, the rawinsonde stations listed in Table I. Data of particular interest in air pollution matters include daily listings of heating degree-days, sunshine, resultant wind direction and speed, average wind speed, precipitation, occurrence of smoke, haze, and fog, in addition to the complete weather observations at 3-hour intervals. An LCD *Annual Summary with Comparative Data* includes a brief narrative climatological summary, normals and extremes, and a history of the station and instrument locations. Such a history can be very useful in evaluating the data for a station, e.g., the effect on wind speed of changes in anemometer elevation. *Climatological Data, National Summary* is also published monthly and annually. It contains a narrative summary of weather conditions over the United States, climatological tables (in both English and metric units) for selected stations, mean monthly upper air data for the 1200 GMT observation, solar radiation data, and total ozone data at a few locations. A number of maps are also presented giving the distribution of temperatures, precipitation, solar radiation, and surface winds, the tracks of cyclones and anticyclones, and upper air maps for the 1200 GMT observation.

Additional specialized climatological publications are described in the *Selective Guide to Published Climatic Data Sources* (2). Unpublished climatological data and summaries, many for military posts, exist in great quantity in the files of the National Climatic Center (NCC), Asheville, North Carolina, and are described in the *Guide to Standard Weather Summaries and Climatic Services (3)*. Frederick (*4*) has written an excellent review of sources and utilization of weather data for air pollution studies, and a report by Stanford Research Institute (*5*), although prepared primarily to develop information on the air pollution climate of California, is also a useful guide to data reservoirs and their exploitation for similar studies elsewhere. For special data summaries or tabulations which are not routinely prepared, the NCC can perform analyses from original records, charging only for their services. The NCC can also supply punched cards and magnetic tapes of meteorological information for local processing.

Climatological publication practices in other countries are quite varied, ranging from monthly national summaries to bulletins of individual observatories. Consultation with the National Weather Service is recommended to obtain reliable information on data sources as well as on previous climatological analyses, in order to avoid duplication of expensive efforts. On a worldwide basis, the following are published in the United States by the National Oceanic and Atmospheric Administration but are accessible in most countries through their National Weather Service:

Monthly Climatic Data for the World [see (*2*)] gives mean surface and upper air data for hundreds of locations throughout the globe. Its publication is sponsored by the World Meteorological Organization.

World Weather Records [see (*2*)] contain monthly and annual tables of mean temperature, mean sea level pressure, and total precipitation for 1951–1960.

Northern Hemisphere Data Tabulations, giving daily surface and upper-air synoptic data, are available on 35-mm microfilm since January, 1964 and in printed booklet form prior to that time. These tabulations lag the observations by roughly 1 year.

3. Air Quality Data

Sources of air quality data throughout the world are quite heterogeneous. In general, measurements of air pollution are made by one or more agencies at several levels of government—national, provincial, town, and borough—as well as by public and private laboratories, and universities in some instances. At the national level the primary responsibility does not adhere to an institutional pattern as it may be a health ministry, regulatory agency, or a department of science and technology. A comprehensive inventory of worldwide air monitoring practices was published by the Smithsonian Institution (*6*) in 1970 and was being updated in 1974.

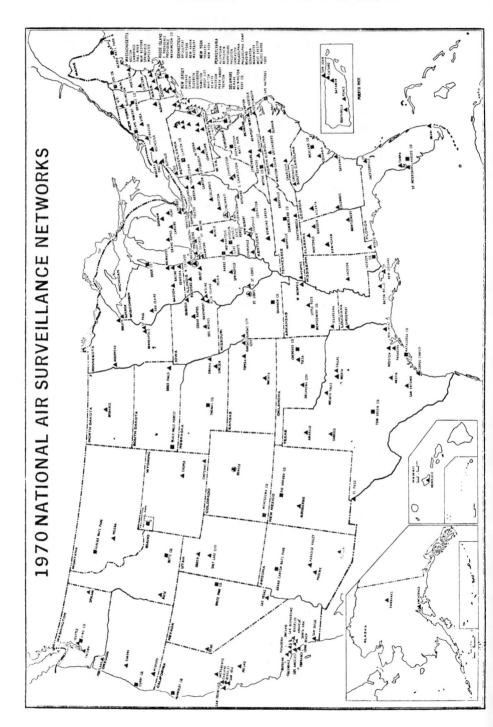

1970 NATIONAL AIR SURVEILLANCE NETWORKS

In the United States a ". . . comprehensive analysis and interpretation of data and information collected from Federal, State, and local air quality and emissions surveillance activities . . ." has been published by the Environmental Protection Agency (EPA) (7). The basic Federal program is the National Aerometric Surveillance Network (NASN). It is operated in cooperation with state and local agencies. Samples of suspended particulates at over 200 sites, urban and rural, and of several gaseous pollutants, e.g., SO_2 and NO_2, at over 50 of those stations are collected over a 24-hour period once every 2 weeks. Figure 1 shows the extent of the NASN.

NASN data are supplemented by a continuous air monitoring program (CAMP) in a few major urban areas. The program continuously monitors the concentrations of a number of important pollutants including those for which National Ambient Air Quality Standards have been set.

There are also many other state and local networks in operation in the United States, some quite elaborate and providing telemetered data to control authority headquarters. Many of these activities are in response to the requirements of the Federal Clean Air Act which requires adequate monitoring to establish compliance with air quality standards.

State air quality data are now required to be submitted to the National Aerometric Data Bank (NADB) in the format of the United States Environmental Protection Agency's Storage and Retrieval of Aerometric Data (SAROAD) (8) system on a quarterly basis. The NADB contains some data going back to 1958 but currently receives reports from on the order of 3000 individual sites over the United States. An inventory of the EPA environmental data systems, including those for air, to which reference can be made for complete information is available (9).

There are two international air quality monitoring activities underway sponsored by the World Health Organization (WHO) and the World Meteorological Organization (WMO), respectively. The aim of the WHO program (10) is to facilitate generation of internationally comparable data on levels and trends of air pollution in urban and industrial areas to enhance planning and assessment of health effects studies. It is implemented through the two WHO International Reference Centers (IRC) in London and Washington and through collaborating laboratories on air pollution in several of the member countries. Data from the network are available upon application to the WHO headquarters in Geneva, Switzerland.

The WMO program is concerned primarily with documenting trends

---

Figure 1. United States National Air Sampling Networks—1970. ▲: Urban site; ■: nonurban site; ○: continuous air monitoring station (CAMP).

in air composition which could effect large scale or even global climatic change, e.g., carbon dioxide and aerosols. Two networks are involved—"baseline" stations at isolated sites to detect changes in the constituents of "clean" air, and "regional" stations at rural sites to determine variations in air pollution which might be attributed to changes in regional land-use practices. When in full operation, there are expected to be 10 to 15 baseline, and over 150 regional stations participating in the program. The United States has established baseline sites at

> Mauna Loa, Hawaii
> Point Barrow, Alaska
> Tutuila, American Samoa
> South Pole, Antarctica

and regional stations at

> Alamosa, Colorado
> Atlantic City, New Jersey
> Bishop, California
> Caribou, Maine
> Huron, South Dakota
> Meridian, Mississippi
> Pendleton, Oregon
> Raleigh, North Carolina
> Salem, Illinois
> Victoria, Texas

The WMO network data for atmospheric turbidity and the composition of precipitation are published jointly by the United States National Oceanographic and Atmospheric Administration (NOAA) and the EPA.

B. Representativeness of Aerometric Data

Aside from questions of precision and resolution of aerometric (i.e., meteorological and air quality) measurements, discussed elsewhere in these volumes, the matter of "representativeness" is always of concern. The term representativeness shall be taken to connote ". . . the effect of individual sensor placement on the usability of its readings" (*11*), hence attention must be focused on the *purpose* of the measurements. Aerometric observations are commonly made for one of three objectives, (a) to survey or document the state of the atmosphere for record purposes, (b) to collect experimental data for research studies, and (c) to monitor the air pollution from a particular source or source complex for possible control action. Data collected for one of these reasons are unlikely to be representative for either of the others. As it is the first reason which produces the information from which air pollution climatologies can be deduced, considerations of sensor placements for that purpose only will be discussed.

The climatology of air pollution is largely restricted to the climatology of the zones of highest pollutant concentrations in urban areas. The most usual meteorological data are collected at rural—i.e., airport—sites, but in the latter half of the 1960's Environmental Meteorological Support Units were established by NOAA and EPA at more central locations in a number of major cities in the United States (*12*). These units obtain vertical sounding data on wind and temperature on a twice-daily basis for utilization by air pollution control agencies. However, even these data are not always usable to depict the "state of the atmosphere" unless the disturbing "noise" has been eliminated or minimized on all scales (in time and space) different from those intended, e.g., daily averages. Normally the problem is to eliminate local effects which are transient or unique. The most obvious and common sense precautions are to avoid a nearby source of primary pollutants which would bias air quality measurements, and in measuring airflow, to minimize local aerodynamic influences at the point of sensor exposure. The source-sampling point relationship is not so critical for secondary pollutants insofar as distance is concerned since more time is available for adequate mixing to take place over the length of fetch. The height of the sensors above the ground is probably the most important meteorological consideration, and it can be a serious one for air quality too, particularly for primary pollutants. Scorer and Barrett (*13*) have suggested the concept of a basic skyline in a city defined as the surface above which three-quarters of the area is unobstructed, below which ". . . the pollution is uniformly mixed. . . ." Presumably, meteorological measurements for climatological purposes should be made at an elevation not less than the basic skyline. For meteorological data, the concept is probably a very good one in principle, and the practice is recommended, provided that only a negligible portion of the obstructing area is closer to the sensor than at least four times the height of the obstruction (*14*). Pollution distributions around buildings and in street canyons have been discussed and investigated by McCormick (*15*) and Johnson et al. (*16*), respectively. Differences of factors of 2 or more are shown to be observable with street canyons depending on wind direction, building geometry, and sensor placement. No generalized formula has been developed to avoid aerodynamic anomalies in sensor placements, but Halitsky (*17*) has presented a scheme whereby an estimate can be made in many cases to avoid contamination from air exhausts or stacks for rooftop sitings.

Guidelines for the establishment of air quality surveillance networks have been published by the United States Environmental Protection Agency (*18*). These apply to measurements made near the ground in order to sample the air at about the breathing level of most of the population. Because of convenience and availability, instrument siting in park areas is common, but the results could be misleading because of absorption and/or deposition of pollutants on the vegetation (one of the rationales for setting up "green" belts or areas). To presume that the pollution is uniformly mixed below the basic skyline can be seriously in error, depending upon the vertical distribution of the source(s) and the static stability of the urban boundary layer. Evidence from London, England; Paris, France; and Cincinnati, Ohio (*19–22*) indicates that in the lower 300 m or so, gaseous pollutants, i.e., SO_2 and CH_4, are distributed fairly uniformly with height in the daytime, but decrease markedly in the vertical at night. On the other hand, aerosol materials appear to decrease in concentration with altitude in all stability conditions. More factual data must be obtained to resolve the effect of height, by pollutant, on the siting of air quality instrumentations particularly with respect to the interpretation of climatological data in terms of compliance with air quality standards.

Still to be resolved is the areal significance of an air quality measurement at individual sites in a city. Two studies dealing with the question, in Sheffield, England (*23*) and Nashville, Tennessee (*24*), indicate that a station spacing of the order of $\frac{1}{2}$ mile (0.8 km) is required to estimate mean daily levels of SO_2 over a city with reasonable accuracy. A subsequent study in West Germany (*25*) generally confirms that estimate, but these results do not seem to have gained general credibility, or at least acceptability.

II. Weather Elements

Air pollution climatology is interwoven with urban climatology because most sources and receptors of air pollution are located in cities. Nevertheless, the fundamental elements and mechanisms making up the air pollution climatology of a region are everywhere existent, differing only in degree of importance of one or more of the contributing factors.

A. Temperature

1. Ambient Value

The most obvious significance of ambient temperatures as an air pollution climatic factor is with respect to their influence on space heating requirements and the attendant discharge of pollutants into the atmosphere. This influence is often expressed in terms of the number of heating degree-days Q defined as the magnitude of the difference between the average daily temperature, T_m, and 65°F when T_m is below 65°F, i.e.,

$$Q \equiv 65 - T_m \tag{1}$$

Besides daily values, heating degree-days are also expressed as monthly (Fig. 2), seasonal, and annual totals. On a much finer time scale, Turner (26) developed empirical relationships between hourly temperatures and fuel usage by residences and commercial establishments during the winter in St. Louis, Missouri. His study showed that in addition to temperature effects, the diurnal variation of fuel usage was also influenced by socioeconomic factors. Besides its influence on space heating requirements, temperature may also have an impact on space cooling requirements and fuel usage for electrical generation. An interesting consideration of this matter in terms of cooling degree-days has been given by Thom (27).

The maximum and minimum atmospheric temperatures, over periods of about 6 hours, experienced by vegetation during the growing season have a very significant influence on the subsequent sensitivity of the plants to phytotoxic air pollutants. While temperature effects have not yet been isolated from effects of light intensity, the temperatures during daylight hours of air pollution episodes affect the sensitivity of plants (28). Available data indicate that air temperature is also an important factor in the formation of photochemical air pollution, and that seasonal and diurnal temperature changes can cause significant changes in photo-oxidation rates (29).

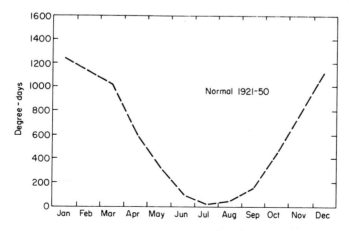

Figure 2. Heating requirements—Muskegon, Michigan.

2. Spatial Variations

The urban "heat island" expresses the fact that ordinarily cities are warmer than their nearby rural surroundings, particularly at night during light winds and clear skies. Such conditions favor the formation of wind patterns in which the air near ground level converges toward the center of the heat island (*30–32*). Furthermore, as the air moves across the city and is warmed, it becomes less stable. Thus, the urban heat island can have a profound effect on the transport and dilution of pollutants.

Urban areas are characteristically enveloped in a mass of warm air created by thermal processes related to the existence of the city itself and independent of its topographic setting (*30–35*). These processes include release of sensible heat by combustion, back radiation from urban structures of high thermal capacity, back radiation from air pollution, etc. Osaka, Japan is somewhat of an anomaly in that temperatures in the downtown section tend to be cooler than in surrounding districts. This has been attributed (*36*) to large heat sources at the edge of the city, an absence of heating in downtown buildings at night, and influences of sea and land breezes. On an annual basis mean urban–rural temperature differences for major cities are not large (Table II) (*37*). The urban heat island commonly reaches its greatest magnitude at night or around sunrise, possibly because the volume of air heated by urban processes is ordinarily smaller at night than in the daytime. Urban–rural contrasts in maximum temperatures are not often very large although an excellent example has been given for Toronto, Canada by Findlay and Hirt (*38*).

Extreme contrasts in minimum temperatures of 8.9°C have been reported for London, England (*39*) and 11.1°C for San Francisco, California (*40*). Mitchell (*34*) has presented an equation attributed to Sundborg to approximate the nocturnal urban–rural temperature difference D of the form

$$D = (a - bN)/V \qquad (2)$$

where N is the percent cloud cover, V is wind speed, and a and b are empirically determined constants. Although developed for Uppsala,

Table II   Annual Mean Urban–Rural Temperature Differences (°C) of Cities[a]

Berlin, Federal Republic of Germany	1.0	Los Angeles, California	0.7	Paris, France	0.7
Chicago, Illinois	0.6	Moscow, USSR	0.7	Philadelphia, Pennsylvania	0.8
London, England	1.3	New York, New York	1.1	Washington, D.C.	1.0

[a] Based on data given by Peterson (*37*).

Sweden, Mitchell suggests that the formula might be adapted to other cities as well. Also Ludwig and Kealoha (*41*) have made extensive investigations of the urban heat island. For nighttime they have developed equations for the magnitude of the urban heat island as a function of the variation of temperature with height in the nearby nonurban area (Table III). Oke (*42*) has considered data for additional cities and, limiting his analysis to "ideal" calm and clear conditions, offers equations for the maximum value of the urban heat island, $\Delta T_{u-r(max)}$ (°C), as a function of population, P:

$$\Delta T_{u-r(max)} = 2.96 \log P - 6.41 \qquad \text{for North American settlements} \qquad (3)$$

$$\Delta T_{u-r(max)} = 2.01 \log P - 4.06 \qquad \text{for European settlements} \qquad (4)$$

The correlation coefficients are 0.96 and 0.74, and root-mean-square errors are ±0.7 and ±0.9°C, respectively, for Equations (3) and (4). Oke points out that city population densities tend to be lower in Europe than in North America.

3. Atmospheric Stability

The vertical variation in temperature over a region is one of the two most important elements of air pollution climatology—the other being the rate of airflow or ventilation. The relevant parameter is the variation with height of potential temperature, $\delta\theta/\delta Z$, rather than of temperature, $\delta T/\delta Z$, since the former determines the stability condition of the atmosphere. To achieve static stability, $\delta\theta/\delta Z$ must be positive and since

$$\frac{\delta\theta}{\delta Z} \simeq \frac{\delta T}{\delta Z} + \Gamma \qquad (5)$$

where Γ is the dry adiabatic lapse rate (e.g., 1°C/100 m), it is apparent

Table III   Nocturnal Urban–Rural Temperature Difference (ΔT, °C) Equations and Supporting Statistics[a]

City population	Equation	Correlation coefficient	Root-mean-square error (°C)
$\frac{1}{2}$ million	$\Delta T = 1.3 - 6.78\nu$	−0.95	±0.66
$\frac{1}{2}$–2 million	$\Delta T = 1.7 - 7.24\nu$	−0.80	±1.0
2 million	$\Delta T = 2.6 - 14.8\nu$	−0.87	±0.96

[a] After Ludwig and Kealoha (*41*), where ν is the vertical temperature change with pressure (C°/mbar).

that a temperature inversion is not necessary for a layer to be stable. However, as vertical temperature data are usually the kind of information most readily available, determinations of inversion frequencies are quite common in assessing the frequency of stable conditions. Such data underestimate the actual occurrence of stable conditions, but are nevertheless informative.

Continuous data on atmospheric stability over urban areas are not numerous but DeMarrais (43) has prepared an excellent report on this matter based on temperature data taken on a television tower in downtown Louisville, Kentucky. He showed (Fig. 3) that the average temperature difference between 60 and 524 ft (18 and 160 m) became superadiabatic roughly 3 hours after sunrise and remained superadiabatic until about sunset, much like similar data for open country. However, after sunset and before sunrise, unlike open country, the average temperature difference in downtown Louisville rarely showed an inversion. Such conditions are expected from theoretical considerations of the effect of a city's heat island on its vertical structure of temperature (44). In addition, theory suggests that at night at higher levels over a city the temperature structure should become similar to that over the nearby rural areas, e.g., inversion conditions. Such conditions have been described by Clarke (45) for Cincinnati, Ohio.

Climatological analyses of low-level temperature data obtained by radiosonde techniques are relatively common although there are several limitations on such data. First, some loss of resolution of the temperature profile occurs because the response of the temperature sensor is relatively slow with respect to the ascent rate of the balloon-borne sonde, so that small excursions in the temperature profile are not necessarily reported. Second, the sonde-launching sites are usually at airports in rural or suburban surroundings so that low-level data are not necessarily representative of urban conditions. Finally, the scheduled times of daily soundings are synoptic by international agreement (currently 0000 and 1200 GMT), which may not be the most desirable times in terms of local time. The local times of the soundings are especially important when comparing data for different locations. This has been clearly demonstrated by Munn et al. (46) in their analysis of the frequency of ground-based inversions in Canada. In spite of these limitations, radiosonde observations can shed much light on general stability conditions when properly interpreted. Besides, they are often the only data available on the vertical variation of stability.

One of the noteworthy detailed studies of radiosonde data has been made by Szepesi (47, 48) at Budapest, Hungary to provide part of a climatological foundation for the forecasting of air pollution. Based on

	August	September	October	November	December	January	February	March	April	May	June	July
Midnight	+1.5	-0.2	-0.5	-0.4	-1.3	-1.0	-2.2	-1.8	-1.0	-0.4	-1.8	-1.1
2	+2.1	-0.5	+1.3	-0.3	-1.4	-1.5	-2.2	-2.0	-0.9	+0.1	-1.5	-0.6
4	+2.2	-0.4	+0.9	0.0	-1.0	-0.8	-2.2	-2.1	-0.3	-0.3	-1.2	-0.5
6	+1.9	-0.6	+0.9	+0.2	-1.0	-1.1	-1.8	-2.1	+0.5	-1.1	-2.0	-1.2
8	-1.8	-2.4	-1.1	-1.0	-1.5	-1.3	-2.1	-2.2	-1.6	-2.4	-3.0	-3.6
10	-2.9	-3.2	-3.0	-2.6	-2.4	-2.4	-2.5	-2.8	-3.3	-3.2	-3.3	-3.8
Noon	-3.4	-3.4	-3.5	-3.1	-2.7	-3.0	-3.0	-3.1	-3.6	-3.7	-3.7	-4.0
14	-3.3	-3.5	-3.2	-2.9	-3.0	-2.9	-2.8	-3.0	-3.3	-3.8	-3.8	-4.3
16	-3.2	-3.4	-3.0	-2.8	-2.5	-2.6	-2.8	-2.7	-3.2	-3.7	-3.7	-4.0
18	-2.5	-2.6	-2.2	-1.8	-1.9	-2.2	-2.4	-2.6/-2.7	-2.9	-3.2	-3.3	-4.0
20	-1.0	-1.4	-0.2	-1.0	-1.7	-1.3	-2.3	-2.2	-1.3	-2.1	-2.6	-2.9
22	+0.3	-0.4	+1.1	+0.2	-1.4	-0.9	-2.3	-1.9	-0.8	-0.5	-1.7	-1.9
Midnight	+1.5	-0.2	-0.7	-0.4	-1.3	-1.0	-2.2	-1.8	-1.0	0.0	-1.8	-1.1

Figure 3. Semimonthly average hourly temperature difference (°F), Louisville Kentucky, August 23, 1957–July 15, 1958. $T_{524\,ft} - T_{60\,ft}$. Adiabatic temperature change (524–60 ft) = −2.5°F (43).

4 to 5 years of soundings for each of six different observation hours, analyses were made of the monthly distribution of stability conditions in the lower 300 m by time of day, by cumulative distribution, and by frequency of inversion duration.

The monthly *Summary of Meteorological Data* (Table IV), issued by

Table IV   Summary of Meteorological Data, Los Angeles, California, July 1968

1. Number of days inversion base height in the following ranges (feet above mean sea level):

Height	1966	1967	1968	Av (1950–1968)
Surface	6	8	7	4
Less than 1500 feet	21	22	22	19
Less than 2500 feet	28	30	29	28

2. Number of days with maximum mixing height (Calculated maximum inversion base height) equal to or less than 3500 feet:

1966	1967	1968	Av (1950–1968)
31	29	29	27

3. Number of days average 0600–1200 PST wind speed equal to or less than 5.0 mph:

1966	1967	1968	Av (1950–1968)
25	20	24	26

4. Number of days inversion base height less than 1500 feet, maximum mixing height 3500 feet or less and average 0600–1200 PST wind speed 5.0 mph or less:

1966	1967	1968	Av (1950–1968)
16	12	18	14

5. Temperature:

	Normal	1968
a. Monthly average	73.0°F	73.3°F
b. Daily average maximum	83.3°F	82.5°F
c. Daily average minimum	62.6°F	64.1°F

	1940–1968	1968
d. Record highest	103°F (1959)	93°F (10th)
e. Record lowest	54°F (1952)	58°F (2nd)

the Los Angeles County Air Pollution Control District, is an example of the type of presentation of atmospheric stability which can be worked up from routine observations. In Table IV the conditions specified in item number 4 are generally considered undesirable in terms of air quality in Los Angeles.

Although based on only twice-daily radiosonde ascents (0300 and 1500 GMT), an excellent study of the frequency and duration of stable conditions over Munich and Erlangen, Federal Republic of Germany for a period of record of about 8 years has been published (49). Table V shows the seasonal percentage of days with a surface-based inversion at Munich, Erlangen, and Los Angeles, California. The low frequencies at Los Angeles, especially in summer, are misleading. Surface-based inversions are usually eliminated in the forenoon but subsidence inversions, which are very common during the warmer part of the year in Los Angeles and which have an average base height of about 450 m, persist both night and day, and are important factors in the smog problem.

Numerous analyses of radiosonde data have been made to describe low-level stability. Hosler's study (50) for the contiguous United States is based mainly on radiosonde observations at four different synoptic times. He presents seasonal maps of the percentage frequency of isothermal-inversion conditions within 500 ft (150 m) of the surface, with respect to both total hours during the data period and total observations for the observation time with the maximum occurrence (Fig. 4). It has been pointed out (51) that the analyses shown in Figure 4 for southern Florida were based on data for Key West and Tampa, and that data for Miami were inadvertently omitted. Values for Miami in winter, spring, summer, and fall were 60, 47, 65, and 75%, respectively. Hosler's study also includes supplementary information on quantities that bear on atmospheric stability, i.e., nights with little cloudiness and slow winds, and a climatic geographic delineation of low-level inversion frequency. Figure 4 may be interpreted as the percent frequency of all nights with a low-level inversion, since the data given in the appendix to Hosler's report indicate that the observation time with the greatest frequency of inversions was

Table V   Percentage Frequency of Days with Surface-Based Inversions

	Winter	Spring	Summer	Fall
Munich (1953–1960)	55.5	65.3	68.5	72.5
Erlangen (1949–1956)	44.4	70.6	80.4	66.0
Los Angeles (1950–1965)	69.0	22.9	13.1	40.7

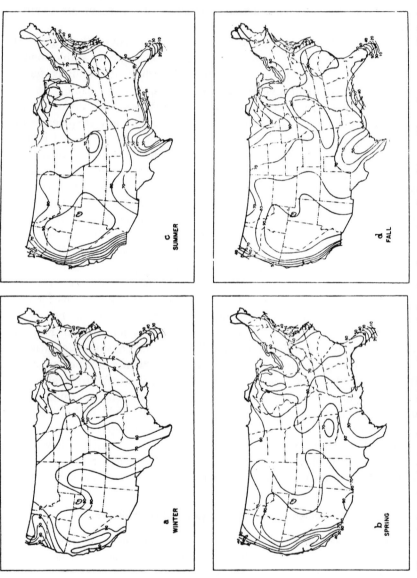

Figure 4. Low-level inversion frequency (percent) for the observation time with maximum frequency, United States (50); (a) winter; (b) spring; (c) summer; (d) fall.

almost invariably at night or near sunrise. Figure 4 indicates that in each season, over at least three-fourths of the contiguous United States, low-level inversions throughout the night occur at least one-half the time; in some areas, especially in summer and fall, on more than 80% of all nights.

A study by Munn et al. (46) presents seasonal maps of the percent frequency of ground-based inversions throughout Canada at each of two synoptic (radiosonde) observation times. The maps include isopleths of solar time, and illustrate that, in general, ground-based inversions are more common at night than in daytime, especially at inland locations. More detailed information on inversions at eleven arctic and subarctic radiosonde stations in Canada, Greenland, and Alaska have been prepared by Bilello (52). He gives data on inversions with respect to frequency, base height, thickness, base temperature, and temperature gradient. The 0000 GMT observations are summarized for all eleven stations but the 1200 GMT observations are summarized for only two stations. It should be pointed out that during the periods of record used by Bilello the radiosonde observation times were changed by 3 hours, although the data were not broken down accordingly. A very comprehensive analysis of inversion conditions has been prepared for the United Kingdom (53). It includes data on inversion thickness and magnitude, inversions by wind speed, and widespread inversions.

A basic problem in summarizing inversion conditions involves the handling of complicated inversions, i.e., successive layers of inversion, lapse, inversion. In order to keep summaries manageable it is desirable to simplify such complicated situations. One system (54) that has been used successfully in analyzing radiosonde observations defines the "inversion base" as the base of the lowest inversion and the "inversion top" as the maximum temperature that is associated with any inversion (below 700 m-bars). Isothermal layers are treated as in the examples of Figure 5. Table VI (55) summarizes certain conditions of simplified inversions

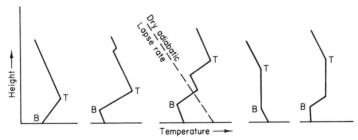

Figure 5. Examples of a system for simplifying complicated inversions. B: base, T: top of simplified inversion (54).

Table VI   Seasonal (Summer) Inversion Frequency Data, Santa Monica, California, June 1957–June 1962 (55)

Inversion base height (ft, mean sea level)	Frequency %	Frequency No.	ΔT (°C) (Top minus bottom) 0.0 to 3.5	3.6 to 6.5	6.6 to 9.5	>9.5	Thickness (ft) 1 to 500	501 to 1000	1001 to 1500	1501 to 2000	2001 to 2500	2501 to 3000	>3000
						0400 PST							
3001 to 5000	5	24	4	7	10	3	3	4	4	8	2	1	2
2501 to 3000	7	31	4	10	9	8	2	4	11	5	6	3	0
2001 to 2500	7	34	3	8	12	11	4	1	6	8	5	8	2
1501 to 2000	21	95	8	25	31	31	2	6	13	25	18	19	12
1001 to 1500	22	100	6	19	31	44	0	4	16	27	27	17	9
501 to 1000	16	73	1	19	26	27	0	1	5	22	13	16	16
126 to 500	4	17	1	5	8	3	0	1	3	4	2	7	0
Surface (125)	16	73	7	23	19	24	0	1	3	11	12	22	24
						1600 PST							
3001 to 5000	3	12	8	2	2	0	2	5	1	0	2	0	2
2501 to 3000	2	7	2	4	1	0	0		1	1	3	1	0
2001 to 2500	3	14	1	6	3	4	1	3	2	1	1	3	3
1501 to 2000	9	43	11	16	9	7	3	17	9	3	1	4	6
1001 to 1500	37	169	11	41	61	56	2	28	48	33	25	16	17
501 to 1000	36	165	10	37	58	60	0	17	46	51	23	10	18
126 to 500	9	41	3	14	14	10	1	1	5	17	10	7	0
Surface (125)	a	2	2	0	0	0	0	0	0	1	1	0	0

a Less than 0.5%.

at Santa Monica (Los Angeles), California. Notice in this table for summer that surface-based inversions are rare, even in the morning before sunrise, but strong inversions with bases between 500 and 2000 ft (150 and 600 m) are quite common both in the morning and afternoon. This is the subsidence inversion that plagues much of the United States west coast during the summer; in winter at night surface-based, radiation-type inversions predominate (Fig. 4).

One should keep in mind that because of heat-island effects, much of the data presented here dealing with radiosonde observations do not strictly apply to urban boundary layers at night. It would probably be more realistic to assume that at night urban boundary layers are in neutral static stability, i.e., $\delta\theta/\delta Z = 0$, at least over downtown sections of large cities.

The lower layers of the atmosphere normally undergo a marked diurnal variation in thermal structure and height through which *relatively vigorous* convective and turbulent mixing takes place. This height can range from virtually zero at night to several kilometers in the afternoon. Holzworth (*56*) has prepared seasonal and annual maps showing the distribution of mean estimated morning and afternoon mixing heights (MH) over the contiguous United States. Daily afternoon MH's were estimated by extending the dry adiabatic lapse rate from the afternoon maximum surface temperature to its intersection with the 1200 GMT radiosonde observation of that morning. Estimates associated with precipitation, when the dry adiabatic assumption might not be valid, were allowed for statistically in an arbitrary manner (*56*). Figure 6 shows isopleths of seasonal mean afternoon MH's based on 5 years of data for 62 upper air observing stations in the United States. [These data are considered more representative than those of an earlier study (*57*).] These maps represent the mean limiting conditions on the vertical extent of daytime mixing available for the dispersion of air pollutants. The significance of the values may be judged from the fact that one of the criteria for forecasting episodes of high air pollution potential (*58*) was that afternoon MH's be no greater than 1500 m. Some values of afternoon MH's for other parts of the world, estimated similar to the technique of (*56*), are listed in Table VII. The monthly variations at Ankara, Turkey (*59*) and Budapest, Hungary (*60*) with maxima in midsummer are much like those for most locations in the United States, but Tel Aviv, Israel (*61*) with a minimum in midsummer is similar to locations along the California coast. At Bombay, India (*62*) the peak value in November is unusual; relatively low values in March and October are perhaps due to conditions associated with the rainy southwest monsoon, for which MH's were not estimated.

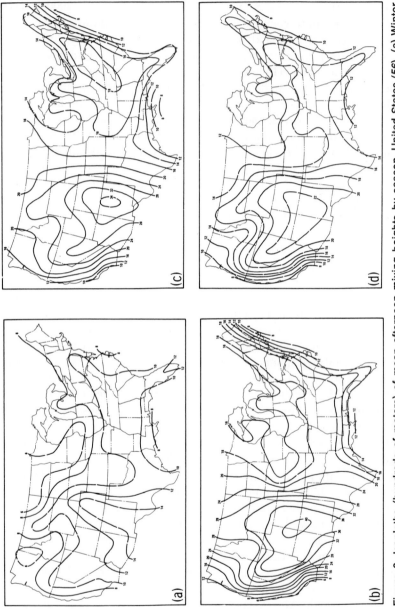

Figure 6. Isopleths (hundreds of meters) of mean afternoon mixing heights by season, United States (56). (a) Winter, (b) spring, (c) summer, (d) autumn.

Table VII Average Afternoon Mixing Heights[a]

	Jan.	Feb.	Mar.	Apr.	May	June	July	Aug.	Sept.	Oct.	Nov.	Dec.
Ankara, Turkey (59)	860	1090	1690	2130	2480	2600	2630	2620	2290	1840	1160	870
Budapest, Hungary (60)	560	950	1460	1750	2140	2200	2270	2100	1800	940	750	490
Tel-Aviv, Israel (61)	1630	1600	1560	1260	1010	920	760	820	1360	1700	1820	1600
Bombay, India (62)	1503	1182	1033							1238	1707	1578

[a] Meters above the surface.

Morning MH's have been estimated for the United States (*56*) in the same manner as for afternoons, except the afternoon maximum temperature was replaced by the rural minimum surface temperature plus 5°C. The "plus 5°C" was incorporated partly to allow for urban heat-island effects. Strictly speaking, morning MH's calculated by this technique apply to the time and place where the surface temperature is 5°C greater than the rural minimum temperature. Although nighttime and early morning mixing heights over cities are likely to vary over different parts of a city (*45, 63*), the method described here is useful for general comparative purposes. Figure 7 shows isopleths of annual mean morning MH's. Unlike afternoon MH's, the morning values do not vary greatly on a seasonal basis. Most of the variation occurs over inland regions where the morning MH's tend to be roughly 100–200 m higher in spring, about 100 m lower in summer and autumn, and about the same in winter as on an annual basis.

It should be noted that in the absence of more direct measurements of atmospheric stability, indirect indices such as the daily temperature range are sometimes utilized. The diurnal temperature range from early morning low to afternoon maximum on clear or partly cloudy days is related to the strength of the inversion at sunrise. In one study in the USSR (*64*) a correlation coefficient of 0.73 was reported between the mag-

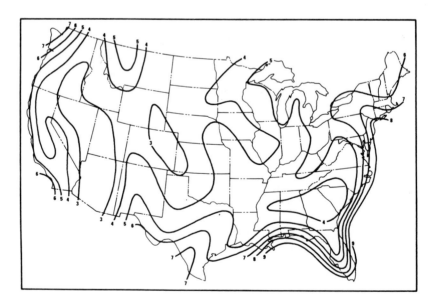

Figure 7. Isopleths (hundreds of meters) of mean annual morning mixing heights, United States (*56*).

nitude of the inversion (in the 2–100-m layer) and the daily air temperature amplitude at the 2-m level.

B. Ventilation—Transport and Diffusion of Air Pollutants

1. Atmospheric Stagnation

It is well known that many major air pollution episodes have been associated with the presence of quiet, quasi-stationary anticyclonic conditions over the afflicted regions (*65–69*). This is no wonder, since such anticyclones characteristically contain regions of calm or light winds, low-level radiation-type inversions at night, and sinking or subsidence of the air aloft, the latter sometimes markedly enhancing the low-level stability (*70*). Hence, when an anticyclone becomes nearly stationary so that light wind and great stability conditions persist in an area over a period of days, atmospheric stagnation or a high meteorological potential for air pollution is said to occur. Under such conditions the only requirement for an air pollution episode is a sufficient emission of pollutants. A synoptic climatology of stagnating anticyclones in the United States east of the Rocky Mountains, based on an analysis of daily weather maps, has been prepared by Korshover (*71*). His stagnation criteria were

(a) Surface geostrophic wind speed 15 knots (7.7 m/second) or less
(b) An absence of weather fronts and precipitation
(c) Persistence of (a) and (b) for at least 4 days

The greatest total number of stagnation days in 35 years at any grid point was 368, roughly one day in 35, and was centered over Georgia. But the values fell to zero in the Great Plains and to almost zero over southern Canada. October was the month with most stagnation cases, followed closely by September.

Korshover could not extend his study to the western United States because of unrealistic pressure gradients that arise from the reduction of pressure to sea level in high and irregular terrain. However, based on daily morning and afternoon values of MH and average wind speed within the mixing layer, the total number of episodes and episode-days in 5 years with a high meterological potential for air pollution have been determined for the contiguous United States (*56*). Figure 8 shows the total number of episode-days meeting the following criteria:

(a) Mixing heights 1500 m or less
(b) Average wind speed 4.0 m/second or less
(c) No significant occurrence of precipitation
(d) Persistence of (a), (b), and (c) for at least 2 days

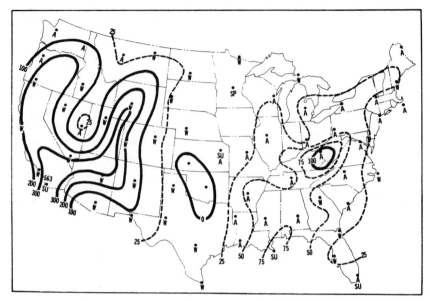

Figure 8. Isopleths of total number of episode-days in 5 years with mixing heights ≤1500 m, wind speeds ≤4.0 m/second, and no significant precipitation—for episodes lasting at least 2 days. Season with greatest number of episode-days indicated as winter (W), spring (SP), summer (SU), or autumn (A), United States (56). Note: Isopleths for data at San Diego, California omitted for clarity. See text.

These criteria are similar to those used in a national forecasting program (58), but episode data for other MH, wind speed, and persistence values have also been presented (56). In the eastern United States (Fig. 8) the greatest number of episode days in 5 years was just over 100, roughly 2 days in 35, which is not in bad disagreement with Korshover's study, considering the different criteria. In the western United States (Fig. 8) 100 episode-days in 5 years are exceeded at most stations and 200 days are exceeded over a large area. The disparity between values for San Diego and nearby Los Angeles, California (a factor of about 2) arises mainly because the wind speeds at San Diego are slightly slower. The season with the greatest number of episode-days in the eastern United States generally is autumn, while in the west it is winter. It has been noted (56) that many of the specific episodes comprising Figure 8 occurred simultaneously at adjacent stations in association with slow-moving anticyclones.

The occurrence of stagnation conditions throughout the USSR has been studied by Bezuglaja (72), based mainly on an analysis of 5 years of surface wind speeds at 220 stations in the midseason months of January,

April, July, and October. Stagnation was defined as a day on which the wind speeds did not exceed 1 m/second for 24 hours. It was noted that such conditions usually occurred simultaneously at several locations in association with the central part of anticyclones or weak pressure gradient fields (73). The greatest average number of stagnation days/month occurred in the vast area of eastern Siberia (except coastlines), where the January values generally exceeded 5 days but reached 25 days at some locations; in the other midseason months the values reached 10–13 days/month. Stagnation was very rare, occurring perhaps once in 3–4 years, along sea coasts and in a large region that is very roughly south of 55° latitude and between 35° and 85° east longitude. In other regions of the USSR, except mountainous areas, stagnation generally averaged 1–5 days/month in all midseason months. The frequencies in mountainous areas were found to be very complicated. Bezuglaja points out that since meteorological processes that are associated with undesirable concentrations of pollutants at the surface are often different for tall and short sources, air pollution potential should be estimated separately for such sources.

A study (74) of persistent surface wind speeds of 7 miles/hour (3.1 m/second) or less has been made for Canada, based on observations at 1-hour intervals at 111 weather stations. Persistence was defined in two categories: 24–47 hours and 48 hours or longer. Overall, the 10-year seasonal frequencies varied from zero to slightly over 100 occurrences, roughly three occurrences per month, for each persistence category. The frequencies were greatest in British Columbia, the Yukon, and northern Alberta. At most locations they occurred most often in winter and least often in spring. It was pointed out that the synoptic feature most often associated with persistent light winds was the stagnating anticyclone.

In Japan considerable effort has gone into the specification of synoptic pressure patterns that are associated with air pollution episodes (36, 75, 76). While high pressure is usually involved, the synoptic situations seem to be more variable than in other parts of the world. In one situation that appears fairly commonly, a quasi-stationary front is located along the southern coast of Japan so that a frontal inversion occurs at low levels over the mainland. Such a situation in New York, New York has been described by Nudelman and Frizola (77).

2. Wind Direction and Speed

Some of the most useful expressions of the climatology of airflow are in terms of wind roses. They usually depict the relative frequency, often with a breakdown by speed classes, with which the wind blows from the various sectors around the compass. Diurnal wind roses on a monthly or seasonal basis are desirable to show systematic variations. Features other than direction can also be summarized. As shown in Figure 9, wind roses for unstable and stable conditions may differ markedly, especially in uneven terrain. The figure, for a station along the Ohio River, also

LAPSE WIND ROSE

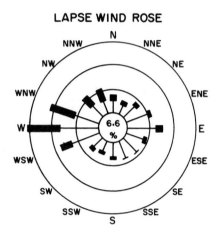

SCALE

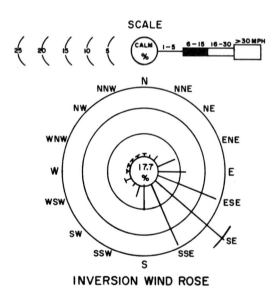

INVERSION WIND ROSE

Figure 9. Lapse and inversion wind roses, June 1955–May 1957, Shippingport, Pennsylvania. D. H. Pack, C. R. Hosler, and T. B. Harris, "A Meteorological Survey of the PWR Site at Shippingport, Pennsylvania." U.S. Weather Bureau, Washington, D.C., 1957.

indicates major differences in the wind features between daytime (mostly lapse) and nighttime (mostly inversion). It is not difficult to visualize that if these two wind roses were combined, the overall prevailing direction would not be very significant. Minimum dilution roses are informative, and pollution roses can be invaluable in identifying sources and evaluating their impact on air quality. Figure 10 (78) shows average SO_2 and hydrocarbon concentrations in downtown Philadelphia, Pennsylvania for each wind direction during winter and summer, and the corresponding wind roses. For hydrocarbons the concentrations were rather independent of wind direction and apparently nearby emissions did not vary much by season, as expected for motor vehicle sources. On the other hand, SO_2 emissions were greater in winter than summer due largely to

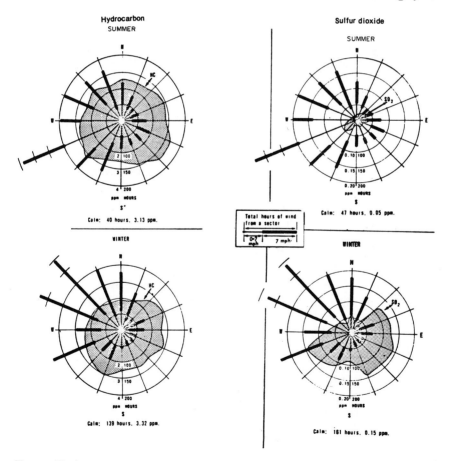

Figure 10. Average concentrations of hydrocarbons and sulfur dioxide by wind direction, Philadelphia, Pennsylvania, 1963 (78).

space heating requirements. In winter relatively high SO_2 concentrations with winds from northeasterly clockwise through southwesterly directions were attributed variously to industrial and space heating operations (*78*).

In using wind observations where the human eye has been involved, e.g., watching a pointer or evaluating strip chart recordings, one should be alert for observer bias for certain directions. For example, for a 16-point compass it is not uncommon to find a preference for the eight primary directions (N, NE, E, etc.) relative to secondary directions (NNE, ENE, etc.). Such bias can be largely eliminated by a fairly simple technique (*79*).

Where wind data for a network of stations are available it is often desirable to prepare streamline maps, indicating the transport of pollution at particular times or time periods, as has been done for Denver, Colorado (*80*) and southern California (*55*). Such transport can be determined more specifically by the construction of trajectories. In Los Angeles, California automatic data processing techniques have been used to construct a large number of trajectories in order to determine the source areas of pollutants associated with high oxidant concentrations (*81*).

In estimating time-averaged concentrations of pollutants (e.g., for time periods specified in air quality standards) at particular receptor points with respect to a particular emission point, the frequency that the wind blows from particular directions during the specified averaging periods is of obvious importance. Along this line Singer and Nagle (*82*) have prepared maps of recurrence intervals for wind direction persistence of 2, 4, and 6 days, based on hourly wind data at 35 locations in the United States. In general they found that the regimes of most persistent directions occurred along coastlines and were associated with strong winds; lowest persistence was in mountainous regions. Other analyses of wind direction persistence have been described elsewhere (*83*).

Experience has shown that the persistence of surface wind speeds less than about 7 miles/hour (3.1 m/second) is often conducive to the accumulation of air pollutants in cities. In connection with his work on low-level inversions Hosler (*50*) included seasonal maps for the United States of the percentage frequency of nighttime wind speeds 7 miles/hour or less. Table VIII gives the seasonal breakdown of similar light wind speed data for some major cities of the world. A tendency for light wind regimes to occur often during the autumn transition season is apparent, and the remarkable paucity of conditions of good ventilation at Milan, Italy would indicate high air pollution potential conditions there through most of the year.

Winds within urban areas are generally retarded relative to rural sur-

Table VIII   Percentage Frequency of Surface Winds < 3.6 m sec⁻¹ (7.9 mph)

City[a]	Winter	Spring	Summer	Fall
Berlin, Fed. Rep. Germany	52.9	57.2	64.1	64.1
Frankfurt, Fed. Rep. Germany	56.4	60.4	64.0	66.0
Rome, Italy	41.5	50.8	55.5	51.4
Milan, Italy	93.2	85.3	90.6	95.1
Madrid, Spain	66.8	62.5	67.7	74.2
Tokyo, Japan	45.5	55.3	77.6	74.8
Paris, France	43.5	47.3	46.6	62.7
London, England	42.3	38.2	43.7	49.7
Budapest, Hungary	81.9	75.9	77.4	83.4
Aleksandrovsk, USSR	56.1	53.8	64.8	47.5
Salt Lake City, Utah[b]	56.2	37.0	36.1	49.1
Pittsburgh, Pennsylvania[b]	37.5	38.9	59.8	50.5
Chicago, Illinois[b]	24.0	26.7	47.0	35.3
New York, New York[b]	16.3	17.3	27.0	23.3
Los Angeles, California[b]	69.3	59.7	63.3	73.7

[a] Data from 6-hour observations at airport stations in vicinity of city. Periods of record vary from 5 to 15 years.
[b] From hourly observations of winds ≤ 3.1 m sec⁻¹ (7 mph).

roundings by the braking effect of the building obstacles, and the structure of the flow is more erratic, i.e., turbulent. In a comparison of nonurban airport wind data with a 30-station wind measurement network in Nashville, Tennessee, Frederick (84) found that the urban wind averaged 60–70% of that observed at the airport. Nashville lies in a region of moderate topographic relief with height variations of as much as 500–600 ft (150–180 m). Frederick also found that the 24-hour wind movement at the airport was highly correlated with the daily average for the city network; the correlations were over 0.9 on most of the winter days and were somewhat dependent upon the actual wind speed. This indicates that, except perhaps for very light winds, airport data can well be quantitatively usable to estimate the airflow over an adjacent city.

During conditions of weak pressure gradient on the macroscale the description of airflow in cities cannot be spelled out in detail because of the heat-island effect, the channeling of flows by city "street canyons," and local thermodynamic and aerodynamic influences (32, 85). True calm conditions probably seldom exist over any appreciable length of time over a significant area, although it is not unusual to find a high frequency of calms in some wind summaries. These are often attributed to the starting speeds of the anemometers. An interesting analysis of this problem has been made by Truppi (86), who offers a method for detecting anomalous frequencies of calms in wind summaries due to anemometer starting speeds. Because of this difficulty it is not advisable generally to compare frequencies of calms at different locations; the frequencies of all speeds less than about 2–3 m/second would be more reliable. As previously noted, during conditions of weak pressure gradient and for cities in flat terrain, there is often a general drift of air toward the city center. Theoretical estimates for London, England (87) show that this drift can be on the order of 5 miles/hour (2.2 m/second) [see also Findlay and Hirt (38), Ariel and Kliuchnikova (88), and Pooler (89)].

The effects of irregular topography on the general properties of airflow, e.g., mountain–valley winds and land–sea breezes, over a region have been discussed in Chapter 8. In some regions of the globe, e.g., Los Angeles, California (90) and Sydney, Australia (91), nature has combined these effects in a fashion to most adversely bias the air pollution potential. Their situations with respect to circulations in semipermanent high pressure belts leads to light winds and persistent low-level inversions, particularly in summer. High terrain behind the cities tends to enhance the occurrence of both sea and land breezes, and thereby some recirculation of pollutants. Although air parcels seldom remain within the Los Angeles Basin longer than 24 hours (81), the daily back and forth flow of air relative to the automotive emission cycle plays an important role in the photochemical oxidant problem in Los Angeles (92). Recirculation of air has also been shown to be an important factor in air pollution problems of Denver, Colorado (80) and generally along the shores of Lake Michigan (93).

3. Atmospheric Turbulence and Diffusion

While wind and stability or temperature structure data are of interest in themselves, they are not independent. They jointly affect the intensity and structure of the air turbulence which diffuses or dilutes air pollutants. Their interrelationships are neither simple nor linear. For instance, the coexistence of marked vertical thermal stratification and high wind speeds is not possible in situations of practical concern. Thorough mechanical mixing will simply cause a dry adiabatic lapse rate (i.e., neutral stability, $\delta\theta/\delta Z = 0$) to be established in the layer. On the other hand, extremely stable ($\delta\theta/\delta Z \gg 0$) or unstable ($\delta\theta/\delta Z \ll 0$) stratification can exist in conditions of very light and variable winds.

The climatology of atmospheric turbulence and diffusion can be most directly based on measurements of the three-dimensional fluctuations or eddying motions of the air and their spectral distribution as a function of meteorological and geometric factors, e.g., height above the ground.

In general, however, such measurements are seldom made routinely and continuously so that a climatology can be revealed. The nearest approach is the well-known subjective classification of horizontal wind direction fluctuations at Brookhaven National Laboratory (*94*), which has been found to be an effective scheme for climatological studies. It has also been demonstrated to be a useful tool for the development of a diffusion climatology (Table IX) (*95*). As developed at Brookhaven, the classifications are based on the wind direction traces recorded by an "Aerovane" wind sensor at an elevation of 350 ft (106 m) above the ground, and are defined in the following tabulation:

Classification	*Description*
A	Fluctuations of wind direction exceeding 90°
B_2	Fluctuations ranging from 40° to 90°
B_1	Similar to A and B_2 with fluctuations confined to 15° and 45° limits
C	Distinguished by the unbroken solid core of the trace, through which a straight line can be drawn for the entire hour, without touching "open space"; the fluctuations must reach 15° but no upper limit is imposed
D	The trace approximates a line; short-term fluctuations do not exceed 15°

The relation of these classes to wind speed and vertical differences of temperature is shown in Table X (*96*).

Where direct measurements of wind fluctuations are not available, as is often the case, Pasquill (*97*) has proposed six "stability" categories to describe the diffusive potential of the lower atmosphere in estimating the dispersion of air pollutants. The categories are specified in terms of wind speed and, as affected by cloudiness, incoming solar and outgoing terrestrial radiation intensities. The categories vary from A (very unstable) to D (neutral) to F (moderately stable) (Table XI), and while originally developed to apply to diffusion in open or rural terrain, have been successfully applied, with slight modification, to the urban environment (*98, 99*). In the United States in response to a large number of requests, the National Climatic Center, Asheville, North Carolina has developed a standard computer program (called the "Star Program") to determine Pasquill stability categories based on the regular hourly surface observations of the National Weather Service. The Star Program has been run for over 250 different locations in the United States. Table

Table IX   Monthly Percentage Frequency of Gustiness Classes by SO₂ Concentration, Nashville, Tennessee *(95)*

SO_2 Concentration (ppm)		Total	Gustiness classes				
			A	B_2	B_1	C	D
			October				
	Total	100.0	4.8	9.2	39.4	6.8	39.8
Less than 0.025		75.3	4.2	8.3	32.8	3.4	26.6
0.025–0.049		19.5	0.6	0.6	6.3	2.0	10.0
0.050–0.074		4.3		0.3	0.3	1.1	2.6
0.075–0.099		0.9				0.3	0.6
0.100 and over		—					
			November				
	Total	100.0	2.8	5.6	52.2	12.6	26.8
Less than 0.025		53.1	1.9	3.1	38.1	6.0	4.0
0.025–0.049		23.2	0.3	1.6	9.9	6.0	5.4
0.050–0.074		11.4	0.3	0.6	2.8	0.6	7.1
0.075–0.099		6.8			0.8		6.0
0.100 and over		5.5	0.3	0.3	0.6		4.3
			December				
	Total	100.0		5.6	51.3	8.9	34.2
Less than 0.025		10.7		1.1	8.3	0.8	0.5
0.025–0.049		37.1		2.4	29.0	1.3	4.4
0.050–0.074		20.8		1.1	7.3	2.7	9.7
0.075–0.099		13.6		0.7	4.3	2.2	6.4
0.100 and over		17.8		0.3	2.4	1.9	13.2
			January				
	Total	100.0	0.3	6.4	60.7	15.4	17.2
Less than 0.025		27.1		1.9	16.1	9.1	
0.025–0.049		41.7		3.2	32.0	4.3	2.2
0.050–0.074		16.1	0.3	0.5	8.6	0.5	6.2
0.075–0.099		9.2		0.5	2.7	1.2	4.8
0.100 and over		5.9		0.3	1.3	0.3	4.0
			February				
	Total	100.0	1.7	8.7	63.4	7.4	18.8
Less than 0.025		47.8	1.1	3.6	37.5	5.0	0.6
0.025–0.049		30.1	0.6	4.5	18.2	1.8	5.0
0.050–0.074		14.0		0.6	7.1	0.3	6.0
0.075–0.099		2.7			0.3	0.3	2.4
0.100 and over		5.4			0.3	0.3	4.8
			March				
	Total	100.0	1.3	4.8	54.0	11.8	28.1
Less than 0.025		57.6	1.0	4.1	40.6	6.1	5.8
0.025–0.049		22.3	0.3	0.7	10.8	4.7	5.8
0.050–0.074		12.4			2.0	0.3	10.1
0.075–0.099		6.7			0.3	0.7	5.7
0.100 and over		1.0			0.3		0.7

Table X   Relation of Gustiness Classes to Meteorological Conditions—Brookhaven

Gustiness class	Mean wind speed ($m\ sec^{-1}$)	$T_{410'} - T_{37'}$ [a] (°C)	% Frequency of occurrence
A	1.8	−1.25	1
B_2	3.8	−1.60	3
B_1	7.0	−1.20	42
C	10.4	−0.64	14
D	6.4	+2.00	40

[a] Adiabatic temperature change $\simeq$ −1.1°C over this interval (96).

XII shows an example of the printout, for Birmingham, Alabama based on 5 years of hourly observations for the month of August. Notice that the table is for the D stability category, which occurred 26.1% of the time, and was most frequent with winds from the SSW at 7–10 knots (3.6–5.1 m/second). From climatological records at 21 stations the frequency of occurrence of the various Pasquill categories has been determined for January, April, July, and October in Great Britain (100). Similarly, the prevalence of these categories during each season over the Netherlands has been derived from data for 20 stations (Fig. 11) (101). Again, it should be noted that where the data are primarily for rural or suburban locations, they may overestimate the occurrence of more stable conditions in cities. Care should also be exercised in applying the Pasquill stability categories to tall sources, e.g., chimneys, where the stability regimes may be different than near ground level.

A climatology of theoretical city-wide average concentrations, $\bar{\chi}(\mu g/m^3)$, normalized for an average area emission rate, $\bar{Q}(\mu g/m^2$ second),

Table XI   Pasquill Stability Categories (97)

Surface wind speed at 10 m ($m\ sec^{-1}$)	Insolation			Night	
	Strong	Moderate	Slight	Thinly overcast or $\geq 4/8$ low cloud	$\leq 3/8$ cloud
2	A	A–B	B	—	—
2–3	A–B	B	C	E	F
3–5	B	B–C	C	D	E
5–6	C	C–D	D	D	D
>6	C	D	D	D	D

Table XII   Star Program Relative Frequency Distribution; Birmingham, Alabama; Month of August 1960–1964[a,b]

Direction	Speed (knots)						
	0–3	4–6	7–10	11–16	17–21	Greater than 21	Total
N	0.0013	0.0043	0.0051	0.0051	0.0002	0.0000	0.0161
NNE	0.0024	0.0032	0.0077	0.0013	0.0005	0.0000	0.0153
NE	0.0005	0.0021	0.0013	0.0010	0.0000	0.0002	0.0053
ENE	0.0009	0.0032	0.0091	0.0053	0.0002	0.0000	0.0189
E	0.0005	0.0024	0.0129	0.0056	0.0000	0.0002	0.0217
ESE	0.0007	0.0045	0.0102	0.0056	0.0000	0.0000	0.0212
SE	0.0006	0.0056	0.0075	0.0018	0.0002	0.0000	0.0159
SSE	0.0019	0.0067	0.0142	0.0029	0.0002	0.0000	0.0261
S	0.0020	0.0051	0.0104	0.0013	0.0000	0.0000	0.0189
SSW	0.0013	0.0067	0.0147	0.0048	0.0000	0.0000	0.0276
SW	0.0006	0.0032	0.0088	0.0034	0.0000	0.0000	0.0162
WSW	0.0009	0.0034	0.0051	0.0037	0.0000	0.0000	0.0133
W	0.0007	0.0045	0.0096	0.0040	0.0000	0.0000	0.0190
WNW	0.0002	0.0024	0.0018	0.0010	0.0000	0.0000	0.0056
NW	0.0002	0.0026	0.0016	0.0005	0.0005	0.0000	0.0056
NNW	0.0007	0.0040	0.0061	0.0024	0.0000	0.0002	0.0136
	0.0161	0.0645	0.1269	0.0505	0.0021	0.0008	

[a] Relative frequency of occurrence of D stability = 0.2610.
[b] Relative frequency of calms distributed above with D stability = 0.0077.

has been prepared for the United States (56). In this work a simple diffusion model was used in which the variables were mixing height, wind speed, and city size (i.e., distance across the city). The climatology includes isoplethed seasonal and annual maps of morning and afternoon $\bar{\chi}/\bar{Q}$ (second/m) values that are exceeded 10, 25, and 50% of the time for various city sizes (Fig. 12). (In Fig. 12 if $\bar{Q} = 1$ μg/m² second, then isopleths are of $\bar{\chi}$ in μg/m³.) In applying such data careful attention should be paid to the model and the assumptions.

C. Solar Radiation

The interactions of solar radiation and air pollutants have several consequences which are of both local and geophysical importance. According to Sheppard (102), backscatter of the solar beam by pollutant aerosols can result in as much as a 20% loss of the energy available to heat the earth-atmosphere system. Ångström (103) reaches about the same conclusion on the basis of theoretical estimates of the effect of atmospheric turbidity on the planetary albedo of the earth. Roach (104) suggested

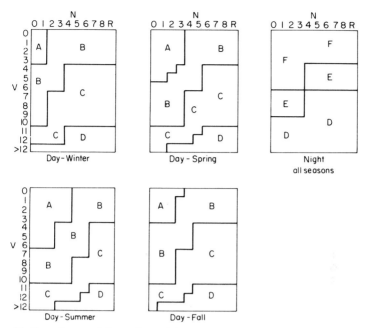

Figure 11. Prevalence of Pasquill stability categories over the Netherlands (*101*). V, wind speed (knots); N, total cloudiness (eighths); R, obscured.

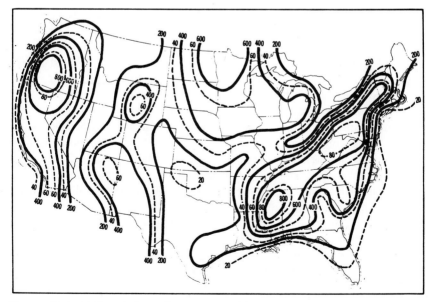

Figure 12. Isopleths of theoretical concentrations (see text) exceeded on 10% of mornings annually, for city sizes of 10 (dashed) and 100 km (solid), United States (*56*).

that absorption by gases and aerosols in polluted areas is capable of producing atmospheric heating rates in excess of 5°C per day which could not be counteracted by infrared cooling. On the other hand, Rasool and Schneider (105) conclude that the net effect of backscatter of solar radiation by aerosols could be to reduce the surface temperature of the earth. Further data are needed on the optical properties of aerosols *in situ* and on surface albedo distributions as input to the theoretical models to resolve the current uncertainty. However, it has been clearly shown that at least in one city, London, England, a decrease in the peak intensity of solar radiation in summer during the decade of the 1960's relative to the 1950's was undoubtedly caused by a general increase in background air pollution (106, 107), ". . . in spite of any effects of the (1956) Clean Air Act, etc. . . . ."

It is well known now that solar radiation can be instrumental in the creation of air pollution problems. The intensity of visible light during the exposure of plants to phytotoxic air pollutants has been reported to exert a marked effect on plant response (28). A threshold light intensity is suspected which varies with species and other growth and exposure conditions. The importance of ultraviolet radiation in the presence of suitable air contaminants in producing photochemical "smog" is discussed elsewhere in this volume.

Some meteorological services over the world have established solar radiation networks of climatological value. The records available for 88 stations with at least 3 years of record of measurements of total solar radiation on a horizontal surface have been summarized on a monthly basis by Black (108). It is notable that in October the United States receives, on the average, nearly twice the solar radiation per day as most of Europe (Fig. 13). While the relative potential for photochemical air pollution is not necessarily in direct proportion, it is obviously indicative.

More detailed radiation maps are available in many countries (e.g., 109, 110), also on a monthly basis. In general, national networks for the measurement of ultraviolet radiation comparable to that for total solar radiation have not been established. However, in the United States NOAA and EPA have a network of about 40 stations which measure the atmospheric turbidity at a wavelength of 0.380 μm. The program also extends to the order of 20 additional stations over the globe in cooperation with foreign countries and the United States Air Force Air Weather Service. Short-term, ad hoc networks are also set up from time to time such as that begun in October 1973 in the United States, sponsored by the Department of Transportation and administered by Temple University, Philadelphia, Pennsylvania to measure solar radiation intensities of erythematic importance (111). MacDonald (112) has made an esti-

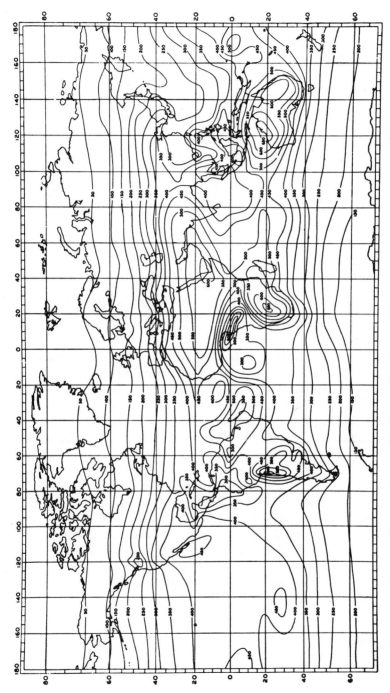

Figure 13. Isopleths of solar radiation (from sun and sky) on a horizontal surface in gm cal/cm² day, October *(108)*.

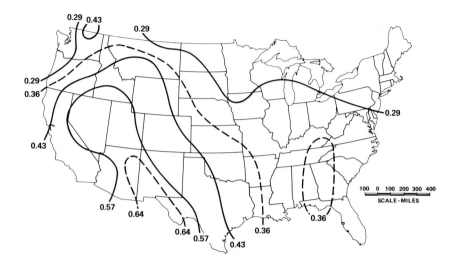

Figure 14. Estimated annual mean daily radiation (gm cal/cm²) for wavelengths ≤0.3192 μm on a horizontal plane at the earth's surface, United States (112).

mate of the annual mean daily radiation of wavelengths ≤0.3192 μm reaching a horizontal surface on the ground (Fig. 14). This figure is based on analysis of measurements of solar ultraviolet radiation taken by Stair and Coblentz [published by Koller (113)], and by the United States Weather Bureau. Although somewhat longer wavelengths may be of more photochemical significance, the distribution shown by the figure is probably indicative of the geographic availability of radiant energy throughout the ultraviolet region. Figure 15 presents similar isopheths to Figure 14 but for total solar radiation. (One langley is equal to 1 gm cal/cm².) Note that aside from the general schematic similarity in the maps, the range in intensity of the ultraviolet radiation is appreciably greater than for total radiation. Quasi-theoretical estimates of the monthly and annual global distribution of ultraviolet radiation at wavelengths of 0.350 and 0.3075 μm have been prepared by Schulze (114). The estimates are for conditions of normal cloudiness as well as for clear skies, and are extrapolations of limited observational data.

D. Visibility

The primary climatological interest in visibility in relation to air pollution is in analysis of secular trends in "seeing," which may provide some

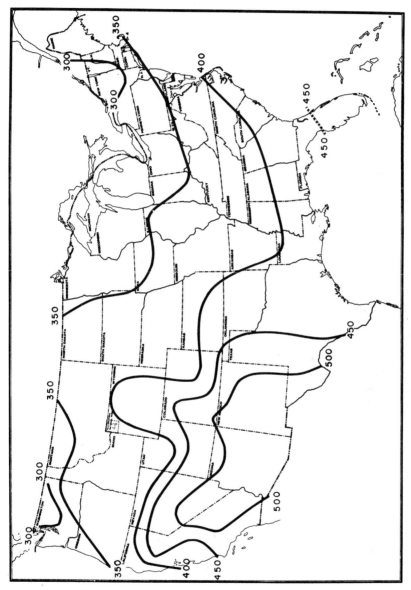

Figure 15. Mean daily total solar radiation (langleys), annual, United States.

insight into air quality trends. Weather service records of visibility are usually much longer than those available for air pollutants. The circumstances under which one can serve as an index of the other is of major consideration, and are discussed in Chapter 11, this volume and Chapter 1, Vol. II. Relatively short-period climatological variations, such as seasonal ones, are complicated by effects of moisture. When the visibility is good, the air is usually dry, but when it is not good the air tends to be relatively moist (*115, 116*). It is often presumed that over a period of years the moisture effect can be discounted, and any trend revealed will be the consequence of changes in atmospheric pollutants, i.e., aerosols. On the other hand, visibility data can be restricted to those observations when the relative humidity was less than some specified value, e.g., 70%. But since some meteorological conditions that are conducive to high humidities, e.g., radiation-type fogs, are also conducive to limited dispersion, it is clear that in some locations such a procedure eliminates some of the very data being sought.

Summer evening visibility over London, England in the period 1950–1959 did not show ". . . the slightest indication of any secular trend . . ." (*117*). On the other hand, a study (*118*) of visibility data at Sacramento (1935–1958) and Bakersfield (1948–1957), both in the California Central Valley, showed a worsening of visibility over the years, attributed to growing air pollution in the study area. This report, incidently, presents an interesting technique for the analysis of visibility trends in terms of the shift with time in percentage frequency of observed visibilities from one range of values to another. Application (*118*) of this method to Weather Bureau observations in downtown Los Angeles, California (1932–1959) indicated declining visibilities prior to about 1947 when certain emissions controls went into effect, but a negligible trend from 1948 to 1959. The latter was considered particularly significant in light of the rapid population growth that was occurring in the Los Angeles area during that period. Evaluation of the visibility trend in downtown Los Angeles has been extended through 1969 (*119*) on the basis of average noontime visibilities. It was concluded that although the overall trend was downward during 1933–1969, some improvement occurred during 1951–1959 over the previous 9-year period. Visibility was clearly highest during 1933–1941. Yet another study (*120*) for Los Angeles, considering only low visibility days [i.e., minimum visibility <3 miles (4.8 km) with relative humidity $\leq 60\%$] at eight stations in the basin, found that the number of such days had increased slowly during the period 1950–1961. It was also concluded that this increase implied ". . . an increase in pollution rather than a change in weather conditions."

In an effort (*121*) to determine the visibility trend for the United

States in general the percent frequencies, by months, of all hourly visibilities less than 7 miles (11.3 km) at 28 airports across the country was examined for two periods separated by an interval of roughly 15–20 years (Table XIII). The study was limited to stations whose location had not changed significantly. Where the frequency of visibilities less than 7 miles was greater in the later period (lower row) than in the earlier period (upper row), the figures were boxed in. It can be seen that low visibilities are more frequent in the later period in less than 26% of the comparisons, and these cases represent an average increase of less than 10% in the occurrence of low visibilities. Apparently, in most of the locations that were studied the visibility had generally improved in the 1950's compared to what it had been in the 1930's. Notice that the improvement tended to be greatest in winter, and in larger cities like Chicago, Illinois and St. Louis, Missouri. In general, this improvement was attributed mainly to a reduction in smoke emissions. Beebe (*122*) arrived at much the same conclusion based on a study of visibilities less than 7 miles due to various combinations of smoke and haze. He examined the hourly weather observations at eight major airports throughout the United States and compared data for January 1945 with that for January 1965.

On the other hand, in a very thorough study Miller *et al.* (*123*) found that summertime visibilities less than 7 miles at Akron, Ohio; Lexington, Kentucky; and Memphis, Tennessee increased significantly from the period 1962–1965 to the period 1966–1969. Similarly, Pritchard and Chopra (*116*) determined on an annual basis that visibilities at Norfolk, Virginia were somewhat lower (i.e., poorer) in the period 1965–1970 than during 1960–1964, and were lowest in the summer months of the 1965–1970 period. These findings for Akron, Lexington, Memphis, and Norfolk are not in disagreement with Table XIII, however, for the nearby stations at Columbus, Ohio; Nashville, Tennessee; and Richmond, Virginia, there is a declining visibility trend *during the summer.*

E. Moisture

Atmospheric and terrestrial moisture in its various phases is the final weather element of notable import to air pollution. It is generally appreciated that precipitation, i.e., rain and snow, can be effective in cleansing the atmosphere and at the same time that it can result in contamination of the earth's surface. Much of the early work on precipitation scavenging was done in connection with airborne radioactive materials (*83, 124*), but the principles involved are generally applicable to current problems. More recently there has been considerable interest in the occurrence of

Table XIII Percent Frequencies of All Visibilities Less than 7 Miles at Weather Bureau Airport Station[a]

		Jan.	Feb.	Mar.	Apr.	May	Jun.	Jul.	Aug.	Sept.	Oct.	Nov.	Dec.
Bakersfield, California	1/29–12/38	23	9	3	0	1	1	2	3	7	8	11	29
	7/55–6/59	47	29	6	5	2	3	2	5	9	15	40	58
Burbank, California	10/31–12/38	18	24	24	29	32	48	13	42	43	38	22	21
	5/50–4/55	25	14	21	32	37	42	45	54	51	47	34	26
Caribou, Maine	1/38–12/41	36	30	24	19	15	16	18	16	19	20	31	39
	7/59–3/61	21	30	28	21	13	14	10	13	10	14	23	18
Chicago, Illinois	1/30–12/38	72	70	65	62	54	48	47	52	52	58	56	73
	1/59–12/60	57	37	48	32	34	30	32	38	33	40	29	39
Columbia, Missouri	6/30–12/38	31	34	29	22	12	7	4	8	13	18	23	36
	7/59–3/61	21	28	24	4	2	6	6	4	2	14	5	15
Columbus, Ohio	1/30–12/38	63	57	45	28	19	16	16	23	23	35	53	68
	7/58–6/61	46	37	29	19	16	20	26	32	24	25	23	41
Des Moines, Iowa	4/33–12/38	54	46	41	29	21	14	9	20	20	26	32	52
	7/58–6/61	23	30	28	11	9	9	10	8	9	12	11	17
El Paso, Texas	7/30–12/38	3	5	9	10	4	2	1	0	2	1	1	3
	1/49–12/53	4	4	4	4	2	1	1	0	1	1	1	3
Grand Island, Nebraska	9/31–12/38	25	23	20	23	13	4	3	8	10	10	15	20
	7/58–6/61	15	23	20	10	10	5	3	2	5	3	6	9
Greensboro, N. Carolina	1/30–12/38	31	24	23	18	10	6	10	11	16	15	26	34
	7/58–6/61	23	24	19	11	12	17	22	28	22	24	14	13
Indianapolis, Indiana	7/29–12/38	61	56	50	38	26	18	19	31	32	38	46	65
	7/58–6/61	35	40	33	14	14	18	22	30	19	21	18	31
Lake Charles, Louisiana	3/39–4/42	11	12	17	11	6	3	3	5	6	12	11	19
	7/59–3/61	20	24	16	9	3	4	8	10	9	15	15	20
Medford, Oregon	7/29–12/38	35	12	5	4	2	1	3	2	9	13	23	37
	8/50–7/55	40	23	4	5	2	0	1	3	5	29	49	46
Milwaukee, Wisconsin	1/30–12/38	50	45	40	34	32	27	22	30	32	38	36	43
	7/58–6/61	29	30	28	17	18	11	11	22	12	19	18	21

City	Period												
Moline, Illinois	1/30–12/38	63	55	51	40	25	21	18	37	35	45	51	66
	7/58–6/61	31	34	34	14	11	14	20	19	14	21	12	26
Nashville, Tennessee	7/37–12/41	45	41	32	11	10	8	13	16	12	23	43	44
	7/58–6/61	26	27	17	9	7	10	15	17	17	19	16	25
Oakland, California	7/29–12/38	26	19	9	4	2	3	7	11	25	28	41	30
	5/50–4/55	28	26	8	9	8	5	11	10	18	27	37	30
Peoria, Illinois	3/36–12/38	68	56	40	29	20	16	17	28	30	37	41	62
	7/58–6/61	36	36	33	12	11	11	16	20	12	20	17	26
Richmond, Virginia	6/29–12/38	39	33	27	22	19	18	20	22	26	26	34	38
	7/58–6/61	22	24	20	13	18	22	26	41	27	23	18	18
Sacramento, California	1/30–12/38	42	22	9	4	2	1	1	2	13	16	37	55
	3/50–2/55	39	30	9	9	3	2	4	4	8	19	44	46
Salem, Oregon	1/34–12/38	35	29	15	6	2	3	6	11	35	51	46	37
	7/55–6/59	25	19	9	3	3	1	1	1	10	24	27	35
San Diego, California	1/30–12/38	90	85	86	92	91	87	88	87	81	77	86	88
	8/49–7/54	79	78	88	82	88	86	81	75	67	71	73	82
Seattle, Washington	1/30–12/38	38	36	19	16	8	8	10	30	44	53	46	36
	7/55–6/59	24	27	13	10	7	6	10	13	28	34	38	28
Sioux City, Iowa	1/47–12/51	16	20	17	10	5	7	3	7	4	7	11	17
	7/58–6/61	22	24	25	8	8	5	4	4	5	6	9	11
South Bend, Indiana	6/30–12/38	65	56	51	33	21	18	14	20	24	37	45	67
	7/58–6/61	51	49	37	21	13	15	21	23	20	24	27	41
St. Louis, Missouri	1/35–12/41	50	45	39	32	18	14	14	17	22	33	40	52
	1/58–12/60	34	18	28	10	12	8	13	11	9	20	13	22
Tulsa, Oklahoma	3/30–12/36	26	24	17	16	7	3	2	4	6	11	14	22
	7/58–6/61	14	19	12	2	3	5	3	2	6	8	4	12
Winslow, Arizona	2/31–12/38	3	2	2	2	0	0	0	0	0	0	0	4
	7/55–6/59	3	2	2	1	0	0	0	0	0	2	1	0

a Observations at hourly intervals. Table from Holzworth (121). Data sources: U.S. Weather Bureau Local Climatological Data, Climatography of the U.S. No. 30 (for selected stations), and Normal Flying Weather for the U.S. (New Orleans, 1945).

acid precipitation, especially in northern Europe (*125*), although in this case the main effort to date has centered on identifying the sources through consideration of horizontal transport (*126, 127*). Precipitation scavenging commonly distinguishes two processes:

(a) In-cloud scavenging by cloud elements and precipitation, referred to as "rainout" or "snowout"

(b) Below-cloud scavenging by precipitation elements, referred to as "washout"

Although these processes (see Chapter 8) have been treated theoretically and experimentally (*83, 128–131*), their general application is often difficult because of unknown variables, e.g., size distributions of rainfall and of particulate materials, solubility of gases, density of larger particles, electrical charge, wettability, airborne chemical reactions, etc. For climatological purposes a more practical approach, at least for particulate materials, may be to empirically determine washout ratios W:

$$W = k/\chi \tag{6}$$

where k is the concentration of particulate materials in precipitation ($\mu g/gm$) and χ is the concentration of particulate materials in air ($\mu g/gm$) at ground level. However, measurements (*132*) have shown that W tends to vary according to the chemical composition of the particulate material, and this may be related to characteristic sizes of different materials. The role of pollutants acting as condensation nuclei for water vapor is discussed in Chapter 1, Vol. II, but it is important climatologically from its impact on the occurrence of fog. In this connection the study of fog at Munich and Nürnberg, Federal Republic of Germany (*49*) may be of interest. The occurrence of fog in London, England, has been investigated rather extensively (*133, 135*), and its decreasing trend is generally believed to be connected with changes in smoke emissions.

Without moisture in the atmosphere there would be no corrosion of materials, even in the most heavily polluted air. Moisture is also a key parameter for the formation of pollutant materials such as sulfuric acid mists and perhaps also for sulfates (*136*), which are physiologically damaging to animal tissues. Relative humidity and soil moisture have been noted to exert a marked effect upon the sensitivity of plants to phytotoxic air pollutants. Plants grown under drought conditions are reported to be less sensitive than when the moisture supply is more nearly normal (*28*).

Along with temperature, precipitation is one of the best-documented weather elements, so that weather records on climatological distributions are quite readily available for most communities. Although relative humidity is a standard (derived) element of the observations at most weather stations, climatological summaries of it are not too common. Hence, significant features such as diurnal and seasonal variations may have to be worked up from original records. In extrapolation of rural data to city areas it is often assumed that the humidity, both relative and absolute, is generally less in towns than in the country (*35*), mainly due to the immediate runoff of precipitation from impervious surfaces into storm sewers. Nighttime traverses of London, England, for example, have shown differences in relative humidity between the central regions of the city and the suburbs on the order of 30% (*137*). More recently, Ackerman (*138*) has made a thorough study of urban–rural differences in dew-point temperature near Chicago, Illinois. She found that average differences were generally positive (higher dew points in the urban area) at night, changed to negative values shortly after sunrise, and remained negative through afternoons in spring and early summer but were positive during winter. She also discussed the possible causes of such variations in terms of dew deposition and evaporation, plant transpiration, surface moisture, combustion, and snow cover. It was concluded, at least in this particular comparison, that characteristically city air is neither drier nor more humid than country air. Soil moisture is not commonly measured at weather posts, and such direct information as may be available is normally collected by agricultural research stations. In the absence of actual measurements, a method for determining soil moisture from climatic data has been developed (*139*).

III. Air Quality

Even though there is increasing appreciation that atmospheric pollution ". . . is a climatological factor in its own right . . ." (*140*), there are many obstacles to be overcome before it will be possible to delineate suitable climatographies of air quality from direct measurements. This stems not only from the relative paucity of lengthy, and representative data, but also from the basic nature of the air pollution problem. On any map of air pollution distribution over a country which one might be able to construct, the data for urban areas would stand out as sharp peaks at levels up to one or two orders of magnitude or so higher than surrounding background or rural levels. This is shown in the following tabulation (*32*) as an ". . . example of the air above a densely populated area compared with the composition of a pure atmosphere . . .":

Component	Clean air	Polluted air
Particulate matter	0.01–0.02 mg/m³	0.07–0.7 mg/m³
Sulfur dioxide	10^{-3}–10^{-2} ppm	0.02–2 ppm
Carbon dioxide	310–330 ppm	350–700 ppm
Carbon monoxide	<1 ppm	5–200 ppm
Oxides of nitrogen	10^{-3}–10^{-2} ppm	10^{-2}–10^{-1} ppm
Total hydrocarbons	<1 ppm	1–20 ppm

In contrast, the rural–urban difference in all other atmospheric elements would rarely be as much as a factor of 2. For this reason the climatological features of urban air pollutants and of "background levels" will be considered separately.

A. Urban Air Pollution

At or near ground level, where the great majority of measurements are made the variations in pollutant concentrations are functions of both meteorological and emission parameters which also vary in time and space (141). The meteorological parameters have two basic periodicities: diurnal and annual, while those of emissions have three: diurnal, weekly, and seasonal. Generally speaking, meteorological conditions at night favor the accumulation of air pollutants, and in the daytime are favorable for rapid dispersion. An annual cycle in this sequence is imposed by the change in relative duration of day and night, as well as by manifestations of characteristic seasonal changes in features of the general atmospheric circulation. The diurnal and weekly variations in pollution emission arise from man's living habits and the accomodations he has made to social, economic, and physiological pressures imposed by his total environment. To some extent these influences are also operative on an annual scale, but here the variations are due primarily to those connected with agricultural pursuits or, in temperate or colder climes, also to the necessities of space heating.

The result of all the above influences is that on a climatological basis a certain amount of order seems to emerge. Much has been written on this subject as well as on analytical techniques to bring out significant features of the daily and seasonal cycles (32, 140–143). Other chapters in this volume should be consulted for details, but to summarize, it can be stated that most primary pollutants—e.g., SO_2, CO, NO_x, particulates—have two diurnal peaks, one in the morning sometime between 0700 and 1000 local standard time (LST), and another in the evening, usually after sunset, 1900–2200 LST. Further, most of these have a larger winter-to-summer concentration ratio, attributed mainly to increased combustion because of space heating demands. The possible mechanisms causing the double daily peaks have been studied in some depth (*140, 144*), and seem to be due to the optimum relative coincidence of changes in emission and dissipation rates of air pollutants during those hours.

Secondary pollutants formed in the atmosphere by photochemical processes, e.g., oxidants and eye irritants, have but one diurnal maximum. As might be expected, this usually occurs in the early afternoon. The annual peak is found in the late summer or autumn season when the favorable combinations of radiation, temperature, and atmospheric stagnations are met.

The horizontal variations in pollutants across a city are of the same order as the temporal variation in mean levels and the deviations about them at individual locations (i.e., a function of time as well as space scales, source distributions, and meteorological factors). As indicated above, there may be two or more orders of magnitude difference in the short-term concentration levels between some city sites and background, but on a monthly or annual basis the difference is usually much less. Although ground-level distribution from source areas reflects prevailing wind direction patterns, the area of maximum concentration for primary pollutants is usually near the core region of cities, but considerably displaced from that region for secondary pollutants [see, e.g., Mosher *et al.* (*145*)]. It has been suggested that a significant fraction of the total sulfate measured at rural sites in the eastern United States is the result of atmospheric transformation of sulfur dioxide transported from urban complexes (*146*).

With respect to long-range trends, the EPA study (*7*) indicates that both total suspended particulates (TSP) and SO_2 levels have ". . . improved considerably over the past 12 years . . ." at NASN urban sites. This is shown in Figure 16 which gives the composite average at 95 urban locations, 1960–1971. The "standards" indicated in the figure are those established by EPA as required under the provisions of the United States Clean Air Act of 1970. It is important to note, however, that another study (*147*) showed that the sulfates levels had ". . . basically stayed unchanged . . ." at a majority of 62 urban sites, with at least 6 years of record, that had experienced significant declines in SO_2 and total suspended particulate matter (TSP). It was suggested that the influence of more distant SO_2 sources ". . . could satisfactorily explain . . ." the differences in the SO_2 and sulfate trends at inner-city locations.

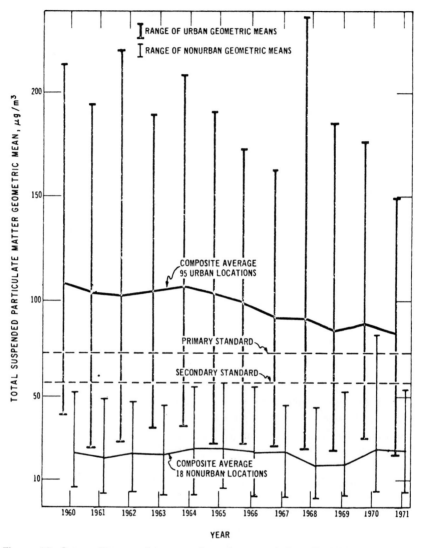

Figure 16. Composite annual means of total suspended particulate at urban and nonurban National Air Surveillance Network stations United States (7).

B. Background Air Pollution

Sufficient data to justify an attempt to delineate atmospheric background levels and trends have been collected for only a few pollutants. In Figures 17 and 18 isopleths (μg/m³) are shown for geometric means of TSP for January–March and July–September, respectively, at NASN

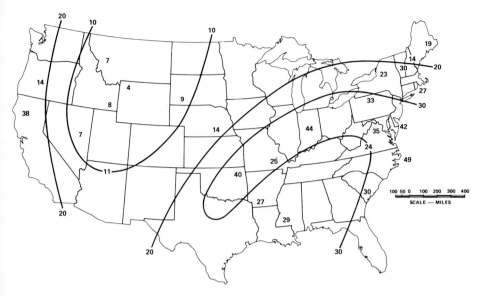

Figure 17. Geometric mean suspended particulate matter concentrations ($\mu g/m^3$)—nonurban National Air Surveillance Network stations—January–March, United States.

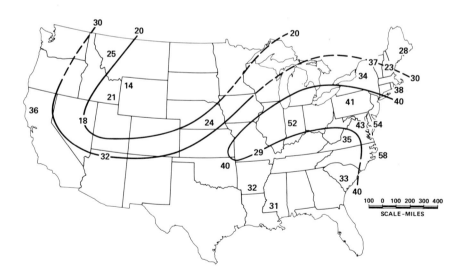

Figure 18. Geometric mean suspended particulate matter concentrations ($\mu g/m^3$)—nonurban National Air Surveillance Network stations—July–September, United States.

stations with at least 10 years of record. There are two particularly notable features of these maps, the near constancy of the geographic distribution of the higher and lower values, and the marked seasonal variation of the loadings over most of the United States with a higher summer/winter ratio. This is in good agreement with the seasonal and geographic distribution of atmospheric turbidity found by Flowers *et al.* (*148*). Since turbidity is a measure of the total vertical particulate loading of the atmosphere, it is clear that the summer maximum of TSP is not merely a low-level phenomenon, although diminished entrainment of ground dust into the air because of persistent snow cover over the mountain regions in the west must be an important factor. As is also apparent in Figure 16 there does not appear to be a significant trend in nonurban TSP at stations with records going back to 1960. On a global scale the picture is uncertain. In terms of atmospheric transmission of solar radiation, or turbidity, there is some evidence (e.g., *107, 149*) to indicate upward trends in the vicinity of urban complexes but this may be purely a low-level phenomenon since study (*150*) on a global scale at high-altitude astronomical stations of the Smithsonian Institution indicated that ". . . there has been no detectable change in the global atmospheric transmission measured from remote, high altitude sites in the last half century."

According to one report (*147*) there is a fairly definite geographical distribution of sulfates over the United States. The highest concentrations, and percentage of TSP, are in the industrial northeastern part of the United States and lowest west of the Mississippi River, in the mountain states. There was ". . . general agreement between levels of particulate sulfate, ambient sulfur dioxide and SO_2 emissions. The highest SO_2 both in ambient concentrations and in emission densities are found in the industrial northeast sector of the country." The data were not sufficient to draw any conclusions with regard to long-term trends.

Measurements of total ozone in vertical columns in the atmosphere are now made at a score or more sites in the northern hemisphere. The data for all of these stations through 1959 have been analyzed by London (*151*) for the whole year and each of the four seasons. There is a marked increase in total ozone with latitude in all seasons (Fig. 19) with a strong overall spring maximum. London points out that the observed seasonal and latitudinal variations are somewhat different from that which might be expected from photochemical theory. Large-scale circulations in the upper troposphere and lower stratosphere transporting ozone from the lower latitudes to polar regions are postulated to account for the difference. No secular trends in total ozone content of the atmosphere have been reported.

There is no question but that the carbon dioxide content of the atmo-

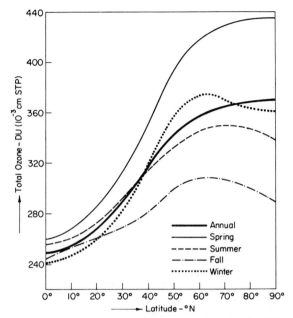

Figure 19. The average variation of total ozone with latitude (151). Ozone values in Dobson units (DU), i.e., 10^{-3} cm at standard temperature and pressure.

sphere is increasing and that, as with atmospheric aerosols, large increases can have a significant effect on global climate (105, 152). The standard surface value for carbon dioxide for 1970 has been accepted as about 322 ppm, Machta (153) has predicted that by the year 2000 the level will be 380 ppm. The effect of this increase is not precisely known, but a group of experts have estimated (154) that an increase of that order could cause a warming of the surface atmospheric layer of about 0.5°C.

REFERENCES

1. "World Meteorological Bulletin (Quarterly)." World Meteorol. Organ., Geneva, Switzerland.
2. United States Dept. of Commerce, "Selective Guide to Climatic Data Sources." U.S. Govt. Printing Office, Washington, D.C., 1969.
3. United States Naval Weather Service Command, "Guide to Standard Weather Summaries and Climatic Services." National Climatic Center, Asheville, North Carolina, 1973.
4. R. H. Frederick, J. Air Pollut. Contr. Ass. 14, 60 (1964).
5. Stanford Research Institute, "The Uses of Meteorological Data in Large Scale Air Pollution Surveys." State of California, Dept. of Public Health, Berkeley, California, 1958.

6. Smithsonian Institution, "National and International Environmental Monitoring Activities." Washington, D.C., 1970.
7. United States Environmental Protection Agency, "The National Air Monitoring Program: Air Quality and Emissions Trends," Annual Report, Vol. 1, EPA-450/1-73-001-a. Research Triangle Park, North Carolina, 1973.
8. United States Dept. of Health, Education and Welfare, "Storage and Retrieval of Air Quality Data, System Description and Data Coding Manual," Publ. No. APTD-68-8. Pub. Health Serv., Cincinnati, Ohio, 1968.
9. United States Environmental Protection Agency," "Environmental Data Systems Directory." Washington, D.C., 1973.
10. World Health Organization, "WHO International Air Pollution Monitoring Network, Data Users Guide." Geneva, Switzerland, 1972.
11. D. P. Petersen and D. Middleton, *Tellus* **15**, 387 (1963).
12. B. H. Kirschner, "Environmental Meteorological Support Units: A New Weather Bureau Program Supporting Urban Air Quality Control," Proc. 2nd Int. Clean Air Congr. (H. M. England and W. T. Berry, eds.), p. 987. Academic Press, New York, New York, 1971.
13. R. S. Scorer and C. F. Barrett, *Int. J. Air Water Pollut.* **6**, 49 (1962).
14. B. H. Evans, *Tex., Eng. Exp. Sta., Res. Rep.* **59**, (1957).
15. R. A. McCormick, *Trans. Roy. Soc. London* **269**, 515 (1971).
16. W. B. Johnson, Jr., F. L. Ludwig, W. F. Dabbert, and R. J. Allen, *J. Air Pollut. Contr. Ass.* **23**, 490 (1973).
17. J. Halitsky, *Amer. Ind. Hyg. Ass., J.* **26**, 106 (1961).
18. United States Environmental Protection Agency, "Guidelines: Air Quality Surveillance Networks." Washington, D.C., 1971.
19. R. C. Braun and M. J. G. Wilson, *Int. J. Air Water Pollut.* **5**, 1 (1961).
20. J. Pelletier, *Int. J. Air Water Pollut.* **7**, 973 (1963).
21. G. A. Cleeves, T. J. Lemmons, and C. A. Clemens, *J. Air. Pollut. Contr. Ass.* **16**, 212 (1966).
22. R. A. McCormick and D. M. Baulch, *J. Air Pollut. Contr. Ass.* **12**, 492 (1962).
23. M. Clifton, D. Kerridge, W. Moulds, J. Pemberton, and J. K. Donoghue, *Int. J. Air Pollut.* **2**, 188 (1959).
24. W. W. Stalker, R. C. Dickerson, and G. D. Kramer, *J. Air Pollut. Contr. Ass.* **12**, 361 (1962).
25. H. von Stratmann, *Staub* **25**, 341 (1965).
26. D. B. Turner, *Atmos. Environ.* **2**, 339 (1968).
27. E. C. Thom, *Air Cond., Heat., Vent.* **55**, 65 (1958).
28. W. W. Heck, J. A. Dunning, and I. J. Hindawi, *J. Air Pollut. Contr. Ass.* **15**, 511 (1965).
29. J. J. Bufalini and A. P. Altschuller, *Int. J. Air Water Pollut.* **7**, 769 (1963).
30. T. J. Chandler, *Meteorol. Mag.* **91**, 146 (1962).
31. T. Okita, *Int. J. Air Water Pollut.* **9**, 323 (1965).
32. H. W. Georgii, *Bull. W. H. O.* **40**, 624 (1969).
33. F. Linke, *in* "Biologie der Grossstadt, Frankfurter Konferenzen fur Naturwissenschaliche Zusammenarbeit," Vol. 4, p. 75. Steinkopf, Dresden and Leipsig, Germany, 1940.
34. J. M. Mitchell, Jr., *Westherwise* **14**, 224 (1961).
35. H. E. Landsberg, *in* "Man's Role in Changing the Face of the Earth" (W. L. Thomas, Jr., ed.), p. 584. Univ. of Chicago Press, Chicago, Illinois, 1956.
36. Osaka Environmental Pollution Control Center, "Air Pollution Situations and Meteorological Elements in Osaka," 2nd Report. Osaka, Japan, 1970.

37. J. T. Peterson, "The Climate of Cities: A Survey of Recent Literature," Nat. Air Pollut. Contr. Admin. Publ. No. AP-59. U.S. Govt. Printing Office, Washington, D.C., 1969.

38. B. F. Findlay and M. S. Hirt, *Atmos. Environ.* **3**, 537 (1969).

39. T. J. Chandler, *Int. J. Air Water Pollut.* **7**, 959 (1963).

40. F. S. Duckworth and J. S. Sandberg, *Bull. Amer. Meteorol. Soc.* **35**, 198 (1954).

41. F. L. Ludwig and J. H. S. Kealoha, "Urban Climatological Studies," Project MU 6300-140. Stanford Research Institute, Stanford, California, 1968 (available from Clearinghouse for Federal Scientific and Technical Information, Springfield, Virginia).

42. T. R. Oke, "City Size and the Urban Heat Island," Prepr. Conf. Urban Environ. and 2nd Conf. Biometeorol. Amer. Meteorol. Soc., Boston, Massachusetts, 1972.

43. G. A. DeMarrais, *Bull. Amer. Meteorol. Soc.* **42**, 548 (1961).

44. P. W. Summers, *in* "Proceedings of the First Canadian Conference on Micrometeorology" (R. E. Munn, ed.). Dept. of Transport, Meteorol. Branch, Toronto, Ontario, Canada, 1967.

45. J. F. Clarke, *Mon. Weather Rev.* **97**, 582 (1969).

46. R. E. Munn, J. Tomlain, and R. I. Titus, *Atmosphere* **8**, 52 (1970).

47. D. Szepesi, *Idojaras* **1**, 10 (1964).

48. D. Szepesi, "Meteorological Conditions of the Turbulent Diffusion of Atmospheric Pollutants in Hungary," Vol. 32. Official Publications of the National Meteorological Institute, Budapest, Hungary, 1967.

49. H. Herb, *Staub* **24**, 182 (1964).

50. C. R. Hosler, *Mon. Weather Rev.* **89**, 319 (1961).

51. C. R. Hosler, personal communication (1973).

52. M. A. Bilello, "Survey of Arctic and Subarctic Temperature Inversions," Tech. Rep. No. 161. United States Army Memorial Command Cold Regions and Research and Engineering Laboratory, Hanover, New Hampshire, 1966.

53. Meteorological Office, "The Incidence of Inversions over the United Kingdom," Memo No. 74. Meteorol. Office, Invest. Div., United Kingdom, 1962 (unpublished).

54. G. C. Holzworth, G. B. Bell, and G. A. DeMarrais, "Temperature Inversion Summaries of U.S. Weather Bureau Radiosonde Observations in California." California Dept. of Public Health, Berkeley, California, 1963 (unpublished).

55. G. A. DeMarrais, G. C. Holzworth, and C. R. Hosler, "Meteorological Summaries Pertinent to Atmospheric Transport and Dispersion over Southern California," U.S. Weather Bureau Tech. Pap. No. 54. U.S. Govt. Printing Office, Washington, D.C., 1965

56. G. C. Holzworth, "Mixing Heights, Wind Speeds, and Potential for Air Pollution in the Contiguous United States," Office of Air Programs Publ. No. AP-101. United States Environmental Protection Agency, Washington, D.C., 1972.

57. G. C. Holzworth, *Mon. Weather Rev.* **92**, 235 (1964).

58. E. M. Gross, "The National Air Pollution Potential Forecast Program," Tech. Memo WBTM NMC 47. United States Dept. of Commerce, Environ. Sci. Serv. Admin., Washington, D.C., 1970.

59. T. Tuna, *in* "Proceedings of the Third Meeting of the Expert Panel on Air Pollution Modeling," Rep. No. 14. NATO Committee on Challenges of Modern Society, United States Environmental Protection Agency, Research Triangle Park, North Carolina, 1972.

60. M. Popovics and D. J. Szepesi, "Diffusion Climatological Investigations in

Hungary," Proc. 2nd Int. Clean Air Congr. (H. M. England and W. T. Berry, eds.), p. 1073. Academic Press, New York, New York, 1971.

61. M. Rindsburger, "Analysis of Mixing Depth over Tel Aviv." Israel Meteorological Service, Bet Dagan, Israel (unpublished).

62. C. R. V. Raman and R. R. Kelkar, "Urban Air Pollution Potential over Bombay, India Derived from Mixing Depths and Mean Layer Wind," Prepr. Conf. Urban Environ. and 2nd Conf. Biometeorol. Amer. Meteorol. Soc., Boston, Massachusetts, 1972.

63. B. Davidson, *J. Air Pollut. Contr. Ass.* **17,** 154 (1967).

64. G. B. Mashkova, *in* "Investigation of the Bottom 300-meter Layer of the Atmosphere" (N. L. Byzova, ed.), p. 43. Izv. Akad. Nauk SSR, Moscow, U.S.S.R., 1963.

65. R. Jalu, *Meteorologie* [4] **37,** 247 (1955).

66. H. H. Schrenk, H. Heimann, G. D. Clayton, W. M. Gafafer, and H. Wexler, *U.S., Pub. Health Serv., Pub. Health Bull.* **306** (1949).

67. P. J. Meade, *Int. J. Air Pollut.* **2,** 87 (1959).

68. D. A. Lynn, B. J. Steigerwald, and J. H. Ludwig, "The November-December 1962 Air Pollution Episode in the Eastern United States," Publ. No. 999-AP-7. Pub. Health Serv., Cincinnati, Ohio, 1964.

69. J. C. Fensterstock and R. K. Fankhauser, "Thanksgiving 1966 Air Pollution Episode in the Eastern United States," Publ. No. AP-45. Nat. Air Pollut. Contr. Admin., Durham, North Carolina, 1968.

70. G. C. Holzworth, *Mon. Weather Rev.* **100,** 445 (1972).

71. J. Korshover, "Climatology of Stagnating Anticyclones East of the Rocky Mountains, 1936–1970," Nat. Oceanogr. Atmos. Admin., Tech. Memo ERL ARL-34. Air Resources Laboratories, Silver Spring, Maryland, 1971.

72. E. Ju. Bezuglaja, *Tr. Gl. Geofiz. Observ.* **234,** 69 (1968).

73. L. R. Son'kin, *Tr. Gl. Geofiz. Observ.* **207,** 56 (1968); also *in* "American Institute of Crop Ecology Survey of USSR Air Pollution Literature" (M. Y. Nuttonson, ed.), Vol. I, p. 58. Amer. Inst. Crop Ecology, Silver Spring, Maryland, 1969.

74. R. W. Shaw, M. S. Hirt, and M. A. Tilley, "Persistence of Light Surface Winds in Canada," Pap. No. 71-AP-11 (presented at the Annual Meeting of the Air Pollution Control Association Pacific Northwest International Section, Calgary, Alberta, 1971). Atmospheric Environmental Service, Dounsview, Ontario.

75. K. Shiozawa, A. Ootaki, and S. Okamoto, *Rep. Waseda Univ. Phys. Sci. Res. Inst.* No. 53, p. 131 (1971).

76. M. Nakamo, *Environ. Pollut. Countermeasures* **7,** 205 (1971).

77. H. S. Nudelman and J. A. Frizzola, *J. Air Pollut. Contr. Ass.* **24,** 140 (1974).

78. United States Department of Health, Education and Welfare "Continuous Air Monitoring Projects in Philadelphia 1962–1965," Publ. No. APTD 69-14. Nat. Air Pollut. Contr. Admin., Washington, D.C., 1969.

79. B. Ratner, *Mon. Weather Rev.* **78,** 185 (1950).

80. H. Riehl and D. Herkhof, "Weather Factors in Denver Air Pollution." Dept. of Atmospheric Science, Colorado State University, Fort Collins, Colorado, 1970.

81. J. R. Taylor, "Normalized Air Trajectories and Associated Pollution Levels in the Los Angeles Basin," Air Quality Rep. No. 45. Air Pollution Control District, Los Angeles County, California, 1962.

82. I. A. Singer and C. M. Nagle, *Nucl. Safety* **11,** 34 (1970).

83. D. H. Slade, "Meteorology and Atomic Energy," Library of Congress Catalog

Card No. 68-60097, see pp. 54–55. U.S. At. Energy Comm., Oak Ridge, Tennessee, 1968.

84. R. H. Frederick, *Int. J. Air Water Pollut.* **8,** 11 (1964).
85. F. H. Schmidt and J. H. Boer, *Ber. Deut. Wetterdienstes* **91,** 28 (1962).
86. L. E. Truppi, *Mon. Weather Rev.* **96,** 325 (1968).
87. E. Gold, *Weather* **11,** 230 (1956).
88. N. Z. Ariel' and L. A. Kliuchnikova, *Tr. Gl. Geofiz. Observ.* **94,** 29 (1960).
89. F. Pooler, Jr., *J. Appl. Meteorol.* **2,** 416 (1963).
90. M. Neiburger and J. G. Edinger, *Air Pollut. Found., Rep.* **1** (1954).
91. J. L. Sullivan, *J. Air Pollut. Contr. Ass.* **12,** 431 (1962).
92. E. K. Kauper and C. J. Hopper, *J. Air Pollut. Contr. Ass.* **15,** 210 (1965).
93. W. A. Lyons and L. E. Olsson, *J. Air Pollut. Contr. Ass.* **22,** 876 (1972).
94. I. A. Singer and M. E. Smith, *J. Meteorol.* **10,** 121 (1953).
95. D. M. Baulch, *J. Air Pollut. Contr. Ass.* **12,** 539 (1962).
96. M. E. Smith, *AMA Arch. Ind. Health* **14,** 56 (1956).
97. F. Pasquill, *Meteorol. Mag.* **90,** 33 (1961).
98. D. B. Turner, *J. Air Pollut. Contr. Ass.* **11,** 483 (1961).
99. D. B. Turner, *J. Appl. Meteorol.* **3,** 83 (1964).
100. J. K. Bannon, L. Dods, and P. J. Meade, "Frequencies of Various Stabilities in the Surface Layer," Memo No. 88. Meteorol. Office, Invest. Div., United Kingdom, 1962 (unpublished).
101. F. H. Schmidt, personal communication (1965).
102. P. A. Sheppard, *Int. J. Air Pollut.* **1,** 31 (1958).
103. A. Ångström, *Tellus* **14,** 435 (1962).
104. W. T. Roach, *Quart. J. Roy. Meteorol. Soc.* **87,** 346 (1961).
105. S. I. Rasool and S. H. Schneider, *Science* **173,** 138 (1971).
106. E. N. Lawrence, *Weather* **26,** 164 (1971).
107. E. N. Lawrence, *Weather* **27,** 320 (1972).
108. J. N. Black, *Arch. Meteorol., Geophys. Bioklimatol., Ser. B.* **7,** 165 (1956).
109. G. J. Day, *Meteorol. Mag.* **90,** 269 (1961).
110. U.S. Department of Commerce, "Mean Daily Solar Radiation, Monthly and Annual—1964." U.S. Govt. Printing Office, Washington, D.C., 1964.
111. W. Hass, personal communication (1974).
112. T. H. MacDonald, "Estimated Mean Daily Ultraviolet Radiation for Wavelengths Equal to and Less than .3192 Microns." Office Meteorol. Res., United States Dept. of Commerce, Washington, D.C., 1959.
113. L. R. Koller, "Ultra-violet Radiation." Wiley, New York, New York, 1952.
114. R. Schulze, "Strahelenklima der Erde." Steinkopff, Darmstadt, Federal Republic of Germany, 1970.
115. M. Neiburger and M. G. Wurtele, *Chem. Rev.* **44,** 321 (1949).
116. W. M. Pritchard and K. P. Chopra, "Effect of Air Pollution on Urban Visibility Statistics," Prepr. Conf. Air Pollut. Meteorol., Raleigh, North Carolina. Amer. Meteorol. Soc., Boston, Massachusetts, 1971.
117. L. C. W. Bonacina, *Weather* **15,** 127 (1960).
118. G. C. Holzworth and J. A. Maga, *J. Air Pollut. Contr. Ass.* **10,** 430 (1960).
119. R. W. Keith, "Downtown Los Angeles Noon Visibility Trends 1933–1969," Prepr. Conf. Air Pollut. Meteorol., Raleigh, North Carolina. Amer. Meteorol. Soc., Boston, Massachusetts, 1971.
120. R. W. Keith, "A Study of Low Visibilities in the Los Angeles Basin." Amer. Meteorol. Soc.—Amer. Geophys. Union, Los Angeles, California, 1964.

121. G. C. Holzworth, *Robert A. Taft Sanit. Eng. Cent., Rep.* **SEC TR A62-5** (1962).

122. R. G. Beebe, *Bull. Amer. Meteorol. Soc.* **48**, 348 (1967).

123. M. E. Miller, N. L. Canfield, T. A. Ritter, and C. R. Weaver, *Mon. Weather Rev.* **100**, 67 (1972).

124. W. Bleeker, *World Meteorol. Organ., Tech. Note* **68** (1965).

125. A. Nyberg, *Idojaras* **74**, 145 (1970).

126. E. J. Forland, *Tellus* **25**, 291 (1973).

127. H. Reiquam, *Science* **170**, 318 (1970).

128. J. M. Hales, M. A. Wolf, and M. T. Dana, *AIChE. J.* **19**, 292 (1973).

129. R. G. Semonin and J. R. Adam, "The Washout of Atmospheric Particulates by Rain," Prepr. Conf. Air Pollut. Meteorol., Raleigh, North Carolina. Amer. Meteorol. Soc., Boston, Massachusetts, 1971.

130. J. M. Hales, *Atmos. Environ.* **6**, 635 (1972).

131. United States Atomic Energy Commission "Precipitation Scavenging," Library of Congress Catalog Card No. 70-609397. Washington, D.C., 1970 (available as CONF-700601, National Technical Information Service, Springfield, Virginia).

132. D. F. Gatz, "Washout Ratios in Urban and Non-Urban Areas," Prepr. Conf Urban Environ. and 2nd Conf. Biometeorol., Philadelphia. Amer. Meteorol. Soc., Boston, Massachusetts, 1972.

133. J. H. Brazell, *Meteorol. Mag.* **93**, 129 (1964).

134. T. Kelly, *Meteorol. Mag.* **100**, 257 (1971).

135. I. Jenkins, *Meteorol. Mag.* **100**, 317 (1971).

136. J. Wagman, R. E. Lee, Jr., and C. J. Axt, *Atmos. Environ.* **1**, 479 (1967).

137. T. J. Chandler, *Weather* **17**, 235 (1962).

138. B. Ackerman, "Moisture Content of City and Country Air," Prepr. Conf. Air Pollut. Meteorol., Raleigh, North Carolina. Amer. Meteorol. Soc., Boston, Massachusetts, 1971.

139. J. R. Mather, *Bull. Amer. Meteorol. Soc.* **35**, 63 (1954).

140. R. E. Munn and M. Katz, *Int. J. Air Pollut.* **2**, 51 (1959).

141. P. H. Merz, L. J. Painter, and P. R. Ryason, *Atmos. Environ.* **6**, 319 (1972).

142. United States Public Health Service, "Continuous Air Monitoring Program in Cincinnati, 1962–1963." Div. Air Pollut., Robert A. Taft Sanit. Eng. Cent., Cincinnati, Ohio, 1965.

143. W. J. Hamming, R. D. MacPhee, and J. R. Taylor, *J. Air Pollut. Contr. Ass.* **10**, 7 (1960).

144. E. C. Halliday and E. Kemeny, *Int. J. Air Water Pollut.* **8**, 43 (1964).

145. J. C. Mosher, W. G. MacBeth, M. J. Leonard, T. P. Mullins, and M. F. Brunelle, *J. Air Pollut. Contr. Ass.* **20**, 35 (1970).

146. A. P. Altshuller, *Environ. Sci. Technol.* **7**, 709 (1973).

147. N. H. Frank, unpublished manuscript (1973).

148. E. C. Flowers, R. A. McCormick, and K. R. Kurfis, *J. Appl. Meteorol.* **8**, 955 (1969).

149. R. A. McCormick and J. H. Ludwig, *Science* **156**, 1358 (1967).

150. R. G. Roosen, R. G. Angione, and C. H. Klemcke, *Bull. Amer. Meteorol. Soc.* **54**, 307 (1973).

151. J. London, *Beitr. Phys. Atmos.* **36**, 254 (1963).

152. P. Chylek and J. A. Coakley, Jr., *Science* **183**, 75 (1974).

153. L. Machta, *Bull. Amer. Meteorol. Soc.* **53**, 402 (1972).

154. "Report of the Study of Man's Impact on Climate." Massachusetts Institute of Technology Press, Cambridge, Massachusetts, 1971.

Subject Index

A

Actinometers, solar radiation measurements with, 609–610

Activation products, radionuclides, 206–207

Adiabatic lapse rate
dry air, 338–339
saturated air, 339

Advection, in temperature lapse rate, 342, 344–346

Aeroallergens, 172, 175, *see also* Pollen
effect on health, 190

Aerometric data, representativeness of, 650–652

Aerosols
atmospheric, 83
condensation, 82
dispersion, 82
effect on temperature, 334–336
particle size, 15
photochemical, 255

Air ions, 104–116
concentration, 109–110
effects on health, 111
formation, 105–108
measurement, 108–109
velocity, 105

Air pollution, *see also* Pollutants, Particulates, specific substances, sources
definition, 13–14
effect on agriculture, 20
on health, 19
on materials, 20
global, 289–324
history, 3–22
parameters in, 349–352

Air pollution analysis
analogies in eddy diffusion, 379–381
airflow, 386–390
combinations of perturbation, 387–390
local mechanical perturbation, 386–387
local thermal perturbation, 387
cross-sectional, 382–386

meteorological approaches, 378–396, *see also* Meteorological measurements
urban effects on atmospheric behavior, 390
precipitation, 394–396
temperature and airflow, 390–394
weather categories and actual weather, 381–382

Air pollution climatology, *see* Climatology, air pollution

Air pollution dispersal, 327–396

Air pollution episodes
Donora, Pennsylvania, 8
historical perspective, 7–10
London, England, 7, 8
Los Angeles, California, 7
Poza Rica, Mexico, 9

Air pollution forecasting, 19, 245–249

Air pollution sources
agricultural and forest emissions (decaying farm wastes, fertilizers, pesticides, slash burning, soil erosion), 292
automotive emissions, 11
global emissions, 291–293
concentration patterns, 307–314
secular time trends, 310–312
synoptic events, 307–308
trace substances, 308–310
ecological significance, 320–322
estimating techniques, 293–294
removal processes, 298–301
at earth's surface, 299–301
sink strength estimates, 301–305
synoptic models of transport and diffusion, 314–318
transport and diffusion processes, 305–307
natural emissions, 292–293
blowing dust, 292
evaporation, 293
forest, bush, or grass fires, 292
forest terpenes, 293
sea spray, 293
volcanoes, 293

ENVIRONMENTAL SCIENCES

An Interdisciplinary Monograph Series

EDITORS

DOUGLAS H. K. LEE

National Institute of
Environmental Health Sciences
Research Triangle Park
North Carolina

E. WENDELL HEWSON

Department of
Atmospheric Science
Oregon State University
Corvallis, Oregon

DANIEL OKUN

Department of Environmental
Sciences and Engineering
University of North Carolina
Chapel Hill, North Carolina

ARTHUR C. STERN, editor, AIR POLLUTION, Second Edition, Volumes I–III, 1968; Third Edition, Volume I, 1976, Volumes II–V, in preparation

L. FISHBEIN, W. G. FLAMM, and H. L. FALK, CHEMICAL MUTAGENS: Environmental Effects on Biological Systems, 1970

DOUGLAS H. K. LEE and DAVID MINARD, editors, PHYSIOLOGY, ENVIRONMENT, AND MAN, 1970

KARL D. KRYTER, THE EFFECTS OF NOISE ON MAN, 1970

R. E. MUNN, BIOMETEOROLOGICAL METHODS, 1970

M. M. KEY, L. E. KERR, and M. BUNDY, PULMONARY REACTIONS TO COAL DUST: "A Review of U. S. Experience," 1971

DOUGLAS H. K. LEE, editor, METALLIC CONTAMINANTS AND HUMAN HEALTH, 1972

DOUGLAS H. K. LEE, editor, ENVIRONMENTAL FACTORS IN RESPIRATORY DISEASE, 1972

H. ELDON SUTTON and MAUREEN I. HARRIS, editors, MUTAGENIC EFFECTS OF ENVIRONMENTAL CONTAMINANTS, 1972

RAY T. OGLESBY, CLARENCE A. CARLSON, and JAMES A. McCANN, editors, RIVER ECOLOGY AND MAN, 1972

LESTER V. CRALLEY, LEWIS T. CRALLEY, GEORGE D. CLAYTON, and JOHN A. JURGIEL, editors, INDUSTRIAL ENVIRONMENTAL HEALTH: The Worker and the Community, 1972

MOHAMMED K. YOUSEF, STEVEN M. HORVATH, and ROBERT W. BULLARD, PHYSIOLOGICAL ADAPTATIONS: Desert and Mountain, 1972

DOUGLAS H. K. LEE and PAUL KOTIN, editors, MULTIPLE FACTORS IN THE CAUSATION OF ENVIRONMENTALLY INDUCED DISEASE, 1972

MERRIL EISENBUD, ENVIRONMENTAL RADIOACTIVITY, Second Edition, 1973

JAMES G. WILSON, ENVIRONMENT AND BIRTH DEFECTS, 1973

RAYMOND C. LOEHR, AGRICULTURAL WASTE MANAGEMENT: Problems, Processes, and Approaches, 1974

LESTER V. CRALLEY, PATRICK R. ATKINS, LEWIS J. CRALLEY, and GEORGE D. CLAYTON, editors, INDUSTRIAL ENVIRONMENTAL HEALTH: The Worker and the Community, Second Edition, 1975

A 6
B 7
C 8
D 9
E 0
F 1
G 2
H 3
I 4
J 5

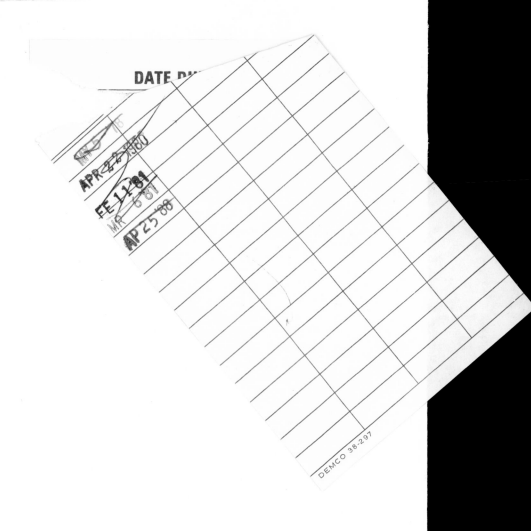

DATE DUE

DEMCO 38-297